Recent Advances in Renewable Energy

(Volume 1)

(Microalgae as a Source of Bioenergy: Products, Processes and Economics)

Edited by

José Carlos Magalhães Pires
Faculty of Engineering, University of Porto, Porto, Portugal

Recent Advances in Renewable Energy

Volume # 1

Microalgae as a Source of Bioenergy: Products, Processes and Economics

Editor: José Carlos Magalhães Pires

ISSN (Online): 2543-2397

ISSN: (Print): 2543-2389

ISBN (Online): 978-1-68108-522-7

ISBN (Print): 978-1-68108-523-4

General:

1. Any dispute or claim arising out of or in connection with this License Agreement or the Work (including non-contractual disputes or claims) will be governed by and construed in accordance with the laws of the U.A.E. as applied in the Emirate of Dubai. Each party agrees that the courts of the Emirate of Dubai shall have exclusive jurisdiction to settle any dispute or claim arising out of or in connection with this License Agreement or the Work (including non-contractual disputes or claims).
2. Your rights under this License Agreement will automatically terminate without notice and without the need for a court order if at any point you breach any terms of this License Agreement. In no event will any delay or failure by Bentham Science Publishers in enforcing your compliance with this License Agreement constitute a waiver of any of its rights.
3. You acknowledge that you have read this License Agreement, and agree to be bound by its terms and conditions. To the extent that any other terms and conditions presented on any website of Bentham Science Publishers conflict with, or are inconsistent with, the terms and conditions set out in this License Agreement, you acknowledge that the terms and conditions set out in this License Agreement shall prevail.

Bentham Science Publishers Ltd.
Executive Suite Y - 2
PO Box 7917, Saif Zone
Sharjah, U.A.E.
Email: subscriptions@benthamscience.org

BENTHAM SCIENCE

CONTENTS

PREFACE

The book "Microalgae as a Source of Bioenergy: Products, Processes and Economics" presents recent advances in biofuel production with microalgae. It is composed of 17 well written chapters by leading researchers in this field.

European Union (EU) defined targets for each Member State regarding climate and energy. Microalgae are considered a promising and sustainable source of energy, due to their biomass productivity and ability to capture CO_2. Several research studies were performed and new procedures to reduce the biomass production cost were developed. In this context, the proposed book have the contributions of relevant researchers in microalgal research area, focusing on the energy product yields, process developments and economics.

Biodiesel is one of the most studied biofuels, which can be produced by several raw materials. Microalgal biodiesel is the third generation biofuel and it is considered the most sustainable route for the production of this renewable form of energy. Microalgae present high areal productivities and their production does not compete with food market. Besides the biodiesel production with extracted oil, the residual biomass can be used for other energetic applications, reducing the cost of each type of produced energy. Chapter 1 presents the integration of biodiesel production and cogeneration and concludes that microalgae is an economic viable energy solution, if the biorefinery concept is applied, taking part from several products that microalgae can offer. Chapter 2 presents the potential of microalgae for bioethanol production, focusing on cultivation strategies to enhance carbohydrate productivity (which is easier to break down to simple reducing sugar), biomass pre-treatment methods, and hydrolysis and fermentation process. Chapter 3 presents different routes to produce bioethanol and presents a review of the research works about bioethanol production from algae. Chapter 4 introduces the basic principles of anaerobic digestion (biogas production), particularly when using microalgae as substrate. The influence of the most important operating parameters is also described. Biohydrogen is the last focused biofuel product. Chapter 5 shows the recent progresses regarding microalgal cellular mechanisms and production processes. As biofuel should present a lower price, several strategies should be implemented to reduce the production costs. Chapter 6 shows the main characteristics of the most used configurations of photobioreactors. Chapter 7 presents CO_2 biofixation from industrial flue gases by microalgae, describing the microalgal metabolism. The use of CO_2 from this source has dual benefit: (i) reduction of biomass production cost; and (ii) mitigation of industrial emissions of CO_2, which is one of the most important greenhouse gas. Harvesting, thickening and dewatering processes represent a significant cost in the production of microalgal biomass. Chapter 8 gives an overview of the related technologies, presenting their advantages. Chapter 9 summarizes recent developments in microalgal oil extraction processes, including drying methods, cell disruption methods, conventional or supercritical solvent extraction methods, and recent approaches for direct biodiesel production.

Concerning the current production and economics of microalgal production, Chapter 10 places international energy policy in the context of the current and projected energy landscape. It gives an overview of the global and commercially installed infrastructure. Some important research projects are also presented. Chapter 11 presents a current view of the commercial production of microalgae cultivation at a large scale worldwide. It also presents the main variables that influence microalgal cultures and compares different types of photobioreactors. Chapter 12 describes the environmental applications of microalgae: CO_2 capture and nutrient recycling. Mechanisms of assimilation of nitrogen and phosphorus are discussed in this chapter. Configurations of photobioreactors are compared in terms of wastewater treatment enhancement (and downstream processes) and improving mass transfer of CO_2 from the gaseous stream to the culture. Chapter 13 presents a wide range of products obtained from microalgal biomass (biorefinery context) that can enhance the economic viability of biofuel production. Chapter 14 performs a techno-economic assessment of microalgal oil production. Process modelling was performed through simulation software Aspen Plus. Chapter 15 reviews biofuel production from different feedstocks, focusing on the techno-economic challenges. Sensitivity analysis was performed to examine the economic parameters are the sales price was highly dependent on algae doubling time. Chapter 16 presents a life cycle assessment of the greenhouse gases of microalgal biofuels. Chapter 17 shows the results of the triple bottom line (for sustainability evaluation) assessment of algal bio-crude production. The considered stages in the analysis are the cultivation of algae, extraction of bio-crude and transport of bio-crude to a refinery. A region of Australia was selected for algae production. The results show that algae bio-crude production is more sustainable than crude oil production.

I would like to thank all the authors for their efforts in writing such excellent chapters. I also acknowledge the entire team of Bentham Science Publishers, particularly Ms. Fariya Zulfiqar (Assistant Manager Publications) due to the important orientations at different stages in the publication of the book. I am confident that this book will attract the attention of researchers and professionals of microalgal biofuel production.

José Carlos Magalhães Pires
Faculty of Engineering
University of Porto
Porto
Portugal

List of Contributors

Adeniyi Lawal	Department of Chemical Engineering and Materials Science, Stevens Institute of Technology, Cole Eye Institute, Cleveland Clinic 9500 Euclid Ave, Hoboken, NJ 07030, United States
Adriano S.A. Henrard	Laboratory of Microbiology and Biochemistry, College of Chemistry and Food Engineering, Federal University of Rio Grande, P.O. Box 474, Av. Itália km 8, 96203-900 Rio Grande, RS, Brazil
Aitor Lekuona-Amundarain	Tecnalia Research and Innovation, Leonardo Da Vinci 11, E-01510, Miñano, (Araba), Spain
Alessandro Marco Lizzul	Department of Environmental Engineering, University College London, Gower Street, WC1E 6BT, London, United Kingdom
Amarjeet Bassi	Department of Chemical and Biochemical Engineering, Faculty of Engineering, Western University, London ON N6A 5B9, Canada
Arunima Malik	ISA, School of Physics A28, The University of Sydney, NSW 2006, Sydney, Australia
Carlos A. Cardona	Group of Chemical, Catalytic and Biotechnological Processes, Institute of Biotechnology and Agroindustry. Department of Chemical Engineering, Universidad Nacional de Colombia – Sede Manizales. Cra. 27 No. 64-60, Manizales, Colombia
Choon Gek Khoo	School of Chemical Engineering, Universiti Sains Malaysia, Engineering Campus, Seri Ampangan, 14300 Nibong Tebal, Pulau Pinang, Malaysia
Christophe Bengoa	Departament d'Enginyeria Química, Universitat Rovira i Virgili, Av. Països Catalans, 26, 43007 Tarragona, Spain
Daissy Lorena Restrepo Serna	Instituto de Biotecnología y Agroindustria, Departamento de Ingeniería Química, Universidad Nacional de Colombia, Manizales campus, Manizales, Colombia
Daniela Parra	Group of Chemical, Catalytic and Biotechnological Processes. Institute of Biotechnology and Agroindustry. Department of Chemical Engineering, Universidad Nacional de Colombia – Sede Manizales. Cra. 27 No. 64-60, Manizales, Colombia
Darren L. Oatley-Radcliffe	Centre for Water Advanced Technologies and Environmental Research (CWATER), College of Engineering, Swansea University, Fabian Way, Swansea SA1 8EN, UK; Membranology Ltd., Charter Court, Phoenix Way, Enterprise Park, Swansea, SA79FS, UK
Dries Vandamme	KU Leuven Campus Kulak, Laboratory for Aquatic Biology, E. Sabbelaan 53, 8500 Kortrijk, Belgium
Eduardo Jacob-Lopes	Food Science and Technology Department, Federal University of Santa Maria, UFSM, Roraima Avenue 1000, 97105-900, Santa Maria, RS, Brazil
Esther Torrens	Departament d'Enginyeria Química, Universitat Rovira i Virgili, Av. Països Catalans, 26, 43007 Tarragona, Spain

Etiele G. Morais	Laboratory of Biochemical Engineering, College of Chemistry and Food Engineering, Federal University of Rio Grande, P.O. Box 474, Av. Itália km 8, 96203-900 Rio Grande, RS, Brazil
Fabiana Regina Xavier Batista	School of Chemical Engineering, Federal University of Uberlandia, Uberlandia/MG, Brazil
Igor S. Gonçalves	Laboratory of Microbiology and Biochemistry, College of Chemistry and Food Engineering, Federal University of Rio Grande, P.O. Box 474, Av. Itália km 8, 96203-900 Rio Grande, RS, Brazil
James L. Manganaro	Department of Chemical Engineering and Materials Science, Stevens Institute of Technology, Hoboken, NJ 07030, United States
Jorge Alberto V. Costa	Laboratory of Biochemical Engineering, College of Chemistry and Food Engineering, Federal University of Rio Grande, P.O. Box 474, Av. Itália km 8, 96203-900 Rio Grande, RS, Brazil
Juan Carlos Higuita Vásquez	Instituto de Biotecnología y Agroindustria, Departamento de Ingeniería Química, Universidad Nacional de Colombia, Manizales campus, Manizales, Colombia
Juliana de Souza Ferreira	School of Chemical Engineering, Federal University of Uberlandia, Uberlandia/MG, Brazil
Keat Teong Lee	School of Chemical Engineering, Universiti Sains Malaysia, Engineering Campus, Seri Ampangan, 14300 Nibong Tebal, Pulau Pinang, Malaysia
Leila Queiroz Zepka	Food Science and Technology Department, Federal University of Santa Maria, UFSM, Roraima Avenue 1000, 97105-900, Santa Maria, RS, Brazil
Lieve M.L. Laurens	National Bioenergy Center, National Renewable Energy Laboratory, Golden, CO 80401, USA
Lin Zhou	Department of Chemical Engineering and Materials Science, Stevens Institute of Technology, Hoboken, NJ 07030, United States
Lucas Reijnders	IBED, University of Amsterdam, Science Park 904, PO box 94248, 10 GE Amsterdam, the Netherlands
Luis G. Ramírez-Mérida	Applied Biotechnology Center, Department of Biology, University of Carabobo, Universidad Avenue, 2002, Valencia, Edo. Carabobo, Venezuela
Luiza Moraes	Laboratory of Biochemical Engineering, College of Chemistry and Food Engineering, Federal University of Rio Grande, P.O. Box 474, Av. Itália km 8, 96203-900 Rio Grande, RS, Brazil
Man Kee Lam	Chemical Engineering Department, Universiti Teknologi PETRONAS, 32610 Bandar Seri Iskandar, Perak, Malaysia
Martin Pablo Caporgno	Departament d'Enginyeria Química, Universitat Rovira i Virgili, Av. Països Catalans, 26, 43007 Tarragona, Spain
Melodie Chen-Glasser	National Bioenergy Center, National Renewable Energy Laboratory, Golden, CO 80401, USA

Meng Wang	Civil and Environmental Engineering, University of South Florida, Tampa, USA
Mengyue Gong	Department of Chemical and Biochemical Engineering, Faculty of Engineering, Western University, London ON N6A 5B9, Canada
Michele G. Morais	Laboratory of Microbiology and Biochemistry, College of Chemistry and Food Engineering, Federal University of Rio Grande, P.O. Box 474, Av. Itália km 8, 96203-900 Rio Grande, RS, Brazil
Robert W. Lovitt	Centre for Water Advanced Technologies and Environmental Research (CWATER), College of Engineering, Swansea University, Fabian Way, Swansea SA1 8EN, UK Membranology Ltd., Charter Court, Phoenix Way, Enterprise Park, Swansea, SA79FS, UK
Sebastián Serna	Group of Chemical, Catalytic and Biotechnological Processes, Institute of Biotechnology and Agroindustry. Department of Chemical Engineering, Universidad Nacional de Colombia – Sede Manizales. Cra. 27 No. 64-60, Manizales, Colombia
Shreyas Yedahalli	Department of Chemical and Biochemical Engineering, Faculty of Engineering, Western University, London ON N6A 5B9, Canada
Suphi S. Oncel	Department of Bioengineering, Engineering Faculty, Ege University, Izmir, Turkey
Thea Ekins-Coward	Centre for Water Advanced Technologies and Environmental Research (CWATER), College of Engineering, Swansea University, Fabian Way, Swansea SA1 8EN, UK
Vicelma Luiz Cardoso	School of Chemical Engineering, Federal University of Uberlandia, Uberlandia/MG, Brazil
Yulin Hu	Department of Chemical and Biochemical Engineering, Faculty of Engineering, Western University, London ON N6A 5B9, Canada

Part I

Energy from Microalgae: Products and Processes

CHAPTER 1

Perspectives of Energy Production from Microalgae: The Biodiesel and Cogeneration Cases

Carlos A. Cardona*, **Daniela Parra** and **Sebastián Serna**

Group of Chemical, Catalytic and Biotechnological Processes, Institute of Biotechnology and Agroindustry, Department of Chemical Engineering, Universidad Nacional de Colombia - Sede Manizales, Cra. 27 No. 64-60, Manizales, Colombia

Abstract: During the last two decades, the use of biofuels has shown rapid growth, driven mostly by policies focused on increasing energy efficiency, and replacing fossil energy by renewable energy. There are different biomass raw materials that have been evaluated for the production of several added-value products. These raw materials have been classified into the first generation (agricultural and edible crops), second generation (inedible agroindustrial residues) and third generation (algae). The interest in the cultivation of microalgae has been increasing due to the high value products that can be obtained. Additionally, the oils present in microalgae are used for the production of biodiesel and the cake resulting after processing can be used for the production of bioethanol, biobutanol or energy. Based on this, this chapter first introduces the current uses and applications of multiple species of microalgae in terms of energy production and describes the technologies being used for the production of bioenergy using micro-algae. Then, two specific cases are analyzed: cogeneration and biodiesel production. The performed analysis serves to conclude that in order to establish microalgae as an energy-producing feedstock, it is necessary to integrate their use to obtain multiple products simultaneously: metabolites due to their high value, oils to produce biodiesel and the dry cake for thermochemical production of energy. The extraction of multiple products can only be made possible if a biorefinery concept is applied.

Keywords: Microalgae, Biorefineries, Bioenergy, Biofuels, Biodiesel, Cogeneration.

INTRODUCTION

During the last two decades, the use of biofuels has shown a rapid growth, driven mostly by the policies focused on increasing energy efficiency, and replacing fossil energy by renewable energy [1, 2]. These policies also aim at achieving

* **Corresponding author Carlos A. Cardona:** Group of Chemical, Catalytic and Biotechnological Processes, Institute of Biotechnology and Agroindustry, Department of Chemical Engineering, Universidad Nacional de Colombia - Sede Manizales, Km. 7 vía al Magdalena, Manizales, Colombia; Tel/Fax: (+57) (6) 8879300 ext 55354; E-mail: ccardonaal@unal.edu.co

José Carlos Magalhães Pires (Ed.)

energy security and mitigation of greenhouse gas (GHG) emissions [3]. There are different biomass raw materials that have been evaluated for the production of several added-value products. These raw materials have been classified into the first generation (agricultural and edible crops), second generation (inedible agroindustrial residues) and third generation (algae) [4].

Microalgae are unicellular or multi-cellular photosynthetic microorganisms. Microalgae use sunlight and convert carbon dioxide into biofuel, food, biological products and biomass. For example, cyanobacteria, green algae and diatoms are the most common microalgae. They are present in different environments with high salinity. There are more than 50000 species, but only around 30000 have been studied and analyzed [5]. The morphological characteristics presented by each species of microalgae are dependent on environmental conditions such as the amount of light, temperature, pH and nutrients in the water. All microalgae usually possess chlorophyll-a, which confers green microalgae pigment and at least one accessory, which can sometimes mask chlorophyll-a. Chlorophyll-d is found in red microalgae [6] and chlorophyll-c can be found in yellow-green microalgae [7]. Microalgae can either be autotrophic or heterotrophic, and some photosynthetic algae are mixotrophic. The carbon source for the autotrophic algae is CO_2. This substrate can be supplied from an upstream power plant or other emission source, while the heterotrophic algae use organic carbon sources [8]. The mixotrophic algae have the ability to perform both photosynthesis and acquire exogenous organic nutrients.

The interest in the microalgal cultivation has been increasing due to the production of high value products. In this regard, different types of bioreactors have been developed for their cultivation. For many years, the algae have grown in natural water sources such as lakes and ponds. This technique is currently being used for wastewater treatment, which can be carried out in natural or artificial lakes, tanks or round ponds. This culture system is a good production method in which the obtained biomass can be used later for other purposes [9]. Another method recently developed in the close crop system, or closed photobioreactors (PBRs), which can be thin panels or horizontal tubes. One of the limitations of such bioreactors is the supply of light and carbon dioxide. Consequently, the flat panel PBR should be thin and the tubes (in tubular PBR) have small diameter [10]. Another closed PBR is the airlift, which is compact, inexpensive and easy to operate. This device offers a pneumatic contact in which the fluid flows in a defined pattern through channels constructed for this purpose [11]. Currently, the use of stirred tank reactors is being studied to increase the productivity of microalgae in their application at large scale; given that this type of reactor is a standard at industry and laboratory scales [12]. The design and scale of these reactors must consider primarily an efficient provision of light, and minimizing carbon dioxide losses. Therefore, the challenge is to develop efficient methods for developing large-scale bioreactors that work equally well as at laboratory scale. The large-scale production of microalgae biomass is continuous throughout the day; at night the broth is mixed only. The only viable large scale methods of production of microalgae are "circuit" type ponds and tubular PBRs [12, 13].

Microalgae for Energy Production

Microalgae have great advantage for obtaining added-value products, since these microorganisms can produce lipids and protein in a short time compared with other crops [14]. Microalgae have been also used to produce biofuels and energy. Recent studies have found that the biomass and lipid content of algae can be increased through changing cultivation conditions such as CO_2 sparging, carbon source, temperature, salinity and nutrient concentration [15]. *Chlorella sp.* can accumulate starch and lipids under various growth conditions. In this process, starch synthesis preceded lipid accumulation. The content of starch is generally from 21 to 45% of the dry weight (DW) and lipids content is from 23 to 60% DW [16]. Microalgae have many advantages such as higher photosynthetic efficiency, higher biomass productivities, a growth rate faster than higher plants, highest CO_2 fixation and O_2 production [17]. They require less water than terrestrial crops, making the process more efficient and environmentally friendly [18]. In addition, macronutrients as nitrogen, phosphorus and sulphur can be obtained from wastewater, allowing the use of this effluent as nutrient source for microalgae culture for bioremediation [19]. The composition of the algae can be modified changing growth conditions. In some cases, the absence or presence of components in the culture medium (*e.g.* nitrogen) increases the accumulation of lipids or starch in the microalgae [20].

The oils present in microalgae are used for the production of biodiesel. Biodiesel is derived from triglycerides through an equilibrium reaction known as transesterification, which produces methyl esters (biodiesel and glycerol) [13]. Mainly, microalgae are used to obtain oil and a solid cake with high content of starch and fibers, which can be used as substrate in fermentation processes [21, 22]. After oil and starch extraction, the microalgae cake can also be used to feed a cogeneration system [23]. The most direct route for obtaining biofuels is *via* anaerobic digestion to biogas. This process has many advantages over other forms of bioenergy production. It is the most energy efficient and environmentally friendly technology [24].

Table **1** shows some microalgae with their respective products, conditions and starch or lipid content. In some cases, microalgae are grown to obtain metabolites.

For example, *Dunaliella tertiolecta* produces β-carotene [25]. *Spirulina maxima* produces β-carotene and a-tocopherol [26].

Table 1. Multiple microalgae strains used for the production of starch and lipids.

Microalgae	Strain	Product	Conditions	Culture Medium	Starch or Lipid Content (DW)	Ref.
Tetraselmis subcordiformis		Starch	Aeration was kept at 0.4 vvm with 3% CO_2, temperature 25±2 °C Illuminated with fluorescent lamps (200 µmm-2s-1)	Sea water with nutrients addition	62.1%	[27]
Chlorella vulgaris Beijerinck	CCALA 924	Starch	Irradiance (23 µmol m-2s-1), cycles 12/12 h (dark/light), temperature at 12 - 15 °C.	Mineral medium	60%	[28]
Chlamydomonas reinhardtii	UTEX 2337	Starch	Illuminated with white fluorescent tubes. Temperature 25 °C and stirred at 500 rpm. Aeration.	UTEX medium from Chlamydomonas	58%	[29]
	UTEX 90	Starch	Temperature 25 °C, in presence of 2% CO_2	TAP media	53%	[30]
	UTEX 2247	Starch	Temperature 25 °C, air with 0.5% of CO_2 at a rate of 300 ml/min. Illuminated with fluorescent lamps.	Sea water with nutrients addition	45%	[31]
Dunaliella tertiolecta	ATCC 30929	Starch	Temperature 30 °C, 1% CO_2, illumination with fluorescent lamps (10 W m-2)	F/2 medium	18%	[32]
Chlorella sp.	TISTR 8485	Starch	Initial OD1000=0.2. Illumination with cool white fluorescence lamps (60 µmol m-2s-1) Shaken at 150 rpm at 28±1 °C	BG-11 medium	27.0±1.1%	[33]
Chlorococcum sp.	TISTR 8583	Starch	Initial OD1000=0.2. Illumination with cool white fluorescence lamps (60 µmol m-2s-1) Shaken at 150 rpm at 28±1 °C	BG-11 medium	25.28±1.12%	[33]

(Table 1) contd.....

Microalgae	Strain	Product	Conditions	Culture Medium	Starch or Lipid Content (DW)	Ref.
N. muscorum	TISTR 8871	Starch	Initial OD1000=0.2. Illumination with cool white fluorescence lamps (60 μmole photon m-2--1) Shaken at 150 rpm at 28±1 °C	BG-11 medium	33.49±0.89%	[33]
Nostoc sp.	TISTR 8873	Starch	Initial OD1000=0.2. Illumination with cool white fluorescence lamps (60 μmol m-2s-1) Shaken at 150 rpm at 28±1 °C	BG-11 medium	32.85±1.52%	[33]
Green algae	NKG 121701	Starch	Aeration flasks with illumination (40 μE m-2s-1) at room temperature		>50%	[34]
Spirulina fusiformis		Starch	Sea water with different salinity enriched with Zarouk media and autoclaved at 121 °C for 15 min.	Zarouk media with sea water	16.8 - 31.4%	[35]
Skeletonema costatum	CP-12	Lipids	Temperature 16±1 °C, illumination , cycles 12/12 h (dark/light), 0.3 cm3 min-1 CO2	F/2 medium with sea water	65%	[36]
N. gregaria	CP-20	Lipids	Temperature 16 ± 1 °C, illumination , cycles 12/12 h (dark/light), 0.3 cm3 min-1CO2	F/2 medium with sea water	76%	[36]
Nitzschia palea		Lipids	Temperature 22 ± 2 °C under continuous light of 2.789 W m-2	Standard Navicula medium	26.2%	[37]
Scenedesmus sp.		Lipids	Mixotrophic conditions. Temperature 22 ± 2 °C under continuous light of 250 μmol m-2s-1 illumination	10 g/L of technical glycerol	15.1 ± 1.4% 15.1 ± 1.4%	[38]
Nannochloris sp.		Lipids	Mixotrophic conditions. Temperature 22 ± 2 °C under continuous light of 250 μmol m-2s-1 illumination	2 g/L of technical glycerol	16.2±1.6	[38]

(Table 1) contd.....

Microalgae	Strain	Product	Conditions	Culture Medium	Starch or Lipid Content (DW)	Ref.
Isochrysis sp.	T.ISO	Lipids	Temperature 25 °C. Illuminated with white light (80±2 μmol m-2s-1), pH 8.3	f/2 medium	20.7 ± 4.0%	[39]
Cryptomonas	CRFI01	Lipids	Temperature 25 °C. Illuminated with white light (80±2 μmol m-2s-1), pH 8.3	f/2 medium	21.4 ± 4.0%	[39]
Schizochytrium limacinum	OUC88	Lipids	Temperature 23 °C shaking rotationally at 500 rpm	GS medium	~13 g/L	[40]
Crypthecodinium cohnii	ATCC 30772	Lipids	Temperature 27 °C, shaking at 180 rpm, pH 6.5	ATCC Culture Medium 460 A2E6		[41]

DW = Dry Weight

Biofuel Production Processes from Microalgae

It is possible to identify three main types of components for the further conversion of microalgae into biofuels: oils, starch and the dry cake. Each of these components will lead to a different biofuel or energy source (biodiesel, bioethanol, energy, and hydrogen, among others). Therefore, there are conversion options for algal biomass associated to different energy carriers (solid, liquid or gaseous fuels). The production processes can be separated into four main groups: biochemical processes, thermochemical processes, algal biomass to biodiesel and biorefineries [14, 42]. The most significant differences between them are the desired form of the energy (end form of the product) and the state of the feedstock. Just to exemplify these differences, for thermochemical processes the end form of the energy is usable energy and gaseous fuels using the dry cake of the microalgae, while for certain biochemical process the end form is a liquid biofuel (biodiesel) using the oils metabolized by the microalgae. Fig. (**1**) shows the production processes of biofuels from microalgae.

Biochemical Conversion

Biochemical conversion of microalgae is directed generally to the production of gas and liquid biofuels (methane, hydrogen and ethanol). This type of processes implies the usage of the biomass as a feedstock for its processing by other micro-organisms to produce the biofuel. The main processes are anaerobic digestion, alcoholic fermentation and photobiological hydrogen production [14, 44].

Anaerobic Digestion

Anaerobic digestion is a biological process to decompose organic matter through the assortment of microbes in the absence of oxygen to produce biogas [45]. Biogas consists mainly in methane and carbon dioxide with traces of other gases such as hydrogen sulphite [46]. This process has three sequential stages: hydrolysis, fermentation and methanogenesis. The first stage breaks the complex compounds into soluble sugars. Then, fermentative bacteria convert these into alcohols, acetic acid, volatile fatty acids (VFA), and a gas containing H_2 and CO_2. Finally, methanogens metabolizes these compounds mainly into CH_4 (60-70%) and CO_2 (30-40%) [19]. The produced gas has an energy content approximate to 20-40% of the lower heating value of the feedstock [14].

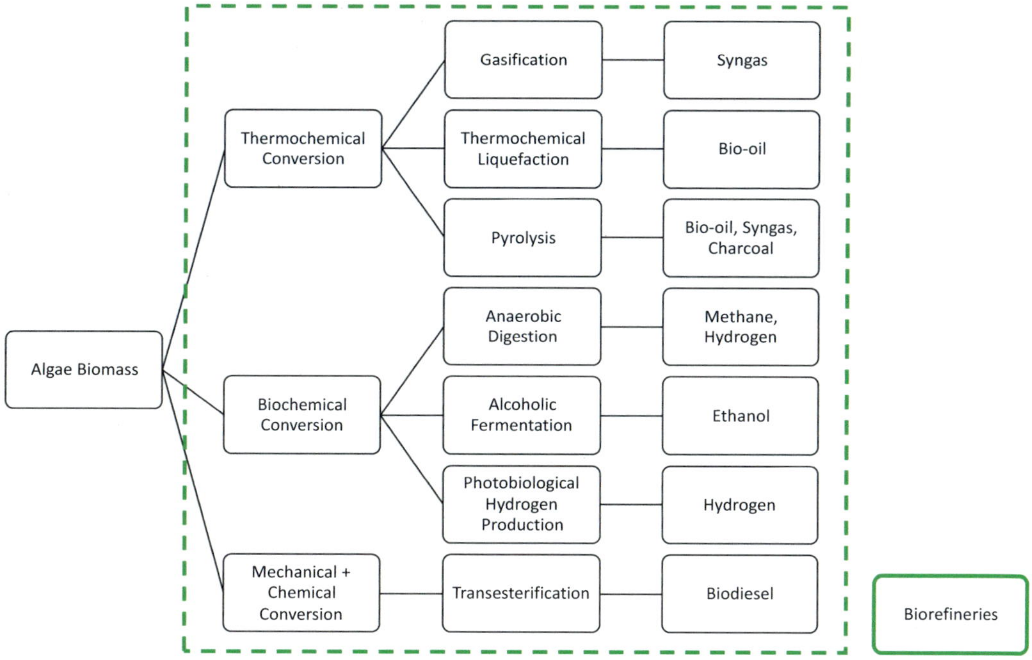

Fig. (1). Production processes of biofuels from microalgae. Adapted from [14, 43].

The first authors to report the anaerobic digestion of microalgae biomass were Golueke *et al.* [47]. These authors investigated the anaerobic digestion of *Chlorella vulgaris* and *Scenedesmus*, microalgae that generally grow as part of a wastewater treatment process. According to Sialve *et al.* [48], the conversion of algal biomass into methane could recover as much energy as that obtained from the extraction of cell lipids and this process can be useful for wet algal biomass given that moisture content must be between 80-90% [49]. Since the early advances performed in this topic, authors identified that the harvesting of microalgae biomass presents a fundamental challenge for the economic viability of an energy system using microalgae biomass as a substrate for anaerobic digestion. Table **2** shows the methane production from the anaerobic digestion of multiple microalgae reported in literature. This table summarizes the research done about several microalgae and highlights the gas potential from microalgae as a viable process for biogas production.

Table 2. Methane production from the anaerobic digestion of multiple microalgae reported in literature. Adapted from [50].

Microalgae Species	*Methane Yield*	*Loading Rate*
Arthrospira maxima	173 mL g^{-1} VS	500 mg/TS/L
Arthrospira platensis	481 mL g^{-1} VS	2000 mg/TS/L
Blue green algae	366 mL g^{-1} VS	281.96 mg/VS/L
Chlamydomonas reinhardtii	587 mL g^{-1} VS	2000 mg/TS/L
Chlorella kessleri	335 mL g^{-1} VS	2000 mg/TS/L
Chlorella sp., Pseudokirchneriella sp. and Chlaqmydomas sp.	0.28-0.60 m^3/kg/VS	402 mg VS
Chlorella sp., Scenedesmus, Euglena and Oscillatoria	300-800 mL g^{-1} VS	N/R
Chlorella sp., Scenedesmus	170-320 mL g^{-1} VS	1.44-2.89 g/VS/L
Chlorella sorokiniana	212 mL g^{-1} VS	N/A
Chlorella vulgaris	403 mL g^{-1} VS	2 g/VS/L
Chlorella vulgaris	286 mL g^{-1} VS	5000 mg/VS/L
Chlorella vulgaris	240 mL g^{-1} VS	1000 mg/VS/L
Chlorella vulgaris	0.40-0.45 L	2677-6714 mg (COD)
Dunaliella	440 mL g^{-1} VS	910 mg/VS/L
Dunaliella salina	505 mL g^{-1} VS	2000 mg/TS/L
Dunaliella tertiolecta	24 mL g^{-1} VS	5000 mg/VS/L
Durvillea Antarctica	492 mL g^{-1} VS	3000 mg/dry/TS/d
Euglena gracilis	485 mL g^{-1} VS	2000 mg/TS/L
Lake Chaohu natural population consortium	295 mL g^{-1} VS	N/R
Macrocystis pyrifera and Durvillaea Antartica(50% blend)	540 mL g^{-1} VS	3000 mg/dry/TS/d
Macrocystis pyrifera	545 mL g^{-1} VS	3000 mg/dry/TS/day
Microcystis sp.	70.33-153.51 mL	1500-6000 mg/VS
Nannochloropsis oculata	204 mL g^{-1} VS	N/R
Nannochloropsis salina (lipid extracted biomass)	130 mL g^{-1} VS	2000 mg/L/VS
Scenedesmus obliquus	287 mL g^{-1} VS	2000 mg/TS/L

(Table 2) contd.....

Microalgae Species	*Methane Yield*	*Loading Rate*
Scenedesmus obliquus	240 mL g^{-1} VS	2000 mg/VS/L
Scenedesmus sp.	170 mLg^{-1} COD	1000 mg/COD/L
Scenedesmus sp. (single stage)	290 mL g^{-1} VS	18,000 mg/VS/L
Scenedesmus sp. (two stage) Note: 46 mL/g/VS Hydrogen	354 mL g^{-1} VS	18,000 mg/VS/L
Scenedesmus sp. and Chlorella sp.	16.3-15.8 ft^3	7.8-9.2 ft^3/lb (VS)
Scenedesmus sp. and Chlorella sp.	143 mL g^{-1} VS	4000 mg/VS/L
Spirulina Leb 18	0.79 g/L	72,000 mg/L/TS
Spirulina maxima	0.35-0.80 m^3	20-100 kg/m^3 (VS)
Spirulina maxima	320 mL g^{-1} VS	910 mg/VS/L
Spirulina maxima	330 mL g^{-1} VS	22,500 mg/VS/L
Spirulina platensis UTEX1926	0.40 m3 kg	N/R
Tetraselmis	0.25-0.31 L g^{-1} VS	2000 mg/VS
Waste water grown community	497 mL g^{-1} TS	2.16 g/L/TS
Zygogonium sp.	344 mL g^{-1} TS	N/R

N/R = Non-reported

Alcoholic Fermentation

Alcoholic fermentation is the conversion of sugars, starch or cellulose into ethanol by a microorganisms. As it was previously observed, microalgae can be a source of both biomass and starch, and hence they can be used as substrate. However, depending on the type of fermenting microorganism, it is necessary to perform a pretreatment stage in order to convert the microalgae into an absorbable substrate. The most common microorganisms used for alcohol fermentation can metabolize glucose, sucrose, xylose, and amylose. Therefore, this is the purpose of the pretreatment stage, to convert the feedstock into a system easily accessible for the enzymes to produce sugars. Generally, after the fermentation stage comes a purification stage that consists of two distillation columns, one for distillation and the other for rectification, and a system of molecular sieves to dehydrate the ethanol.

Ethanol can be used as a vehicle fuel, but it is normally used as an additive that aims to increase the octane rating and improving vehicle emissions [51]. Car petrol engines can operate on petroleum/gasoline blends of up to 15% of bioethanol. An advantage of ethanol is its higher octane rating compared to ethanol-free gasoline, which allows increasing the engine's compression ratio for increased thermal efficiency [51]. The solid residue generated in the process can be used for cattle-feed or for gasification, which helps diminishing feedstock costs, which typically are 55-80% of the final alcohol selling price [14, 49].

C. vulgaris has a high starch content (37% DW), and conversion efficiencies of up to 65% of ethanol have been recorded [31]. Ethanol production has also been reported *via* dark fermentation process with a maximum ethanol productivity of 450 μmol g^{-1} DW at 30 °C [34, 52]. Despite the possibility of using microalgae in alcoholic fermentation, the most efficient microalgae potential is in the context of lipid production, and ethanol production is applied to the waste algae biomass from oil extraction [14].

Photobiological Hydrogen Production

Hydrogen fuel can provide power to multiple types of engines (rockets, cars, boats and airplanes) [53]. Microalgae are able to photoproduce H_2 gas given their genetic, metabolic and enzymatic characteristics under anaerobic conditions [14, 54]. During photosynthesis, microalgae convert water molecules into oxygen and hydrogen ions (H^+), which are then converted by hydrogenase enzymes into H_2 under anaerobic conditions [19].

There are two main approaches for photosynthetic H_2 production. The first approach consists of two stages. In the first stage, algae are grown photosynthetically in normal conditions. During the second stage, the algae are induced to anaerobic conditions and stimulating consistent hydrogen production [54, 55]. The hydrogen production yield maintains high until 60 h of production and does not generate toxic or environmentally harmful products, and allows obtaining value added products as a result of biomass cultivation [56]. The second approach simultaneously produces photosynthetic oxygen and H_2 gas. The electrons that are released in the photosynthetic H_2O oxidation are fed directly into the hydrogenase-mediated H_2-evolution process [14]. The H_2 productivity of the second approach is theoretically superior to the first one, but it presents severe hydrogenase inhibition after a very short period due to the photosynthetic production of oxygen [55]. The theoretical maximum yield of hydrogen by green algae can be approximately 198 kg H_2 ha^{-1} per day [57].

Thermochemical Conversion

Gasification

Gasification is a process that converts solid or liquid hydrocarbons into synthesis gases, which has proved to be a successful option for waste management, chemical production and energy production from non-conventional feeds like forest wastes, agricultural wastes, poultry waste, and municipal sewage, among others [58]. Gasification, as a biomass-to-energy pathway, offers the possibility of producing syngas from multiple potential feedstocks. Syngas is a low calorific gas (typical 4-6 MJ m^{-3}) that can be burnt directly or used as a fuel for gas engines/turbines [14].

Regarding the gasification of microalgae biomass, *Spirulina* has been partially oxidized at temperature ranges between 850-1000 °C, estimating around 1000 °C; the highest theoretical yield of 0.64 g methanol from 1 g of biomass [59]. The authors estimated an energy balance (ratio of methanol produced to the total required energy) of 1.1, therefore obtaining a marginal positive value. According to Minowa and Sawayama [60], it is possible to use the microalgae *C. vulgaris* in systems with nitrogen cycling to obtain methane-rich fuel. In addition, the nitrogen components of the microalgae can be converted into fertilizer ammonia. However, this area needs more research especially into the energy balance of drying the biomass for gasification [14].

Thermochemical Liquefaction

The thermochemical liquefaction is a catalyzed process that takes place at low temperature (300-350 °C) and high pressure (5-20 MPa) in the presence of hydrogen [61]. Despite the expensive reactors and fuel-feed systems [49], it has good advantages for the production of energy from wet biomass [62]. This process uses sub-critical water, taking advantage of its high activity, to decompose biomass materials in shorter molecular materials with higher energy density [63].

According to Brennan and Owende [14], there are processes that can be employed to convert wet algal biomass material into liquid fuel. For example, this process was applied at 300 °C to *B. braunii*, obtaining a maximum yield of 64% DW of oil with higher heating value (HHV) of 45.9 MJ kg^{-1} [64]. Another study obtained an oil yield of 42% DW from *Dunaliella tertiolecta* giving an HHV of 34.9 MJ kg^{-1} [65]. These results indicate that this process is a feasible option for the conversion of algal biomass into liquid fuel [14].

Pyrolysis

Pyrolysis is the conversion of biomass into bio-oil, syngas and charcoal at medium to high temperatures (350-700 °C) in the absence of air. This process is the basis of several methods that are being developed for producing biomass based fuel, including either first or second generation feedstocks [61]. However, there are technical challenges given that the oils obtained through this process are acidic, unstable, viscous, and contain solids and chemically dissolved water [14, 66]. Therefore, the process require further processing to lower the oxygen content and remove alkalis [67].

This technology compared to the others is one of the most studied in terms of the usage of algal biomass. For example, *Chlorella prothothecoides* under heterotrophic growth conditions recorded an oil yield of 57.9% DW with an HHV of 41 MJ kg^{-1}, a value 3.4 times higher than the achieved by phototrophic cultivation [68]. For *C. prothothecoides* and *Microcystis aeruginosa* grown phototrophically, bio-oil yields of 18% with a HHV of 30 MJ kg^{-1} and 24% with a HHV of 29 MJ kg^{-1} were achieved, respectively. According to Demirbas [69], for *C. prothothecoides* it is possible to obtain an HHV of 39.7 MJ kg^{-1} at temperatures ranging from 502-552 °C. According to the information shown in Table **3**, bio-oils obtained from microalgae are of higher quality than those extracted from lignocellulosic materials [68, 69].

Table 3. Comparison of typical properties of bio-oils (pyrolysis of wood and microalgae) and petroleum oils. Taken from [14].

Properties	*Typical Values*		
	Bio-oils		*Petroleum Oil*
	Wood	*Microalgae*	
C (%)	56.4	62.07	83.0-87.0
H (%)	6.2	8.76	10.0-14.0
O (%)	37.3	11.24	0.05-1.5
N (%)	0.1	9.74	0.01-0.7
Density (kg L^{-1})	1.2	1.06	0.75-1.0
Viscosity (Pa s)	0.04-0.20 (at 408C)	0.10 (at 408C)	2-1000
HHV (MJ kg^{-1})	21	29-45.9	42

Finally, for comparison purposes, Table **4** shows some research results obtained for the previous three technologies in the production of energy using microalgae as feedstock.

Table 4. Comparison of gasification, pyrolysis and thermochemical liquefaction for energy production from microalgae. Taken from [14].

Conversion Process	*Microalgae*	*Production*	*Operation Conditions*	*Liquid*		*Gas Content (% dry wt.)*	*Solid Content (% dry wt.)*
				Content (% dry wt.)	*HHV (MJ/kg)*		
Gasification	*Spirulina*	N/A	1000 °C; 0.101 MPa	-	-	64	-
Thermochemical liquefaction	*Botryococus braunii*	N/A	300 °C; 3 MPa	64	45.9	-	-

(Table 4) contd.....

Conversion Process	Microalgae	Production	Operation Conditions	Liquid		Gas Content (% dry wt.)	Solid Content (% dry wt.)
				Content (% dry wt.)	HHV (MJ/kg)		
Thermochemical liquefaction	*Duanliella tertiolecta*	N/A	300 °C; 3 MPa	42	34.9	-	-
Pyrolysis	*Chlorella prothothecoides*	Heterotrophic	450 °C; 0.101 MPa	57.9	41	32	10.1
Pyrolysis	*Chlorella prothothecoides*	Phototrophic	450 °C; 0.101 MPa	16.6	30	-	-
Pyrolysis	*Chlorella prothothecoides*	Phototrophic	500 °C; 0.101 MPa	18	30	-	-
Pyrolysis	*Chlorella prothothecoides*	N/A	502 °C; 0.101 MPa	55.3	39.7	36.3	8.4
Pyrolysis	*Microcystis aeruginosa*	Phototrophic	500 °C; 0.101 MPa	24	29	-	-

Algal Biomass to Biodiesel

Biodiesel is a drop-in diesel alternative, made from domestic, renewable resources such as plant oils, animal fats, used cooking oil, and even new sources such as algae. It is a fuel comprised of mono-alkyl esters of long chain fatty acids derived from vegetable oils or animal fats, meeting ASTM D 6751 [70]. It is produced through transesterification, in which glycerin is separated from the oil. This process has two products, the methyl esters (biodiesel) and glycerin.

As aforementioned, microalgae are known for producing intra-cellular oils that can be used as feedstock in the transesterification for generating biodiesel. These oils have similar physical and chemical properties to petroleum diesel, 1st generation biodiesel from oil crops and compares favorably with the international standard EN14214 (International Biodiesel Standard for Vehicles) [14].

Some of the advantages of algal biodiesel are that it is renewable, biodegradable, and quasi-carbon neutral; it is non-toxic and contains reduced levels of particulates, carbon monoxide, soot, hydrocarbons and SO_x, and the reduced CO_2 emissions compared to petroleum diesel [14, 71]. Algal biodiesel is more suitable in the aviation industry given its low freezing points and high energy densities, compared to 1st generation biodiesel [72].

Biorefineries

For biotechnological process, the most advanced concept up to date is the biorefinery, which is a structure that integrates biomass conversion processes in a given scheme for the production of different products such as biomolecules, biomaterials, bioenergy and biofuels [73]. This methodology applies three concepts: hierarchy, sequencing and integration. Hierarchy refers to the hierarchical classification of the principal elements that conforms a biorefinery system (feedstock, technologies and products). Sequencing involves the analysis of the different routes to obtain the proposed products (chemical, biochemical and thermochemical). Integration involves the relationship between the feedstock, technologies or processes and products through material and energy balances [23, 74].

The first level in a biorefinery is the raw materials. Based on the composition of the raw material, it is only possible to obtain certain chemical building blocks (*i.e.* sugars, fats and oils, proteins, polyphenols, among others) that can be converted in specific products. Moncada [74] proposed five families of products that can be obtained from biomass, which are biofuels, bioenergy (*e.g.* power and heat), food products, biomaterials, biomolecules and chemical products. Fig. (**2**) shows the linkage between the types of biomass, obtained chemical building block and possible products.

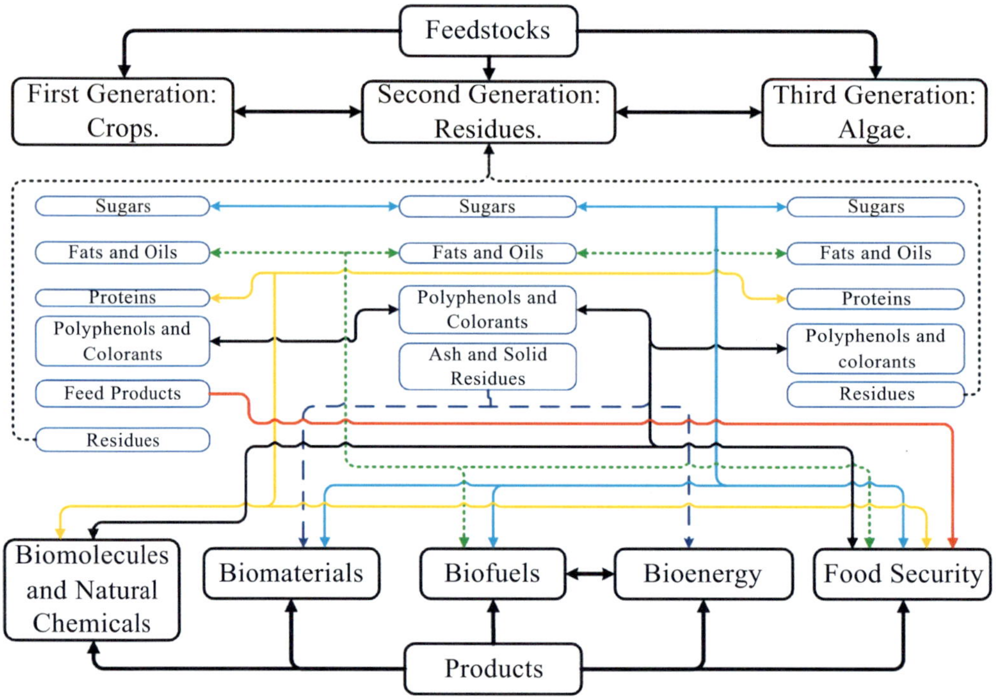

Fig. (2). Hierarchical distribution of types of biomass, obtained chemical building block and possible products [75].

In the specific case of microalgae, biorefineries aim for the integration of multiple processes and technologies to generate different energy blocks out of one single raw material. To give a mere example of this application, it is possible to consider the growth of the microalgae to maximize the oils production. Then, after proceeding to remove the oils, there are two feedstock blocks. One, the oils that can be used to produce biodiesel. The second block is the wet/dry cake of the microalgae. This feedstock can be used for several thermochemical conversion processes or to produce bioethanol, as it was previously mentioned. In this sense, there is a complete and more intensive usage of the microalgae. This process might even consider the extraction of metabolites (high added-value product) before the obtainment of energy production, which will increase the economic performance of the process.

Challenges in the Use of Microalgae for Energy Production

Microalgae biomass has a wide range of possible biotechnological uses, not only in the field of energy production, but also in the field of biomolecules. This raw material can be used for the production of energy and other commodities for several reasons such as high productivity, multiple range of substrates (from sugars, to low quality water and seawater), and that no fertile land is needed [76]. However, these uses have a significant difference in terms of the required amount of biomass to process. The most common use for this raw material is for obtaining added-value products for human applications (health, cosmetics, nutraceutical and foods) and aquaculture [77]. For these applications, the required amount of biomass is very small compared to other markets such as energy or commodities [76].

Although microalgae have a high potential to contribute to the world demand of energy and commodities, the current available technology is not enough to supply the potential world demand. Thus, this raw material and its subsequent production processes have some challenges that must be overcome and it is still necessary to solve several bottlenecks related with biological, engineering, economic and environmental aspects [78].

First, the technology must be scaled-up several orders to significantly contribute to the biofuels market. Specially, it is necessary to reduce the biomass production cost [78]. According to Acién *et al.* [76], this production cost is associated to some specific factors. The cost of the equipment is strongly related to the type of production scheme (PBR or open ponds/raceways), which can vary from 0.1 €/L to 3 €/kg. Another factor is the consumption of raw material, specifically the carbon dioxide, which is the most expensive consumable and whose purity affects the biomass production yields. Finally, water and fertilizers uptake contributes significantly to the production cost. For this, it is recommended to use wastewater containing minerals, which may reduce the production cost, although it is necessary to define the required pretreatment. When optimizing these factors, it is possible to achieve production costs near to 1.8 €/kg, but this cost is still higher than what would be acceptable for the biofuels market [76].

It has been reported that microalgae fuel production is relatively close to economic feasibility [79], but current production technologies request substantial reduction of the current costs and to operate near the optimal photosynthetic yield. Therefore, regarding the production costs, it is necessary to use low-cost carbon dioxide and medium, developing more productive photobioreactor systems and dramatically reducing their cost [76, 80].

Second, as the achieved microalgae density in the medium is low in harvesting stage, the required areas to produce high amounts of biomass are very high. According to Pulz and Gross [81] and Chisti [82], microalgae biomass market produces only 5 ktonne/year, with a cost of 25000 $/tonne, and in order to replace 5% of the US demand of transport fuel, it is necessary to produce between 20000-66000 ktonne/year of biomass at production costs below 400 $/tonne. Therefore, it is required to select an optimized location to increase the productivity, improving the photosynthetic efficiency of microalgae by adequate culture conditions and control, in order to decrease the required areas by enhancing the productivities per area.

Finally, the most important challenge to overcome is associated to the integration of the current and potential uses of the microalgae. This means to evaluate the use of the microalgae to obtain simultaneously added-value products and energy. There is a significant potential in the simultaneous extraction of valuable co-products (*viz.*, β-carotene, polyunsaturated fatty acids, biofertilizers, among others) with biofuel production [14]. The large-scale production of microalgae required for biofuels/energy would increase the net amount of these high added-value products, whose price will boost the economic performance of the process. In addition, the biomass resulting after the extraction of the high added-value products could still be used for energy production.

Based on this, biorefineries can become an excellent option for the integration of the possible processing lines of the microalgae, obtaining in first place the high added-value products, given that they require higher purity and less processing and then using the microalgae biomass cake for energy production. This scheme offers the possibility of producing energy or biofuels, depending on the selected process (based on the multiple ones previously analyzed). However, the most technical, economic and environmental biorefinery scheme requires further research.

SCENARIOS ANALYSIS: SIMULATION FOR COGENERATION AND BIODIESEL CASES

After analyzing and showing the wide potential of microalgae, its current technologies and processes and the challenges still to overcome, this section shows certain specific examples of the use of microalgae for the production of biofuels and energy, specifically biodiesel and cogeneration, respectively.

Aspen Plus software was used to simulate the global processes. The thermodynamic model used was UNIFAC Dortmund for the liquid phase and Soave Redlich Kwong with the Boston Mathias modification (RK&BM) for the vapor phase. Conversely, extraction residues were used as a fuel in a biomass fired cogeneration, included into the simulation according to chemical compositions reported by Phukan *et al.* (Microalgae paste) [83]. The kinetic model for basic catalysis employed in this work was reported by Granjo *et al.* [84]. Economic evaluations were calculated using Aspen Process Economic Analyzer with an annual interest rate of 17%, a tax rate of 33% and the straight-line depreciation method over a 12-years period of analysis. For feedstock prices, the international reports from ICIS pricing were employed. The reported prices were the following: methanol (2.11 USD/kg), sodium hydroxide (0.39 USD/kg), phosphoric acid (0.78 USD/kg). For microalgae oil as a non-commercial oil, the values reported in literature by Campbell *et al.* [85] were employed (1.24 USD/L). Operating charges such as operator and supervisor labor costs were defined for Colombia at 2.14 and 4.29 USD/h, respectively. Electricity, potable water, low and high pressure steam costs were 0.0304 USD/kWh, 1.25 USD/ m^3, 1.57 USD/ton and 8.18 USD/ton, respectively.

The environmental impact was assessed with the waste reduction (WAR) algorithm (developed by the U.S. Environmental Protection Agency) to estimate the potential environmental impact (PEI) generated in the process considering eight environmental impact categories: human toxicity potential by ingestion (HTPI), human toxicity potential by dermal and inhalation exposure (HTPE), terrestrial toxicity potential (TTP), aquatic toxicity potential (ATP), global warming potential (GWP), ozone depletion potential (ODP), photochemical oxidation potential (PCOP), and acidification potential (AP). The mass flow rate of each component in the process streams was multiplied by its chemical potential to determine its contribution to the potential environmental impact categories.

Cogeneration from Extraction Cake Residues

Cogeneration technologies are an option for the production of mechanical and thermal energy with significant energy and cost savings. Operationally, it has also a higher efficiency compared to systems that produce the same energy separately.

Fig. (**3**) shows the process flow-sheet diagram for a cogeneration system based on the extraction cake residues. The extraction cake residues are obtained after extracting the oil of the grown microalgae. This residue has high content of starch and fibers. Among the available technologies for biomass-fired cogeneration, the combined-cycle gas turbine (CCGT) is considered as a promising alternative due to its efficiency. The CCGT configurations are composed mainly by three systems: (i) a gas turbine where chemical energy is converted into mechanical work; (ii) a heat steam recovery generator (HSRG), which is a high-efficiency steam boiler that uses hot gases from a gas turbine or a reciprocating engine to generate steam, in a thermodynamic Rankine cycle, generating steam at different pressure levels depending on the requirements of the chemical process; and (iii) a steam turbine where the steam produced in HSRG system is employed to generate additional power [86].

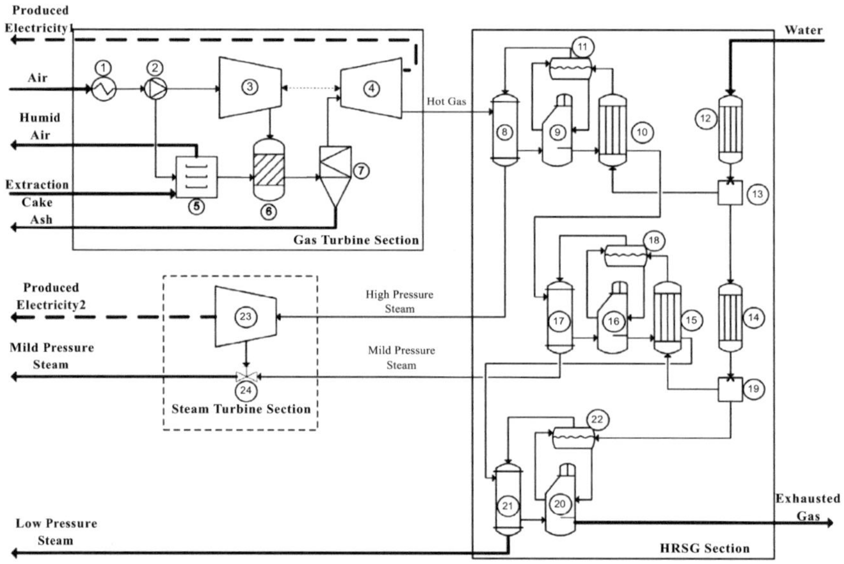

Fig. (3). Process flow-sheet diagram for cogeneration system based on extraction cake residues. 1. Direct use heat exchanger, 2. Air divisor, 3. Compressor, 4. Turbine, 5. Dryer, 6. Combustion chamber, 7. Cyclone, 8. High-pressure superheater, 9. High-pressure evaporator, 10. High-pressure economizer, 11. High pressure drum, 12. Low-pressure economizer I, 13. Flow divisor, 14. Low-pressure economizer II, 15. Intermediate pressure economizer, 16. Intermediate pressure evaporator, 17. Intermediate pressure superheater, 18. Intermediate pressure drum, 19. Flow divisor, 20. Low-pressure evaporator, 21. Low-pressure superheater, 22. Low-pressure drum, 23. Steam turbine, 24. Flow mixer.

Simulation results for the biomass fired cogeneration plant using microalgae paste as fuel reveal an energy production of 33.33 MW and electricity production of 8.22 MW. In this case, 1.17 MW of cogenerating heating are consumed in direct heat use and 0.011 MW of cogenerating electricity are consumed in pumping.

Biodiesel Production Using Basic Catalysis

The biodiesel production has three main sequential stages: Pretreatment, reaction and purification. These stages are shown in Fig. (**4**).

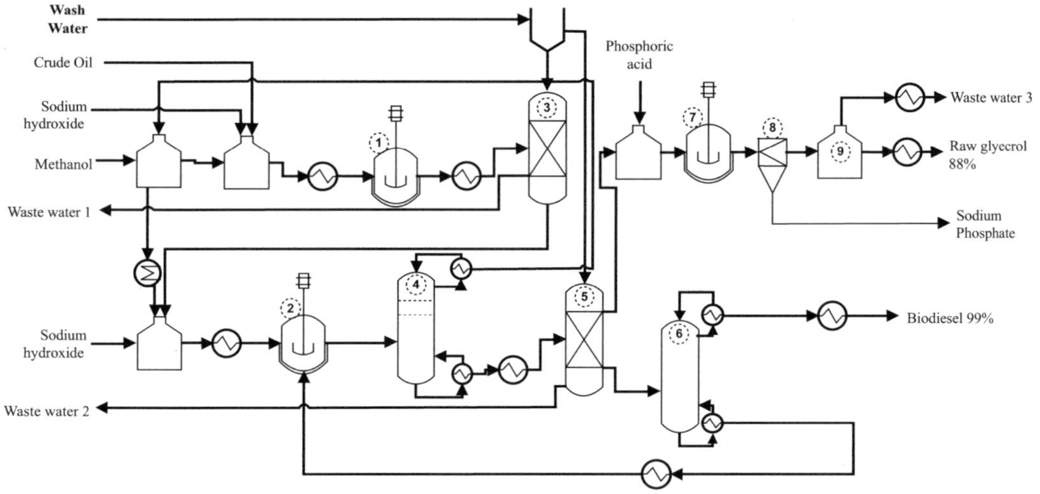

Fig. (4). Process flow diagram for biodiesel production using basic catalysis. 1. Neutralization reactor I, 2. Transesterification reactor, 3. Liquid liquid extraction column I, 4.Distillation tower I, 5. Liquid liquid extraction column II, 6. Distillation tower II, 7. Neutralization reactor II, 8. Solid separator, 9. Flash.

In the pretreatment stage, particles, colloidal mater, pigments, extraction residues and other impurities can be removed using filtration. First, it is necessary to dry the oil, and then eliminate the free fatty acids (FFA) using neutralization or through the pre-esterification of the FFA, to avoid the saponification reaction. Esterification can be performed under batch operation at 200-250 °C. Generally, the reaction is conducted close to the methanol boiling point.

The reaction stage consists in a catalyzed transesterification to reduce the reaction time and costs. Transesterification is a reaction of fatty acids with an alcohol to produce three molecules of alkyl esters and one of glycerol. The catalyst can be basic or acid, the acid catalyst converts free fatty acids into biodiesel esters through esterification, while simultaneously catalyzes the transesterification of triglycerides to biodiesel. However, acid catalysis has slower reaction rates than basic and requires higher rates of methanol.

After the transesterification, the reaction mixture contains alkyl esters, glycerol, non-converted alcohol, catalyst and mono-, di- and tri-glycerides. There are two alternatives to purify this mixture. First, a liquid-liquid extraction can be performed, recovering alkyl esters in the light phase, while glycerin is in heavy phase, and then non-converted alcohol is separated. Second, vacuum distillation is used first to separate the non-converted alcohol and then liquid-liquid extraction to separate glycerin and biodiesel. The second alternative is the one applied in this case because the initial separation of alcohol facilitates the subsequent separation of phases. After this point, both for homogeneous alkaline and acid processes, catalyst must be neutralized, producing salts that could be removed later by filtration or centrifugation. After neutralization, the ester-enriched phase is purified, removing residual alcohol, catalyst, neutralization salts, soaps and residual glycerol. Finally, biodiesel is dried using distillation or vacuum flash.

The economic evaluation results for biodiesel production are summarized in Table **5**. As significant results, raw material costs (including costs of all feedstocks used in biodiesel production such as oil, methanol, catalyst and acids) for basic catalyzed processes (0.497 USD/L) were the highest. The total cost with cogeneration was 0.307 USD/L. For biodiesel from microalgae oil, there are not commercial values reported in the open literature. However, authors such as Demirbas [87] and Campbell *et al.* [85] agree that the production cost of this oil should be lower than diesel oil. In this sense, biodiesel obtained from microalgae oil covers this requirement. When the cogeneration was included, the process cost changed positively.

Table 5. Economic evaluation results for biodiesel production.

	Cost (USD/L)
Raw material cost	0.497
Total utilities cost	0.022
Operating labor	0.008
Maintenance	0.005
Operating charges	0.002
Plant overhead	0.006
General and administrative cost	0.037
Subtotal cost	0.577
Credit by electricity selling	-0.269
Total cost with cogeneration	0.230

The environmental analysis results using WAR algorithm are shown in Fig. (**5**). In Fig. (**5a**), it can be seen that the microalgae used as feedstock processed using basic catalyzed has a high mitigation potential, reflected in the negative PEI. In Fig. (**5b**), it can be seen how all outlet streams contain pollutant elements, which can affect the ecosystem. This means from the generation point of view that all the alternatives have the potential to mitigate the environmental load by transforming feedstocks into added-value products. In this analysis, it can be seen how all outlet streams contain pollutant elements, which can affect the ecosystem. Therefore, it would be desirable for these streams to have a PEI close to zero, in order to reduce the potential environmental impacts of products.

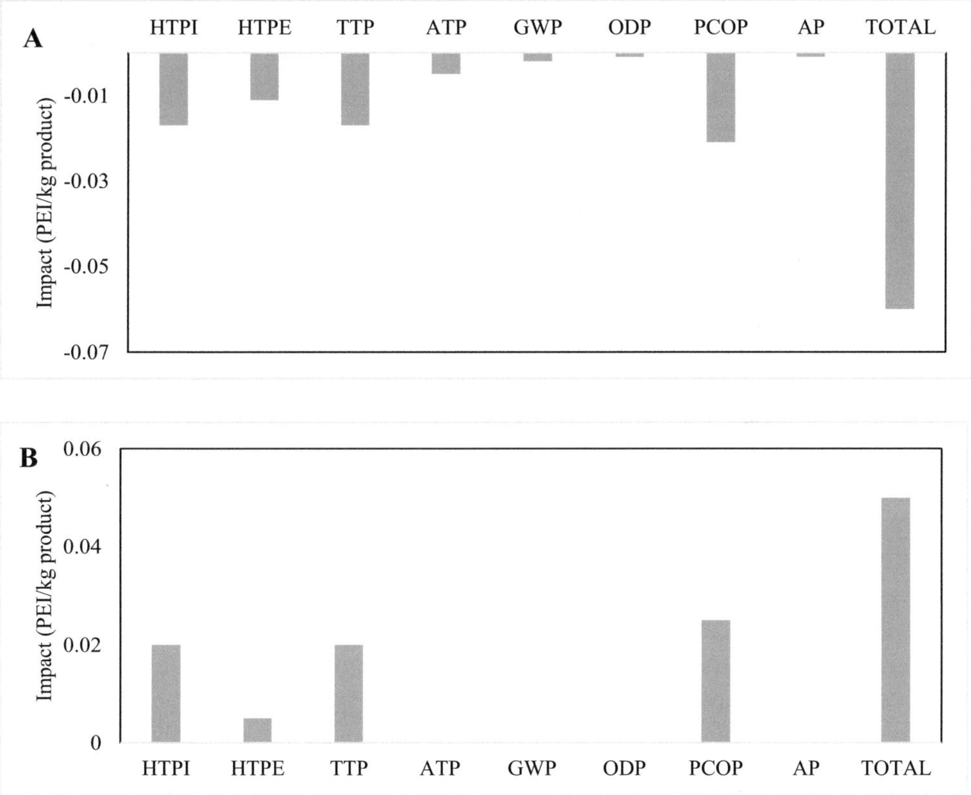

Fig. (5). PEI analysis for biodiesel production from microalgae cake. a) Generated PEI per mass of product. b) Outlet PEI per mass of product. Human Toxicity Potential by Ingestion (HTPI), Human Toxicity Potential by Exposure (HTPE), Aquatic Toxicity Potential (ATP), Terrestrial Toxicity Potential (TTP), Global Warming Potential (GWP), Ozone Depletion Potential (ODP), Photo-chemical Oxidation Potential (PCOP) and Acidification Potential (AP).

Cogeneration results gave an advantage to basic catalyzed processes due to the possibility of employing the extraction residues to generate heat and power. Both higher heating and power potential were released by microalgae paste. The high protein and low fiber content of these feedstocks apparently helped to increase the calorific value of these residues. As long as they have comparative higher calorific values and higher available rates, these residues could generate more steam and consequently, produce more heat and power.

Microalgae oil is more expensive than other common oils used for biodiesel production, but the production volume makes the actual cost lower and makes the process more efficient in the context of biofuel production in Colombia. The oil price would be reduced by the intensification of the cultivation and growth of biomass because the algae have a high photosynthetic capacity and a high duplication rate of cells in a short residence time with lower cultivation area compared to conventional energy crops [88].

The main advantage of microalgae cultivation compared to other crops is the ability to assimilate waste streams, such as greenhouse gases. Intensive cultivation of algae is an attractive alternative for replacing fossil fuels and the production of value-added products. Microalgae have high productivity and yield per planted area, as well as adaptation to diverse media availability of nutrients and water. However, technology does not guarantee low costs in the stages of oil separation due to the large amounts of water. This last step requires more research and development to be introduced in the industry. When microalgae processes obtain oil and cake, they can become much more profitable due to the decline in energy demand.

CONCLUSION

Microalgae are a source of different value-added products. However, obtaining each of these products as a stand-alone process is not very efficient in economic, technical and environmental terms. Microalgae are one of the main candidates for a future source of biofuel production and it has been proven through its usage under different bioenergy and biofuels productions schemes. However, these processes are performed currently only as separate processes that do not integrate the multiple potential of microalgae. The biodiesel is much more efficient coupled to a cogeneration system, which gives value to the microalgae cake, which is one of the wastes of the oil extraction process. It is necessary to continue the development of these schemes to improve the respective technology and reduce production costs.

CONFLICT OF INTEREST

The authors declare no conflict of interest, financial or otherwise.

ACKNOWLEDGEMENTS

The authors express their acknowledgments to the Universidad Nacional de Colombia at Manizales, the research group in Chemical, Catalytic and Biotechnological Processes, the Project entitled "Development of modular small-scale integrated biorefineries to produce an optimal range of bioproducts from a variety of rural agricultural and agroindustrial residues/wastes with a minimum consumption of fossil energy - SMBIO" from ERANET LAC 2015, the project entitled "Evaluación experimental de la gasificación de residuos para la producción de gas de síntesis y electricidad" and the Research Direction of the Research Information System (*Sistema de Información de la Investigación*) HERMES of the Universidad Nacional de Colombia.

REFERENCES

[1] J. P Sorest, *Climatic change: solutions in sight, a Dutch perspective.*, 2000.

[2] Ó.J Sánchez, and C.A Cardona, "Conceptual design of cost-effective and environmentally-friendly configurations for fuel ethanol production from sugarcane by knowledge-based process synthesis", *Bioresour. Technol.*, vol. 104, pp. 305-314, 2012. [http://dx.doi.org/10.1016/j.biortech.2011.08.125] [PMID: 22137752]

[3] International Energy Agency (IEA), "World energy outlook 2007", Paris, France, 2000.

[4] J Moncada, J.J Jaramillo, J.C Higuita, C Younes, and C.A Cardona, "Production of bioethanol using chlorella vulgaris cake: A technoeconomic and environmental assessment in the colombian context", *Ind. Eng. Chem. Res.*, vol. 52, no. 47, pp. 16786-16794, 2013. [http://dx.doi.org/10.1021/ie402376z]

[5] T.M Mata, A.A Martins, and N.S Caetano, "Microalgae for biodiesel production and other applications: A review", *Renew. Sustain. Energy Rev.*, vol. 14, no. 1, pp. 217-232, 2010. [http://dx.doi.org/10.1016/j.rser.2009.07.020]

[6] A Murakami, H Miyashita, M Iseki, K Adachi, and M Mimuro, "Chlorophyll d in an epiphytic cyanobacterium of red algae", *Science.*, vol. 303, no. 5664, p. 1633, 2004. [http://dx.doi.org/10.1126/science.1095459] [PMID: 15016990]

[7] R.R Guillard, and C.J Lorenzen, "Yellow-green algae with chlorophyll C", *Phycology.*, vol. 8, pp. 10-14, 1972.

[8] R Davis, A Aden, and P.T Pienkos, "Techno-economic analysis of autotrophic microalgae for fuel production", *Appl. Energy.*, vol. 88, no. 10, pp. 3524-3531, 2011. [http://dx.doi.org/10.1016/j.apenergy.2011.04.018]

[9] J de la Noüe, G Laliberté, and D Proulx, "Algae and waste water", *J. Appl. Phycol.*, vol. 4, no. 3, pp. 247-254, 1992. [http://dx.doi.org/10.1007/BF02161210]

[10] A.H Scragg, A.M Illman, A Carden, and S.W Shales, "Growth of microalgae with increased calorific values in a tubular bioreactor", *Biomass Bioenergy.*, vol. 23, no. 1, pp. 67-73, 2002. [http://dx.doi.org/10.1016/S0961-9534(02)00028-4]

[11] G Vunjak-novakovic, Y Kim, X Wu, and I Berzin, "Air-Lift Bioreactors for Algal Growth on Flue Gas : Mathematical Modeling and Pilot-Plant Studies", *Ind. Eng. Chem. Res.*, vol. 44, pp. 6154-6163, 2005. [http://dx.doi.org/10.1021/ie049099z]

[12] L Xu, P.J Weathers, X.R Xiong, and C.Z Liu, "Microalgal bioreactors: Challenges and opportunities", *Eng. Life Sci.,* vol. 9, no. 3, pp. 178-189, 2009.
[http://dx.doi.org/10.1002/elsc.200800111]

[13] Y Chisti, "Biodiesel from microalgae beats bioethanol", *Trends Biotechnol.,* vol. 26, no. 3, pp. 126-131, 2008.
[http://dx.doi.org/10.1016/j.tibtech.2007.12.002] [PMID: 18221809]

[14] L Brennan, and P Owende, "Biofuels from microalgae - A review of technologies for production, processing, and extractions of biofuels and co-products", *Renew. Sustain. Energy Rev.,* vol. 14, pp. 557-577, 2010.
[http://dx.doi.org/10.1016/j.rser.2009.10.009]

[15] R Chandra, M.V Rohit, Y.V Swamy, and S Venkata Mohan, "Regulatory function of organic carbon supplementation on biodiesel production during growth and nutrient stress phases of mixotrophic microalgae cultivation", *Bioresour. Technol.,* vol. 165, no. C, pp. 279-287, 2014.
[http://dx.doi.org/10.1016/j.biortech.2014.02.102] [PMID: 24703606]

[16] S Zhu, W Huang, J Xu, Z Wang, J Xu, and Z Yuan, "Metabolic changes of starch and lipid triggered by nitrogen starvation in the microalga Chlorella zofingiensis", *Bioresour. Technol.,* vol. 152, pp. 292-298, 2014.
[http://dx.doi.org/10.1016/j.biortech.2013.10.092] [PMID: 24308944]

[17] L Gouveia, and A.C Oliveira, "Microalgae as a raw material for biofuels production", *J. Ind. Microbiol. Biotechnol.,* vol. 36, no. 2, pp. 269-274, 2009.
[http://dx.doi.org/10.1007/s10295-008-0495-6] [PMID: 18982369]

[18] G.C Dismukes, D Carrieri, N Bennette, G.M Ananyev, and M.C Posewitz, "Aquatic phototrophs: efficient alternatives to land-based crops for biofuels", *Curr. Opin. Biotechnol.,* vol. 19, no. 3, pp. 235-240, 2008.
[http://dx.doi.org/10.1016/j.copbio.2008.05.007] [PMID: 18539450]

[19] K.B Cantrell, T Ducey, K.S Ro, and P.G Hunt, "Livestock waste-to-bioenergy generation opportunities", *Bioresour. Technol.,* vol. 99, no. 17, pp. 7941-7953, 2008.
[http://dx.doi.org/10.1016/j.biortech.2008.02.061] [PMID: 18485701]

[20] J Quin, "Bio - Hydrocarbons from Algae, Barton", *Aust. Rural Ind. Res. Dev. Corp.,* vol. 05/025, no. 5, pp. 1-26, 2005.

[21] R.P John, G.S Anisha, K.M Nampoothiri, and A Pandey, "Micro and macroalgal biomass: a renewable source for bioethanol", *Bioresour. Technol.,* vol. 102, no. 1, pp. 186-193, 2011.
[http://dx.doi.org/10.1016/j.biortech.2010.06.139] [PMID: 20663661]

[22] J Andres Quintero, E Ruth Felix, L Eduardo Rincón, M Crisspón, J Fernandez Baca, Y Khwaja, and C.A Cardona, "Social and techno-economical analysis of biodiesel production in Peru", *Energy Policy.,* vol. 43, pp. 427-435, 2012.
[http://dx.doi.org/10.1016/j.enpol.2012.01.029]

[23] J Moncada, J.A Tamayo, and C.A Cardona, "Integrating first, second, and third generation biorefineries: Incorporating microalgae into the sugarcane biorefinery", *Chem. Eng. Sci.,* vol. 118, pp. 126-140, 2014.
[http://dx.doi.org/10.1016/j.ces.2014.07.035]

[24] M.E Montingelli, S Tedesco, and A.G Olabi, "Biogas production from algal biomass: A review", *Renew. Sustain. Energy Rev.,* vol. 43, pp. 961-972, 2015.
[http://dx.doi.org/10.1016/j.rser.2014.11.052]

[25] Z Shuping, W Yulong, Y Mingde, I Kaleem, L Chun, and J Tong, "Production and characterization of bio-oil from hydrothermal liquefaction of microalgae Dunaliella tertiolecta cake", *Energy.,* vol. 35, no. 12, pp. 5406-5411, 2010.
[http://dx.doi.org/10.1016/j.energy.2010.07.013]

[26] M.S Miranda, R.G Cintra, S.B Barros, and J Mancini Filho, "Antioxidant activity of the microalga Spirulina maxima", *Braz. J. Med. Biol. Res.*, vol. 31, no. 8, pp. 1075-1079, 1998. [http://dx.doi.org/10.1590/S0100-879X1998000800007] [PMID: 9777014]

[27] C Yao, J Ai, X Cao, S Xue, and W Zhang, "Enhancing starch production of a marine green microalga Tetraselmis subcordiformis through nutrient limitation", *Bioresour. Technol.*, vol. 118, pp. 438-444, 2012. [http://dx.doi.org/10.1016/j.biortech.2012.05.030] [PMID: 22717561]

[28] I Brányiková, B Maršálková, J Doucha, T Brányik, K Bišová, V Zachleder, and M Vítová, "Microalgae--novel highly efficient starch producers", *Biotechnol. Bioeng.*, vol. 108, no. 4, pp. 766-776, 2011. [http://dx.doi.org/10.1002/bit.23016] [PMID: 21404251]

[29] N.T Eriksen, F.K Riisgård, W.S Gunther, and J.J Lønsmann Iversen, "On-line estimation of O(2) production, CO(2) uptake, and growth kinetics of microalgal cultures in a gas-tight photobioreactor", *J. Appl. Phycol.*, vol. 19, no. 2, pp. 161-174, 2007. [http://dx.doi.org/10.1007/s10811-006-9122-y] [PMID: 19396354]

[30] M.S Kim, J.S Baek, Y.S Yun, S Jun Sim, S Park, and S.C Kim, "Hydrogen production from Chlamydomonas reinhardtii biomass using a two-step conversion process: Anaerobic conversion and photosynthetic fermentation", *Int. J. Hydrogen Energy.*, vol. 31, no. 6, pp. 812-816, 2006. [http://dx.doi.org/10.1016/j.ijhydene.2005.06.009]

[31] A Hirano, R Ueda, S Hirayama, and Y Ogushi, "CO2 fixation and ethanol production with microalgal photosynthesis and intracellular anaerobic fermentation", *Energy.*, vol. 22, no. 2-3, pp. 137-142, 1997. [http://dx.doi.org/10.1016/S0360-5442(96)00123-5]

[32] A Ike, N Toda, N Tsuji, K Hirata, and K Miyamoto, "Hydrogen photoproduction from CO2-fixing microalgal biomass: Application of halotolerant photosynthetic bacteria", *J. Ferment. Bioeng.*, vol. 84, no. 6, pp. 606-609, 1997. [http://dx.doi.org/10.1016/S0922-338X(97)81921-6]

[33] S Rodjaroen, N Juntawong, A Mahakhant, and K Miyamoto, "High biomass production and starch accumulation in native green algal strains and cyanobacterial strains of Thailand", *Witthayasan Kasetsat Witthayasat.*, vol. 41, no. 3, pp. 570-575, 2007.

[34] M Matsumoto, H Yokouchi, N Suzuki, H Ohata, and T Matsunaga, "Saccharification of marine microalgae using marine bacteria for ethanol production", *Appl. Biochem. Biotechnol.*, vol. 105, no. 108, pp. 247-254, 2003. [http://dx.doi.org/10.1385/ABAB:105:1-3:247] [PMID: 12721490]

[35] I.M Rafiquil, A Hassan, G Sulebele, C.A Orosco, P Roustaian, and K.C Jalal, "Salt stress culture of blue-green algae Spirulina fusiformis", *Pak. J. Biol. Sci.*, vol. 6, no. 7, pp. 648-650, 2003. [http://dx.doi.org/10.3923/pjbs.2003.648.650]

[36] C.A Popovich, C Damiani, D Constenla, and P.I Leonardi, "Lipid quality of the diatoms Skeletonema costatum and Navicula gregaria from the South Atlantic Coast (Argentina): Evaluation of its suitability as biodiesel feedstock", *J. Appl. Phycol.*, vol. 24, no. 1, pp. 1-10, 2012. [http://dx.doi.org/10.1007/s10811-010-9639-y]

[37] M.I Abdel-Hamid, D.a El-Refaay, M Abdel-Mogib, and Y.a Azab, "Studies on biomass and lipid production of seven diatom species with special emphasis on lipid composition of <I>Nitzschia palea</I> (Bacillariophyceae) as reliable biodiesel feedstock", *Arch. Hydrobiol. Supplbd. Algol. Stud.*, vol. 143, no. 1, pp. 65-87, 2013.

[38] V Andruleviciute, V Makareviciene, V Skorupskaite, and M Gumbyte, "Biomass and oil content of Chlorella sp., Haematococcus sp., Nannochloris sp. and Scenedesmus sp. under mixotrophic growth conditions in the presence of technical glycerol", *J. Appl. Phycol.*, vol. 26, no. 1, pp. 83-90, 2014. [http://dx.doi.org/10.1007/s10811-013-0048-x]

[39] S.M Renaud, L-V Thinh, G Lambrinidis, and D.L Parry, "Effect of temperature on growth, chemical composition and fatty acid composition of tropical Australian microalgae grown in batch cultures", *Aquaculture.,* vol. 211, no. 1-4, pp. 195-214, 2002. [http://dx.doi.org/10.1016/S0044-8486(01)00875-4]

[40] L Zhu, X Zhang, L Ji, X Song, and C Kuang, "Changes of lipid content and fatty acid composition of Schizochytrium limacinum in response to different temperatures and salinities", *Process Biochem.,* vol. 42, no. 2, pp. 210-214, 2007. [http://dx.doi.org/10.1016/j.procbio.2006.08.002]

[41] C Ratledge, K Kanagachandran, A.J Anderson, D.J Grantham, and J.C Stephenson, "Production of docosahexaenoic acid by Crypthecodinium cohnii grown in a pH-auxostat culture with acetic acid as principal carbon source", *Lipids.,* vol. 36, no. 11, pp. 1241-1246, 2001. [http://dx.doi.org/10.1007/s11745-001-0838-x] [PMID: 11795857]

[42] J Moncada, C.A Cardona, and L.E Rincón, "Design and analysis of a second and third generation biorefinery: The case of castorbean and microalgae", *Bioresour. Technol.,* vol. 198, pp. 836-843, 2015. [http://dx.doi.org/10.1016/j.biortech.2015.09.077] [PMID: 26457832]

[43] K Tsukahara, and S Sawayama, "Liquid fuel production using microalgae", *J. Jpn. Petrol. Inst.,* vol. 48, no. 5, pp. 251-259, 2005. [http://dx.doi.org/10.1627/jpi.48.251]

[44] USDOE, "Roadmap for biomass technologies in the United States", 2002.

[45] C Rodriguez, A Alaswad, J Mooney, T Prescott, and A.G Olabi, "Pre-treatment techniques used for anaerobic digestion of algae", *Fuel Process. Technol.,* vol. 138, pp. 765-779, 2015. [http://dx.doi.org/10.1016/j.fuproc.2015.06.027]

[46] EU, "Biomass conversion technologies: achievements and prospects for heat and power generation", 1999.

[47] C.G Golueke, W.J Oswald, and H.B Gotaas, "Anaerobic digestion of Algae", *Appl. Microbiol.,* vol. 5, no. 1, pp. 47-55, 1957. [PMID: 13403639]

[48] B Sialve, N Bernet, and O Bernard, "Anaerobic digestion of microalgae as a necessary step to make microalgal biodiesel sustainable", *Biotechnol. Adv.,* vol. 27, no. 4, pp. 409-416, 2009. [http://dx.doi.org/10.1016/j.biotechadv.2009.03.001] [PMID: 19289163]

[49] P McKendry, "Energy production from biomass (Part 2): Conversion technologies", *Bioresour. Technol.,* vol. 83, no. 1, pp. 47-54, 2002. [http://dx.doi.org/10.1016/S0960-8524(01)00119-5] [PMID: 12058830]

[50] A.J Ward, D.M Lewis, and F.B Green, "Anaerobic digestion of algae biomass: A review", *Algal Res.,* vol. 5, no. 1, pp. 204-214, 2014. [http://dx.doi.org/10.1016/j.algal.2014.02.001]

[51] M.N Bin Hossain, J.K Basu, and M Mamun, "“The Production of Ethanol from Micro-Algae Spirulina,” Procedia Eng., vol. 105, no", *Icte.,* vol. 2014, pp. 733-738, 2015.

[52] Y Ueno, N Kurano, and S Miyachi, "Ethanol production by dark fermentation in the marine green alga, Chlorococcum littorale", *J. Ferment. Bioeng.,* vol. 86, no. 1, pp. 38-43, 1998. [http://dx.doi.org/10.1016/S0922-338X(98)80031-7]

[53] W.G Colella, "Switching to a U.S. hydrogen fuel cell vehicle fleet: The resultant change in emissions, energy use, and greenhouse gases", *J. Power Sources.,* vol. 150, no. 1/2, pp. 150-181, 2005. [http://dx.doi.org/10.1016/j.jpowsour.2005.05.092]

[54] E Greenbaum, "Energetic efficiency of hydrogen photoevolution by algal water splitting", *Biophys. J.,* vol. 54, no. 2, pp. 365-368, 1988. [http://dx.doi.org/10.1016/S0006-3495(88)82968-0] [PMID: 19431729]

[55] M.L Ghirardi, L Zhang, J.W Lee, T Flynn, M Seibert, E Greenbaum, and A Melis, "Microalgae: a green source of renewable H(2)", *Trends Biotechnol.,* vol. 18, no. 12, pp. 506-511, 2000. [http://dx.doi.org/10.1016/S0167-7799(00)01511-0] [PMID: 11102662]

[56] A Melis, "Green alga hydrogen production: progress, challenges and prospects", *Int. J. Hydrogen Energy.,* vol. 27, no. 11-12, pp. 1217-1228, 2002. [http://dx.doi.org/10.1016/S0360-3199(02)00110-6]

[57] A Melis, and T Happe, "Hydrogen production. Green algae as a source of energy", *Plant Physiol.,* vol. 127, no. 3, pp. 740-748, 2001. [http://dx.doi.org/10.1104/pp.010498] [PMID: 11706159]

[58] T.K Patra, and P.N Sheth, "Biomass gasification models for downdraft gasifier: A state-of-the-art review", *Renew. Sustain. Energy Rev.,* vol. 50, pp. 583-593, 2015. [http://dx.doi.org/10.1016/j.rser.2015.05.012]

[59] A Hirano, K Hon-Nami, S Kunito, M Hada, and Y Ogushi, "Temperature effect on continuous gasification of microalgal biomass: theoretical yield of methanol production and its energy balance", *Catal. Today.,* vol. 45, no. 1-4, pp. 399-404, 1998. [http://dx.doi.org/10.1016/S0920-5861(98)00275-2]

[60] T Minowa, and S Sawayama, "A novel microalgal system for energy production with nitrogen cycling", *Fuel.,* vol. 78, no. 10, pp. 1213-1215, 1999. [http://dx.doi.org/10.1016/S0016-2361(99)00047-2]

[61] H.B Goyal, D Seal, and R.C Saxena, "Bio-fuels from thermochemical conversion of renewable resources: a review", *Renew. Sustain. Energy Rev.,* vol. 12, no. 2, pp. 504-517, 2008. [http://dx.doi.org/10.1016/j.rser.2006.07.014]

[62] J Clark, and F Deswarte, "Introduction to chemicals from biomass", In: *Wiley series in renewable resources.,* C.V Stevens, Ed., John Wiley & Sons, Inc., 2008. [http://dx.doi.org/10.1002/9780470697474]

[63] V Patil, K.Q Tran, and H.R Giselrød, "Towards sustainable production of biofuels from microalgae", *Int. J. Mol. Sci.,* vol. 9, no. 7, pp. 1188-1195, 2008. [http://dx.doi.org/10.3390/ijms9071188] [PMID: 19325798]

[64] Y Dote, S Sawayama, S Inoue, T Minowa, and S.Y Yokoyama, "Recovery of liquid fuel from hydrocarbon-rich microalgae by thermochemical liquefaction", *Fuel.,* vol. 73, no. 12, pp. 1855-1857, 1994. [http://dx.doi.org/10.1016/0016-2361(94)90211-9]

[65] T Minowa, S.y Yokoyama, M Kishimoto, and T Okakura, "Oil production from algal cells of Dunaliella tertiolecta by direct thermochemical liquefaction", *Fuel.,* vol. 74, no. 12, pp. 1735-1738, 1995. [http://dx.doi.org/10.1016/0016-2361(95)80001-X]

[66] D Chiaramonti, A Oasmaa, and Y Solantausta, "Power generation using fast pyrolysis liquids from biomass", *Renew. Sustain. Energy Rev.,* vol. 11, no. 6, pp. 1056-1086, 2007. [http://dx.doi.org/10.1016/j.rser.2005.07.008]

[67] A Demirbas, "Biomass resource facilities and biomass conversion processing for fuels and chemicals", *Energy Convers. Manage.,* vol. 42, no. 11, pp. 1357-1378, 2001. [http://dx.doi.org/10.1016/S0196-8904(00)00137-0]

[68] X Miao, and Q Wu, "High yield bio-oil production from fast pyrolysis by metabolic controlling of Chlorella protothecoides", *J. Biotechnol.,* vol. 110, no. 1, pp. 85-93, 2004. [http://dx.doi.org/10.1016/j.jbiotec.2004.01.013] [PMID: 15099908]

[69] A Demirbas, "Oily products from mosses and algae via pyrolysis", *Energy Sources A Recovery Util. Environ. Effects.,* vol. 28, no. 10, pp. 933-940, 2006. [http://dx.doi.org/10.1080/009083190910389]

[70] National Biodiesel Board, "Introduction to Biodiesel and Glossary of Terms", In: *Bioenergy - Biomass to Biofuels*. Academic Press: USA, 2015, pp. 37-40.

[71] J Sheehan, V Camobreco, J Duffield, and M Graboski, "An overview of biodiesel and petroleum diesel life cycles," 1998.

[72] National Renewable Energy Laboratory, "Jet fuel from microalgal lipids," USA, 2006.

[73] J Moncada, M.M El-Halwagi, and C.A Cardona, "Techno-economic analysis for a sugarcane biorefinery: Colombian case", *Bioresour. Technol.*, vol. 135, pp. 533-543, 2013. [http://dx.doi.org/10.1016/j.biortech.2012.08.137] [PMID: 23021947]

[74] J Moncada, *Design and Evaluation of Sustainable Biorefineries from Feedstock in Tropical Regions*. Universidad Nacional de Colombia - Sede Manizales, 2012.

[75] J.M Botero, Design and Evaluation of Sustainable Biorefineries from Feedstocks in Tropical Regions., 2012.

[76] F.G Acién, J.M Fernández, J.J Magán, and E Molina, "Production cost of a real microalgae production plant and strategies to reduce it", *Biotechnol. Adv.*, vol. 30, no. 6, pp. 1344-1353, 2012. [http://dx.doi.org/10.1016/j.biotechadv.2012.02.005] [PMID: 22361647]

[77] M.A Borowitzka, "Commercial production of microalgae: ponds, tanks, tubes and fermenters", *J. Biotechnol.*, vol. 70, pp. 313-321, 1999. [http://dx.doi.org/10.1016/S0168-1656(99)00083-8]

[78] A Richmond, "Microalgal biotechnology at the turn of the millennium: a personal view", *J. Appl. Phycol.*, vol. 12, pp. 441-451, 2000. [http://dx.doi.org/10.1023/A:1008123131307]

[79] E Stephens, I.L Ross, Z King, J.H Mussgnug, O Kruse, C Posten, M.A Borowitzka, and B Hankamer, "An economic and technical evaluation of microalgal biofuels", *Nat. Biotechnol.*, vol. 28, no. 2, pp. 126-128, 2010. [http://dx.doi.org/10.1038/nbt0210-126] [PMID: 20139944]

[80] N.H Norsker, M.J Barbosa, M.H Vermuë, and R.H Wijffels, "Microalgal production--a close look at the economics", *Biotechnol. Adv.*, vol. 29, no. 1, pp. 24-27, 2011. [http://dx.doi.org/10.1016/j.biotechadv.2010.08.005] [PMID: 20728528]

[81] O Pulz, and W Gross, "Valuable products from biotechnology of microalgae", *Appl. Microbiol. Biotechnol.*, vol. 65, no. 6, pp. 635-648, 2004. [http://dx.doi.org/10.1007/s00253-004-1647-x] [PMID: 15300417]

[82] Y Chisti, "Biodiesel from microalgae", *Biotechnol. Adv.*, vol. 25, no. 3, pp. 294-306, 2007. [http://dx.doi.org/10.1016/j.biotechadv.2007.02.001] [PMID: 17350212]

[83] M.M Phukan, R.S Chutia, B.K Konwar, and R Kataki, "Microalgae Chlorella as a potential bio-energy feedstock", *Appl. Energy.*, vol. 88, no. 10, pp. 3307-3312, 2011. [http://dx.doi.org/10.1016/j.apenergy.2010.11.026]

[84] J.F Granjo, B.P Duarte, and N.M Oliveira, "Kinetic Models for the Homogeneous Alkaline and Acid Catalysis in Biodiesel Production", *Computer-Aided Chem. Eng.*, vol. 27, pp. 483-488, 2009. [http://dx.doi.org/10.1016/S1570-7946(09)70301-3]

[85] P.K Campbell, T Beer, and D Batten, "Life cycle assessment of biodiesel production from microalgae in ponds", *Bioresour. Technol.*, vol. 102, no. 1, pp. 50-56, 2011. [http://dx.doi.org/10.1016/j.biortech.2010.06.048] [PMID: 20594828]

[86] L.E Rincón, J.J Jaramillo, and C.A Cardona, "Comparison of feedstocks and technologies for biodiesel production: An environmental and techno-economic evaluation", *Renew. Energy.*, vol. 69, pp. 479-487, 2014. [http://dx.doi.org/10.1016/j.renene.2014.03.058]

[87] A Demirbas, and M Fatih Demirbas, "Importance of algae oil as a source of biodiesel", *Energy Convers. Manage.,* vol. 52, no. 1, pp. 163-170, 2011.
[http://dx.doi.org/10.1016/j.enconman.2010.06.055]

[88] J.J Jaramillo, J.M Naranjo, and C.A Cardona, "Growth and Oil Extraction from Chlorella vulgaris: A Techno- Economic and Environmental Assessment", *Ind. Eng. Chem. Res.,* vol. 51, pp. 10503-10508, 2012.
[http://dx.doi.org/10.1021/ie300207x]

CHAPTER 2

Energetic Products from Microalgae: Bioethanol

Choon Gek Khoo[1], **Man Kee Lam**[2,*] and **Keat Teong Lee**[1]

[1] *School of Chemical Engineering, Universiti Sains Malaysia, Engineering Campus, 14300 Nibong Tebal, Pulau Pinang, Malaysia*

[2] *Chemical Engineering Department, Universiti Teknologi PETRONAS, 32610 Seri Iskandar, Perak, Malaysia*

Abstract: Cultivation of microalgae for bioethanol production emerged as one of the potential solutions to drive green and sustainable biofuel production. Microalgae are photosynthetic microorganisms that can grow rapidly and are able to accumulate high content of carbohydrate within their cells. The carbohydrate is usually stored in the form of starch, which is easier to breakdown to simple reducing sugar compared to lignocellulosic biomass. In the present chapter, the process route to produce bioethanol from microalgae biomass is discussed, including the cultivation strategies to enhance microalgae carbohydrate productivity, biomass pre-treatment methods, hydrolysis and fermentation process.

Keywords: Microalgae, Bioethanol, Carbohydrate, Hydrolysis, Fermentation, Starch, Distillation, Biofuel, Pre-treatment, Energy.

OVERVIEW ON THE RENEWABLE ENERGY DEVELOPMENT

As reviewed in 2014, the global energy demand increased with a slower rate than in 2013 (1.1% versus 2.5%) [1]. Fig. (**1**) shows the distribution of Total Primary Energy Supply (TPES) with respect to different energy sources, in which 13.8% or 1,894 Mtoe was contributed by renewable energy in 2014 [1]. As shown in Fig. (**1a**), fossil fuels remained as the largest source of energy consumption, which in turn resulted in substantial increment in greenhouse gases (GHGs) emission to the atmosphere. Thus, utilization of renewable energy is encouraged to fulfil the requirement of COP21 (2015 United Nations Climate Change Conference), an international environmental agreement to keep the global average temperature below 2 °C [2, 3].

* **Corresponding author Man Kee Lam:** Chemical Engineering Department, Universiti Teknologi PETRONAS, 32610 Seri Iskandar, Perak, Malaysia; Tel: +605-3687548; Fax: +605-3656176; E-mails: lam.mankee@utp.edu.my and lam.mankee@petronas.com.my

José Carlos Magalhães Pires (Ed.)

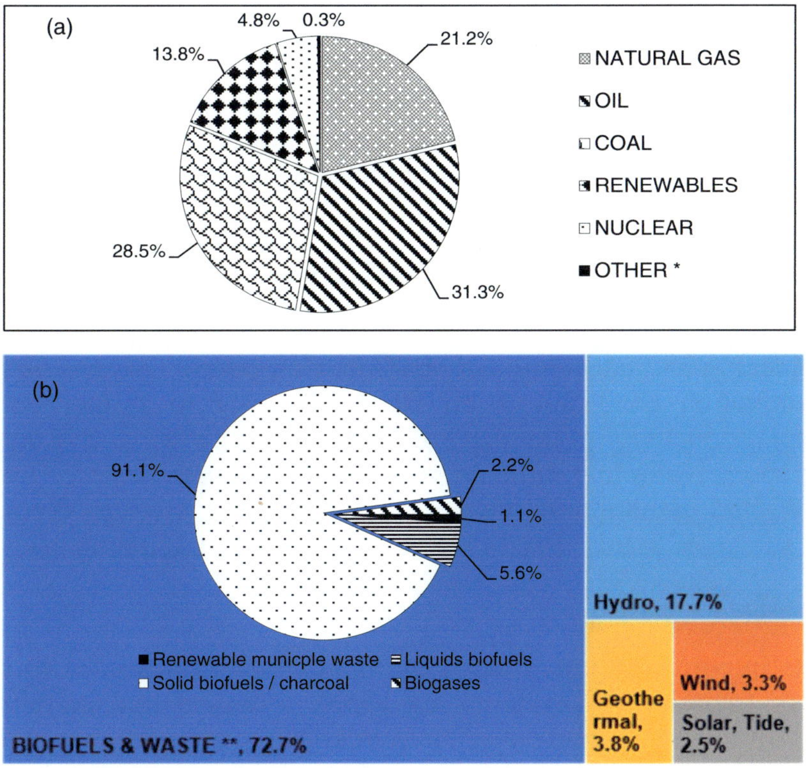

Fig. (1). Distribution of energy supply by 2014: (a) Fuel shares in world total primary energy supply; and (b) Product shares in world renewable energy supply [1].
Note:
*Others include energy sources such as non-renewable wastes, peat, oil shale and chemical heat.
**Details distribution of Biofuels & Waste energy supply according to categories is represented by the pie chart.

BIOETHANOL FROM MICROALGAE

"Ethanol's value as an octane booster was underscored in 2015. Even with the drop in oil prices, we saw record demand for U.S. ethanol because it remained the most cost effective- and cleanest-source of octane in the world."- by RFA Vice Chairman Mick Henderson, Commonwealth Agri-Energy, LLC [4].

Bioethanol is a green biofuel that emits less GHGs than conventional gasoline [5]. However, using sugarcane, coarse grains, wheat and molasses as the main feedstock for bioethanol production has attributed to two major controversy issues [6]: (i) the imbalance of global food market due to "food versus fuel" scenario [7]; and (ii) the use of large agricultural land [8]. Therefore, microalgae which can grow rapidly, able to fix CO_2 from flue gases and bio-treat wastewater, are recommended as a potential feedstock for sustainable bioethanol production [9].

Microalgae are unicellular microorganisms with size ranging from a few to hundred micrometers (μm). They are commonly found in freshwater or marine water and exist either in the form of individual or group [10]. They are also photosynthetic microorganisms that utilize CO_2 as carbon source and release O_2 to the atmosphere [11]. The photosynthetic process (or light-driven redox reaction) converts the absorbed CO_2 into carbohydrates and lipid within the microalgae cells [12].

EFFECT OF CULTIVATION CONDITIONS TO IMPROVE CARBOHYDRATE PRODUCTION

Besides screening and choosing carbohydrate-rich microalgae species, inducing specific cultivation condition is an alternative way to further enhance the carbohydrate productivity of microalgae. In fact, this strategy is also applicable to microalgae species with low carbohydrate content, where they easily grow and are able to withstand extreme environment. The following sections discuss several cultivation strategies to increase the carbohydrate content in microalgae cell.

Effect of Sulfur

Sulfur is an important element in all living cells, as it is needed to synthesize amino acids (methionine, cysteine, cysteine), vitamins and sulfolipids [13]. Deprivation of the sulfur concentration in microalgae cultivation was carried out to produce bio-H_2 from *Chlamydomonas reinhardtii* through direct biophotolysis process [14]. It was found that when microalgae were cultivated under limited sulfur concentration, synthesis of the PS II D1 polypeptide chain (32 kDa) was impeded, which thus, interfered in the repair of PS II from photo-oxidative damage [15, 16]. As a result, the microalgae cells were able to maintain high respiration activity that exceeded the rate of photosynthesis, rendering an anaerobic condition in the cultivation system that allowed bio-H_2 production for 4 to 5 days [15, 17].

During the bio-H_2 production, protein and starch content in *Chlamydomonas reinhardtii* were found increased to 40% and 600%, respectively, for the first 30 h of sulfur deprivation, but gradually declined thereafter [18]. This observation clearly shows that substantial protein and starch catabolism occurs under sulfur limited condition, and subsequently leading to utilization of these endogenous substrates by microalgae to sustain the production of bio-H_2 [18]. For the interest of bioethanol production, once the microalgae cells have accumulated optimum starch concentration under sulfur deprivation condition, the microalgae are harvested immediately instead of extending the cultivation time for bio-H_2 production.

Apart from that, sulfur deprivation method could attain more promising result than deprivation of other macronutrients, such as nitrogen and phosphorus [19]. In a study reported by Brányiková *et al.*, when *Chlorella vulgaris* was cultivated under limited nitrogen or phosphorus nutrients, the starch accumulation in the microalgae cells could reach up to 38-55%; however, after 20 h of cultivation, the microalgae stopped to synthesize the starch [19]. On the other hand, under sulfur limited condition, the *Chlorella vulgaris* was able to maintain 60% of starch content even after 32 h (stationary phase). In addition, the study also revealed positive result with sulfur limitation method in a scaled-up open door thin-layer solar photobioreactor (working volume was 250 L). Under outdoor condition, *Chlorella vulgaris* was able to accumulate 50% of starch content within the cells after exposing to sunlight for 11 h.

Effect of Nitrogen

Nitrogen is an essential element of photosynthesis component and usually participating in DNA, proteins, amino acids, enzymes and pigments synthesis [20, 21]. Hence, sufficient of nitrogen concentration is always necessary to accelerate the growth of microalgae and to enhance their biomass productivity. However, recent studies have shown that when microalgae are cultivated under nitrogen deficient condition, higher accumulation of lipid or carbohydrate could be attained [22 - 27]. This is because protein synthesis pathway in the microalgae cell is diverted to form lipid or carbohydrate as the main energy reserve components. Nevertheless, the main drawback of this cultivation method is low biomass yield, predominantly due to the decrease of photosynthesis pigments (chlorophyll) that inhibits the growth of microalgae [20, 21]. Thus, two-stage cultivation method is recommended to enhance the microalgae biomass productivity with high lipid or carbohydrate content [24, 28 - 30]. In the first stage, microalgae are cultivated in nutrients-rich medium until sufficient biomass are produced, and then transferred to nitrogen-deficient medium to trigger the lipid or carbohydrate accumulation.

Apparently, the competition between lipid and carbohydrate accumulation in microalgae cell under nitrogen deficient condition is still unclear as these two compounds share similar synthesis pathway [24, 26, 31, 32]. However, some recent studies have reported that cultivation time played a significant role in manipulating the accumulation of lipid or carbohydrate in microalgae cell [24, 31 - 33]. During initial phase of cultivation under nitrogen-deficient condition, microalgae preferred to accumulate carbohydrate as a quick response to environmental stress; however, if the cultivation time is prolonged, the carbohydrate synthesis pathway will be diverted to form lipid as a long-term energy storage [32]. For example, the carbohydrate content in *Scenedesmus obliquus* was found to increase from 20.90% to 49.36% after two days of nitrogen starvation, while the protein content decreased from 50.43% to 27.38% [33]. Nevertheless, for the subsequent days of cultivation, the carbohydrate content decreased to 46.98% whereas the lipid content started to increase from 11.72% to 13.66%. Similar results were also observed for *Chlorella vulgaris* [24] and *Chlorella zofingiensis* [32].

Effect of Phosphorus

Phosphorus is an essential macronutrient that regulates microalgae growth and metabolism. It is mainly incorporated into nucleic acid and phospholipids, which serve as important modifiers of protein function and are involved in generating metabolic energy [34, 35]. Up to now, the effect of phosphorus on carbohydrate accumulation in microalgae cells is not well understood and it is believed to be species-specific [36]. Nevertheless, it is well known that carbohydrate accumulation started to increase when intracellular phosphorus content decreased to a critical level [34, 37]. In a recent study by Brányiková *et al.*, *Chlorella vulgaris* tended to accumulate higher starch content (55%) under phosphorus starvation condition instead of nitrogen starvation condition (38%) [19]. In addition, an added advantage of inducing phosphorus starvation instead of nitrogen starvation to microalgae cultivation is that higher biomass productivity could be attained resulting in higher carbohydrate yield [19, 38]. However, in some studies, phosphorus starvation only caused lipid accumulation with no significant effect on carbohydrate accumulation. For example, cultivation of *Tetraselmis subcordiformis* under phosphorus starvation did not increase the carbohydrate content in the microalgae and the concentration remained constant throughout 8 days of cultivation [39].

Effect of Carbon Source

Inorganic Carbon

It is a common practice to cultivate phototrophic microalgae by supplying CO_2 as the main carbon source. Through photosynthesis process, CO_2 is absorbed by microalgae cells to support their growth by converting the carbon to build proteins, nucleic acids, carbohydrate and lipids [40]. However, it is interesting to note that the carbohydrate content in microalgae cells usually does not increase linearly with the CO_2 concentration. Based on several recent studies, the increment of carbohydrate content in *Chlorella*, *Scenesdusmus obliquus* and *Chlamydomonas reinhardtii* did not increase significantly although high CO_2 concentration was supplied [24, 33, 41, 42]. In a study carried out by Izumo *et al.*, the carbohydrate content in *Chlorella* reduced when switching the CO_2 supply from 0.04% (atmospheric air) to 3% [41]. It was observed that when supplying low concentration of CO_2, the location of starch shifted from stroma to pyrenoid.

As a result, the expression of granule-bound starch synthase (GBSS) increased, leading to higher accumulation of starch content under low concentration of CO_2 [41]. However, it is well established that supplying high concentration of CO_2 to microalgae cultivation could accelerate the biomass productivity and shorten the overall cultivation time [43 - 46]. Subsequently, by inducing appropriate stress conditions to the microalgae cultivation, such as nitrogen depletion, the overall carbohydrate yield could further be increased [24, 33].

Organic Carbon

Besides CO_2, specific microalgae species could utilize organic substrate (*e.g.* glucose, acetate, glycerol and *etc.*) as carbon and energy source to grow [47]. Usually, this is referred as heterotrophic cultivation (independent of light). The consumption of organic substrates by microalgae is achieved through several mechanisms, such as phosphorylation (glucose), simple diffusion (glycerol) or using membrane transporter proteins (organic acids) [20, 48]. Up to now, several microalgae species such as *Chlorella protothecoide* [49 - 51], *Chlorella vulgaris* [52], *Crypthecodinium cohnii* [53], *Neochloris oleoabundans* [54] and *Schizochytrium limacinum* [55] have been studied under heterotrophic growth conditions to achieve high biomass productivity.

However, majority of the heterotrophic cultivation studies are focused on lipid productivity and studies related to carbohydrate production are rarely reported in the literature. This is because the organic carbon source used in growing heterotrophic microalgae is glucose, which is a derivative of carbohydrate after hydrolysis process. In other words, the net production of carbohydrate in heterotrophic cultivation could significantly be lower than phototrophic cultivation. In a recent study by Morales-Sánchez *et al.*, 54% (wt) of carbohydrate content (comprised of 90% glucose) was observed when *Neochloris oleoabundans* was cultivated under fed-batch heterotrophic condition (glucose as carbon source) [54]. Nevertheless, in terms of glucose balance, the overall glucose produced by *Neochloris oleoabundans* was estimated to be 7.67 g/L (based on biomass yield of 14.2 g/L), while the total glucose consumed by the microalgae to grow was 50 g/L. Thus, it is clear that high carbohydrate content attained in microalgae under heterotrophic cultivation does not directly indicate the potential of this approach, but the overall mass balance of carbohydrate should be taken into consideration.

Light Intensity

Light is the main energy driver for photosynthetic microalgae [20]. By providing suitable light intensity to microalgae cultivation system [56, 57], the carbohydrate and lipid content in microalgae cells could further be enhanced [19, 58, 59].

Recent study has shown that by supplying light intensity between 200 to 420 μmol m^{-2} s^{-1}, the carbohydrate content of *Scenedesmus obliquus* could be enhanced [60]. At light intensity of 420 μmol m^{-2} s^{-1}, *Scenedesmus obliquus* accumulated the highest carbohydrate content of 38%; which was 2.3 times higher compared to the cultivation supplied with light intensity of 60 μmol m^{-2} s^{-1}. Similar observation was also reported by Brányiková *et al.*, in which the starch content in *Chlorella* increased from 8.5% to 40% when the light intensity switched from 215 μmol m^{-2} s^{-1} to 330 μmol m^{-2} s^{-1}. However, it should be noted that providing extremely high light intensity could reduce the biomass productivity of microalgae, which is generally known as photo-inhibition [61, 62]. The carbohydrate productivity in *Scenedesmus obliquus* was reported to reduce by about 45% when high intensity increased from 420 μmol $m^{-2}s^{-1}$ to 540 μmol $m^{-2}s^{-1}$ [60].

CARBOHYDRATE IN MICROALGAE

Microalgae cells are composed by different polysaccharides that can be broken down to monosaccharides through pre-treatments and hydrolysis processes. The produced monosaccharides (*e.g.* glucose) can be converted into bioethanol through fermentation process [63, 64]. Microalgae such as *Chlorella, Chlamydomonas, Dunaliella*, and *Scenedesmus* consist of high carbohydrates content (Table **1**), which is suitable to be cultivated for bioethanol production [65].

Table 1. Carbohydrates content of selected species of microalgae under various growth conditions.

Microalgae	Growth Conditions	Biomass Accumulation	Carbohydrates Content (wt %)		Ref.
			Total Carbohydrates	Specific Carbohydrate	
Chlamydomonas mexicana (YSL008)	Phototrophic cultivation: 27±2 °C; light intensity of 60 μmol m^{-2} s^{-1}; 0.6 LPM sterile air; 10% (v/v) CO_2	n.a.	52.6	Glucose: 33.5 Mannose: 12.8 Galactose: 4.6 Glucosamine: 1.8	[66]

(Table 1) contd.....

Microalgae	Growth Conditions	Biomass Accumulation	Carbohydrates Content (wt %)		Ref.
Chlamydomonas reinhardtii (UTEX 90)	Phototrophic cultivation: 23 °C; continuous illumination of 450 μmol m^{-2} s^{-1}; pH 7.0-7.4; agitation rate of 130 rpm; feeding 1.0M acetic acid; inoculum size of 10%	Yield: 12.4 g/L	59.7±0.5	Starch: 43.6±1.4 Glucose: 44.7±0.8 Fucose: 0.40±0.01 Rhamnose: 0.90±0.02 Arabinose: 1.90±0.04 Galactose: 2.70±0.04 Mannose: 1.40±0.03	[67]
Chlorella vulgaris (KMMCC-9; UTEX 26)	Heterotrophic cultivation: Nutrient stress cultivation; 21±1 °C; light intensity of 150 μmol m^{-2} s^{-1}; 16:8 h light-dark cycle; air aeration	n.a.	22.4±0.4	Glucose: 10.8±0.2 Galactose: 5.3±0.1 Mannose: 1.4±0.1 Rhamnose: 3.1±0.1 Arabinose: 0.9±0.0 Xylose: 0.9±0.0	[68]
Chlorella sp. (KR-1)	Phototrophic cultivation: 30 °C; light intensity of 80 μmol m^{-2} s^{-1}; 10% (v/v) CO_2	n.a.	36.1	Glucose: 29.9 Galactose/xylose: 2.3 Rhamnose: 1.9 Arabinose: 2.0	[69]
Dunaliella tertiolecta (LB999)	Phototrophic cultivation: 20-25 °C; light intensity of 60 μmol m^{-2} s^{-1}; 2% (v/v) CO_2; inoculum size of 0.3 g/L	Yield: 9.1 g/L	51.9	n.a.	[70]
Scenedesmus obliquus (CNW-N)	Phototrophic cultivation: Two-stage cultivation (from nutrient rich medium switched to a nutrient deficient condition); 28 °C; light intensity of 210-230 μmol m^{-2} s^{-1}; pH 6.2; agitation rate of 300 rpm; 2.5% CO_2; inoculum size of 19-21 mg L^{-1}	Productivity: 681.4±26.5 mg/L/day	51.7±0.5	Glucose: 41.5±4.2	[23]

(Table 1) contd.....

Microalgae	Growth Conditions	Biomass Accumulation	Carbohydrates Content (wt %)		Ref.
Scenedesmus sp. (CCNM 1077)	Heterotrophic cultivation: Nitrogen starvation (nitrate free); 25±2 °C; 12:12 h light dark cycle; 10% (v/v) inoculum size	Productivity: 8.48 mg/L/day	45.74	n.a.	[71]

Cellulose

Microalgae phyla that have cellulose storage are Chlorophyceae (Chlorophyta), Dinophyta, Phaeophyta (Ochrophyta), Prymnesiophyceae, Rhodophyceae (Rhodophyta), and Xanthophyceae (Ochrophyta) [73]. Cellulose is the major component of plant fibres found in the secondary cell wall, embedded in a matrix of other polysaccharides (hemicelluloses, pectins, and lignins) to form biocomposite. The molecular constitution of cellulose is β-1,4-linked polymer of $_{D}$-glucose (Fig. **2**). Cellulose is easy to degrade *via* cellulolytic enzymes. These enzyme complexes (cellulosomes) consist of *endo-* and *exo*-cellulase, β-glucosidases, and carbohydrate-binding proteins, and are capable to degrade the cellulose for bioethanol production [74].

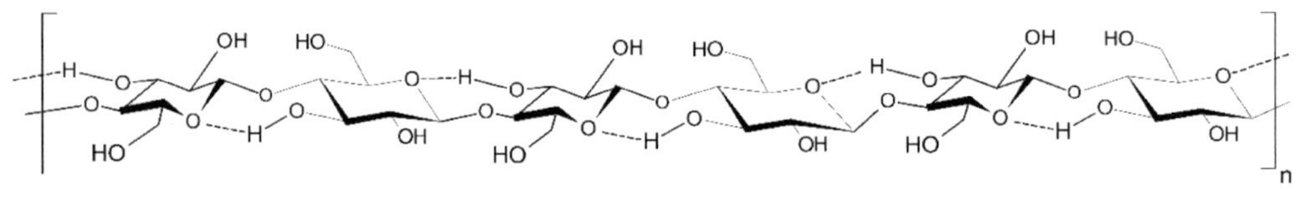

Fig. (2). Molecular constitution of cellulose (β-1,4-linked polymer of $_{D}$-glucose) [74].

Starch

Starch is a complex carbohydrate that found inside cytoplasm as insoluble and semi-crystalline granules and functioned as energy reservoir [73]. Starch granules comprise of α-glucans that are formed by amylose and amylopectin (Fig. **3**). Amylose is a linear polymer consists of several thousand α-1,4-linked $_{D}$-glucose residues; whereas amylopectin is a branched polymer consists of α-1,4-linked $_{D}$-glucose residues and α-1,6-linked branch for each 24-30 residues [74].

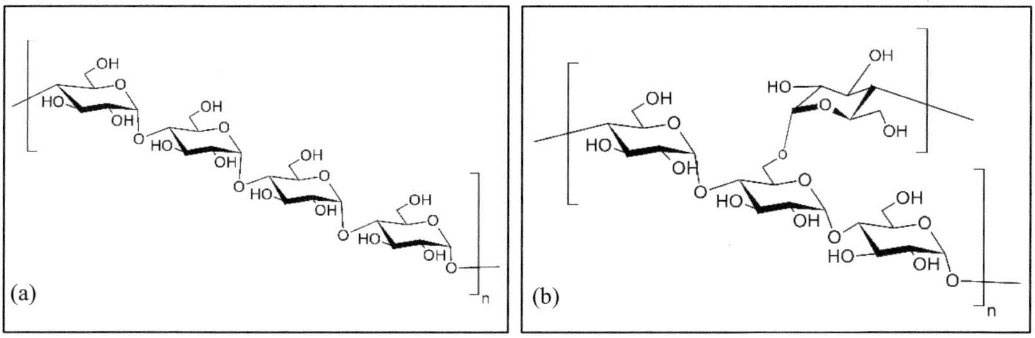

Fig. (3). Molecular constitution of starch granules: (**a**) Amylose: linear polymer of α-1,4-linked $_{D}$-glucose residues; and (**b**) Amylopectin: branched polymer consists of α-1,4-linked D-glucose residues and α-1,--linked branch [74].

Similar to cellulose, starch can be degraded by enzymatic hydrolysis to produce bioethanol. Amylases are among the most commonly used enzymes to hydrolyse starch [72].

MICROALGAE BIOETHANOL PRODUCTION

Microalgae based bioethanol can be produced *via* the following chemical equation, in which the sugar produced from microalgae biomass is fermented into bioethanol:

$$C_nH_{2n}O_n(sugar) \rightarrow \frac{n}{3}C_2H_5OH + \frac{n}{3}CO_2 + Heat \qquad \textbf{(1)}$$

The primary carbohydrates ($(C_nH_{2n}O)_n$) source is usually the simple sugar, such as glucose, starch, and cellulose available in microalgae biomass. Generally, bioethanol production from microalgae involved several processes, such as pre-treatment of biomass, hydrolysis or saccharification of complex carbohydrates, fermentation of bioethanol, and bioethanol recovery, as shown in Fig. (**4**) [65].

Up to date, various researches have been conducted to enhance the microalgae based bioethanol production. The performance of fermentation process greatly depends on the biomass pre-treatment method and carbohydrates hydrolysis process [65]. Basically, the pre-treatment process is aimed to disrupt the microalgae biomass cell wall to release the entrapped complex carbohydrates for the subsequent hydrolysis process. However, it should be noted that the formation of undesired by-products, such as formic acid, acetic acid, and furan compounds during carbohydrate hydrolysis process will affect the subsequent fermentation efficiency [75].

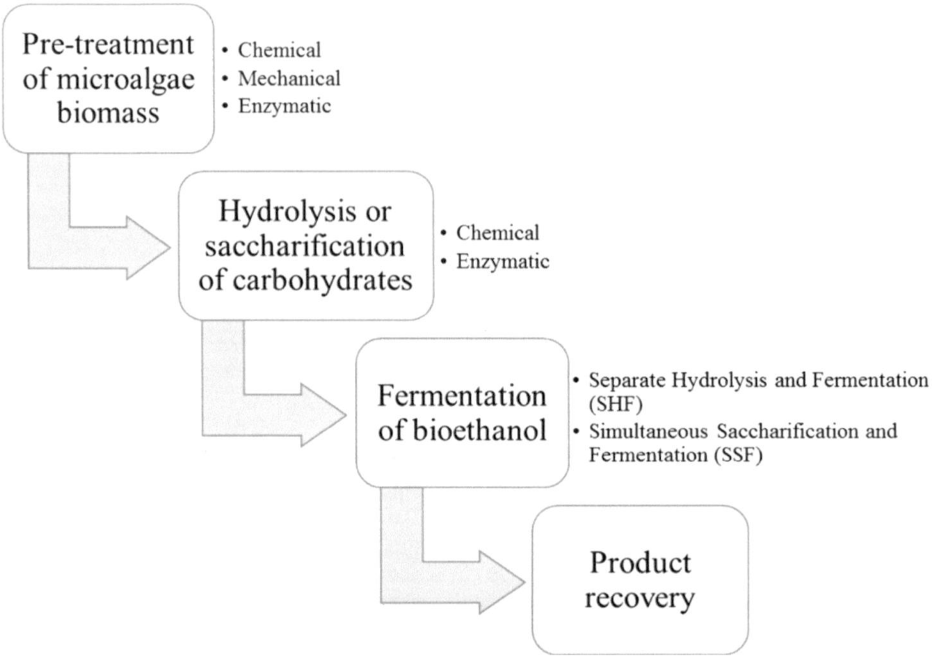

Fig. (4). Routes for microalgae based bioethanol production.

PRE-TREATMENT & HYDROLYSIS

Since pre-treatment and hydrolysis of microalgae biomass play a vital role in bioethanol production, it is important to optimize the operating conditions prior to the subsequent fermentation process. There are a number of important factors to classify the effectiveness of microalgae biomass pre-treatment method: (i) quantifying the carbohydrates and sugars contents of the pre-treated samples; (ii) choosing the correct fermentation source, either from liquid hydrolysate or water insoluble solids phase; (iii) analysing the inhibitory effects of pre-treated samples before fermentation process; and (iv) evaluating the potential of value-added products from the pre-treated samples [76]. The pre-treatment of microalgae biomass can be categorized into: (i) physical; (ii) chemical; and (iii) enzymatic. The following sub-sections discuss the microalgae biomass pre-treatment method and carbohydrate hydrolysis process.

Physical Pre-treatment

Physical pre-treatments, such as high-pressure homogenization, thermal heating, and ultra-sonication, are able to disrupt the microalgae biomass cell wall to release the carbohydrate [64]. The ultra-sonication method employs cavitation process to destabilize the cells, in which the pressures created by the ultrasonic bubbles are collapsed violently on the microalgae cell wall. During the sonication process, excessive heat is dissipated to the surrounding and therefore, it is essential to carry out the pre-treatment process under ice cold condition [77].

Ultrasonic assisted extraction (UAE) is an effective biomass pre-treatment for carbohydrates extraction, as compared with conventional solvent extraction (CSE) and fluidized bed extraction (FBE) method [78]. The glucose extraction by UAE showed an increment of almost 4-fold (36.94 g/100 g biomass), as compared with the CSE and FBE methods. In addition, yeast fermentation process can further be improved by joint sonication and heating during biomass pre-treatment [79]. Filamentous microalgae are partially disrupted by sonication, and complete cell wall destruction is achieved by thermal heating with enzymes.

Chemical Pre-Treatment

The chemical hydrolysis method is a cheaper method compared to enzymatic hydrolysis. Usually, acids or alkalis are used for the hydrolysis process; acids are more commonly employed. Acid hydrolysis is capable to break the intra- and inter-molecular hydrogen bonds across the glycosidic link which leading to the breakdown of carbohydrates [72]. Among the acids, sulfuric acid (H_2SO_4) is the most attractive option for the carbohydrate hydrolysis process. This is due to the advantage of released anions (SO_4^{2-}) as nutrient source for yeast fermentation process [80]. The optimization parameters of chemical pre-treatment include time, temperature and acid or alkaline concentration.

By referring to the acid hydrolysis performed by Miranda *et al.* [77], the optimum hydrolysis conditions for *Scenedesmus obliquus* biomass could be achieved by using 2 M of H_2SO_4, at temperature of 120 °C for 30 min. The sugar yield attained was 95.6%, indicating that chemical pre-treatment method is an alternative low cost method to release the sugar from microalgae biomass efficiently.

Besides, alkaline-peroxide also demonstrated a significantly increment on sugar extraction (maximum 87%) from microalgae biomass. This is due to the enhancement of oxidation of sugar to form organic acids during fermentation process [81].

Enzymatic Pre-treatment

Enzymatic pre-treatment degrades the cellulose for subsequent fermentation process. As compared with the previous discussed pre-treatment (physical and chemicals) methods, enzymatic pre-treatment has the advantages of less energy consumption, mild operating conditions and does not incur corrosion issue to the fermenter. However, the disadvantages of enzymatic pre-treatment are: too slow enzymatic rate for commercial application (10-14 days) [76] and high enzymes specificity towards different microalgae species [82]. The current commercial enzymes are cellulase and amylase, which are used to hydrolyze cellulose and starch, respectively [72].

Kim *et al.* used cellulase and pectinase to saccharify *Chlorella vulgaris* [68]. From the attained results, *Chlorella vulgaris* cultivated under nitrogen stress condition was able to increase the carbohydrates content within the cells. By using *Aspergillus* pectinase enzyme, 79% of carbohydrates were successfully extracted during the saccharification process. The study also evaluated the activity of immobilized yeast fermentation and attained 89% of conversion. It can be concluded that immobilization of enzymes on support can enhance the bioethanol production yield.

FERMENTATION PROCESS

Free sugars extracted from microalgae biomass can be fermented into bioethanol through some ethanologen microorganisms (bacteria, filamentous fungi, yeast). There are two different types of fermentation process: (i) separate hydrolysis and fermentation (SHF); and (ii) simultaneous saccharification and fermentation (SSF). The comparison of both SHF and SSF is tabulated in Table **2** [83 - 85].

Table 2. Comparison of Separate Hydrolysis and Fermentation (SHF) and Simultaneous Saccharification and Fermentation (SSF).

Fermentation	Benefit	Drawback
SHF	■ Easier to optimize the hydrolysis and fermentation operating conditions ■ Simpler equipment design ■ Shorter residence time ■ Using cheaper chemicals	■ Higher contamination rates ■ Inhibitory effects
SSF	■ Higher bioethanol yield ■ Higher rates of hydrolysis ■ Less inhibitory effects ■ Lower operating cost ■ Lower contamination rate ■ Required less sterile conditions ■ Smaller amount of enzyme intake ■ Shorter residence time	■ Only can be performed under optimal operating conditions for either hydrolysis or fermentation process ■ Difficulty in process control

Depending on the microalgae species and cultivation conditions, different hydrolysis and fermentation (SSF and SHF) approaches are introduced for bioethanol production, as summarized in Table **3**

Table 3. Microalgae based bioethanol production from SHF and SSF.

Fermentation	Feedstock	Hydrolysis		Fermentation		Bioethanol Accumulation	Ref.
		Source	Operating Conditions	Source	Operating Conditions		
Separate Hydrolysis and Fermentation (SHF)	*Chlamydomonas reinhardtii* (UTEX 90)	2 stages enzymatic: liquefaction by α-amylase (Termamyl 120 L) and saccharification by amyloglucosidase (AMG 300L)	Liquefaction: 0.0001-0.02% enzyme, pH 6; 10% H_2SO_4; 70-100°C; 30 min Saccharification: pH 4.5-5.5; 50-65 °C; 0.001-0.3% enzyme; 5-60 min	Yeast: *Saccharomyces cerevisiae* (S288C)	30 °C; 40 h; 160 rpm	0.235 g ethanol / g dry algal biomass	[67]
	Chlorella sp. (KR-1)	Enzymatic (Polygalacturonase, amyloglucosidase, cellulase, and Viscozyme L); Chemical (dilute acid)	Enzymatic: pH 3.5-6.5; 35-55 °C; 0.08-2.4 mL/g enzyme Chemical: 5% (w/v) biomass, 0.1-1 N of HCl or H_2SO_4, 121 °C, 15 min	Yeast: *Saccharomyces cerevisiae* (KCTC 7931)	30 °C; 20 h; 180 rpm	0.16 g ethanol / g dry algal biomass	[69]
	Scenedesmus bijugatus	Acid hydrolysis	1-3% v/v H_2SO_4; 100-180 °C, 45 min	Yeast: *Saccharomyces cerevisiae*	24-144 h	0.158 g ethanol / g dry algal biomass	[86]
Simultaneous Saccharification and Fermentation (SSF)	*Chlamydomonas fasciata* (Ettl 437)	Ultrasonic	30 W and 20 kHz; 0-40 min	Yeast: *Saccharomyces cerevisiae* (AM12)	40 °C; 30 h; 100 rpm	0.194 g ethanol / g dry algal biomass	[87]
	Chlorella vulgaris (FSP-E)	Enzymatic (cellulose-hydrolyzing bacterium); Chemical (acid)	Enzymatic: pH 6; slurry concentration of 10-40 g/L; sonicated 10 min; 45 °C; 200 rpm Chemical: 0.1-5.0% (v/v) acid; 121 °C; 20 min	Bacterium: *Zymomonas mobilis*	30 °C in dark	0.233 g ethanol / g dry algal biomass	[88]
	Scenedesmus abundans (PKUAC 12)	Chemical (acid); Enzymatic (amylase)	Chemical: 3% H_2SO_4; 170 rpm; 5 min Enzymatic: pH 5.5; 15 U (α-amylase)/g algae or 0.4 U (glucoamylase) / g algae; 30 min; 55 °C	Yeast: *Saccharomyces cerevisiae Hansen*	30 °C; 48 h; 200 rpm	0.103 g ethanol / g dry algal biomass	[89]

Separate Hydrolysis and Fermentation (SHF)

During SHF, enzymatic hydrolysis and fermentation process take place in two different units of reactors. Hence, optimizing the operating conditions for each stage can be performed separately. However, the inhibitory effect from the accumulation of hydrolysis products (especially degradation of glucose and cellobiose), will suppress the activity of cellulases and results in slower hydrolysis rate [90].

Simultaneous Saccharification and Fermentation (SSF)

In SSF, enzymatic hydrolysis and fermentation take place in a single reactor. Therefore, the hydrolysis products (glucose and cellobiose) are consumed by the yeast in fermentation process, which greatly reduces the inhibitory effects [66]. In addition, higher concentration of bioethanol could result in lower degree of contamination, due to the lower yield of solid content (> 15% dry weight basis) [83]. Since both the enzymatic hydrolysis and fermentation are operated under the same conditions, it would rather be difficult to attain an optimal temperature for SSF operation.

Microorganism for Fermentation of Microalgae Biomass

The performance parameters of fermentation process include: temperature, pH, microorganism growth rate, bioethanol productivity, alcohol tolerance, osmotic tolerance, inhibitor tolerance, and genetic stability [90]. It is well known that *Saccharomyces cerevisiae* is the most efficient yeast in fermentative bioethanol process. The yeast has shown good tolerance towards alcohol and inhibitor effect, associated with high bioethanol productivity [84]. *Saccharomyces cerevisiae* was utilized to produce bioethanol *via* SHF and SSF processes [66]. The yeast was able to maximize the sugar consumption efficiency (91-98%), achieving 0.5 g/g or an equivalent of 88.2% (theoretical) of bioethanol yield from SSF. However, energy recovery was higher in the SHF mode (85.96%) than SSF (70%).

Other than that, other microorganisms such as bacteria, are also used in microalgae biomass fermentation process. These bacteria include *Bacillus stearothermophilus*, *Clostridium thermohydrosulfuricum*, *Clostridium thermocellum*, *Escherichia coli*, *Klebsiella oxytoca*, *Thermoanaerobacterium thermosaccharolyticum*, *Thermoanaerobacter ethnolicus*, and *Zymomonas mobilis* [72]. Recently, *Z. mobilis* was reported to be an efficient bacterium in sugar fermentation, in which the bacterium had high tolerance to ethanol content (~ 11% v/v) and was able to withstand high osmotic pressure caused by high sugar concentrations [91]. In addition, Yang *et al.* also concluded that *Z. mobilis* was able to utilize glucose, sucrose, and fructose to produce bioethanol [92].

PRODUCTS RECOVERY

Commercial production of bioethanol requires product recovery process [90]. The bioethanol can be recovered from distillation process, where the bioethanol distillate is concentrated (~37%) below the azeotrope condition in the distillation column. The bottom product is discharged to stripping column to remove excessive water, and the remaining bioethanol is recycled back to the feed. The recovery of bioethanol process can be controlled at 99.6% to minimize the lost. On the other hand, solid products obtained from the fermentation process are separated by a centrifuge, followed by drying process in rotary dryer. Then, the solid products are recovered in an evaporator as clean condensate and concentrated syrup (15-20% of total solids weight).

CONCLUSIONS

Microalgae biomass is the third generation feedstock for biofuel production. Besides the accumulation of lipid within the microalgae cells for biodiesel production, the stored carbohydrate can also be further utilized for bioethanol production. It is foreseen that bioethanol derived from microalgae biomass could pave a sustainable route to substitute the gasoline in the global market. However, more researches are still required to justify the commercial feasibility of microalgae bioethanol production, especially in terms of life cycle impact assessment and techno-economic analysis.

CONFLICT OF INTEREST

The authors declare no conflict of interest, financial or otherwise.

ACKNOWLEDGEMENTS

The authors would like to acknowledge the financial support received from the Ministry of Higher Education (MOHE) Malaysia (Fundamental Research Grant Scheme (FRGS) with cost center 0153AB-L25 and MyBrain15-MyPhD scholarship) and Universiti Sains Malaysia (Research University Grant No. 814147). Technical support from Green Technology Mission Oriented Research and Centre for Biofuel and Biochemical Research (CBBR) of Universiti Teknologi PETRONAS is highly appreciated.

REFERENCES

[1] *"Excerpt from: Renewables information," IEA Publishing.* Statistics: United States of America, 2016.

[2] *What is COP21?,* 2015. Available at: http://www.cop21paris.org/about/cop21, [Accessed: 10 Sept. 2016].

[3] C.H. Ra, and S-K. Kim, "Bioethanol Production from Macroalgae and Microbes," In: *Marine Bioenergy: Trends and Developments*, S.-K. Kim and C.-G. Lee, Eds., ed United States of America:

CRC Press, Taylor & Francis Group, 2014, pp. 257-271.

[4] *Fueling a High Octane Future.* RFA Publishing: United States of America, 2016.

[5] *Ethanol Vehicle Emissions,* 2016. Available at: http://www.afdc.energy.gov/vehicles/flexible_fuel_emissions.html, [Accessed: 10 Sept. 2016].

[6] *OECD-FAO Agricultural Outlook 2016-2025,* .

[7] S. Lim, and L.K. Teong, "Recent trends, opportunities and challenges of biodiesel in Malaysia: An overview", *Renew. Sustain. Energy Rev.,* vol. 14, pp. 938-954, 2010. [http://dx.doi.org/10.1016/j.rser.2009.10.027]

[8] A. Kirkels, "Biomass boom or bubble? A longitudinal study on expectation dynamics", *Technol. Forecast. Soc. Change,* vol. 103, pp. 83-96, 2016. [http://dx.doi.org/10.1016/j.techfore.2015.11.013]

[9] J.P. Maity, J. Bundschuh, C-Y. Chen, and P. Bhattacharya, "Microalgae for third generation biofuel production, mitigation of greenhouse gas emissions and wastewater treatment: Present and future perspectives – A mini review", *Energy,* 2014. [http://dx.doi.org/10.1016/j.energy.2014.04.003]

[10] T. Suganya, M. Varman, H.H. Masjuki, and S. Renganathan, "Macroalgae and microalgae as a potential source for commercial applications along with biofuels production: A biorefinery approach", *Renew. Sustain. Energy Rev.,* vol. 55, pp. 909-941, 2016. [http://dx.doi.org/10.1016/j.rser.2015.11.026]

[11] R.A. Andersen, "The Microalgal Cell", In: *Handbook of Microalgal Culture: Applied Phycology and Biotechnology.,* A. Richmond, Q. Hu, Eds., 2nd ed. John Wiley & Sons: USA, 2013, pp. 3-20. [http://dx.doi.org/10.1002/9781118567166.ch1]

[12] J. Masojídek, G. Torzillo, and M. Koblížek, "Photosynthesis in Microalgae", In: *Handbook of Microalgal Culture: Applied Phycology and Biotechnology.,* A. Richmond, Q. Hu, Eds., 2nd ed. John Wiley & Sons: USA, 2013, pp. 21-36. [http://dx.doi.org/10.1002/9781118567166.ch2]

[13] E.W. Becker, *Microalgae: Biotechnology and Microbiology.* Cambridge University Press, 1994.

[14] A. Melis, L. Zhang, M. Forestier, M.L. Ghirardi, and M. Seibert, "Sustained photobiological hydrogen gas production upon reversible inactivation of oxygen evolution in the green alga Chlamydomonas reinhardtii", *Plant Physiol.,* vol. 122, no. 1, pp. 127-136, 2000. [http://dx.doi.org/10.1104/pp.122.1.127] [PMID: 10631256]

[15] A. Melis, "Photosynthetic H2 metabolism in Chlamydomonas reinhardtii (unicellular green algae)", *Planta,* vol. 226, no. 5, pp. 1075-1086, 2007. [http://dx.doi.org/10.1007/s00425-007-0609-9] [PMID: 17721788]

[16] D.D. Wykoff, J.P. Davies, A. Melis, and A.R. Grossman, "The regulation of photosynthetic electron transport during nutrient deprivation in Chlamydomonas reinhardtii", *Plant Physiol.,* vol. 117, no. 1, pp. 129-139, 1998. [http://dx.doi.org/10.1104/pp.117.1.129] [PMID: 9576782]

[17] M.L. Ghirardi, A. Dubini, J. Yu, and P.C. Maness, "Photobiological hydrogen-producing systems", *Chem. Soc. Rev.,* vol. 38, no. 1, pp. 52-61, 2009. [http://dx.doi.org/10.1039/B718939G] [PMID: 19088964]

[18] L. Zhang, T. Happe, and A. Melis, "Biochemical and morphological characterization of sulfur-deprived and H2-producing Chlamydomonas reinhardtii (green alga)", *Planta,* vol. 214, no. 4, pp. 552-561, 2002. [http://dx.doi.org/10.1007/s004250100660] [PMID: 11925039]

[19] I. Brányiková, B. Maršálková, J. Doucha, T. Brányik, K. Bišová, V. Zachleder, and M. Vítová, "Microalgae--novel highly efficient starch producers", *Biotechnol. Bioeng.,* vol. 108, no. 4, pp. 766-

776, 2011.
[http://dx.doi.org/10.1002/bit.23016] [PMID: 21404251]

[20] G. Markou, I. Angelidaki, and D. Georgakakis, "Microalgal carbohydrates: an overview of the factors influencing carbohydrates production, and of main bioconversion technologies for production of biofuels", *Appl. Microbiol. Biotechnol.,* vol. 96, no. 3, pp. 631-645, 2012.
[http://dx.doi.org/10.1007/s00253-012-4398-0] [PMID: 22996277]

[21] J.A. Berges, D.O. Charlebois, D.C. Mauzerall, and P.G. Falkowski, "Differential effects of nitrogen limitation on photosynthetic efficiency of photosystems I and II in microalgae", *Plant Physiol.,* vol. 110, no. 2, pp. 689-696, 1996.
[http://dx.doi.org/10.1104/pp.110.2.689] [PMID: 12226211]

[22] Y. Li, D. Han, G. Hu, M. Sommerfeld, and Q. Hu, "Inhibition of starch synthesis results in overproduction of lipids in Chlamydomonas reinhardtii", *Biotechnol. Bioeng.,* vol. 107, no. 2, pp. 258-268, 2010.
[http://dx.doi.org/10.1002/bit.22807] [PMID: 20506159]

[23] S.H. Ho, S.W. Huang, C.Y. Chen, T. Hasunuma, A. Kondo, and J.S. Chang, "Bioethanol production using carbohydrate-rich microalgae biomass as feedstock", *Bioresour. Technol.,* vol. 135, pp. 191-198, 2013.
[http://dx.doi.org/10.1016/j.biortech.2012.10.015] [PMID: 23116819]

[24] S.H. Ho, S.W. Huang, C.Y. Chen, T. Hasunuma, A. Kondo, and J.S. Chang, "Characterization and optimization of carbohydrate production from an indigenous microalga Chlorella vulgaris FSP-E", *Bioresour. Technol.,* vol. 135, pp. 157-165, 2013.
[http://dx.doi.org/10.1016/j.biortech.2012.10.100] [PMID: 23186680]

[25] L. Gouveia, A.E. Marques, T.L. da Silva, and A. Reis, "Neochloris oleabundans UTEX #1185: a suitable renewable lipid source for biofuel production", *J. Ind. Microbiol. Biotechnol.,* vol. 36, no. 6, pp. 821-826, 2009.
[http://dx.doi.org/10.1007/s10295-009-0559-2] [PMID: 19377896]

[26] Y. Li, M. Horsman, B. Wang, N. Wu, and C.Q. Lan, "Effects of nitrogen sources on cell growth and lipid accumulation of green alga Neochloris oleoabundans", *Appl. Microbiol. Biotechnol.,* vol. 81, no. 4, pp. 629-636, 2008.
[http://dx.doi.org/10.1007/s00253-008-1681-1] [PMID: 18795284]

[27] A. Converti, A.A. Casazza, E.Y. Ortiz, P. Perego, and M. Del Borghi, "Effect of temperature and nitrogen concentration on the growth and lipid content of Nannochloropsis oculata and Chlorella vulgaris for biodiesel production", *Chem. Eng. Process. Process Intensif.,* vol. 48, pp. 1146-1151, 2009.
[http://dx.doi.org/10.1016/j.cep.2009.03.006]

[28] A. Widjaja, C.C. Chien, and Y.H. Ju, "Study of increasing lipid production from fresh water microalgae Chlorella vulgaris", *J. Taiwan Inst. Chem. Eng.,* vol. 40, pp. 13-20, 2009.
[http://dx.doi.org/10.1016/j.jtice.2008.07.007]

[29] S.H. Ho, W.B. Lu, and J.S. Chang, "Photobioreactor strategies for improving the CO2 fixation efficiency of indigenous Scenedesmus obliquus CNW-N: statistical optimization of CO2 feeding, illumination, and operation mode", *Bioresour. Technol.,* vol. 105, pp. 106-113, 2012.
[http://dx.doi.org/10.1016/j.biortech.2011.11.091] [PMID: 22177736]

[30] G. Mujtaba, W. Choi, C.G. Lee, and K. Lee, "Lipid production by Chlorella vulgaris after a shift from nutrient-rich to nitrogen starvation conditions", *Bioresour. Technol.,* vol. 123, pp. 279-283, 2012.
[http://dx.doi.org/10.1016/j.biortech.2012.07.057] [PMID: 22940330]

[31] M. Siaut, S. Cuiné, C. Cagnon, B. Fessler, M. Nguyen, and P. Carrier, *Oil accumulation in the model green alga Chlamydomonas reinhardtii: Characterization, variability between common laboratory strains and relationship with starch reserves.* vol. Vol. 11. BMC Biotech, 2011.

[32] S. Zhu, W. Huang, J. Xu, Z. Wang, J. Xu, and Z. Yuan, "Metabolic changes of starch and lipid

triggered by nitrogen starvation in the microalga Chlorella zofingiensis", *Bioresour. Technol.*, vol. 152, pp. 292-298, 2014.
[http://dx.doi.org/10.1016/j.biortech.2013.10.092] [PMID: 24308944]

[33] S.H. Ho, A. Kondo, T. Hasunuma, and J.S. Chang, "Engineering strategies for improving the CO2 fixation and carbohydrate productivity of Scenedesmus obliquus CNW-N used for bioethanol fermentation", *Bioresour. Technol.*, vol. 143, pp. 163-171, 2013.
[http://dx.doi.org/10.1016/j.biortech.2013.05.043] [PMID: 23792755]

[34] C.H. Yao, J.N. Ai, X.P. Cao, and S. Xue, "Characterization of cell growth and starch production in the marine green microalga Tetraselmis subcordiformis under extracellular phosphorus-deprived and sequentially phosphorus-replete conditions", *Appl. Microbiol. Biotechnol.*, vol. 97, no. 13, pp. 6099-6110, 2013.
[http://dx.doi.org/10.1007/s00253-013-4983-x] [PMID: 23685550]

[35] J. Moseley, and A.R. Grossman, *Phosphate Metabolism and Responses to Phosphorus Deficiency*, 2009.
[http://dx.doi.org/10.1016/B978-0-12-370873-1.00014-9]

[36] C. González-Fernández, and M. Ballesteros, "Linking microalgae and cyanobacteria culture conditions and key-enzymes for carbohydrate accumulation", *Biotechnol. Adv.*, vol. 30, no. 6, pp. 1655-1661, 2012.
[http://dx.doi.org/10.1016/j.biotechadv.2012.07.003] [PMID: 22820270]

[37] B.J. Cade-Menun, and A. Paytan, "Nutrient temperature and light stress alter phosphorus and carbon forms in culture-grown algae", *Mar. Chem.*, vol. 121, pp. 27-36, 2010.
[http://dx.doi.org/10.1016/j.marchem.2010.03.002]

[38] L. Rodolfi, G. Chini Zittelli, N. Bassi, G. Padovani, N. Biondi, G. Bonini, and M.R. Tredici, "Microalgae for oil: strain selection, induction of lipid synthesis and outdoor mass cultivation in a low-cost photobioreactor", *Biotechnol. Bioeng.*, vol. 102, no. 1, pp. 100-112, 2009.
[http://dx.doi.org/10.1002/bit.22033] [PMID: 18683258]

[39] C.F. Ji, X.J. Yu, Z.A. Chen, S. Xue, J. Legrand, and W. Zhang, "Effects of nutrient deprivation on biochemical compositions and photo-hydrogen production of Tetraselmis subcordiformis", *Int. J. Hydrogen Energy,* vol. 36, pp. 5817-5821, 2011.
[http://dx.doi.org/10.1016/j.ijhydene.2010.12.138]

[40] L.L. Beer, E.S. Boyd, J.W. Peters, and M.C. Posewitz, "Engineering algae for biohydrogen and biofuel production", *Curr. Opin. Biotechnol.*, vol. 20, no. 3, pp. 264-271, 2009.
[http://dx.doi.org/10.1016/j.copbio.2009.06.002] [PMID: 19560336]

[41] A. Izumo, S. Fujiwara, Y. Oyama, A. Satoh, N. Fujita, and Y. Nakamura, "Physicochemical properties of starch in Chlorella change depending on the CO2 concentration during growth: Comparison of structure and properties of pyrenoid and stroma starch", *Plant Sci.*, vol. 172, pp. 1138-1147, 2007.
[http://dx.doi.org/10.1016/j.plantsci.2007.03.001]

[42] C. Thyssen, R. Schlichting, and C. Giersch, "The CO2-concentrating mechanism in the physiological context: lowering the CO2 supply diminishes culture growth and economises starch utilisation in Chlamydomonas reinhardtii", *Planta,* vol. 213, no. 4, pp. 629-639, 2001.
[http://dx.doi.org/10.1007/s004250100534] [PMID: 11556796]

[43] A. Kumar, S. Ergas, X. Yuan, A. Sahu, Q. Zhang, J. Dewulf, F.X. Malcata, and H. van Langenhove, "Enhanced CO(2) fixation and biofuel production via microalgae: recent developments and future directions", *Trends Biotechnol.*, vol. 28, no. 7, pp. 371-380, 2010.
[http://dx.doi.org/10.1016/j.tibtech.2010.04.004] [PMID: 20541270]

[44] S.H. Ho, C.Y. Chen, D.J. Lee, and J.S. Chang, "Perspectives on microalgal CO2-emission mitigation systems--a review", *Biotechnol. Adv.*, vol. 29, no. 2, pp. 189-198, 2011.
[http://dx.doi.org/10.1016/j.biotechadv.2010.11.001] [PMID: 21094248]

[45] D. Tang, W. Han, P. Li, X. Miao, and J. Zhong, "CO2 biofixation and fatty acid composition of

Scenedesmus obliquus and Chlorella pyrenoidosa in response to different CO2 levels", *Bioresour. Technol.,* vol. 102, no. 3, pp. 3071-3076, 2011. [http://dx.doi.org/10.1016/j.biortech.2010.10.047] [PMID: 21041075]

[46] M.K. Lam, and K.T. Lee, "Effect of carbon source towards the growth of Chlorella vulgaris for CO2 bio-mitigation and biodiesel production", *Int. J. Greenh. Gas Control,* vol. 14, pp. 169-176, 2013. [http://dx.doi.org/10.1016/j.ijggc.2013.01.016]

[47] T.M. Mata, A.A. Martins, and N.S. Caetano, "Microalgae for biodiesel production and other applications: A review", *Renew. Sustain. Energy Rev.,* vol. 14, pp. 217-232, 2010. [http://dx.doi.org/10.1016/j.rser.2009.07.020]

[48] O. Perez-Garcia, F.M. Escalante, L.E. de-Bashan, and Y. Bashan, "Heterotrophic cultures of microalgae: metabolism and potential products", *Water Res.,* vol. 45, no. 1, pp. 11-36, 2011. [http://dx.doi.org/10.1016/j.watres.2010.08.037] [PMID: 20970155]

[49] Y. Cheng, W. Zhou, C. Gao, K. Lan, Y. Gao, and Q. Wu, "Biodiesel production from Jerusalem artichoke (Helianthus Tuberosus L.) tuber by heterotrophic microalgae Chlorella protothecoides", *J. Chem. Technol. Biotechnol.,* vol. 84, pp. 777-781, 2009. [http://dx.doi.org/10.1002/jctb.2111]

[50] W. Xiong, X. Li, J. Xiang, and Q. Wu, "High-density fermentation of microalga Chlorella protothecoides in bioreactor for microbio-diesel production", *Appl. Microbiol. Biotechnol.,* vol. 78, no. 1, pp. 29-36, 2008. [http://dx.doi.org/10.1007/s00253-007-1285-1] [PMID: 18064453]

[51] H. Xu, X. Miao, and Q. Wu, "High quality biodiesel production from a microalga Chlorella protothecoides by heterotrophic growth in fermenters", *J. Biotechnol.,* vol. 126, no. 4, pp. 499-507, 2006. [http://dx.doi.org/10.1016/j.jbiotec.2006.05.002] [PMID: 16772097]

[52] Y. Liang, N. Sarkany, and Y. Cui, "Biomass and lipid productivities of Chlorella vulgaris under autotrophic, heterotrophic and mixotrophic growth conditions", *Biotechnol. Lett.,* vol. 31, no. 7, pp. 1043-1049, 2009. [http://dx.doi.org/10.1007/s10529-009-9975-7] [PMID: 19322523]

[53] R.M. Couto, P.C. Simões, A. Reis, T.L. Da Silva, V.H. Martins, and Y. Sánchez-Vicente, "Supercritical fluid extraction of lipids from the heterotrophic microalga Crypthecodinium cohnii", *Eng. Life Sci.,* vol. 10, pp. 158-164, 2010.

[54] D. Morales-Sánchez, R. Tinoco-Valencia, J. Kyndt, and A. Martinez, *Heterotrophic growth of Neochloris oleoabundans using glucose as a carbon source.* vol. Vol. 6. Biotechnol. Biofuels, 2013.

[55] M.B. Johnson, and Z. Wen, "Production of biodiesel fuel from the microalga Schizochytrium limacinum by direct transesterification of algal biomass", *Energy Fuels,* vol. 23, pp. 5179-5183, 2009. [http://dx.doi.org/10.1021/ef900704h]

[56] Q. Hu, "Environmental Effects on Cell Composition," In: *Handbook of Microalgal Culture*, ed: Blackwell Publishing Ltd, 2007, pp. 83-94.

[57] C. Yao, J. Ai, X. Cao, S. Xue, and W. Zhang, "Enhancing starch production of a marine green microalga Tetraselmis subcordiformis through nutrient limitation", *Bioresour. Technol.,* vol. 118, pp. 438-444, 2012. [http://dx.doi.org/10.1016/j.biortech.2012.05.030] [PMID: 22717561]

[58] J.M. Lv, L.H. Cheng, X.H. Xu, L. Zhang, and H.L. Chen, "Enhanced lipid production of Chlorella vulgaris by adjustment of cultivation conditions", *Bioresour. Technol.,* vol. 101, no. 17, pp. 6797-6804, 2010. [http://dx.doi.org/10.1016/j.biortech.2010.03.120] [PMID: 20456951]

[59] A. Blanchemain, and D. Grizeau, "Eicosapentaenoic acid content of Skeletonema costatum as a function of growth and irradiance; relation with chlorophyll a content and photosynthetic capacity", *J.*

Exp. Mar. Biol. Ecol., vol. 196, pp. 177-188, 1996. [http://dx.doi.org/10.1016/0022-0981(95)00129-8]

[60] S.H. Ho, C.Y. Chen, and J.S. Chang, "Effect of light intensity and nitrogen starvation on CO2 fixation and lipid/carbohydrate production of an indigenous microalga Scenedesmus obliquus CNW-N", *Bioresour. Technol.*, vol. 113, pp. 244-252, 2012. [http://dx.doi.org/10.1016/j.biortech.2011.11.133] [PMID: 22209130]

[61] Y. Chisti, "Biodiesel from microalgae beats bioethanol", *Trends Biotechnol.*, vol. 26, no. 3, pp. 126-131, 2008. [http://dx.doi.org/10.1016/j.tibtech.2007.12.002] [PMID: 18221809]

[62] A.P. Carvalho, S.O. Silva, J.M. Baptista, and F.X. Malcata, "Light requirements in microalgal photobioreactors: an overview of biophotonic aspects", *Appl. Microbiol. Biotechnol.*, vol. 89, no. 5, pp. 1275-1288, 2011. [http://dx.doi.org/10.1007/s00253-010-3047-8] [PMID: 21181149]

[63] L. Barsanti, and P. Gualtieri, "Algae Utilization", In: *Algae: Anatomy, Biochemistry, and Biotechnology*, 2nd ed. ed USA: CRC Press, 2014, pp. 267-308. [http://dx.doi.org/10.1201/b16544-8]

[64] D. Hernández, B. Riaño, M. Coca, and M.C. García-González, "Saccharification of carbohydrates in microalgal biomass by physical, chemical and enzymatic pre-treatments as a previous step for bioethanol production", *Chem. Eng. J.*, vol. 262, pp. 939-945, 2015. [http://dx.doi.org/10.1016/j.cej.2014.10.049]

[65] R. Harun, J.W. Yip, S. Thiruvenkadam, W.A. Ghani, T. Cherrington, and M.K. Danquah, "Algal biomass conversion to bioethanol - a step-by-step assessment", *Biotechnol. J.*, vol. 9, no. 1, pp. 73-86, 2014. [http://dx.doi.org/10.1002/biot.201200353] [PMID: 24227697]

[66] M.M. El-Dalatony, M.B. Kurade, R.A. Abou-Shanab, H. Kim, S. Salama, and B-H. Jeon, "Long-term production of bioethanol in repeated-batch fermentation of microalgal biomass using immobilized Saccharomyces cerevisiae", *Bioresour. Technol.*, vol. 219, pp. 98-105, 2016. [http://dx.doi.org/10.1016/j.biortech.2016.07.113] [PMID: 27479800]

[67] S.P. Choi, M.T. Nguyen, and S.J. Sim, "Enzymatic pretreatment of Chlamydomonas reinhardtii biomass for ethanol production", *Bioresour. Technol.*, vol. 101, no. 14, pp. 5330-5336, 2010. [http://dx.doi.org/10.1016/j.biortech.2010.02.026] [PMID: 20219355]

[68] K.H. Kim, I.S. Choi, H.M. Kim, S.G. Wi, and H-J. Bae, "Bioethanol production from the nutrient stress-induced microalga Chlorella vulgaris by enzymatic hydrolysis and immobilized yeast fermentation", *Bioresour. Technol.*, vol. 153, pp. 47-54, 2014. [http://dx.doi.org/10.1016/j.biortech.2013.11.059] [PMID: 24333701]

[69] O.K. Lee, Y-K. Oh, and E.Y. Lee, "Bioethanol production from carbohydrate-enriched residual biomass obtained after lipid extraction of Chlorella sp. KR-1", *Bioresour. Technol.*, vol. 196, pp. 22-27, 2015. [http://dx.doi.org/10.1016/j.biortech.2015.07.040] [PMID: 26218538]

[70] O.K. Lee, A.L. Kim, D.H. Seong, C.G. Lee, Y.T. Jung, J.W. Lee, and E.Y. Lee, "Chemo-enzymatic saccharification and bioethanol fermentation of lipid-extracted residual biomass of the microalga, Dunaliella tertiolecta", *Bioresour. Technol.*, vol. 132, pp. 197-201, 2013. [http://dx.doi.org/10.1016/j.biortech.2013.01.007] [PMID: 23411448]

[71] I. Pancha, K. Chokshi, B. George, T. Ghosh, C. Paliwal, R. Maurya, and S. Mishra, "Nitrogen stress triggered biochemical and morphological changes in the microalgae Scenedesmus sp. CCNM 1077", *Bioresour. Technol.*, vol. 156, pp. 146-154, 2014. [http://dx.doi.org/10.1016/j.biortech.2014.01.025] [PMID: 24495540]

[72] C. Simas-Rodrigues, H.D. Villela, A.P. Martins, L.G. Marques, P. Colepicolo, and A.P. Tonon, "Microalgae for economic applications: advantages and perspectives for bioethanol", *J. Exp. Bot.*, vol.

66, no. 14, pp. 4097-4108, 2015.
[http://dx.doi.org/10.1093/jxb/erv130] [PMID: 25873683]

[73] L. Barsanti, and P. Gualtieri, "General Overview," In: *Algae: Anatomy, Biochemistry, and Biotechnology*, 2nd ed. ed USA: CRC Press, 2014, pp. 1-48.
[http://dx.doi.org/10.1201/b16544-2]

[74] R.V. Stick, and S.J. Williams, "Chapter 9 - Disaccharides, Oligosaccharides and Polysaccharides," In: *Carbohydrates: The Essential Molecules of Life*, 2nd ed. ed Amsterdam: Elsevier, 2009, pp. 321-342.

[75] P.I. Hargreaves, C.A. Barcelos, A.C. da Costa, and N. Pereira Jr, "Production of ethanol 3G from Kappaphycus alvarezii: evaluation of different process strategies", *Bioresour. Technol.,* vol. 134, pp. 257-263, 2013.
[http://dx.doi.org/10.1016/j.biortech.2013.02.002] [PMID: 23500583]

[76] V.B. Agbor, N. Cicek, R. Sparling, A. Berlin, and D.B. Levin, "Biomass pretreatment: fundamentals toward application", *Biotechnol. Adv.,* vol. 29, no. 6, pp. 675-685, 2011.
[http://dx.doi.org/10.1016/j.biotechadv.2011.05.005] [PMID: 21624451]

[77] J.R. Miranda, P.C. Passarinho, and L. Gouveia, "Pre-treatment optimization of Scenedesmus obliquus microalga for bioethanol production", *Bioresour. Technol.,* vol. 104, pp. 342-348, 2012.
[http://dx.doi.org/10.1016/j.biortech.2011.10.059] [PMID: 22093974]

[78] G. Zhao, X. Chen, L. Wang, S. Zhou, H. Feng, W.N. Chen, and R. Lau, "Ultrasound assisted extraction of carbohydrates from microalgae as feedstock for yeast fermentation", *Bioresour. Technol.,* vol. 128, pp. 337-344, 2013.
[http://dx.doi.org/10.1016/j.biortech.2012.10.038] [PMID: 23196255]

[79] J-H. Hwang, A.N. Kabra, M-K. Ji, J. Choi, M.M. El-Dalatony, and B-H. Jeon, "Enhancement of continuous fermentative bioethanol production using combined treatment of mixed microalgal biomass", *Algal Res.,* vol. 17, pp. 14-20, 2016.
[http://dx.doi.org/10.1016/j.algal.2016.03.029]

[80] J.R. Miranda, P.C. Passarinho, and L. Gouveia, "Bioethanol production from Scenedesmus obliquus sugars: the influence of photobioreactors and culture conditions on biomass production", *Appl. Microbiol. Biotechnol.,* vol. 96, no. 2, pp. 555-564, 2012.
[http://dx.doi.org/10.1007/s00253-012-4338-z] [PMID: 22899495]

[81] J. Martín Juárez, A. Lorenzo Hernando, R. Muñoz Torre, S. Blanco Lanza, and S. Bolado Rodríguez, "Saccharification of microalgae biomass obtained from wastewater treatment by enzymatic hydrolysis. Effect of alkaline-peroxide pretreatment", *Bioresour. Technol.,* vol. 218, pp. 265-271, 2016.
[http://dx.doi.org/10.1016/j.biortech.2016.06.087] [PMID: 27372005]

[82] T. Yuan, X. Li, S. Xiao, Y. Guo, W. Zhou, J. Xu, and Z. Yuan, "Microalgae pretreatment with liquid hot water to enhance enzymatic hydrolysis efficiency", *Bioresour. Technol.,* vol. 220, pp. 530-536, 2016.
[http://dx.doi.org/10.1016/j.biortech.2016.08.117] [PMID: 27614155]

[83] R. Harun, B. Liu, and M.K. Danquah, "Analysis of process configurations for bioethanol production from microalgal biomass," In: *Progress in biomass and bioenergy production*, S. S. Shaukat, Ed., ed Croatia: Intech Science, Technology and Medicine, 2011, pp. 978-53.

[84] E. Tomás-Pejó, J.M. Oliva, M. Ballesteros, and L. Olsson, "Comparison of SHF and SSF processes from steam-exploded wheat straw for ethanol production by xylose-fermenting and robust glucose-fermenting Saccharomyces cerevisiae strains", *Biotechnol. Bioeng.,* vol. 100, no. 6, pp. 1122-1131, 2008.
[http://dx.doi.org/10.1002/bit.21849] [PMID: 18383076]

[85] M. Devarapalli, and H.K. Atiyeh, "A review of conversion processes for bioethanol production with a focus on syngas fermentation", *Biofuel Res. J.,* vol. 7, pp. 268-280, 2015.
[http://dx.doi.org/10.18331/BRJ2015.2.3.5]

[86] V. Ashokkumar, Z. Salam, O.N. Tiwari, S. Chinnasamy, S. Mohammed, and F.N. Ani, "An integrated approach for biodiesel and bioethanol production from Scenedesmus bijugatus cultivated in a vertical tubular photobioreactor", *Energy Convers. Manage.*, vol. 101, pp. 778-786, 2015. [http://dx.doi.org/10.1016/j.enconman.2015.06.006]

[87] C. Asada, K. Doi, C. Sasaki, and Y. Nakamura, "Efficient extraction of starch from microalgae using ultrasonic homogenizer and its conversion into ethanol by simultaneous saccharification and fermentation", *Natural Resources,* vol. 3, pp. 175-179, 2012. [http://dx.doi.org/10.4236/nr.2012.34023]

[88] S-H. Ho, S-W. Huang, C-Y. Chen, T. Hasunuma, A. Kondo, and J-S. Chang, "Bioethanol production using carbohydrate-rich microalgae biomass as feedstock", *Bioresour. Technol.,* vol. 135, pp. 191-198, 2013. [http://dx.doi.org/10.1016/j.biortech.2012.10.015] [PMID: 23116819]

[89] H. Guo, M. Daroch, L. Liu, G. Qiu, S. Geng, and G. Wang, "Biochemical features and bioethanol production of microalgae from coastal waters of Pearl River Delta", *Bioresour. Technol.,* vol. 127, pp. 422-428, 2013. [http://dx.doi.org/10.1016/j.biortech.2012.10.006] [PMID: 23138065]

[90] M. Balat, H. Balat, and C. Öz, "Progress in bioethanol processing", *Pror. Energy Combust. Sci.,* vol. 34, pp. 551-573, 2008. [http://dx.doi.org/10.1016/j.pecs.2007.11.001]

[91] U. Kalnenieks, A. Pentjuss, R. Rutkis, E. Stalidzans, and D.A. Fell, "Modeling of Zymomonas mobilis central metabolism for novel metabolic engineering strategies", *Front. Microbiol.,* vol. 5, p. 42, 2014. [http://dx.doi.org/10.3389/fmicb.2014.00042] [PMID: 24550906]

[92] S. Yang, Q. Fei, Y. Zhang, L.M. Contreras, S.M. Utturkar, S.D. Brown, M.E. Himmel, and M. Zhang, "Zymomonas mobilis as a model system for production of biofuels and biochemicals", *Microb. Biotechnol.,* vol. 9, no. 6, pp. 699-717, 2016. [http://dx.doi.org/10.1111/1751-7915.12408] [PMID: 27629544]

CHAPTER 3

Bioethanol Production Process

Fabiana Regina Xavier Batista*, **Juliana de Souza Ferreira** and **Vicelma Luiz Cardoso**

School of Chemical Engineering, Federal University of Uberlandia, Uberlandia/MG, Brazil

Abstract: Ethanol has increasingly attracted attention of researchers. Currently, ethanol is synthesized by different routes, such as production from the synthesis gas, ethylene hydration and fermentation of sugars. In addition, microalgae have been recognized as the potential sources for biofuel production, since they reach high biomass productivities. Recently, the use of microalgal biomass as feedstock for bioethanol production became an interest for researchers and then for the general public. Therefore, the aim of this chapter is to introduce the basics of algal-ethanol production and the current status of this emerging bioethanol source.

Keywords: Bioethanol, Biomass, *Chlamydomonas reinhardtii*, Purple non sulfur bacterium, Hybrid system.

DIFFERENT ROUTES TO ETHANOL PRODUCTION

Bioethanol is a renewable fuel whose production has increased worldwide. This is due to environmental care and oil depletion since it is possible to maintain sustainable production and reduce pollution. Besides, the combustion of bioethanol does not result in high emissions of greenhouse gases.

Originally, ethanol was synthetized from ethylene or synthesis gas (syngas). Ethanol could be manufactured either by the catalytic hydration of ethene in the presence of steam, using phosphoric acid adsorbed on the surface of silica or by a catalytic process in which the syngas (a mixture of primarily CO, H_2 and CO_2) is converted in a variety of fuels, including ethanol [1]. However, syngas and ethylene are obtained either from fossil resources and from the processes that require extreme operational conditions. Besides, the oil crisis has made serious competitors of petrochemical methods. Therefore, it was demanded to develop technology that is sustainable and based on mild operating parameters. A viable

* **Corresponding author Fabiana Regina Xavier Batista:** School of Chemical Engineering, Federal University of Uberlandia, Uberlandia/MG, Brazil; Tel : +55 (34) 32309401; E-mail: frxbatista@ufu.br

José Carlos Magalhães Pires (Ed.)

alternative is the production of bioethanol by fermentation using renewable source such as sugar, starch or cellulosic biomass.

The biological process, in which bioethanol is obtained from sugar cane using yeast fermentation system, is successful mainly in the South America countries. Brazil was the pioneer in using ethanol as a motor fuel and, nowadays, it is the world's largest sugarcane bioethanol producer (in 2015/16, its production reached 8 billion gallons) [2]. In addition, the sugarcane industry provides the biggest energy system of biomass of the world, since it uses bagasse and straw to generate electricity [3]. In the United States, 90% of ethanol production is from corn that contains 70% of starch. Besides, other carbon sources can also be used for bioethanol production by fermentation pathways including industrial waste [4 - 6].

It is important to note there are different strategies for ethanol production by biological routes: first-generation (1G), second-generation (2G) and third-generation (3G). From the biochemical point of view, 1G bioethanol processes may be categorized by fermentable sugar conversion in bioethanol by yeast. The biomass can either be sugarcane or beet, both providing sucrose that is directly fermented. For the specific case of bioethanol obtained from sucrose, the enzymatic activity of invertase breaks sucrose into glucose and fructose. During the transport into the cells, the sugar is phosphorylated in the glycolysis pathway forming pyruvic acid. The theoretical yield is 0.511 g of bioethanol per gram of the metabolized sugar (hexose) and 0.489 grams of CO_2 per gram of the metabolized sugar (hexose). Two ATP (adenosine triphosphate) molecules generated from glucose are used in the cell metabolism to conduct the yeast cell reactions. Therefore, the bioethanol production might be favored by the cell growth.

The 2G technology is required when the lignocellulosic biomass is used. The 2G bioethanol is produced in four steps including: (i) biomass pre-treatments to open fibers; (ii) the enzymatic hydrolysis of hemicellulose and cellulose providing fermentable sugars; (iii) sugar fermentation to bioethanol; and (iv) recovery and purification of the target-product [7 - 9]. In the 3G bioethanol, algal biomass is used either (1) as a substrate by yeast to the target-product evolution or (2) as a responsible strain for the intracellular bioethanol production and secreted to the supernatant [10]. Microalgae are capable to accumulate polysaccharides such as starch into the cell wall [11]. Based on their metabolic pathway, the starch might be converted into pyruvate which is a precursor for bio-products synthesis; such as acetic and formic acids, bioethanol, CO_2, biohydrogen, among others [12]. The enhancement of bioethanol production may be observed by the increase of production or storage of starch, as well as by feeding sugar to the algae cultures, and subsequent dark, anaerobic fermentation of carbon sources. Nevertheless, a detailed knowledge of the bioethanol production by algae needs to be researched in parallel.

MICROALGAE CONTRIBUTIONS TO THE BIOETHANOL PRODUCTION

General Algae Information

Algae can be unicellular or multi-cellular organisms, found in terrestrial and aquatic environments. They convert CO_2, water and sunlight into biomass such as carbohydrates, proteins and lipids, and obtain energy *via* photosynthesis. They can be prokaryotic (cyanobacteria) or eukaryotic microorganisms including seaweed and phytoplankton [13, 14]. The eukaryotic algae are classified in four different groups depending on the pigmentation, life cycle and basic cellular structure: green algae (*Chlorophyta*), red algae (*Rhodophyta*), golden algae (*Chryosophyta*) and diatoms (*Bacillariophyta*) [15 - 17].

In respect to their composition, algal biomass has a varied composition that favors the application as feedstock in the production of 3G biofuels *via* thermochemical and biochemical technologies [13]. Macroalgae (or seaweeds or brown algae) and microalgae (or green algae) differ in relation to the content of lipids, carbohydrates and proteins. In general, microalgae present lipids (7-23%), carbohydrates (5-23%) and proteins (6-52%). The high content of lipids benefits their application in biodiesel production [18] and the high level of proteins allows them to be used as co-substrates [19]; whilst seaweeds do not present lipids, usually contain abundant carbohydrates (80%) and low levels of proteins (15%) [15].

Specifically to bioethanol, it is produced mainly *via* algal biomass fermentation from both, micro- and macroalgae. In both cases, more economic harvesting methods must be developed to make production of alga-based bioethanol commercially viable.

Brown and Green Algae Used as Feedstock

Algae present several advantages as feedstock to 3G bioethanol. In comparison to 1G land-based crops, they do not demand arable land, fertilizers or irrigation water, so do not compete for food resources or potable water [20, 21]. Besides, marine microalgae species can also reduce freshwater consumption [22]. The areal productivities are higher, for instance, brown algae (seaweeds) can be harvested 4 - 6 times in a year [23]. The production of macroalgae cultures achieved around 14.5 million tons (wet weight) in 2007 [24].

In relation to 2G bioethanol biomass based on lignocellulosic feedstock, the process applying algae is simplified since the pre-treatment of the lignocellulosic biomass is complex. Algae have low or no content of lignin or hemicellulose and the hydrolysis of the cell wall to sugar is simple [19, 25, 26].

In the biophotolysis of water by microalgae, H^+ and e^- are extracted and stored in the form of carbohydrates during photosynthesis and Calvin cycle, in which the metabolism of carbon fixation occurs. These carbohydrates can be either starch accumulated in the plastids as reserve material or cellulose, pectin and sulfated polysaccharides cellulose that are part of the cell wall structure [27]. The composition of these sugars can vary among the species. The bioethanol production requires the selection of strains with high productivity of sugars. The species of microalgae mostly used for bioethanol fermentation are *Chlorella, Dunaliella, Chlamydomonas* and *Scenedesmus* that accumulate more than 50% of starch based on their dry cell weight [22, 28]. With respect to brown algae (macroalgae), the main produced sugars are glucan and mannitol that are easily assimilated by microbes during bioethanol fermentation. In addition, brown algae also produce alginate which demands microorganisms that can simultaneously degrade, uptake and metabolize this sugar [29, 30].

Concerning the saccharification of algal carbohydrates previously to fermentation by yeast, this procedure is performed to cleave the β-1, 4-glycosidic and α-1,4-glycosidic linkages between the subunits of glucose in cellulose and in starch, respectively. Among the methods, they can be chemical (mineral acids, alkaline, hot compressed water, supercritical CO_2) or enzymatic. Enzymatic hydrolysis is the most promising procedure due to elimination of sugar degradation in toxic by-products (furfural and 5-hydroxymethylfurfural) that inhibits fermentation step. In addition, this process presents lower costs of equipment and waste treatment, since the technique is based on mild temperature and pressure conditions. Besides, there is no addition of chemicals [31 - 33].

Besides saccharification, different pre-treatments using a variety of microalgae were also investigated [32]. Depending on the species, the yield of bioethanol production comprehended ranges from 0.240 to 0.888 g bioethanol/g of substrate. For instance, in the study of bioethanol production from biomass of *Chlorella vulgaris* (IAM C-534), the starch (37% in dry weight) was extracted after cell disruption by ultrasonic radiation with 95% (v/v) methanol and deionized water. These authors used enzymatic hydrolysis by incubation with α-amylase at 100 °C in pH 6 and glucoamylase at 60 °C in pH 4.5. The hydrolysate was fermented by *Saccharomyces cerevisiae*. A production of 0.24% (w/w) of bioethanol was attained corresponding to 65% of the theoretical concentration based on the 37% (w/w) of starch [33].

The co-fermentation of *Laminaria digitata* (macroalgae, rich in carbohydrates) and *Arthrospira platensis* (microalgae, rich in protein) for biohydrogen, bioethanol and organic acids production was evaluated [19]. The algal biomass was pre-treated with 2.5% dilute H_2SO_4 at 135 °C for 15 min resulting in a total yield of carbohydrate monomers of 0.268 g/g volatile solids. In the best condition, up to 2.24 g/L of bioethanol was obtained and maximum energy conversion efficiency of 54.5% was achieved producing biohydrogen (5.7%), bioethanol (15.6%), acetic acid (9.6%) and butyric acid (18.5%).

A direct conversion without algal biomass pre-treatment was studied [29]. A thermophilic bacterium *Defluviitalea phaphyphila* Alg1 was used to ferment un-pretreated brown algal biomass at 60 °C. *D. phaphyphila* Alg1 could simultaneously degrade mannitol, glucose and alginate to produce bioethanol. The target product was accumulated (10 g/L) within 72 h with a bioethanol yield of 0.25 g/g.

Self Fermentation or Intracellular Bioethanol Production Excreted to Supernatant

Microalgae biomass can also be used for bioethanol production through the conversion of the starch accumulated in the cells into the alcohol under dark and anaerobic condition [34]. Intracellular bioethanol production is an effective way for CO_2 fixation coupled with energy production and is simpler and less energy intensive than the conventional bioethanol-fermentation (algal feedstock fermentation) [33].

Concerning the type of carbon source, algae can be autotrophic or heterotrophic, and even mixotrophic, having the ability to perform photosynthesis and acquire exogenous organic sources [35]. Green algae such as *Chlamydomonas reinhardtii*, *Chlamydomonas moewusii*, *Chlorogonoium elongatum*, *Chlorella fusca* and *Scenedesmus* D3, metabolize their starch to different products that are acetate, bioethanol, formate, glycerol, lactate, H_2 and CO_2, and the ratio among the excreted products varies depending on the species or even on the strain [36]. Although microalgae are oxygenic phototroph, these photosynthetic microbes can proceed fermentative pathways as well, since they can resist extended periods of anoxia [37]. The initial stage of the microalgae metabolism is similar to any other heterotroph eukaryote, that is, glycolysis occurs producing pyruvate, ATP and NADH (nicotinamide adenine dinucleotide) that is the Embden-Meyerhof-Parnas (EMP) pathway [36]. Among the algae species, the anoxic metabolism pathways from the starch to the end products of *Chlamydomonas reinhardtii* are the most explored, following the same routes as in strict anaerobic microorganisms and parallel metabolic processes occur in the mitochondria, cytosol and the chloroplast [37, 38].

Bioethanol is produced either from acetaldehyde produced *via* pyruvate decarboxylase (PDC) or from acetyl-CoA produced from pyruvate *via* pyruvate dehydrogenase (PDH), pyruvate formate-lyase (PFL) and pyruvate: ferredoxin oxidoreductase (PFO). The conversion of acetaldehyde or acetyl-CoA to bioethanol occurs *via* bifunctional aldehyde/alcohol dehydrogenase (ADHE). This metabolic route to bioethanol production allows the reoxidation of NADH and sustain glycolysis (shown in Fig. **1**) [36, 37, 39].

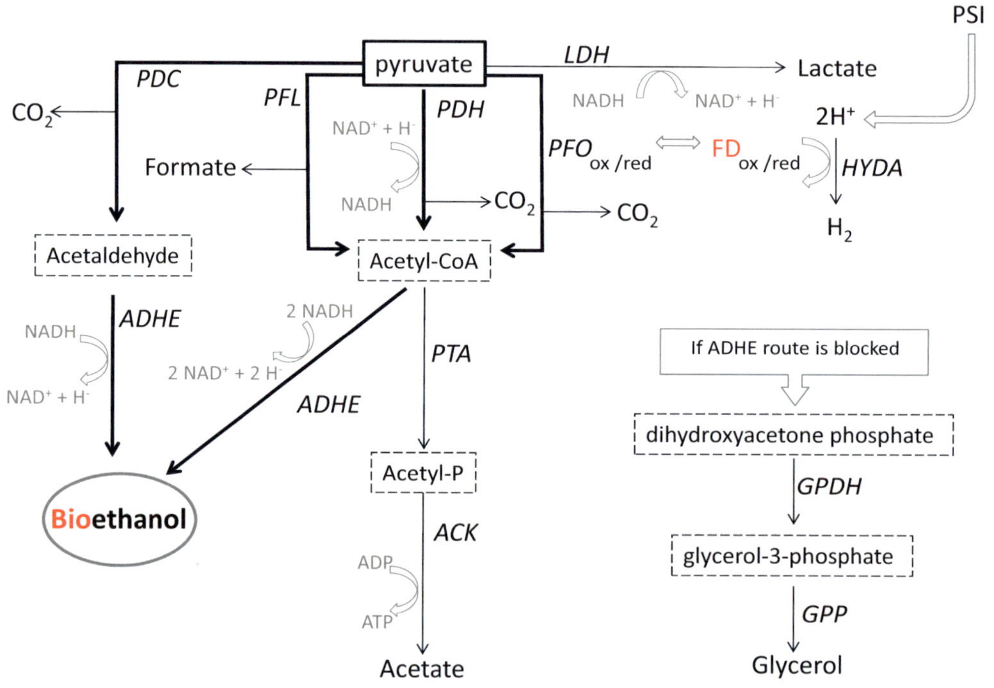

Fig. (1). Production of metabolites by green algae (Adapted from [37]). Bioethanol, acetate, lactate, formate, glycerol, CO_2 and H_2. Pyruvate dehydrogenase (PDH). Pyruvate: ferredoxin oxidoreductase (PFO). Pyruvate formate-lyase (PFL). (FeFe)-hydrogenase (HYDA). Pyruvate decarboxylase (PDC). Alcohol/aldehyde dehydrogenase (ADHE). Lactate dehydrogenase (LDH). Acetate kinase (ACK). Phosphotransacetylase (PTA). Photosystem I (PSI). Glycerol-3-phosphate dehydrogenase (GPDH). Glycerol-3-phosphatase (GPP).

Further metabolites are produced in dark and anaerobic fermentation by the green alga *C. reinhardtti*. PFL and PFO (both sensitive to oxygen) enable the oxidation of pyruvate to acetyl-CoA, yielding, respectively, formate and CO_2 with the reduction of ferredoxin (FD), which can be used by the [FeFe]-hydrogenase (HYDA) to reduce protons to H_2 [37]. During hydrogen production and under sulfur deprivation, H^+ and e^- are fed into the plastoquinone pool and HYDA *via* the photosystem I (PSI). Furthermore, acetate can also be produced from acetyl-

CoA by phosphotransacetylase (PTA) and acetate kinase (ACK) with ATP production [40]. With respect to PFL system, fermentation of dark-adapted algae produces a ratio of formate, bioethanol and acetate of 2:1:1. While the cells are illuminated, formate and bioethanol are also synthesized along with H_2 production, under deprivation of sulfur that leads anaerobic medium [41 - 43]. Other metabolite is the lactate that is produced from pyruvate *via* lactate dehydrogenase (LDH). Besides bioethanol, formate, acetate, lactate, bioH_2 and CO_2, glycerol is also one of the metabolites in a dark fermentation of green algae and its formation occurs when aldehyde/alcohol dehydrogenase (ADHE) route is blocked. The synthesis of glycerol involves a two-step reaction; the intermediate dihydroxyacetone phosphate is reduced to glycerol-3-phosphate by NAD^+-dependent glycerol-3-phosphate dehydrogenase (GPDH) and then glycerol-3-phosphate is hydrolyzed by a glycerol-3-phosphatase (GPP) in glycerol [36, 41, 44].

The formation of these fermentation products is controlled at the level of pyruvate concentration and the ratios of the fermentative products may change as a consequence of culture conditions in order to increment bioethanol production [38].

One of the culture medium (with sulfur deprivation) exerts a distinct effect on the cellular activities of photosynthesis and respiration. It could result in the net consumption of oxygen by the cells causing anaerobiosis in the growth medium; a condition that automatically elicited H_2 production by the algae [45]. However, microalgae cells usually cannot tolerate anaerobiosis for a long period of time since the photosystem II (PSII), responsible for photosynthesis in microalgae, could be suffer irreversible damage.

The fermentation of intracellular starch of *Chlamydomonas reinhardtii* F-60 was investigated [41]. This work tested under the influence of the dark and light and in the presence of 3-(3,4-dichlorophenyl)-1,1-dimethylurea (DCMU) and carbonyl cyanide-p-trifluoromethoxyphenyl hydrazine (FCCP). The results indicated that bioethanol production was inhibited by light. In dark fermentation, the ratio of formate, acetate and bioethanol was 2.07:1.07:0.91. Finally, glycerol, lactate and CO_2 were detected in minor amounts.

The intracellular bioethanol production of different marine species was tested [33]. The highest bioethanol productivity and relatively high conversion percentages of 30-40% (theoretical yield is 0.56 g of bioethanol/1 g of starch) were observed using *Chlamydomonas reinhardtii* UTEX2247. Besides, approximately 0.27% (w/w) of bioethanol was produced and lactic (0.22%, w/w) and acetic (0.11%, w/w) acids were detected. This study showed that the agitation has a little effect on the bioethanol production. In addition, pH and temperature should be controlled around 7-8 and 25-35 °C, respectively.

The fermentation *via Chlamydomonas sp* YA-SH-1 was evaluated [46]. The process resulted in a growth rate of 30 g-dry biomass/(m^2d) with a content of starch of 30% (dry base), conversion rate of 50% and the bioethanol concentration increased during initial 44 h and kept constant until the end of the process (60 h) achieving a concentration of approximately 1.25%.

The production of H_2 by the green alga *Chlamydomonas reinhardtii* cultivated in sulfur-deprived medium was also the focus [42]. The initial pH and the effect of anaerobiose/aerobiose condition were investigated. It was observed that, in addition to H_2, formate, bioethanol and acetate were accumulated in the extracellular medium and the highest concentrations were attained at low pH. At initial pH 7.3 and 7.7, bioethanol was not detect. However, bioethanol was produced at initial pH of 6.5, 6.9 and 8.2, being the concentration of 1.48, 1.07 and 0.39 mmol, respectively. The findings showed that the reductants generated during the initial steps of the glycolytic oxidation of starch were consumed by bioethanol production, preventing their re-direction to NAD(P)H-plastoquinone oxidoreductase and production of H_2.

The production of bioethanol by dark fermentation using the marine green algae *Chlorococcum littorale* was studied [34]. In this case, cellular starch was decomposed immediately after the dark anaerobic incubation. Simultaneously with starch decomposition, formation of end-products was observed, such as bioethanol, acetate, hydrogen, carbon dioxide and glycerol. It suggested that green algae that produce high starch contents could increase bioethanol evolution. According to these authors, *C. littorale* performs dark fermentation by degrading intracellular starch *via* the EMP pathway.

RESEARCH AND DEVELOPMENT OF THE USE OF ALGAE RELATED TO BIOETHANOL PRODUCTION

Currently, microalgae are considered one of the most important sources of biomass, because they can provide a significant contribution to liquid fuels. Algae can be used to generate energy in several ways; for instance, the algal oils to produce biodiesel. Biohydrogen gas and bioethanol under specialized growth conditions might be also observed. In addition, the algal biomass can also be burned to generate heat and electricity. Nevertheless, challenges should be overcome to enhance biofuel production by algae.

According to the microalgal metabolism as described above, bioethanol can be produced from pyruvate decomposition through acetyl-CoA or acetaldehyde. Therefore, knowledge of the metabolic capabilities of green algae provides the foundation for increments in bioethanol production. In general, a balance between starch and end-products must be established [47].

Nevertheless, besides metabolic performance, it is important to note the ability of the starch accumulation and temperature or agitation conditions that can strongly influence the bioethanol production. Research challenges for algal-bioethanol production may be summarized in terms of providing conditions for biomass propagation and understanding the essential biological route to the target-product synthesis. Several strategies might be adopted to reach this purpose. Initially, it is important to define an initial cell density to green algae cultivation. Previous performed assay has already shown that algal inoculum lower than 0.05 g/L did not provide satisfactory bioethanol production. Anaerobic conditions also showed essential to the ethanol evolution. Organic acid such as acetic acid should be supplemented into *C. reinhardtii* cultures to provide a short cell population doubling time, around 30 hours. Nitrogen or argon gases are indicated to processes using green algae to keep the medium anaerobic. Nevertheless, in the case of nitrogen-fixing strains, that is, in the case of cyanobacteria, only argon gas is recommended, since nitrogenase is the enzyme directly related to this biofuel production and it is inhibited by nitrogen. If the enzyme is hydrogenase, both gases may be used.

By using microalgae to intracellular bioethanol production excreted to supernatant, they may also be related to bioethanol in different strategies. Maximum bioethanol content (19.94 g/L) was observed by applying a hybrid system in which the effluent (metabolized supernatant) from *C. reinhardtii* containing organic acids was used as substrate to *R. capsulatus*; a purple non-sulfur bacterium [10]. This investigation was performed in order to enhance the comprehension of the effect of the medium composition and operational conditions on bioethanol production by *C. reinhardtii* and *R. capsulatus*.

Another alternative to approach the production of bioethanol by using algae is concerned to the identification of potential strains. In Brazil, several research efforts on the isolation and characterization of highly productive algal strains from Brazilian biodiversity are currently under way. Microalgae research and production in Brazil started in 1970 [48]. Currently, the only commercial production is for aquiculture, in several small-scale companies mostly in the northeastern region of the country. Nonetheless, Brazil has a great potential for large-scale microalgae production, due to large tropical coastal area.

Further strategy is the application of microalgae to 2G bioethanol production. The Brazilian Agricultural Research Corporation (Ministry of Agriculture, Livestock, and Food Supply) is dedicated to the enzyme (beta-glycosidases) production by microalgae. These enzymes are responsible for the final stage of "breaking" of cellulose; a complex carbohydrate used for bioethanol production. In this case, the cellulose materials must be "broken" in glucose which is then fermented by *Saccharomyces cerevisiae* to give bioethanol. In general, beta-glycosidases are produced by filamentous *fungi,* but Brazilian researchers believe that these enzymes obtained from genetically modified microalgae should result in lower cost than from filamentous *fungi.*

An improved system of bioethanol production has been developed in Europe. The project DEMA (Direct Ethanol from MicroAlgae) aims to develop, demonstrate and license a technology for the direct production of bioethanol from microalgae with low-cost scalable photobioreactors. The concept involves the catalytic conversion of solar energy, water and CO_2 in ethanol [49]. For that, metabolically engineered strain of the cyanobacterium can be used. Furthermore, bioethanol will be extracted from the culture media *via* a membrane technology process, exploiting the existing EU expertise and technology.

In addition, the use of mutants is one of the alternatives that must be investigated to obtain further increase in bioethanol yields [15, 32]. A strain of genetically modified cyanobacteria was designed to be capable of producing bioethanol [50]. It was created as a novel pathway by cloning the coding sequences of pyruvate decarboxylase (pdc) and alcohol dehydrogenase II (adh) from the bacterium. *Zymomonas mobilis* was cloned into the shuttle vector pCB4 and then was used to transform the cyanobacterium *Synechococcus sp.* strain PCC 7942. The pdc and adh genes were expressed at high levels by control of the promoter from the rbcLS operon encoding the cyanobacterial ribulose-1,5-bisphosphate carboxylase/oxygenase. The maximum bioethanol production was 1,710 μM (cell line pCB4-LR(TF)pa).

Finally, it is also important to note the risk associated with inconsistent target-biofuel production caused by bacterial infections. In general, bacterial infection results in the fast decline of algal cultures [51]. Differential centrifugation, sonication and vortexing are suggested as procedures of the purification methods for microalgae [52]. However, the most recent works using antibiotics to the elimination of bacteria from microalgal culture is important to be highlighted. The use of the *Penicillin* G, *Dihydrostreptomycin* and *Gentamycin* antibiotics in microalgal cultures to avoid bacterium strains development was also discussed [52].

CONCLUDING REMARKS

Economic incentives can potentially support several kinds of investment for the bioethanol production by algae cultivation. However, bioethanol produced by algal system is a challenge, mainly when it is secreted into the supernatant by cells. The technology is successful if the cellular metabolism is geared to the bioethanol production. In general, the starch or other carbohydrate found in the cell might be converted into pyruvate or acetaldehyde and afterwards, bioethanol contents can be observed in the broth of the photofermentation. It is undeniable that the cell growth has a direct influence on the biofuel evolution. Nevertheless, the selection of right operation conditions, such as initial cell density, anaerobic statement, light intensity, among others is essential. Finally, several configurations of hybrid systems (green algae/ photosynthetic bacteria/fermentative bacteria/yeast) also show attractive.

CONFLICT OF INTEREST

The authors declare no conflict of interest, financial or otherwise.

ACKNOWLEDGEMENTS

The authors would like to thank Minas Gerais State Foundation for Research Development (FAPEMIG), National Counsel of Technological and Scientific Development (CNPq), Coordination for the Improvement of Higher Education Personnel (CAPES) and Vale S.A. for financial support.

REFERENCES

[1] K.T. Klasson, M.D. Ackerson, and E.C. Clausen, "Bioconversion of syngas into liquid or gaseous fuels", *Enzyme Microb. Technol.,* vol. 14, pp. 602-608, 1992. [http://dx.doi.org/10.1016/0141-0229(92)90033-K]

[2] Sugar cane, Available at: http://sugarcane.org/sugarcane-products/ethanol, [Accessed Sept. 20, 2016].

[3] L. Wu, Y. Li, M. Arakane, M. Ike, M. Wada, Y. Terajima, S. Ishikawa, and K. Tokuyasu, "Efficient conversion of sugarcane stalks into ethanol employing low temperature alkali pretreatment method", *Bioresour. Technol.,* vol. 102, no. 24, pp. 11183-11188, 2011. [http://dx.doi.org/10.1016/j.biortech.2011.09.081] [PMID: 22000967]

[4] S. Ozmihci, and F. Kargi, "Utilization of cheese whey powder for ethanol fermentation: effects of operating parameters", *Enzyme Microb. Technol.,* vol. 38, pp. 711-718, 2006. [http://dx.doi.org/10.1016/j.enzmictec.2005.11.006]

[5] M.O. Dias, T.L. Junqueira, O. Cavalett, M.P. Cunha, C.D. Jesus, P.E. Mantelatto, C.E. Rossell, R. Maciel Filho, and A. Bonomi, "Cogeneration in integrated first and second generation ethanol from sugarcane", *Chem. Eng. Res. Des.,* vol. 91, pp. 1411-1417, 2013. [http://dx.doi.org/10.1016/j.cherd.2013.05.009]

[6] L. Thomas, C. Larroche, and A. Pandey, "Current developments in solid-state fermentation", *Biochem. Eng. J.,* vol. 81, pp. 146-161, 2013. [http://dx.doi.org/10.1016/j.bej.2013.10.013]

[7] P. Binod, M. Kuttiraja, M. Archana, J.K. Usha, R. Sindhu, R.K. Sukumaran, and A. Pandey, "High temperature pretreatment and hydrolysis of cotton stalk for producing sugars for bioethanol production", *Fuel,* vol. 92, pp. 340-345, 2012. [http://dx.doi.org/10.1016/j.fuel.2011.07.044]

[8] Y.Y. Tye, K.T. Lee, W.N. Wan Abdullah, and C.P. Leh, "Potential of Ceiba pentandra (L.) Gaertn. (kapok fiber) as a resource for second generation bioethanol: effect of various simple pretreatment methods on sugar production", *Bioresour. Technol.*, vol. 116, pp. 536-539, 2012. [http://dx.doi.org/10.1016/j.biortech.2012.04.025] [PMID: 22595099]

[9] J.P. Silva, S.I. Mussatto, I.C. Roberto, and J.A. Teixeira, "Fermentation medium and oxygen transfer conditions that maximize the xylose conversion to ethanol by Pichia stipitis", *Renew. Energy,* vol. 37, pp. 259-265, 2012. [http://dx.doi.org/10.1016/j.renene.2011.06.032]

[10] R.L. Costa, T.V. Oliveira, J.S. Ferreira, V.L. Cardoso, and F.R.X. Batista, "Prospective technology on bioethanol production from photofermentation", *Bioresour. Technol.,* vol. 181, pp. 330-337, 2015. [http://dx.doi.org/10.1016/j.biortech.2015.01.090] [PMID: 25678298]

[11] D.O. Hall, and K.K. Rao, *Photosynthesis.* 5th ed. Cambridge University Press: Cambridge, 1994.

[12] G. Philipps, D. Krawietz, A. Hemschemeier, and T. Happe, "A pyruvate formate lyase-deficient *Chlamydomonas reinhardtii* strain provides evidence for a link between fermentation and hydrogen production in green algae", *Plant J.*, vol. 66, no. 2, pp. 330-340, 2011. [http://dx.doi.org/10.1111/j.1365-313X.2011.04494.x] [PMID: 21219510]

[13] Zh.R. Oyang, X.B. Wen, Y.H. Geng, H. Mei, H.J. Hu, G.Y. Zhang, and Y.G. Li, *The Effects of Light Intensities,Temperatures,pH and Salinities on Photosynthesis of Chlorella. Journal of Wuhan Botanical Research,* vol. 28, p. 49, 2010.

[14] L. Wagner, *Biodiesel from Algae Oil.* Mora Associates Ltd.: UK, 2007, pp. 1-7.

[15] A. Singh, P.S. Nigam, and J.D. Murphy, "Renewable fuels from algae: an answer to debatable land based fuels", *Bioresour. Technol.,* vol. 102, no. 1, pp. 10-16, 2011. [http://dx.doi.org/10.1016/j.biortech.2010.06.032] [PMID: 20615690]

[16] S.A. Khan, Rashmi, Mir. Z. Hussain, S. Prasad, and U.C. Banerjee, "Prospects of biodiesel production from microalgae in India", *Renew. Sustain. Energy Rev.,* vol. 13, pp. 2361-2372, 2009. [http://dx.doi.org/10.1016/j.rser.2009.04.005]

[17] M. Packer, "Algal capture of carbon dioxide; biomass generation as a tool for greenhouse gas mitigation with reference to New Zealand energy strategy and policy", *Energ. Pol.,* vol. 37, pp. 3428-3437, 2009. [http://dx.doi.org/10.1016/j.enpol.2008.12.025]

[18] W.S. Borges, B.S. Araújo, L.G. Moura, U. Coutinho Filho, M.M. de Resende, and V.L. Cardoso, "Bio-oil production and removal of organic load by microalga *Scenedesmus sp.* using culture medium contaminated with different sugars, cheese whey and whey permeate", *J. Environ. Manage.,* vol. 173, pp. 134-140, 2016. [http://dx.doi.org/10.1016/j.jenvman.2015.11.015] [PMID: 26948140]

[19] A. Xia, A. Jacob, M.R. Tabassum, C. Herrmann, and J.D. Murphy, "Production of hydrogen, ethanol and volatile fatty acids through co-fermentation of macro- and micro-algae", *Bioresour. Technol.,* vol. 205, pp. 118-125, 2016. [http://dx.doi.org/10.1016/j.biortech.2016.01.025] [PMID: 26820925]

[20] N. Wei, J. Quarterman, and Y-S. Jin, "Marine macroalgae: an untapped resource for producing fuels and chemicals", *Trends Biotechnol.,* vol. 31, no. 2, pp. 70-77, 2013. [http://dx.doi.org/10.1016/j.tibtech.2012.10.009] [PMID: 23245657]

[21] R. Harun, M.K. Danquah, and G.M. Forde, "Microalgal biomass as a fermentation feedstock for bioethanol production", *J. Chem. Technol. Biotechnol.,* vol. 85, pp. 199-203, 2010.

[22] C-Y. Chen, X-Q. Zhao, H-W. Yen, S-H. Ho, C-L. Cheng, D-J. Lee, F-W. Bai, and J-S. Chang, "Microalgae-based carbohydrates for biofuel production", *Biochem. Eng. J.,* vol. 78, pp. 1-10, 2013. [http://dx.doi.org/10.1016/j.bej.2013.03.006]

[23] G. Roesijadi, A.E. Copping, M.H. Huesemann, J. Forster, and J.R. Benemann, "Techno-Economic

Feasibility Analysis of Offshore Seaweed Farming for Bioenergy and Biobased Products; Report Number PNWD-3931. Battelle PacificNorthwest Division",

[24] Food and Agriculture Organization of the United Nations (FAO), "Use of algae and aquatic macrophytes as feed in small scale aquaculture - a review. FAO Fisheries and Aquaculture Technical Paper. No. 531. Rome, FAO. 123 p",

[25] J.F. Rong, X.G. Huang, and X.H. Zhou, "CN Patent 101 870 950 A",

[26] Ch.M. Han, J.Q. Yang, H.T. Chen, H. Zhang, and M.Sh. Liu, "CN Patent 101 735 948 A",

[27] S.H. Ho, C.Y. Chen, D.J. Lee, and J.S. Chang, "Perspectives on microalgal CO_2-emission mitigation systems--a review", *Biotechnol. Adv.,* vol. 29, no. 2, pp. 189-198, 2011. [http://dx.doi.org/10.1016/j.biotechadv.2010.11.001] [PMID: 21094248]

[28] T.Zh. Liu, W. Zhang, Y. Chen, X.W. Peng, X.L. Chen, and L. Chen, "CN Patent 102 206 570 A",

[29] S-Q. Ji, B. Wang, M. Lu, and F-L. Li, "Direct bioconversion of brown algae into ethanol by thermophilic bacterium *Defluviitalea phaphyphila*", *Biotechnol. Biofuels,* vol. 9, p. 81, 2016. [http://dx.doi.org/10.1186/s13068-016-0494-1] [PMID: 27042210]

[30] A.J. Wargacki, E. Leonard, M.N. Win, D.D. Regitsky, C.N. Santos, P.B. Kim, S.R. Cooper, R.M. Raisner, A. Herman, A.B. Sivitz, A. Lakshmanaswamy, Y. Kashiyama, D. Baker, and Y. Yoshikuni, "An engineered microbial platform for direct biofuel production from brown macroalgae", *Science,* vol. 335, no. 6066, pp. 308-313, 2012. [http://dx.doi.org/10.1126/science.1214547] [PMID: 22267807]

[31] S.P. Choi, M.T. Nguyen, and S.J. Sim, "Enzymatic pretreatment of *Chlamydomonas reinhardtii* biomass for ethanol production", *Bioresour. Technol.,* vol. 101, no. 14, pp. 5330-5336, 2010. [http://dx.doi.org/10.1016/j.biortech.2010.02.026] [PMID: 20219355]

[32] J. Zhu, J. Rong, and B. Zong, "Factors in mass cultivation of microalgae for biodiesel", *Chin. J. Catal.,* vol. 34, pp. 80-100, 2013. [http://dx.doi.org/10.1016/S1872-2067(11)60497-X]

[33] A. Hirano, R. Ueda, S. Hirayama, and Y. Ogushi, "CO_2 fixation and ethanol production with microalgal photosynthesis and intracelular anaerobic fermentation", *Energy,* vol. 22, pp. 137-142, 1997. [http://dx.doi.org/10.1016/S0360-5442(96)00123-5]

[34] Y. Ueno, N. Kurano, and S. Miyachi, "Ethanol production by dark fermentation in the marine green alga, Chlorococcum littorale", *J. Ferment. Bioeng.,* vol. 86, pp. 38-43, 1998. [http://dx.doi.org/10.1016/S0922-338X(98)80031-7]

[35] R.E. Lee, *Phycology.* Cambridge University Press: New York, 1980.

[36] A. Atteia, R. van Lis, A.G. Tielens, and W.F. Martin, "Anaerobic energy metabolism in unicellular photosynthetic eukaryotes", *Biochim. Biophys. Acta,* vol. 1827, no. 2, pp. 210-223, 2013. [http://dx.doi.org/10.1016/j.bbabio.2012.08.002] [PMID: 22902601]

[37] A.R. Grossman, M. Croft, V.N. Gladyshev, S.S. Merchant, M.C. Posewitz, S. Prochnik, and M.H. Spalding, "Novel metabolism in *Chlamydomonas* through the lens of genomics", *Curr. Opin. Plant Biol.,* vol. 10, no. 2, pp. 190-198, 2007. [http://dx.doi.org/10.1016/j.pbi.2007.01.012] [PMID: 17291820]

[38] F. Mus, A. Dubini, M. Seibert, M.C. Posewitz, and A.R. Grossman, "Anaerobic acclimation in *Chlamydomonas reinhardtii*: anoxic gene expression, hydrogenase induction, and metabolic pathways", *J. Biol. Chem.,* vol. 282, no. 35, pp. 25475-25486, 2007. [http://dx.doi.org/10.1074/jbc.M701415200] [PMID: 17565990]

[39] A. Hemschemeier, J. Jacobs, and T. Happe, "Biochemical and physiological characterization of the pyruvate formate-lyase Pfl1 of *Chlamydomonas reinhardtii*, a typically bacterial enzyme in a eukaryotic alga", *Eukaryot. Cell,* vol. 7, no. 3, pp. 518-526, 2008. [http://dx.doi.org/10.1128/EC.00368-07] [PMID: 18245276]

[40] O. Kruse, and B. Hankamer, "Microalgal hydrogen production", *Curr. Opin. Biotechnol.,* vol. 21, no. 3, pp. 238-243, 2010. [http://dx.doi.org/10.1016/j.copbio.2010.03.012] [PMID: 20399635]

[41] R.P. Gfeller, and M. Gibbs, "Fermentative Metabolism of *Chlamydomonas reinhardtii*: I. Analysis of Fermentative Products from Starch in Dark and Light", *Plant Physiol.,* vol. 75, no. 1, pp. 212-218, 1984. [http://dx.doi.org/10.1104/pp.75.1.212] [PMID: 16663574]

[42] S. Kosourov, M. Seibert, and M.L. Ghirardi, "Effects of extracellular pH on the metabolic pathways in sulfur-deprived, H_2-producing *Chlamydomonas reinhardtii* cultures", *Plant Cell Physiol.,* vol. 44, no. 2, pp. 146-155, 2003. [http://dx.doi.org/10.1093/pcp/pcg020] [PMID: 12610217]

[43] M. Winkler, A. Hemschemeier, C. Gotor, A. Melis, and T. Happe, "[Fe]-hydrogenases in green algae: photo-fermentation and hydrogen evolution under sulfur-deprivation", *Int. J. Hydrogen Energy,* vol. 27, pp. 1431-1439, 2002. [http://dx.doi.org/10.1016/S0360-3199(02)00095-2]

[44] L. Magneschi, C. Catalanotti, V. Subramanian, A. Dubini, W. Yang, F. Mus, M.C. Posewitz, M. Seibert, P. Perata, and A.R. Grossman, "A mutant in the ADH1 gene of *Chlamydomonas reinhardtii* elicits metabolic restructuring during anaerobiosis", *Plant Physiol.,* vol. 158, no. 3, pp. 1293-1305, 2012. [http://dx.doi.org/10.1104/pp.111.191569] [PMID: 22271746]

[45] A. Melis, L. Zhang, M. Forestier, M.L. Ghirardi, and M. Seibert, "Sustained photobiological hydrogen gas production upon reversible inactivation of oxygen evolution in the green alga *Chlamydomonas reinhardtii*", *Plant Physiol.,* vol. 122, no. 1, pp. 127-136, 2000. [http://dx.doi.org/10.1104/pp.122.1.127] [PMID: 10631256]

[46] S. Hirayama, R. Ueda, Y. Ogushi, A. Hirano, Y. Samejima, K. Hon-Nami, and S. Kunito, "Ethanol production from carbon dioxide by fermentative microalgae", *Stud. Surf. Sci. Catal.,* vol. 114, pp. 657-660, 1998. [http://dx.doi.org/10.1016/S0167-2991(98)80845-8]

[47] DOE (U.S. Department of Energy), National Algal Biofuels Technology Review. U.S. Department of Energy, Office of Energy Efficiency and Renewable Energy, Bioenergy Technologies Office, 2016. [Online] Available at: http://energy.gov/sites/prod/files/2016/06/f33/national_algal_biofuels_technology_review.pdf, [AccessedSept. 20, 2016].

[48] M.C. Damaso, C.M. Machado, D.S. Rodrigues, S.G. Belem, and T.F. Salum, "Bioprocesses for biofuels: an overview of the Brazilian case", *Chem. Biol. Technol. Agric.,* vol. 1, p. 6, 2014. [http://dx.doi.org/10.1186/s40538-014-0006-0]

[49] Direct Ethanol from MicroAlgae (DEMA), Report Summary, Research and Innovation, Europian Comission, pp. 1-3, 2015 . [Online] Available at: http://cordis.europa.eu/result/rcn/163444_en.html, [AccessedSept. 20, 2016].

[50] M.D. Deng, and J.R. Coleman, "Ethanol synthesis by genetic engineering in cyanobacteria", *Appl. Environ. Microbiol.,* vol. 65, no. 2, pp. 523-528, 1999. [PMID: 9925577]

[51] E.E. Powell, and G.A. Hill, "Economic assessment of an integrated bioethanol-biodiesel-microbial fuel cell facility utilizing yeast and photosynthetic algae", *Chem. Eng. Res. Des.,* vol. 87, pp. 1340-1348, 2009. [http://dx.doi.org/10.1016/j.cherd.2009.06.018]

[52] R.R. Guillard, "Purification Methods for Microalgae", In: *Algal Culturing Techniques.,* R.A. Andersen, Ed., , 2005, pp. 117-133.

CHAPTER 4

Methane Production Process for Microalgae Conversion

Martin Pablo Caporgno, Esther Torrens and **Christophe Bengoa***

Departament d'Enginyeria Química, Universitat Rovira i Virgili, Tarragona, Spain

Abstract: Anaerobic digestion (AD) is a biological process in which various types of biomasses are converted mainly into gaseous product called biogas. The importance of biogas lies in its composition: methane is the major compound, being used as a biofuel.

Although the history of biogas dates back to more than two thousand years ago, the first biogas plant was built in the middle of the 19th century. AD started to be applied for wastewater treatment and biogas production in the early 1900s, and soon after, some researchers realised that biogas could be produced from the microalgae generated during the secondary treatment of wastewater. Afterwards, an increasing number of studies about the AD of microalgae have been published worldwide.

This chapter introduces the basic principles of AD and describes the influence of some of the most important operating parameters during biogas production from microalgae.

Keywords: Anaerobic Digestion, Biogas, Methane, Microalgae.

INTRODUCTION

Anaerobic digestion (AD) is a process in which biomass, is converted (in absence of oxygen) into biogas [1]. This biological process is made up by diverse microbial populations, involving several metabolic reactions. Thanks to its versatility, the process makes possible the conversion of various types of biomass as into biogases, *i.e.*, wastewater from food industry, sewage sludge from wastewater treatment plants (WWTPs), animal manure and the organic fraction derived from municipal solid waste amongst others [2, 3].

Due to the dual benefit of waste stabilisation and renewable energy production, AD is gaining more attention worldwide. Several government policies in developed countries in Europe, US and Australia are promoting the utilisation of

* **Corresponding author Christophe Bengoa:** Departament d'Enginyeria Química, Universitat Rovira i Virgili, Av. Països Catalans 26, 43007 Tarragona, Spain; Tel/Fax: +34 977 55 96 18, +34 977 55 96 67; E-mail: christophe.bengoa@urv.cat

José Carlos Magalhães Pires (Ed.)

this technology [4, 5]. Another reason for the expansion of AD is the environmental-friendly characteristics of this process, which has minimal effects on the environment [6, 7]. In fact, three main benefits are usually associated with the process:

1. *No adverse impacts on food security.* Biogas can be generated from residues, non-food crops or crops cultivated in lands unsuitable for food production; as a consequence, AD is classified as second generation technology for biofuels. The agricultural products are basic biomass used for food production and for a wide variety of everyday products. The production of crops for biofuels has changed the use of lands, resulting in a competition: cultivation for food *vs.* cultivation for biofuels [8]. In this context, AD allows the production of biogas for energy needs without risking food supply.
2. *The biogas produced is carbon-neutral.* By using renewable sources of biomass, the release of carbon dioxide from fossil fuels is prevented; there is no increase in the carbon dioxide in the atmosphere when producing and burning biogas [9]. In addition, the conversion of biogas into power and heat prevents methane emissions. Methane is usually released to the atmosphere during the decomposition of untreated waste, and it is 25 times more potent as a greenhouse gas than carbon dioxide [10].
3. *The disposal of waste is greatly reduced.* Only a small fraction of the biomass is converted into cells during AD [1], thus the disposal of waste is reduced. This is probably one of the major advantages of this technology. The generated digestate is the second most abundant product, and it contains liquid and solid fractions. Since this by-product is rich in nitrogen and phosphorus, considerable efforts have been made in order to minimise possible environmental impacts. The directly used as bio-fertiliser in farms [5] or in microalgae cultivation [11 - 13] seems to be some of the best alternatives to recycle the nutrients contained in the digestate.

The amount of biogas generated during digestion and its composition depends on several factors, *i.e.*, the population of microorganisms in the digester, the type of biomass and the operating parameters in the digester. However, the biogas is primarily a mixture of methane (CH_4) and carbon dioxide (CO_2), containing minor fractions of nitrogen (N_2), oxygen (O_2) and water (H_2O), and traces of hydrogen sulphide (H_2S), ammonia (NH_3), volatile organic compounds and siloxanes amongst other impurities [9, 14]. Table **1** summarizes the composition of biogas generated from three different type of biomass.

Regarding the impurities in biogas, some of them can adversely affect its further utilisation. As seen in Table **1**, the major impurity in biogas is CO_2. Although CO_2

is not dangerous, it reduces the calorific value of biogas at high concentration due to the methane dilution. Fortunately, modern co-generation systems for biogas combustion can operate with gas mixtures containing up to 50% in volume of CO_2 and do not require a reduction of the CO_2 content. The biogas is usually saturated with moisture when recovered from the digesters. The presence of water in biogas solubilises H_2S which derives from amino-acids and proteins, being the main cause of corrosion in the biogas system. To prevent corrosion, water can be condensed by decreasing the temperature of biogas. Processes for the removal of H_2S include the use of activated carbon or hydrated ferric oxide filters, scrubber towers and dosing of iron salts like ferric chloride into the digester [15]. Ammonia also derives from amino-acids and proteins. Oxidation of NH_3 during biogas combustion increases the emission of NO_x to the environment. A separate cleaning step for ammonia elimination is usually unnecessary since most technologies for carbon dioxide and hydrogen sulphide removal are also effective to remove ammonia. Volatile organic compounds like benzene or toluene can be found in biogas, but mainly in biogas from WWTPs or landfills [9]. These compounds are components of crude oil and are used in the industry for a large number of products; part of these products end in WWTPs and landfills. Finally, siloxanes constitute another kind of impurities. These organic silicon polymers derive from the degradation of materials commonly found in soaps and detergents, thus they are also problematic in the biogas produced from landfill and WWTPs [9]. During combustion, siloxanes produce solid waste similar to sand that accumulates and causes problems in equipment and pipelines.

Table 1. Average composition and characteristics of biogas from different biomass. Adapted from [14].

Gas Composites/Features	Formula	Sewage Gas	Agricultural Gas	Landfill Gas
Methane (% by vol.)	CH_4	65 – 75	45 – 75	45 – 55
Carbon dioxide (% by vol.)	CO_2	20 – 35	25 – 55	25 – 30
Carbon monoxide (% by vol.)	CO	< 0.2	< 0.2	< 0.2
Nitrogen (% by vol.)	N_2	3.4	0.01 – 5.00	10 – 25
Oxygen (% by vol.)	O_2	0.5	0.01 – 2.00	1 – 5
Hydrogen (% by vol.)	H_2	traces	0.5	0.00
Hydrogen sulphide (mg/Nm³)	H_2S	< 8000	10 – 30.000	< 8000
Ammonium (mg/Nm³)	NH_3	traces	0.01 – 2.50	traces
Siloxanes (mg/Nm³)	-	< 0.1 – 5.0	traces	< 0.1 – 5.0
Benzene, Toluene, Xylene (mg/Nm³)	-	< 0.1 – 5.0	0.00	< 0.1 – 5.0
Chlorofluorocarbons (mg/Nm³)	-	0	20 – 1000	n.a,
Oil (mg/Nm³)	-	traces	traces	0

BIOGAS FROM MICROALGAE: THE HISTORY

The history of biogas dates back to more than two thousand years ago [14, 16]; anecdotal evidence indicated that biogas was used to warm water for bath during the 10th century BC in Assyria. Marco Polo mentioned the use of covered sewage tanks to produce biogas in China. However, the first plant to capture biogas in modern times was built in Mumbai, in 1859. This idea, later brought to the UK in 1895, was used for gas lighting street and homes. Soon after, in the early 1900s, the technologies for wastewater treatment and for biogas production were developed in the UK and Germany.

AD of microalgae dates back to 1950s, when Golueke *et al.* thought about the possibility of converting the microalgal biomass generated during the secondary treatment of wastewater into biogas [17]. According to the authors, AD seemed to be an attractive possibility to turn the solar energy captured by microalgae into usable energy. The results revealed that, in terms of the volume of gas produced per organic matter destroyed, microalgae were able to produce comparable amounts of biogas than the sewage sludge normally digested. In those early experiments, the authors had already identified one of the drawbacks for microalgae digestion: the resistance of microalgae cell walls to the microorganisms attack. While the destruction of the organic matter was close to 60% in digesters fed with sewage sludge under mesophilic conditions, the destruction was about 44% with microalgae. In any case, although a temperature increase was beneficial for microalgae degradability, digesters fed with microalgae were not superior in performance. However, the popularity of AD waned over the following years, the excess of oil in 1955 caused a fall in the price of crude and make biofuels less attractive. Moreover, the utilisation of mineral fertilisers increased, decreasing the utilisation of the digestate for this purpose [14]. As a consequence, only a few publications went out in later years.

A further oil crisis in the early 1970s arouse interest in the possibility of replacing the fossil fuels by biofuels. The production of first generation biofuels was considered as the best option to deal with the situation, *i.e.*, sugarcane, corn, soybean, rapeseed cultivation for biofuels [18]. At that time, the U.S. National Renewable Energy Laboratory (NREL), through the Aquatic Species Program (ASP), started promoting the research in microalgae utilisation for biofuels; the program lasted for almost 20 years [19]. Foree and McCarty were some of the researchers that restarted the investigation on the conversion of microalgae into methane in the 1970s [20]. The main objective of their research was to evaluate the decomposition of biomass, but not the production of methane. Once again, the results evidenced the high resistance of the microalgae to AD; 40% on average of the biomass remained undecomposed even after 200 days. The conversion of microalgae into biogas caught the interest of Samson and Leduy, who contributed significant advances in the field. Over a couple of years and several experiments, the authors identified some of the drawbacks for microalgae digestion which are still being investigated and some possible alternatives to overcome them [21, 22]. After observing that approximately half of the biomass resisted degradation, the authors attributed this fact to the resistance of the cell walls. Aimed to enhance the degradation, the authors investigated the effects of mechanical and thermochemical pre-treatments on the disintegration and degradation of microalgae [21]. The imbalance carbon to nitrogen ratio in microalgae was also suggested as another cause of the low degradation [22]. In this occasion, the authors evaluated the possibility of improving the methane productivity by mixing microalgae and carbon-rich substrates, *i.e.*, co-digestion.

The utilisation of macroalgae for methane production was also investigated as a consequence of the oil crises [23, 24]. However, the interest in the topic did not continue for a long time. The more recent research concerning the AD of macroalgae derived from the possibility of processing the biomass harvested from eutrophicated lagoons and coasts, with environmental and economic benefits [25].

From 2000 onwards, the rising oil prices have reawakened the interests in algal biofuels. A large number of research studies have been conducted, including different kinds of biofuels and microalgae species. Some of the most important findings regarding the AD of microalgae are highlighted in the following sections.

THE ANAEROBIC DIGESTION PROCESS

A wide variety of microbial communities have been reported to be involved in the AD process, playing an indispensable role in the consecutive stages which converts the complex biomass into biogas [6]. These stages, *i.e.*, hydrolysis, acidogenesis, acetogenesis/dehydrogenation and methanogenesis are schematised in Fig. (**1**).

The first stage is the **hydrolysis**, sometimes referred to as solubilisation. The initial attack on the large-molecules of organic compounds take place in this stage. Most of the insoluble particulate matter is decomposed through the action of extracellular enzymes produced and secreted by obligate or facultative anaerobic bacteria. These enzymes make contact with particulate and colloidal substrates, and breakdown or "cut" specific bonds in large-molecules into smaller molecules. The small molecules are solubilised, thus becoming more digestible. Some examples of hydrolytic enzymes are protease, cellulase, amylase, pectinase and lipase. Some of the large-molecules which are decomposed and the products generated are (i) proteins, breakdown into amino acids; (ii) carbohydrates like cellulose, hemicellulose and starch, breakdown into cellobiose, xylose and glucose; and (iii) lipids, breakdown into fatty acids and glycerol.

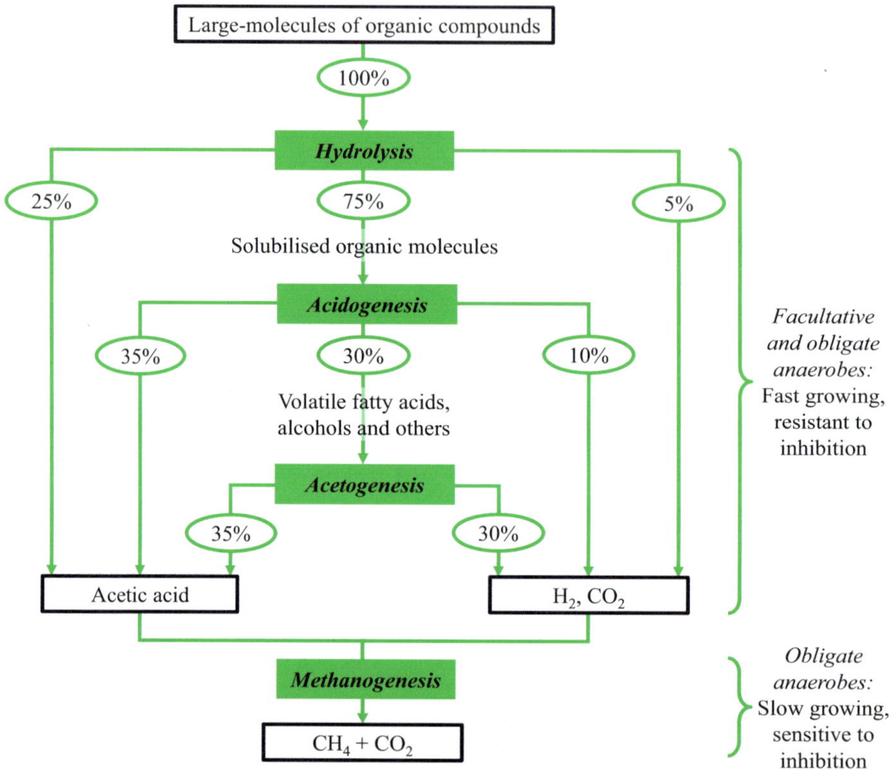

Fig. (1). Scheme of the different stages involved in AD and COD conversion in the stages.

Since not all bacteria produce extracellular enzymes and each enzyme can degrade only a specific group of substrates, hydrolysis is possible thanks to a large and diverse community of bacteria. The conversion of complex organic compounds into methane proceeds as rapidly as these compounds are decomposed into substrates assimilable by the microorganisms responsible for the following stages. For poorly biodegradable substrates, as it occurs with microalgae, hydrolysis becomes the rate-limiting stage [26].

Small amounts of hydrogen, acetic acid and carbon dioxide are generated during hydrolysis; approximately 5% and 20% of the energy stored in the large-molecules is now in the form of hydrogen and acetic acid. A 75% of the energy remains in the smaller molecules generated (see Fig. **1**) [1, 27].

Another group of bacteria, mainly facultative anaerobes, are responsible for the second stage called **acidogenesis**. These bacteria convert the solubilised organic molecules into VFA (*e.g.,* formic, acetic, propionic, butyric, and valeric acid), into alcohols (*e.g.,* methanol and ethanol), and also some other compounds like aldehydes, carbon dioxide, hydrogen, organic-nitrogen and organic-sulphur compounds. Approximately 10% and 35% of the energy from large-molecules is additionally reserved in hydrogen and acetic acid after acidogenesis, whereas a 30% remains in VFA (see Fig. **1**) [1, 27]. Although the facultative anaerobic bacteria use the remaining oxygen and create favourable conditions for obligatory anaerobes like methanogens, they also produce ammonia and hydrogen sulphide which are toxic for methanogens.

The major part of the compounds produced during acidogenesis cannot be converted directly into methane. For this reason, many of the acids and alcohols are degraded to acetate in this third stage denoted **acetogenesis**. At this stage, 17% and 13% of the energy in the molecules pass to acetic acid and hydrogen (see Fig. **1**) [1, 27]. Since the major part of methane comes from acetate, the bacteria involved in acetogenesis constitute one of the most important groups of microorganisms in AD. Acetogens are obligate hydrogen producers; it means that hydrogen is simultaneously produced with acetate. When not removed and maintained at low concentration in media, hydrogen inhibits acetogens. The equilibrium is established by the conversion of hydrogen into methane in the following stage, thus there is a symbiotic relationship with acetogens and methanogens.

The final stage (**methanogenesis**) is characterised by the methane production. It is promoted by microorganisms classified as Archaea or Archaeabacteria, but it is worth pointing out that methanogens are usually mentioned as methanogenic bacteria in the literature. The methanogens are responsible for converting a relatively small number of simple substrates into methane, the acetate, carbon dioxide and hydrogen being their most frequent substrates; approximately 70% of the methane is produced from acetate and less than 30% from hydrogen. Secondary substrates are formate, methanol, carbon monoxide, and methylamine. There are a large diversity of methanogens in digesters. According to the used substrates, methanogens are mainly classified into three principal groups, *i.e.*, hydrogenotrophic, acetotrophic and methylotrophic methanogens.

The hydrogenotrophic methanogens use hydrogen to convert carbon dioxide into methane (Eq. 1). As mentioned before, they are responsible for maintaining low values of hydrogen partial pressure to avoid the inhibition of acetogens. Some of them can convert carbon monoxide too (Eq. 2).

$$CO_2 + 4\,H_2 \rightarrow CH_4 + 2\,H_2O \qquad \textbf{(1)}$$

$$4\,CO + 2\,H_2O \rightarrow CH_4 + 3\,CO_2 \qquad \textbf{(2)}$$

The acetotrophic methanogens convert acetate into methane and carbon dioxide (Eq. 3). Part of the carbon dioxide generated is further converted by hydrogenotrophic methanogens into methane dioxide (Eq. 3). As acetogens, they are adversely affected by the hydrogen partial pressure.

$CH_3COOH \rightarrow CH_4 + CO_2$**(3)**

The third group (methylotrophic methanogens) converts substrates with methyl groups into methane, *i.e.*, methanol (Eq. 4) and methylamines (Eq. 5-7).

$$4CH_3OH \rightarrow 3\ CH_4 + CO_2 + 2\ H_2O \quad \textbf{(4)}$$

$$4(CH_3)_3 - N + 6\ H_2O \rightarrow 9\ CH_4 + 3\ CO_2 + 4\ NH_3 \quad \textbf{(5)}$$

$$2(CH_3)_2 - NH + 2\ H_2O \rightarrow 3\ CH_4 + CO_2 + 2\ NH_3 \quad \textbf{(6)}$$

$$4(CH_3) - NH_2 + 2\ H_2O \rightarrow 3\ CH_4 + CO_2 + 4\ NH_3 \quad \textbf{(7)}$$

Besides being strict anaerobes, methanogens are characterised by a very slow reproduction. Methanogens reproduction requires at least 12 days retention time for a large population in anaerobic digesters [27]. This slow reproduction is the result of the small amount of energy that methanogens can obtained from the substrates; these microorganisms must degrade a large quantity of substrate for doubling their population. A direct consequence, which is also an advantage, is the small quantity of solids produced in anaerobic digesters. A simple comparison of the amount of bacterial cells or sludge generated in anaerobic and aerobic degradation processes shows that approximately 50% of the organic carbon degraded in the aerobic digesters is converted into new bacterial cells, whereas in anaerobic digesters the percentage is close to 5% [28].

Parameters Affecting AD of Microalgae

Optimum operating conditions in anaerobic digesters are required for the satisfactory conversion of organic matter into methane. The methanogens are the most sensitive population in the digesters [6]; changes in pH, alkalinity, hydraulic retention times and temperature can significantly reduce their activity. Under stable digester conditions, the products generated by acidogens and acetogens are converted into methane at roughly the same velocity. If the methanogens do not degrade these products, they accumulate and acidify the medium, being the most common cause of failure in AD.

Over the years, AD has been widely used for the large-scale conversion of sewage sludge, municipal solid waste, manures and agricultural waste into biogas, and for this reason, the optimum conditions for the main operating parameters in digesters are well known [1, 28 - 30]. On the contrary, the major part of the information available concerning AD of microalgae derives from laboratory-scale digesters, which usually operate in batch mode. Nevertheless, the experiences gained from the AD of waste can be applied to microalgae digestion despite the lack of large-scale digesters processing microalgae. Some of the parameters that affect AD are described below.

Substrate Characteristics

When degraded, the substrate may provide the nutrients required by the microorganisms for all their activities. Nevertheless, some intermediate products generated during the degradation process can adversely affect microorganisms and consequently, interfere with the continuity of the process. The degradation of proteins in microalgae is a clear example. Proteins are usually the major constituent in microalgae, up to 60% of the organic fraction [31]; the ammonia and hydrogen sulphide released during their degradation are both potential inhibitors of methanogens [1].

The substrates have complex compositions and structures that are usually unknown. For this reason, some simple parameters are used to give an idea of the fraction of the biomass that can be potentially degraded, *i.e.*, the COD (g O_2/L) and the volatile solids content (VS, in g VS/L or %). Both, the COD and the VS are determined in the laboratory following standard methods [32].

According to Angelidaki and Sanders [33], when the composition of the substrate is known, it is possible to calculate the theoretical methane and ammonium yields for the digestion using a stoichiometric relationship (Eq. 8).

$$C_aH_bO_cN_d + \left(a - \frac{b}{4} - \frac{c}{2} + \frac{3\,d}{4}\right) H_2O \rightarrow$$
$$\left(\frac{a}{2} + \frac{b}{8} - \frac{c}{4} - \frac{3\,d}{8}\right) CH_4 + \left(\frac{a}{2} - \frac{b}{8} + \frac{c}{4} + \frac{3\,d}{8}\right) CO_2 + \, c\, NH_3 \quad \textbf{(8)}$$

Similarly, the oxygen consumed in the oxidation of the organic substrate or theoretical COD (Eq. 9) can be calculated.

$$C_aH_bO_cN_d + \left(a + \frac{b}{4} - \frac{c}{2} - \frac{3\,d}{4}\right) O_2 \rightarrow$$
$$n\, CO_2 + \left(\frac{b}{2} - \frac{3\,d}{2}\right) H_2O + \, c\, NH_3 \quad \textbf{(9)}$$

Both Eq. 8 and Eq. 9 allow the calculation of the theoretical methane yields based on the COD (Eq. 10) or VS (Eq. 11) in the substrate; using the molecular weights of C, H, O, N, and the volume of 1 mol of gas at normal conditions (22.4 L/mol).

$$CH_4 \text{ yield } (L_{CH4}/g_{COD}) = \frac{\left(\frac{a}{2}+\frac{b}{8}-\frac{c}{4}-\frac{3\,d}{8}\right) 22.4}{\left(a+\frac{b}{4}-\frac{c}{2}-\frac{3\,d}{4}\right) 32} \quad \textbf{(10)}$$

$$CH_4 \text{ yield } (L_{CH4}/g_{VS}) = \frac{\left(\frac{a}{2}+\frac{b}{8}-\frac{c}{4}-\frac{3\,d}{8}\right) 22.4}{12\,a+1\,b+16\,c+14\,d} \quad \textbf{(11)}$$

The calculation of the theoretical methane can be simplified by multiplying the lipid, protein and carbohydrate content of the substrate by the theoretical methane productions of these fractions. It has been estimated an average composition of $C_{57}H_{104}O_6$ for lipid, $C_5H_7O_2N$ for proteins and $(C_6H_{10}O_5)_n$ for carbohydrates, thus theoretical methane productions are 1.014, 0.496 and 0.415 NL_{CH4}/g_{VS}, respectively [33]. As can be seen, the lipid fraction has the greatest influence on the theoretical methane production. This methodology can be used to determine the theoretical methane potential in microalgae. Although the composition of microalgae depends on several factors such as strain, cultivation conditions, etc., the lipid content in microalgae is between 5-60%, the protein content between 40-60% and the carbohydrate content between 8-30% [31]. According to these values, the theoretical methane production varies between 0.5 and 0.8 NL_{CH4}/g_{VS}.

Theoretical values of methane production are overestimated, since their calculation assumes that all the substrate is digested and converted into methane. However, a portion of substrate is converted into new cells, typically 5-10% of the organic material degraded. Moreover, some fractions are not degraded and others are inaccessible to microorganisms due to the structure of the substrates.

Characteristics of the substrate also affect its conversion into biogas. A high surface area improves the contact between substrate and microorganism, essential for enhancing degradation. In microalgae, the presence of strong cell walls hampers this contact and reduces the attack of microorganisms. In highly particulate or structural substrates such as microalgae, conversions between 30% and 60% can be expected [33]. Microalgae drying can be also detrimental to AD [34 - 37]; drying can leads to aggregates formation that hamper the interaction between microorganisms and substrate, or to loss of volatile organic compounds.

The concentration of solids or organic fraction in substrates affects their conversion into biogas too. It has been reported that 60-80% of water in substrates, *i.e.*, 20-40% solids, favours methane production [2]. From an engineering point of view, the optimum solid content ranges from 12% to 30% [14]. A high solid content makes pumping and mixing operations more difficult, retards any cell growth and limits the mass transfer in the digester. On the other hand, a low solid content is not economically feasible due to the excessive amount of water. In WWTPs, a typical example of facilities operating anaerobic digesters, the solid content in the sewage sludge fed into the digesters ranges between 3-6% [30]. As discussed in the following section, the solid content in microalgal cultures is considerable lower than in other substrates commonly digested like sewage sludge, being one of the major limitations for microalgae digestion.

Mixing

A smooth agitation ensures that each microorganism is supplied with nutrients and that the products generated are removed. On the contrary, a strong agitation can alter the symbiosis between acetogens and methanogens, cause foam formation and increase the energy consumption [30].

Retention Time

There are two types of retention time: (i) the solids retention time (SRT, in days); and (ii) the hydraulic retention time (HRT, in days). The first one gives the time that solids remain in the digester, whereas the second one gives the time that the liquid influent (including solids) remains in the digester; both give an idea about the length of time that bacteria have to break down the substrate. In digesters without recycle, both parameters have the same value. Another parameter related to the STR and the HRT is the solids loading rate or organic loading rate (OLR, in $kg_{VS}/m^3/d$ or $kg_{COD}/m^3/d$). This parameter depends on the concentration of solid in the influent.

Because of the long generation time for methanogens [16], the retention time must be high enough to ensure cells reproduction and to prevent the washout of methanogens. Typical retention times for sewage sludge digestion are between 12-14 days [30]. Nevertheless, these values are unsuitable for microalgae digestion. The high resistance of the substrate requires a much longer time, usually between 10 and 30 days [38, 39]. When the optimum value is exceeded, the organic fraction destruction achieved in the additional time is insignificant, thus the production capacity of the plant is reduced.

As a consequence of the difficulties to produce high amounts of microalgae, the majority of the experiments reported in the literature have been performed in batch mode. The use of batch mode reactors is more operational in order to perform several experiments simultaneously, *i.e.*, different microalgae species, pre-treatments or amount of substrate in the reactor amongst others. However, large-scale production of biogas from microalgae requires continuous or semi-continuous digesters.

Table 2. Comparison of different results obtained during the AD of some microalgae species reported in the last years.

Reactor Operating Mode	Substrate	T (°C)	OLR (kg_{VS}/m^3/d)	HRT (d)	CH_4 yield (L_{CH4}/kg_{VS})	Ref.
Semi-continuous	*Chlorella vulgaris*[a]	35	0.75	16 28	147 240	[40]
Batch	*Scenedesmus sp.*[b]	35	0.6	23	50	[41]
Semi-continuous	*Scenedesmus sp.*[b] Thermal pre-treated *Scenedesmus sp.*[b]	35	n.a.	33 33	46 102	
Batch	*Scenedesmus obliquus* *Phaeodactylum tricornutum*	33	n.a.	30	210 350	[42]
Semi-continuous	*Scenedesmus obliquus* *Phaeodactylum tricornutum*	33	1.9 2.7	2.1 2.2	270 130	
	Scenedesmus obliquus *Phaeodactylum tricornutum*	54	2.0 2.8	2.3 2.2	290 170	
Semi-continuous	*Scenedesmus sp.*[c]	35	0.6	15	55	[43]
	Thermal pre-treated *Scenedesmus sp.*[c]	35	0.6 1.5	15	160 184	
Batch	*Nannochloropsis salina* Thermal pre-treated *N. salina*	40	n.a.	49	200 570	[44]
Semi-continuous	*Nannochloropsis salina* Thermal pre-treated *N. salina*	38	1.96	120	130 270	
Batch	*Chlorella sp.* Thermochemical pre-treated *Chlorella sp.*	35	n.a.	95	340 459	[45]
	Nannochloropsis sp. Thermochem. pre-treated *Nannochloropsis*	35	n.a.	95	360 511	
	Thalassiosira weissflogii Thermochem. pre-treated *T. weissflogii*	35	n.a.	95	380 440	
	Tetraselmis sp. Thermochem. pre-treated *Tetraselmis sp.*	35	n.a.	95	420 470	
	Pavlova_cf sp. Thermochem. pre-treated *Pavlova_cf sp.*	35	n.a.	95	510 546	

(Table 3) contd.....

Reactor Operating Mode	Substrate	T (°C)	OLR (kg_{VS}/m^3/d)	HRT (d)	CH_4 yield (L_{CH4}/kg_{VS})	Ref.
Batch	LE *Nannochloropsis sp.* Washed LE *Nannochloropsis sp.*	35	n.a.	39 77	194 171	[46]
Semi-continuous	Washed LEW *Nannochloropsis sp.*	35	0.5 1.5 2.0 3.0	183 61 46 30	174 149 147 128	
Semi-continuous		55	1.5 2.0	61 46	220 155	
Batch	LE *Chlorella vulgaris*	35	n.a.	15	195	[47]
Two-stage	Methanogenesis LE *Chlorella vulgaris*	35	2-12 [d]	24-4	336-319	
Batch	*Nannochloropsis oculata* Thermal pre-treated *N. oculata*	37	n.a.	12	237 329	[35]
Batch	Microalgae B (*Scenedesmus and C. vulgaris*)	37 55	n.a.	57	367 315	[36]
	Microalgae C (*Scenedesmus sp.*)	37 55	n.a.	57	178 140	
Semi-continuous	Microalgae grown in a pilot WWTP Thermal pre-treated microalgae	37	0.71 0.68	20	180 300	[48]
Semi-continuous	Microalgae grown in a pilot WWTP	35	0.75	15 20	130 170	[49]
	Microwave pre-treated microalgae	35	0.75	15 20	170 270	
Semi-continuous	PE *Scenedesmus*	37	3.85 1.85	20 40	292 291	[50]
Batch	Microalgae mixture (*Scenedesmus sp.*)	35	n.a.	90	220	[51]
Semi-continuous		35	2.0 2.75 3.5	12.4	139 133 130	
Semi-continuous	LE *Auxenochlorella protothecoides*	35	0.97 2.05	20	210 180	[52]
		35	0.95 1.54	40	250 170	
Batch	*Acutodesmus obliquus* Freeze *A. obliquus* Freeze-dried *A. obliquus*	36	n.a.	30-38	229 350 291	[53]
	Acutodesmus obliquus Enzymatic pre-treated *A. obliquus*	36	n.a.	30-38	191 300	

(Table 3) contd.....

Reactor Operating Mode	Substrate	T (ºC)	OLR (kg_{VS}/m^3/d)	HRT (d)	CH_4 yield (L_{CH4}/kg_{VS})	Ref.
Batch	*Spirulina sp.* Mechanical pre-treated *Spirulina sp.*	37	n.a.	64	113 166	[54]
	Scenedesmus sp. Mechanical pre-treated *Scenedesmus sp.*	37	n.a.	48	54 97	
Semi-continuous	*Chlorella vulgaris*[e] Enzymatic pre-treated *C. vulgaris*[e]	35	0.98	15 20	77 196	[55]
Semi-continuous	*Chlorella vulgaris*[f] Thermally pre-treated *C. vulgaris*[f]	35	0.8 0.9	15	160 207	[56]
Semi-continuous	Microalgae grown in a pilot WWTP Thermal pre-treated microalgae	37	0.70 0.71	20	120 170	[57]
Batch	*Spirulina platensis*	35	n.a.	16	370	[58]
Semi-continuous		35	2.0 2.7 4.0	25 15 10	313 271 163	
Batch	*Chlorella vulgaris* Chemically pre-treated *C. vulgaris*	33	n.a.	27	251 231	[37]
	Nannochloropsis oculata Chemically pre-treated *N. oculata*	33	n.a.	27	238 339	
Batch	*Chlorella vulgaris* LE *C. vulgaris*	33	n.a.	27	251 231	[59]
	Nannochloropsis oculata LE *N. oculata*	33	n.a.	27	238 339	
Batch	*Phaeodactylum tricornutum* LE *P. tricornutum*	33	n.a.	28	258 257	[60]
Batch	*Chlorella sp.* Anaerobic pre-treated *Chlorella sp.*	37	n.a.	20	339 416	[61]
Batch	*Spirulina platensis*	36	n.a.		290	[62]
Two-stage	Acidogenesis *S. platensis* Methanogenesis *S. platensis*	36	3.75 [g]	4 10	26 308	

LE: Lipid-extracted
PE: Protein-extracted
[a] COD/VS = 1.33 according to the authors
[b] COD/VS = 0.6 according to the authors
[c] COD/VS = 1.65 according to the authors
[d] in kg_{COD}/m^3/d.
[e] COD/VS = 1.53 according to the authors
[f] COD/VS = 1.87 and 1.64 for raw and pre-treated biomass according to the authors
[g] kg_{VS}/d fed to the system.

Temperature

The presence of different bacterial groups, with different optimal temperature ranges, make it difficult for the operation of digesters [6, 28]. Acidogens have an optimum temperature between 32 ºC and 42 ºC (mesophilic range), and between 48 ºC and 55 °C (thermophilic range). On the other hand, most of the methanogens are mesophilic; only a minor fraction is psychrophilic or thermophilic.

The higher the temperature in the reactor, the higher the energy needs. Regardless of the biogas yield, it is usually suggested that the energy balance is favoured under mesophilic conditions [30]. However, other aspects may influence the decision of the operating temperature, for example, the stability of the process. The thermophilic conditions increase the inhibition risk caused by an ammonia concentration, especially during the AD of protein-rich substrates like microalgae. Zamalloa *et al.* studied the AD of two different microalgae species under mesophilic and thermophilic conditions, using a hybrid flow-through reactor [42]. Although the methane production and consequently the biodegradability of *Scenedesmus obliquus* was increased under thermophilic conditions, *Phaeodactylum tricornutum* digestion was almost unaffected by the increase in temperature (Table **2**). Kinnunen *et al.* evaluated the effects of the temperature increase when digesting *Nannochloropsis sp.* waste after lipid-extraction [46]. As seen in Table **2**, the highest methane production increase was obtained at low OLR (1.5 $kg_{VS}/m^3/d$); thermophilic digestion resulted in 48% more methane than mesophilic digestion. On the contrary, the methane production was similar in both reactors after increasing the OLR up to 2 $kg_{VS}/m^3/d$. At these OLR, the thermophilic reactor started showing inhibition signs, *i.e.*, VFA accumulation, attributed to ammonia. Mesophilic digestion did not show signs of inhibition when the OLR was increased up to 3 $kg_{VS}/m^3/d$. Opposed effects of the temperature increase on the methane production were reported during *Isochrysis galbana* and *Selenastrum capricornutum* digestion [63]. Although the inhibition signs were not evidenced probably due to the batch mode operation, the highest biogas production were observed under mesophilic conditions for both species. The biodegradability decreased from 55% to 35% and from 30% to 22% for *I. galbana* and *S. capricornutum* after increasing the digestion temperature. However, thermophilic conditions increased the methane content in biogas, as observed by Zamalloa *et al.* [42]. Olsson *et al.* digested some microalgal biomass under mesophilic and thermophilic conditions in batch mode [36]; as reported in Caporgno *et al.*, the temperature increase did not improve the AD [63].

It seems that the effect of temperature is highly dependent on the characteristics of the digested microalgae. As matter of fact, the increase of toxicity of ammonia when the temperature increases is problematic for microalgae digestion. Furthermore, high temperature during acetogenesis can affect the concentration of some intermediate products generated during microalgae decomposition and cause inhibition [64]. The sensitivity and the higher energy consumption under thermophilic conditions seem to be the main reasons why the majority of the experiments using microalgae are performed under mesophilic conditions. On the other hand, it is worth mentioning that thermophilic conditions may be beneficial for other substrates; for example, satisfactory pathogen elimination sludge can be achieved with thermophilic digestion in WWTPs [28, 30].

An alternative to the one-stage digestion is the two-stage process; the hydrolysis/acids production is driven under mesophilic conditions and methanogenesis under thermophilic conditions. Although the two-stage AD of microalgae has been reported some years ago [65], it was not widely investigated. As seen in Table **2**, some researchers have recently reported high efficiency and the absence of inhibition by ammonia in two-stage digesters fed with microalgae [62] and microalgae waste [47].

Alkalinity and pH

The alkalinity prevents fluctuations of pH in the digester. The fluctuations in pH affects the activity of the microorganisms and can decrease degradation. Similarly to temperature, there are optimum pH ranges for the different groups of microorganisms. Acidogens can perform reactions when the pH is above 5.0, whereas methanogens require a more limiting pH range, from 6.8-7.2; a decrease below 6.5 can reduce the activity of methanogens and can lead to accumulation of VFA and digestion failure. This acidification is prevented by carbon dioxide/hydrogen carbonate/carbonate ($CO_2/HCO_3^-/CO_3^{2-}$) buffer system:

$$CO_2 + H_2O \leftrightarrow H_2CO_3 \leftrightarrow H^+ + HCO_3^- \leftrightarrow 2\,H^+ + CO_3^{2-} \quad \textbf{(12)}$$

When the pH is low, CO_2 is present as free molecules. With rising pH values, the dissolved CO_2 forms ionised carbonic acid and release hydrogen ions. An alkalinity of 1000-5000 mg $CaCO_3$/L ensures optimum conditions for the methanogens in the digesters [28].

During the anaerobic digestion of substrates with high protein content such as microalgae, the ammonia/ammonium (NH_3/NH_4^+) buffer system significantly contributes to the alkalinity. Ammonia released during proteins degradation reacts with carbon dioxide and water to produce ammonium carbonate:

$$NH_3 + CO_2 + H_2O \leftrightarrow NH_4^+ + HCO_3^- \quad \textbf{(13)}$$

$$NH_4HCO_3 + R–COOH \leftrightarrow R–COONH_4 + H^+ + HCO_3^- \quad \textbf{(14)}$$

Where R represents the non-carboxyl (–COOH) part of the volatile acid.

This system is a result of the release of amino groups (Eq. 5-7) and production of ammonia when amino acids and proteins are degraded. With falling pH value, ammonium ions are formed and hydroxyl ions released. On the contrary, rising pH values promote the formations of free ammonia.

The ratio VFA:alkalinity can be used as a control parameter in digesters; values between 0.07-0.08 indicate stable operation, 0.1-0.2 are acceptable, but a ratio >0.5 predicts a possible failure. This parameter is particularly important under mesophilic conditions. When the temperature decreases below 32°C, the activity of methanogens decrease but the production of VFA continues at a rapid rate and can cause VFA accumulation [28].

Nutrients

As mentioned before, the formation of new cells during digestion is low. For this reason, the requirement of nutrients is also low. However, these nutrients must be balanced. A nutrient ratio C:N:P:S of 500-1000:15-20:5:3 is considered sufficient for microorganisms [2]. The C:N ratio is probably the most important one, and it may be at least 25:1 for an optimal gas production from the substrates. When the C:N ratio is low, for example in protein-rich substrates, it leads to high concentration of ammonia and inhibition of methanogens. On the contrary, an overly high C:N ratio causes lack of nitrogen. The C:N ratio in microalgae is, most of the time, close to 10 [31]. The low C:N ratio is one of the main drawbacks for microalgae digestion.

Several micronutrients are required due to enzyme systems present in microorganisms. Methanogens have unique enzyme systems that require cobalt, iron, nickel, and sulphide amongst others.

Other Causes of Inhibition

Some alkaline earth-metal salts containing sodium, potassium, calcium or magnesium can cause inhibition in AD [1]. These salts can be commonly found in high concentration in industrial waste, but in low concentration in urban wastewaters. In microalgae, mainly in marine microalgae, the presence of sodium can affect AD. The sodium concentration in cultures during microalgae cultivation considerably exceeds the threshold reported as inhibitory for AD, 3500 mg/L [1]. Although high concentration of sodium interferes with the metabolism of methanogens, the acclimation of methanogens to high concentrations of sodium can in some cases reduce the detrimental effects of this ion [66 - 68]. Schwede *et al.* observed inhibition signs during the AD of *Nannochloropsis salina* in continuous mode and attributed the inhibition to the concentration of sodium; the concentration of sodium in digester was in the range indicated as inhibitory [44]. However, the authors indicated that microorganisms adapted to the adverse conditions and the reactor maintained stable performance. Kinnunen *et al.* suggested that washing could reduce the concentration of salt in the lipid-extracted waste of the marine microalgae *Nannochloropsis sp* [46]. No benefits of washing were observed, and in fact, the loss of organic matter during washing considerably decreased the methane production (Table **2**).

Methane-forming bacteria contain approximately 2.5% w/w (dry weight) of sulphur [28]. Sulphur is required in digesters, but can also lead to inhibition. Sulphides cause inhibition in AD [1]; they can be introduced in the digesters with some waste, but they are mainly generated from the degradation of proteins in case of microalgae. The concentration of sulphured amino acids is low in microalgae [69], thus the amount of hydrogen sulphide released is low. However, sulphur compounds can be present in saline waters and saline substrates [25, 70], and they are converted into hydrogen sulphide by sulphate-reducing bacteria [1]. In addition to the potential inhibition, the high concentrations of sulphide is problematic for the further use of biogas due to corrosion. The concentration of hydrogen sulphide is affected by changes in the operation of the reactors. Ramos-Suárez *et al.* observed that under unstable operation of the digester, caused by ammonia and VFA accumulation, the development of sulphate-reducing bacteria was favoured and consequently, the concentration of hydrogen sulphide in biogas increased considerably [50].

It has been reported that inhibition can derive from the lipid content in biomass. Zhao *et al.* attributed the inhibition during *Nannochloropsis salina* digestion to the high lipid content in microalgae [71]. In spite of the high theoretical methane production of lipids, the authors suggested that the intermediate products generated during the degradation of long chain fatty acid were responsible for the low methane production.

LIMITATIONS OF MICROALGAE AS SUBSTRATES FOR AD

As mentioned above, researchers had already identified the major limitations of microalgae for digestion in experiments performed almost half a century ago [17, 21, 22, 72]. These limitations are listed and described below.

Resistance of the Cell Walls

The presence of cell walls is one of the major reasons behind microalgae degradation in AD. The cell walls hamper the attack of the microorganism, thus reducing the possibility of degradation [73]. The first experiment on microalgae digestion revealed that cells, apparently undamaged, resisted degradation after several weeks in the digester [17]. Even under thermophilic condition and exceeding the thermal death point of microalgae, intact cells were found in the digesters [72].

The composition of the cell walls can significantly affect the suitability of the biomass for digestion. According to some results reported by Mussgnug *et al.*, the composition of the cell wall plays a key role in degradation. The authors observed that species without cell walls or with protein-based cell walls were more easily degraded than those with carbohydrate-based cell walls [34]. Between 12% and 36% of the total biomass dry weight in microalgae correspond to the cell walls. Depending on the species, between 30% and 80% of the cell wall correspond to carbohydrates, and a minor amount, up to 40%, corresponds to proteins [31]. Cellulose, hemicelluloses and neutral sugars are the most common carbohydrates in cell walls, whereas peptides, proline and hydroxyproline are the main components identified in the protein fraction. The microalgae cell walls have been studied along the years due to their unusually high resistance. Additionally to carbohydrates and proteins, the cell walls contain resistant biopolymers; some of these biopolymers have chains of 30 carbon atoms. In some microalgae species, the structure of the cell walls showed a tri-laminar structure resistant to degradation [74]. In several marine microalgal species, the resistance of the cells was attributed to the presence of a non hydrolysable molecule called algaenans [75].

The implementation of pre-treatments to break the cell walls was first reported by Samson and Leduy in 1983; other authors had already evidenced the low degradability of the organic fraction in microalgae and the need of applying processes to increase their degradability [21]. These first results also suggested that the low degradability of microalgae was not only a matter of cell wall resistance, but also presence of recalcitrant biodegradable compounds. Mechanical pre-treatment application resulted in an effective solubilisation of the organic fraction but without improving the biodegradability of the biomass. The biodegradability of the of pre-treated biomass was similar to the obtained when frozen biomass was digested; in this case, the solubilisation was slight and caused by ice crystals formation during biomass freezing. The accessibility to the internal components of pre-treated microalgae contributes to the methane production; however, the content of recalcitrant organic matter, *i.e.*, indigestible components of the cell wall, plays also an important role in methane production. From the moment when researchers realised about the need of pre-treatments, an increasing number of technologies and conditions have been applied to microalgae biomass before AD. Pre-treatments applied to microalgae before digestion involve mechanical, thermal, chemical and biological pre-treatments. Some of the general aspects about these pre-treatments, which have been reported in the literature [39, 76 - 78], are mentioned below. Some results obtained during the AD of pre-treated microalgae to date can be seen in Table **2**.

- *Mechanical pre-treatments.* Generally, the authors conclude that mechanical pre-treatments are very effective for cell wall disruption and, in some cases, for biogas production enhancement. These type of pre-treatments alter the morphology of the cells by breaking down the structure of cellulose, solubilising hemicellulose or lignin; these changes can increase the specific surface area in the substrate and make hydrolysis easier [39]. On the other hand, these pre-treatments are considered inefficient from an energetic point of view [39, 76]. Ultrasound and microwave are typical examples of mechanical pre-treatments. In both methods, the cell wall disruption is caused by energy transfer through waves [77]. In addition to the investment required and the high operating costs identified as the main disadvantage for large-scale application, the possibilities of scaling-up make these techniques unattractive.
- *Thermal pre-treatments.* The thermal pre-treatments disrupt microalgae cells by the action of heat. Moreover, the heat can alter some compounds like cellulose, hemicellulose and lignin structures, and facilitate their hydrolysis [78]. The temperature and exposure time during the pre-treatment affects the efficiency of the pre-treatment, but the biogas production also depends on the characteristics of the microalgae species. □Several authors identify the temperature as the most influencing parameter. Low temperatures (<70 °C) allow the solubilisation of the exopolymers and extracellular compounds from microalgae, whereas cell disruption and the release of intracellular compounds are promoted at higher temperatures (>80 °C). The main advantage in this range of temperatures is the possibility of using the residual heat from biogas combustion in co-generation systems [76]. Temperatures higher than 100 °C can solubilise part of the organic compounds that constitute the cell walls. These high temperatures can be achieved, for example, by applying steam at high temperature (150-180 °C) and pressure, followed by a sudden pressure drop; this technology is called steam explosion and it is available at a large-scale.
- *Chemical pre-treatments.* The addition of chemicals, usually acids and alkalis, can promote the solubilisation of the hemicellulose and lignin that constitute the cell walls, making these compounds more readily available for microorganisms. One of the main disadvantages of chemical pre-treatments lies in the formation compounds that can be toxic for microorganisms [76]. Moreover, when concentrated acids are used, corrosion may also be taken into account due to the special materials required for constructing reactors [78]. Most of the studies

concerning chemical pre-treatments are usually combined with thermal pre-treatments, as their combination may promote biodegradability. Lastly, the utilisation of organic solvents can be also included in this category of pre-treatment. Organic solvents are used to extract valuable components from biomass. Several organic solvents have been used for lipid extraction in the biodiesel production process, thus AD has been studied to convert the lipid extracted waste into biogas. It has been reported that the action of solvents can affect some characteristics of the biomass. An improvement of the degradability of indigestible components and, in some cases, increase in the biogas production has been reported [79]. The extraction of lipids is further explained.

- *Enzymatic pre-treatments.* Finally, the action of enzymes can increase the degradability of some indigestible components present in the cell walls. Compared to the pre-treatments previously mentioned, enzymatic pre-treatments have been suggested as feasible alternatives with low energy consumption, the low risk of inhibition caused by the toxic compounds generated during the pre-treatment and low capital investment [77]. Nevertheless, one of the major drawbacks is the high costs of enzymes associated to their complex production [76]. Due to the selectivity of enzymes, considerable efforts are being made to find a way to apply enzymes capable to attack the cell wall in a broad range of microalgae, *i.e.*, with different composition and structure.

The major part of the information available in the literature correspond to experiments performed in laboratory-scale reactors, where the energetic balance only gives a general idea about the viability of the process at industrial scale. Passos *et al.* found the benefits of applying microwaves to cause microalgae disruption and methane production increase (Table **2**), but the energy balance revealed that the energy consumption was much higher than the additional energy generated [49]. In another work, Passos *et al.* observed a positive energy balance during the digestion of thermal pre-treated microalgae [48]. Since the energy balance depends on the methane production, the operating conditions of the reactors can directly determine whether the application of certain pre-treatment has positive energy balance or not. The analysis and optimisation of the energy balance is essential for any process in large scale, especially considering that some of the pre-treatments are characterised by high energy requirement inputs, *i.e.*, mechanical and thermal pre-treatment. The effectiveness of these pre-treatments also depends on the concentration of the biomass. Since the processes for biomass concentration are high-energy consuming, the energy balance of AD can be adversely affected when considering the impact of dewatering before the pre-treatment [49].

The disruption of the cell walls is crucial for microalgae processing; the cell disruption affects the efficiency and the energy consumption of the processes to recovery high-value compounds from microalgae [77]. Depending on the microalgae composition and the produced compound, the remaining microalgae fractions or waste after processing can represent around 65% of the organic fraction. The compounds in this remaining fraction are fragmented into smaller pieces and/or solubilised, thus their further conversion into biogas may not require additional pre-treatments and results in high biogas production [79]. Some of the results reported during the AD of microalgae before and after lipid extraction can be found in Table **2**. Although some authors suggested that the combination of biodiesel and biogas production increase the energy production [52], additional aspects such as the lipid-extraction process or the characteristics of the biomass, must be taken into account for the energetic balance since they affect the methane production from the microalgae waste [37, 46, 59].

Microalgae Concentration

Whatever the final product is, the excessive content of water in microalgal cultures is a common and serious problem for its further processing due to the high costs of dewatering [13, 80, 81]. The low concentration of the organic fraction in microalgae cultures because of the high water content, adversely affects the operation of the digester. The concentration of the organic fraction determines the volume of the digester: the higher the concentration, the lower the volume necessary for maintaining constant HRT and consequently, the biogas production. Due to the long retention time required for microalgae digestion, the low concentration of the organic fraction may lead to an excessive size of the digester. Shortening the retention time to decrease the size can decrease the productivity of biogas or wash the methanogens out. The implementation of a settling tank after the digester to recycle part of the microorganisms and solids can prevent microorganism washout and increase the SRT. Furthermore, there are some type of reactors that allow to increase the SRT, while keeping the HRT low [29, 42, 82, 83].

For the treatment of slurries with low solid content, usually between 2 and 10%, Continuously Stirred Tank Reactor (CSTR) are the most common reactors. They are generally used with sewage sludge, manures, agricultural wastes, etc. Digesters in WWTPs processing sewage sludge are designed and recommended for loading rates between 3.2-7.2 $kg_{VS}/m^3/day$ but typical values are 0.5-0.6 $kg_{VS}/m^3/day$ [28, 29]. These loading rates are feasible due to the concentration of solids in sewage sludge. Sewage sludge contains between 1-2% solids when generated in WWTPs, but it can be concentrated up to 5% by simple steps before digestion, *i.e.*, thickening [30]. Unfortunately, the technologies currently use for microalgae cultivation in large-scale cannot reach such a high concentration of solids, being commonly less than 2% [13]. Since the microalgae concentration is problematic and costly, it is still limited to microalgae production for high-value products like omega-3 or astaxanthin [84]. Microalgae cultivation and the technologies available for water elimination are addressed in other chapters of this book.

Table **2** summarises some of the last experiments performed to study microalgae digestion under semi-continuous mode. When observing the results reported by authors who digested microalgae in batch and semi-continuous mode, it is possible to notice that the methane production in batch exceeded the production from semi-continuous digesters [41, 42, 44, 51, 58]. One of the reasons is that the retention time in batch digesters is usually higher than in semi-continuous digesters, thus the microorganisms have more time to degrade the microalgae. Another reason is that the concentration of compounds which can cause inhibition is higher in semi-continuous than in batch digesters due the mode of feeding. Usually, decreasing the OLR and increasing the HRT leads to high methane production [42, 49, 55, 58]; however, the results also depend on the characteristics of the biomass amongst other factors. Some authors reported inhibition at OLR around 3 $kg_{VS}/m^3/d$, whereas other authors did not observe inhibition signs under similar operating conditions but digesting different biomass. Ramos-Suárez *et al.* demonstrated that acclimation of microorganisms can be the key to increase the ORL without affecting the stability of the reactor [50].

Carbon:Nitrogen Ratio

The high content of protein in microalgae creates a problem in digesters due to an imbalanced carbon:nitrogen ratio. As mentioned before, the microorganisms require substrates with a ratio close to 25:1 [2], whereas this ratio is most of the time below 10 in microalgae because of the high protein content [31]. The degradation of protein produces ammonia, which is in equilibrium with ammonium ions. Ammonia is much more toxic than ammonium; it can pass through cell membranes, being responsible for inhibition in microorganisms like methanogens [69]. The equilibrium established between ammonia and ammonium is significantly affected by pH, *i.e.*, a change in the pH from 7 to 9 increases the ratio ammonia to ammonium from 1:99 to 30:70 [14]. The temperature increase also increases the ratio ammonia to ammonium, increasing the risk of inhibition under thermophilic than under mesophilic conditions. Several authors observed inhibition cause by ammonia when digesting microalgae [46, 55, 85]. However, even after the extraction of aminoacids, the digestion of the microalgae waste showed ammonia accumulation [50].

The co-digestion of microalgae with other substrates has been suggested by several researchers along these years as a valuable option to balance the carbon:nitrogen ratio [85 - 88]. Some usual co-substrates are waste, *e.g.*, manures, agricultural waste and sewage sludge amongst others [36, 37, 60, 63, 89]. Recently, some by-products obtained from microalgae processing, as the lipid-extracted biomass or the glycerol generated during biodiesel production, have been suggested as potential co-substrates for digestion [90 - 93]. However, results indicate that the carbon:nitrogen ratio is not the key factor for successful digestion; in some cases, the biogas production in co-digestion did not increase in spite of the higher carbon:nitrogen ratio in the mixtures than in microalgae. Furthermore, significant increase in biogas was reported during co-digestion with carbon:nitrogen ratios lower than the values considered as optimal. In addition to the balanced carbon:nitrogen ratio, the addition of a second substrate can complement the lack of nutrients in other substrate or provide essential nutrients for microorganisms [85]. Due to the low concentration of microalgae after cultivation, the co-substrate can improve the operation of the reactor by increasing the load rate and preventing washout.

Another possibility to increase the carbon:nitrogen ratio of microalgae for AD by the manipulation of the cultivation conditions, as suggested by Markou *et al.* [94]. *Arthrospira platensis* was cultivated under phosphorus limitation conditions to trigger carbohydrate accumulation, from 20% to 65% on a dry biomass basis. This accumulation of carbohydrates increase the carbon:nitrogen ratio and the methane production when digested.

CONFLICT OF INTEREST

The authors declare no conflict of interest, financial or otherwise.

ACKNOWLEDGEMENTS

Martín Pablo Caporgno thanks the Spanish Ministerio de Educación, Cultura y Deporte for his pre-doctoral scholarship (Ref. AP2012-3726).

REFERENCES

[1] P.L. McCarty, "Anaerobic waste treatment fundamentals: Part I Chemistry and microbiology", *Public Works,* vol. 95, pp. 107-112, 1964.

[2] A. Khalid, M. Arshad, M. Anjum, T. Mahmood, and L. Dawson, "The anaerobic digestion of solid organic waste", *Waste Manag.,* vol. 31, no. 8, pp. 1737-1744, 2011. [http://dx.doi.org/10.1016/j.wasman.2011.03.021] [PMID: 21530224]

[3] C. Mao, Y. Feng, X. Wang, and G. Ren, "Review on research achievements of biogas from anaerobic digestion", *Renew. Sustain. Energy Rev.,* vol. 45, pp. 540-555, 2015. [http://dx.doi.org/10.1016/j.rser.2015.02.032]

[4] J. Edwards, M. Othman, and S. Burn, "A review of policy drivers and barriers for the use of anaerobic digestion in Europe, the United States and Australia", *Renew. Sustain. Energy Rev.,* vol. 52, pp. 815-828, 2015.

[http://dx.doi.org/10.1016/j.rser.2015.07.112]

[5] A. Tiwary, I.D. Williams, D.C. Pant, and V.V. Kishore, "Emerging perspectives on environmental burden minimisation initiatives from anaerobic digestion technologies for community scale biomass valorisation", *Renew. Sustain. Energy Rev.,* vol. 42, pp. 883-901, 2015. [http://dx.doi.org/10.1016/j.rser.2014.10.052]

[6] F. Ali Shah, Q. Mahmood, M. Maroof Shah, A. Pervez, and S. Ahmad Asad, "Microbial ecology of anaerobic digesters: the key players of anaerobiosis", *Sci. World J.,* vol. 2014, p. 183752, 2014. [http://dx.doi.org/10.1155/2014/183752] [PMID: 24701142]

[7] A.J. Ward, D.M. Lewis, and F.B. Green, "Anaerobic digestion of algae biomass: A review", *Algal Res.,* vol. 5, pp. 204-214, 2014. [http://dx.doi.org/10.1016/j.algal.2014.02.001]

[8] H. Haberl, "Competition for land: A sociometabolic perspective", *Ecol. Econ.,* vol. 119, pp. 424-431, 2014. [http://dx.doi.org/10.1016/j.ecolecon.2014.10.002]

[9] S. Rasi, A. Veijanen, and J. Rintala, "Trace compounds of biogas from different biogas production plants", *Energy,* vol. 32, pp. 1375-1380, 2007. [http://dx.doi.org/10.1016/j.energy.2006.10.018]

[10] Overview of greenhouse gases. Methane Emissions. U.S. Environmental Protection Agency, Washington, DC, USA. [Online] Available at: www3.epa.gov/climatechange/ghgemissions/gases/ch4.html, [Accessed July. 26, 2016].

[11] E. Uggetti, B. Sialve, E. Trably, and J.P. Steyer, "Integrating microalgae production with anaerobic digestion: a biorefinery approach", *Biofuels Bioprod. Biorefin.,* vol. 8, pp. 516-529, 2014. [http://dx.doi.org/10.1002/bbb.1469]

[12] G. Chen, L. Zhao, and Y. Qi, "Enhancing the productivity of microalgae cultivated in wastewater toward biofuel production: A critical review", *Appl. Energy,* vol. 137, pp. 282-291, 2015. [http://dx.doi.org/10.1016/j.apenergy.2014.10.032]

[13] T.Y. Zhang, H.Y. Hu, Y.H. Wu, L.L. Zhuang, X.Q. Xu, X.X. Wang, and G.H. Dao, "Promising solutions to solve the bottle necks in the large-scale cultivation of microalgae for biomass/bioenergy production", *Renew. Sustain. Energy Rev.,* vol. 60, pp. 1602-1614, 2016. [http://dx.doi.org/10.1016/j.rser.2016.02.008]

[14] D. Deublein, and A. Steinhauser, "Biogas from waste and renewable resources: An introduction. October 2010, WILEY-VCH",

[15] N. Abatzoglou, and S. Boivin, "A review of biogas purification processes", *Biofuels Bioprod. Biorefin.,* vol. 3, pp. 42-71, 2009. [http://dx.doi.org/10.1002/bbb.117]

[16] M. Guo, W. Song, and J. Buhain, "Bioenergy and biofuels: History, status, and perspective", *Renew. Sustain. Energy Rev.,* vol. 42, pp. 712-725, 2015. [http://dx.doi.org/10.1016/j.rser.2014.10.013]

[17] C.G. Golueke, W.J. Oswald, and H.B. Gotaas, "Anaerobic digestion of Algae", *Appl. Microbiol.,* vol. 5, no. 1, pp. 47-55, 1957. [PMID: 13403639]

[18] K. Ullah, M. Ahmad, V.K. Sharma, P. Lu, A. Harvey, M. Zafar, S. Sultana, and C.N. Anyanwu, "Algal biomass as a global source of transport fuels: Overview and development perspectives", *Prog. Nat. Sci.: Materials International,* vol. 24, pp. 329-339, 2014. [http://dx.doi.org/10.1016/j.pnsc.2014.06.008]

[19] T.M. Mata, A.A. Martins, and N.S. Caetano, "Microalgae for biodiesel production and other applications: A review", *Renew. Sustain. Energy Rev.,* vol. 14, pp. 217-232, 2010. [http://dx.doi.org/10.1016/j.rser.2009.07.020]

[20] E.G. Foree, and P.L. McCarty, "Anaerobic decomposition of algae", *J. Environ. Sci. Technol.,* vol. 4, pp. 842-849, 1970. [http://dx.doi.org/10.1021/es60045a005]

[21] R. Samson, and A. Leduy, "Influence of mechanical and thermochemical pretreatments on anaerobic digestion of Spirulina maxima algal biomass", *Biotechnol. Lett.,* vol. 5, pp. 671-676, 1983. [http://dx.doi.org/10.1007/BF01386360]

[22] R. Samson, and A. Leduy, "Improved performance of anaerobic digestion of Spirulina maxima algal biomass by addition of carbon-rich wastes", *Biotechnol. Lett.,* vol. 5, pp. 677-682, 1983. [http://dx.doi.org/10.1007/BF01386361]

[23] S.J. Tarwadi, and V.D. Chauhan, "Seaweed biomass as a source of energy", *Energy,* vol. 12, pp. 375-378, 1987. [http://dx.doi.org/10.1016/0360-5442(87)90107-1]

[24] S. Ghosh, D.L. Klass, and D.P. Chynoweth, "Bioconversion of Macrocystis Pyrifera to Methane", *J. Chem. Technol. Biotechnol.,* vol. 31, pp. 791-807, 1981. [http://dx.doi.org/10.1002/jctb.2803101105]

[25] J. McKennedy, and O. Sherlock, "Anaerobic digestion of marine macroalgae: A review", *Renew. Sustain. Energy Rev.,* vol. 52, pp. 1781-1790, 2015. [http://dx.doi.org/10.1016/j.rser.2015.07.101]

[26] C. González-Fernández, B. Sialve, N. Bernet, and J.P. Steyer, "Impact of microalgae characteristics on their conversion to biofuel. Part II: focus on biomethane production", *Biofuels Bioprod. Biorefin.,* vol. 6, pp. 205-218, 2012. [http://dx.doi.org/10.1002/bbb.337]

[27] J. Cheng, *Biomass to renewable energy processes.*. December 2009, CRC Press.

[28] M.H. Gerardi, *The microbiology of anaerobic digesters.*. August 2003, John Wiley & Sons. [http://dx.doi.org/10.1002/0471468967]

[29] T.Z. de Mes, A.J. Stams, J.H. Reith, and G. Zeeman, *"Methane production by anaerobic digestion of wastewater and solid wastes" in Bio-methane and Bio-hydrogen: status and perspectives of biological methane and hydrogen production.* Dutch Biological Hydrogen Foundation, 2003.

[30] L. Appels, J. Baeyens, J. Degrève, and R. Dewil, "Principles and potential of the anaerobic digestion of waste-activated sludge", *Prog. Energ. Combust.,* vol. 34, pp. 755-781, 2008. [http://dx.doi.org/10.1016/j.pecs.2008.06.002]

[31] H. Tijani, N. Abdullah, and A. Yuzir, "Integration of microalgae biomass in biomethanation systems", *Renew. Sustain. Energy Rev.,* vol. 52, pp. 1610-1622, 2015. [http://dx.doi.org/10.1016/j.rser.2015.07.179]

[32] E.W. Rice, R.B. Baird, A.D. Eaton, and L.S. Clesceri, *Standard Methods for the Examination of Water and Wastewater*, 2012.

[33] "Assessment of the anaerobic biodegradability of macropollutants", *Rev. Environ. Sci. Bio.,* vol. 3, pp. 117-129, 2004. [http://dx.doi.org/10.1007/s11157-004-2502-3]

[34] J.H. Mussgnug, V. Klassen, A. Schlüter, and O. Kruse, "Microalgae as substrates for fermentative biogas production in a combined biorefinery concept", *J. Biotechnol.,* vol. 150, no. 1, pp. 51-56, 2010. [http://dx.doi.org/10.1016/j.jbiotec.2010.07.030] [PMID: 20691224]

[35] M.D. Marsolek, E. Kendall, P.L. Thompson, and T.R. Shuman, "Thermal pretreatment of algae for anaerobic digestion", *Bioresour. Technol.,* vol. 151, pp. 373-377, 2014. [http://dx.doi.org/10.1016/j.biortech.2013.09.121] [PMID: 24189036]

[36] J. Olsson, X.M. Feng, J. Ascue, F.G. Gentili, M.A. Shabiimam, E. Nehrenheim, and E. Thorin, "Co-digestion of cultivated microalgae and sewage sludge from municipal waste water treatment",

Bioresour. Technol., vol. 171, pp. 203-210, 2014. [http://dx.doi.org/10.1016/j.biortech.2014.08.069] [PMID: 25203227]

[37] M.P. Caporgno, M. Olkiewicz, A. Fortuny, F. Stüber, A. Fabregat, J. Font, J. Pruvost, O. Lepine, J. Legrand, and C. Bengoa, "Evaluation of different strategies to produce biofuels from Nannochloropsis oculata and Chlorella vulgaris", *Fuel Process. Technol.,* vol. 144, pp. 132-138, 2016. [http://dx.doi.org/10.1016/j.fuproc.2015.12.026]

[38] E. Kwietniewska, and J. Tys, "Process characteristics, inhibition factors and methane yields of anaerobic digestion process, with particular focus on microalgal biomass fermentation", *Renew. Sustain. Energy Rev.,* vol. 34, pp. 491-500, 2014. [http://dx.doi.org/10.1016/j.rser.2014.03.041]

[39] M.E. Montingelli, S. Tedesco, and A.G. Olabi, "Biogas production from algal biomass: A review", *Renew. Sustain. Energy Rev.,* vol. 43, pp. 961-972, 2015. [http://dx.doi.org/10.1016/j.rser.2014.11.052]

[40] M. Ras, L. Lardon, S. Bruno, N. Bernet, and J.P. Steyer, "Experimental study on a coupled process of production and anaerobic digestion of Chlorella vulgaris", *Bioresour. Technol.,* vol. 102, no. 1, pp. 200-206, 2011. [http://dx.doi.org/10.1016/j.biortech.2010.06.146] [PMID: 20678925]

[41] C. González-Fernández, B. Sialve, N. Bernet, and J.P. Steyer, "Thermal pretreatment to improve methane production of Scenedesmus biomass", *Biomass Bioenergy,* vol. 40, pp. 105-111, 2012. [http://dx.doi.org/10.1016/j.biombioe.2012.02.008]

[42] C. Zamalloa, N. Boon, and W. Verstraete, "Anaerobic digestibility of Scenedesmus obliquus and Phaeodactylum tricornutum under mesophilic and thermophilic conditions", *Appl. Energy,* vol. 92, pp. 733-738, 2012. [http://dx.doi.org/10.1016/j.apenergy.2011.08.017]

[43] C. González-Fernández, B. Sialve, N. Bernet, and J.P. Steyer, "Effect of organic loading rate on anaerobic digestion of thermally pretreated Scenedesmus sp. biomass", *Bioresour. Technol.,* vol. 129, pp. 219-223, 2013. [http://dx.doi.org/10.1016/j.biortech.2012.10.123] [PMID: 23247149]

[44] S. Schwede, Z.U. Rehman, M. Gerber, C. Theiss, and R. Span, "Effects of thermal pretreatment on anaerobic digestion of Nannochloropsis salina biomass", *Bioresour. Technol.,* vol. 143, pp. 505-511, 2013. [http://dx.doi.org/10.1016/j.biortech.2013.06.043] [PMID: 23831893]

[45] P. Bohutskyi, M.J. Betenbaugh, and E.J. Bouwer, "The effects of alternative pretreatment strategies on anaerobic digestion and methane production from different algal strains", *Bioresour. Technol.,* vol. 155, pp. 366-372, 2014. [http://dx.doi.org/10.1016/j.biortech.2013.12.095] [PMID: 24468544]

[46] H.V. Kinnunen, P.E. Koskinen, and J. Rintala, "Mesophilic and thermophilic anaerobic laboratory-scale digestion of Nannochloropsis microalga residues", *Bioresour. Technol.,* vol. 155, pp. 314-322, 2014. [http://dx.doi.org/10.1016/j.biortech.2013.12.115] [PMID: 24462882]

[47] Y. Li, M. Gao, D. Hua, J. Zhang, Y. Zhao, H. Mu, H. Xu, X. Liang, F. Jin, and X. Zhang, "One-stage and two-stage anaerobic digestion of lipid-extracted algae", *Ann. Microbiol.,* vol. 65, pp. 1465-1471, 2015. [http://dx.doi.org/10.1007/s13213-014-0985-x]

[48] F. Passos, and I. Ferrer, "Microalgae conversion to biogas: thermal pretreatment contribution on net energy production", *Environ. Sci. Technol.,* vol. 48, no. 12, pp. 7171-7178, 2014. [http://dx.doi.org/10.1021/es500982v] [PMID: 24825469]

[49] F. Passos, M. Hernández-Mariné, J. García, and I. Ferrer, "Long-term anaerobic digestion of microalgae grown in HRAP for wastewater treatment. Effect of microwave pretreatment", *Water Res.,*

vol. 49, pp. 351-359, 2014.
[http://dx.doi.org/10.1016/j.watres.2013.10.013] [PMID: 24183557]

[50] J.L. Ramos-Suárez, F. García Cuadra, F.G. Acién, and N. Carreras, "Benefits of combining anaerobic digestion and amino acid extraction from microalgae", *Chem. Eng. J.,* vol. 258, pp. 1-9, 2014.
[http://dx.doi.org/10.1016/j.cej.2014.07.086]

[51] K.C. Tran, J.L. Mendoza Martin, S. Heavena, C.J. Banks, F.G. Acien Fernandez, and E. Molina Grima, "Cultivation and anaerobic digestion of Scenedesmus spp. grown in a pilot-scale open raceway", *Algal Res.,* vol. 5, pp. 95-102, 2014.
[http://dx.doi.org/10.1016/j.algal.2014.06.001]

[52] P. Bohutskyi, B. Ketter, S. Chow, K.J. Adams, M.J. Betenbaugh, F.C. Allnutt, and E.J. Bouwer, "Anaerobic digestion of lipid-extracted Auxenochlorella protothecoides biomass for methane generation and nutrient recovery", *Bioresour. Technol.,* vol. 183, pp. 229-239, 2015.
[http://dx.doi.org/10.1016/j.biortech.2015.02.012] [PMID: 25746299]

[53] M.R. Gruber-Brunhumer, J. Jerney, E. Zohar, M. Nussbaumer, C. Hieger, G. Bochmann, M. Schagerl, J.P. Obbard, W. Fuchs, and B. Drosg, "Acutodesmus obliquus as a benchmark strain for evaluating methaneproduction from microalgae: influence of different storage and pretreatment methods on biogas yield", *Algal Res.,* vol. 12, pp. 230-238, 2015.
[http://dx.doi.org/10.1016/j.algal.2015.08.022]

[54] A.E. Inglesby, M.J. Griffiths, S.T. Harrison, and R.P. van Hille, "Anaerobic digestion of Spirulina sp. and Scenedesmus sp.: a comparison and investigation of the impact of mechanical pre-treatment", *J. Appl. Phycol.,* vol. 27, pp. 1891-1900, 2015.
[http://dx.doi.org/10.1007/s10811-015-0669-3]

[55] A. Mahdy, L. Mendez, M. Ballesteros, and C. González-Fernández, "Protease pretreated Chlorella vulgaris biomass bioconversion to methane via semi-continuous anaerobic digestion", *Fuel,* vol. 158, pp. 35-41, 2015.
[http://dx.doi.org/10.1016/j.fuel.2015.04.052]

[56] L. Mendez, A. Mahdy, M. Ballesteros, and C. González-Fernández, "Biomethane production using fresh and thermally pretreated Chlorella vulgaris biomass: A comparison of batch and semi-continuous feeding mode", *Ecol. Eng.,* vol. 84, pp. 273-277, 2015.
[http://dx.doi.org/10.1016/j.ecoleng.2015.09.056]

[57] F. Passos, and I. Ferrer, "Influence of hydrothermal pretreatment on microalgal biomass anaerobic digestion and bioenergy production", *Water Res.,* vol. 68, pp. 364-373, 2015.
[http://dx.doi.org/10.1016/j.watres.2014.10.015] [PMID: 25462743]

[58] N. Aramrueang, J. Rapport, and R. Zhang, "Effects of hydraulic retention time and organic loading rate on performance and stability of anaerobic digestion of Spirulina platensis", *Biosyst. Eng.,* vol. 147, pp. 174-182, 2016.
[http://dx.doi.org/10.1016/j.biosystemseng.2016.04.006]

[59] M.P. Caporgno, M. Olkiewicz, J. Pruvost, O. Lepine, J. Legrand, J. Font, and C. Bengoa, "A novel pre-treatment for the methane production from microalgae by using N-methylmorpholine-N-oxide (NMMO)", *Bioresour. Technol.,* vol. 201, pp. 370-373, 2016.
[http://dx.doi.org/10.1016/j.biortech.2015.11.069] [PMID: 26684667]

[60] M.P. Caporgno, M. Olkiewicz, C. Torras, J. Salvadó, E. Clavero, and C. Bengoa, "Effect of pre-treatments on the production of biofuels from Phaeodactylum tricornutum", *J. Environ. Manage.,* vol. 177, pp. 240-246, 2016.
[http://dx.doi.org/10.1016/j.jenvman.2016.04.023] [PMID: 27107392]

[61] S. He, X. Fan, N.R. Katukuri, X. Yuan, F. Wang, and R.B. Guo, "Enhanced methane production from microalgal biomass by anaerobic bio-pretreatment", *Bioresour. Technol.,* vol. 204, pp. 145-151, 2016.
[http://dx.doi.org/10.1016/j.biortech.2015.12.073] [PMID: 26773949]

[62] A. Varol, and A. Ugurlu, "Biogas production from microalgae (Spirulina platensis) in a two-stage

anaerobic system", *Waste Biomass Valoriz.,* vol. 7, pp. 193-200, 2016.
[http://dx.doi.org/10.1007/s12649-015-9442-8]

[63] M.P. Caporgno, R. Trobajo, N. Caiola, C. Ibáñez, A. Fabregat, and C. Bengoa, "Biogas production from sewage sludge and microalgae co-digestion under mesophilic and thermophilic conditions", *Renew. Energy,* vol. 75, pp. 374-380, 2015.
[http://dx.doi.org/10.1016/j.renene.2014.10.019]

[64] M. Gruhn, J.C. Frigon, and S.R. Guiot, "Acidogenic fermentation of Scenedesmus sp.-AMDD: Comparison of volatile fatty acids yields between mesophilic and thermophilic conditions", *Bioresour. Technol.,* vol. 200, pp. 624-630, 2016.
[http://dx.doi.org/10.1016/j.biortech.2015.10.087] [PMID: 26551650]

[65] A. Vergara-Fernández, G. Vargas, N. Alarcón, and A. Velasco, "Evaluation of marine algae as a source of biogas in a two-stage anaerobic reactor system", *Biomass Bioenergy,* vol. 32, pp. 338-344, 2008.
[http://dx.doi.org/10.1016/j.biombioe.2007.10.005]

[66] Y. Chen, J.J. Cheng, and K.S. Creamer, "Inhibition of anaerobic digestion process: a review", *Bioresour. Technol.,* vol. 99, no. 10, pp. 4044-4064, 2008.
[http://dx.doi.org/10.1016/j.biortech.2007.01.057] [PMID: 17399981]

[67] A. Mottet, F. Habouzit, and J.P. Steyer, "Anaerobic digestion of marine microalgae in different salinity levels", *Bioresour. Technol.,* vol. 158, pp. 300-306, 2014.
[http://dx.doi.org/10.1016/j.biortech.2014.02.055] [PMID: 24632407]

[68] A. Ward, A. Ball, and D. Lewis, "Halophytic microalgae as a feedstock for anaerobic digestion", *Algal Res.,* vol. 7, pp. 16-23, 2015.
[http://dx.doi.org/10.1016/j.algal.2014.11.008]

[69] B. Sialve, N. Bernet, and O. Bernard, "Anaerobic digestion of microalgae as a necessary step to make microalgal biodiesel sustainable", *Biotechnol. Adv.,* vol. 27, no. 4, pp. 409-416, 2009.
[http://dx.doi.org/10.1016/j.biotechadv.2009.03.001] [PMID: 19289163]

[70] P. Peu, J.F. Sassi, R. Girault, S. Picard, P. Saint-Cast, F. Béline, and P. Dabert, "Sulphur fate and anaerobic biodegradation potential during co-digestion of seaweed biomass (Ulva sp.) with pig slurry", *Bioresour. Technol.,* vol. 102, no. 23, pp. 10794-10802, 2011.
[http://dx.doi.org/10.1016/j.biortech.2011.08.096] [PMID: 21982451]

[71] B. Zhao, J. Ma, Q. Zhao, L. Laurens, E. Jarvis, S. Chen, and C. Frear, "Efficient anaerobic digestion of whole microalgae and lipid-extracted microalgae residues for methane energy production", *Bioresour. Technol.,* vol. 161, pp. 423-430, 2014.
[http://dx.doi.org/10.1016/j.biortech.2014.03.079] [PMID: 24736123]

[72] C.G. Golueke, and W.J. Oswald, "Biological conversion of light energy to the chemical energy of methane", *Appl. Microbiol.,* vol. 7, no. 4, pp. 219-227, 1959.
[PMID: 13661866]

[73] M. Dębowski, M. Zieliński, A. Grala, and M. Dudek, "Algae biomass as an alternative substrate in biogas production technologies", *Renew. Sustain. Energy Rev.,* vol. 27, pp. 596-604, 2013.
[http://dx.doi.org/10.1016/j.rser.2013.07.029]

[74] S. Derenne, C. Largeau, C. Berkaloff, B. Rousseau, C. Wilhelm, and P.G. Hatcher, "Non-hydrolysable macromolecular constituents from outer walls of Chlorella fusca and Nanochlorum eucaryotum", *Phytochemistry,* vol. 31, pp. 1923-1929, 1992.
[http://dx.doi.org/10.1016/0031-9422(92)80335-C]

[75] F. Gelin, J.K. Volkman, C. Largeau, S. Derenne, J.S. Sinninghe Damsté, and J.W. De Leeuw, "Distribution of aliphatic, nonhydrolyzable biopolymers in marine microalgae", *Org. Geochem.,* vol. 30, pp. 147-159, 1999.
[http://dx.doi.org/10.1016/S0146-6380(98)00206-X]

[76] F. Passos, E. Uggetti, H. Carrère, and I. Ferrer, "Pretreatment of microalgae to improve biogas production: a review", *Bioresour. Technol.,* vol. 172, pp. 403-412, 2014. [http://dx.doi.org/10.1016/j.biortech.2014.08.114] [PMID: 25257071]

[77] E. Günerken, E. D'Hondt, M.H. Eppink, L. Garcia-Gonzalez, K. Elst, and R.H. Wijffels, "Cell disruption for microalgae biorefineries", *Biotechnol. Adv.,* vol. 33, no. 2, pp. 243-260, 2015. [http://dx.doi.org/10.1016/j.biotechadv.2015.01.008] [PMID: 25656098]

[78] C. Rodriguez, A. Alaswad, J. Mooney, T. Prescott, and A.G. Olabi, "Pre-treatment techniques used for anaerobic digestion of algae", *Fuel Process. Technol.,* vol. 138, pp. 765-779, 2015. [http://dx.doi.org/10.1016/j.fuproc.2015.06.027]

[79] M.P. Caporgno, and C. Bengoa, "Anaerobic digestion of microalgae: the benefits of digesting microalgae waste", *Current Biochemical Engineering, Special Issue Biomethane,* vol. 3, pp. 1-13, 2016.

[80] N. Pragya, K.K. Pandey, and P.K. Sahoo, "A review on harvesting, oil extraction and biofuels production technologies from microalgae", *Renew. Sustain. Energy Rev.,* vol. 24, pp. 159-171, 2013. [http://dx.doi.org/10.1016/j.rser.2013.03.034]

[81] S. Pawar, "Effectiveness mapping of open raceway pond and tubular photobioreactors for sustainable production of microalgae biofuel", *Renew. Sustain. Energy Rev.,* vol. 62, pp. 640-653, 2016. [http://dx.doi.org/10.1016/j.rser.2016.04.074]

[82] A.E. Inglesby, and A.C. Fisher, "Enhanced methane yields from anaerobic digestion of Arthrospira maxima biomass in an advanced flow-through reactor with an intergrated recirculation loop microbial fuel cell", *Energy Environ. Sci.,* vol. 5, pp. 7996-8006, 2012. [http://dx.doi.org/10.1039/c2ee21659k]

[83] C. Zamalloa, J. De Vrieze, N. Boon, and W. Verstraete, "Anaerobic digestibility of marine microalgae Phaeodactylum tricornutum in a lab-scale anaerobic membrane bioreactor", *Appl. Microbiol. Biotechnol.,* vol. 93, no. 2, pp. 859-869, 2012. [http://dx.doi.org/10.1007/s00253-011-3624-5] [PMID: 22005739]

[84] T. Suganya, M. Varman, H.H. Masjuki, and S. Renganathan, "Macroalgae and microalgae as a potential source for commercial applications along with biofuels production: A biorefinery approach", *Renew. Sustain. Energy Rev.,* vol. 55, pp. 909-941, 2016. [http://dx.doi.org/10.1016/j.rser.2015.11.026]

[85] S. Schwede, A. Kowalczyk, M. Gerber, and R. Span, "Anaerobic co-digestion of the marine microalga Nannochloropsis salina with energy crops", *Bioresour. Technol.,* vol. 148, pp. 428-435, 2013. [http://dx.doi.org/10.1016/j.biortech.2013.08.157] [PMID: 24071442]

[86] H.M. El-Mashad, "Kinetics of methane production from the codigestion of switchgrass and Spirulina platensis algae", *Bioresour. Technol.,* vol. 132, pp. 305-312, 2013. [http://dx.doi.org/10.1016/j.biortech.2012.12.183] [PMID: 23416617]

[87] S.K. Prajapati, A. Malik, V.K. Vijayc, and T.R. Sreekrishnand, "Enhanced methane production from algal biomass through short duration enzymatic pretreatment and codigestion with carbon rich waste", *RSC Advances,* vol. 5, pp. 67175-67183, 2015. [http://dx.doi.org/10.1039/C5RA12670C]

[88] C. Herrmann, N. Kalita, D. Wall, A. Xia, and J.D. Murphy, "Optimised biogas production from microalgae through co-digestion with carbon-rich co-substrates", *Bioresour. Technol.,* vol. 214, pp. 328-337, 2016. [http://dx.doi.org/10.1016/j.biortech.2016.04.119] [PMID: 27152773]

[89] M. Wang, A.K. Sahu, B. Rusten, and C. Park, "Anaerobic co-digestion of microalgae Chlorella sp. and waste activated sludge", *Bioresour. Technol.,* vol. 142, pp. 585-590, 2013. [http://dx.doi.org/10.1016/j.biortech.2013.05.096] [PMID: 23770998]

[90] E.A. Ehimen, S. Connaughtonw, Z. Sun, and G.C. Carrington, "Energy recovery from lipid extracted,

transesterified and glycerol codigested microalgae biomass", *Glob. Change Biol. Bioenergy,* vol. 1, pp. 371-381, 2009.
[http://dx.doi.org/10.1111/j.1757-1707.2009.01029.x]

[91] E.A. Ehimen, Z.F. Sun, C.G. Carrington, E.J. Birch, and J.J. Eaton-Rye, "Anaerobic digestion of microalgae residues resulting from the biodiesel production process", *Appl. Energy,* vol. 88, pp. 3454-3463, 2011.
[http://dx.doi.org/10.1016/j.apenergy.2010.10.020]

[92] S. Astals, R.S. Musenze, X. Bai, S. Tannock, S. Tait, S. Pratt, and P.D. Jensen, "Anaerobic co-digestion of pig manure and algae: impact of intracellular algal products recovery on co-digestion performance", *Bioresour. Technol.,* vol. 181, pp. 97-104, 2015.
[http://dx.doi.org/10.1016/j.biortech.2015.01.039] [PMID: 25643955]

[93] P. Neumann, A. Torres, F.G. Fermoso, R. Borja, and D. Jeison, "Anaerobic co-digestion of lipid-spent microalgae with waste activated sludge and glycerol in batch mode", *Int. Biodeter. Biodegr.,* vol. 100, pp. 85-88, 2015.
[http://dx.doi.org/10.1016/j.ibiod.2015.01.020]

[94] G. Markou, I. Angelidaki, and D. Georgakakis, "Carbohydrate-enriched cyanobacterial biomass as feedstock for bio-methane production through anaerobic digestion", *Fuel,* vol. 111, pp. 872-879, 2013.
[http://dx.doi.org/10.1016/j.fuel.2013.04.013]

CHAPTER 5

Microalgae, Taking Over the Role in the Hydrogen Future

Suphi S. Oncel*

Department of Bioengineering, Engineering Faculty, Ege University, Izmir, Turkey

Abstract: In the search of alternative routes for energy, microalgal hydrogen production becomes an important topic as a renewable and sustainable candidate. Following the progresses in the knowledge related with cellular mechanisms and production processes, biohydrogen can have the chance to take a strong role in the future even if the current capacities are not comparable to the conventional fossil fuels. This chapter focuses on the state of the art for biohydrogen production from microalgae.

Keywords: Biohydrogen, Microalgae, Photobioreactor, Economy.

INTRODUCTION

With the industrial progress that is energetically supplied by fossil fuels, humanity found itself in a hurdle to feed the immense need for consumption which results in a dramatic pressure on the environment. Fast production and immediate consumption frenzy started to show side effects as the contamination of water and air, forests destruction, decrease in food stocks, decrease in natural resources and the increase of greenhouse effect (associated to climate change). With the rising concern about the global warming, the need of prevention is becoming a priority for the future of humanity. According to the research studies, the temperature increase should be controlled to an average increase of 2 °C to prevent drastic changes in the global climate, which can be realized by an estimated 80% decrease in the carbon-dioxide (CO_2) levels till 2050 [1]. This is a real challenge in an era of population rising, while the economic aspects still pushing the societies to produce, to compensate the demand. The only way to pass this sharp corner seems to depend on the change in the understandings and habits of the societies towards energy.

* **Corresponding author Suphi S. Oncel:** Department of Bioengineering, Engineering Faculty, Ege University, Izmir, Turkey; Tel: +9 0232 311 58 12; Fax: 09 0232 311 58 80; E-mail: suphi.oncel@ege.edu.tr

José Carlos Magalhães Pires (Ed.)

After facing critical problems like oil crisis in 1970's, Chernobyl nuclear disaster in 1986, Deepwater Horizon oil spill in 2010, Fukushima Daini nuclear disaster in 2011, addition to the numerous wars and chaos, humanity admit the importance of the alternative ways for feeding energy demand. Within this regard one of the prosperous way is the utilization of renewable sources like solar, wind, water (waves, rivers), geothermal and bio-based to fulfill the needs for a sustainable and environmentally friendly future [2, 3].

Considering other alternatives of renewables, bio-based sources have important qualifications which make them good candidates for energy especially with a focus on fuels. Bio-based sources like plants or algae are multifarious and can be integrated with the other energy streams, both renewable or not; thus, sub route alternatives will serve an advantage for sustainability. These sources can be used to produce a variety of conventional fuels (Fig. **1**), like diesel, methane and hydrogen addition to some other basic carbon based fuels like ethanol, methanol or butanol [5]. Also these sources can be processed by using various technologies like fermentation, digestion, combustion or pyrolysis in order to reach the intended fuel product [8].

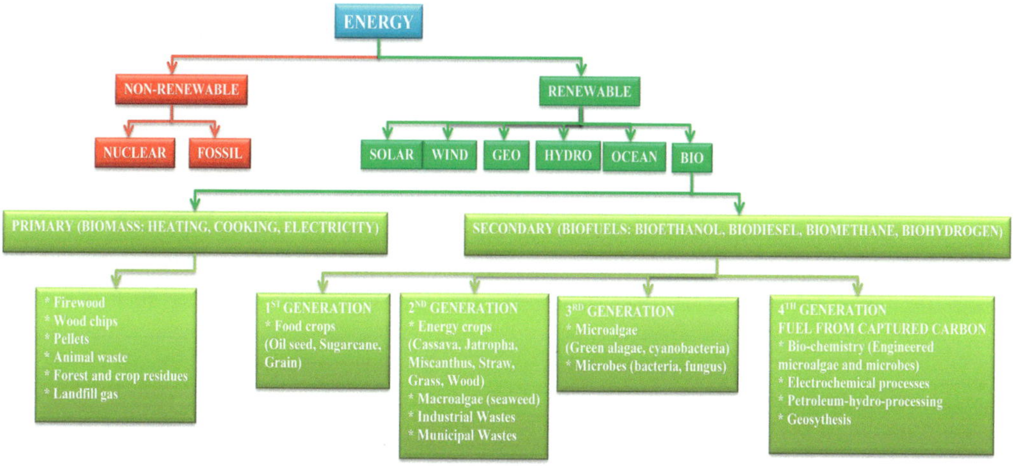

Fig. (1). Major energy production routes: Modified from [4 - 7].

Similar to the other energy technologies, the ultimate target of the bio-based energy approach is to reach a sustainable, environment friendly, efficient, economically feasible, safe, distributable, reliable, scalable and equitable energy future. In order to reach these targets, algae can be considered as a strong alternative among other bio-based sources. Algae are a diverse group of organisms that can be classified into two basic groups: one is the macroalgae and the other is the microalgae. They both are able to be used directly or indirectly in energy systems in order to produce various fuel alternatives [9 - 11].

Macroalgae can be defined simply as the water plants found in aqueous environments under three basic groups based on their pigment colors: red (*Rhodophyta*), brown (*Phaeophyta*) and green (*Chlorophyta*), which can convert solar energy for survival (photosynthesis) [12, 13]. They are conventionally cultured or harvested for years as a source of native food in Far East Asia countries like China, Korea, Philippines or Japan, so they have a quite prevalent usage and awareness [14]. They are also important sources for hydrocolloids like carrageenan, alginate or agar which makes them an important economic value [15].

Microalgae are a group of diversified microscopic organisms which can perform photosynthesis similar to the macroalgae. Taxonomically, microscopic algae can be classified into two major groups: (i) prokaryotic (cyanobacteria formerly termed as *Cyanophyceae*); and (ii) eukaryotic [16].

Cyanobacteria (also known as blue-green algae) are prokaryotic photosynthetic microscopic algae dominating the marine environments that have diversified members related to their morphology or physiology. They are one of the oldest residents of Earth (which can be traced back to 3.5 billion years) and played a major role in the change to the current composition of the atmosphere in terms of oxygen. According to the endosymbiosis theory, they are the predecessor for the chloroplasts of eukaryotic algae and green plants. The large phylum of the cyanobacteria has been divided into *Gloeobacterales*, *Synechococcales*, *Spirulinales*, *Chroococcales*, *Pleurocapsales*, *Oscillatoriales*, *Chroococcidiopsidales* and *Nostocales*, which is a good sign of their potential for various products considering the diversified living environments from moderate to harsh [17 - 19].

Eukaryotic algae (mostly termed as the microalgae) are classified into groups based on different characteristics, morphological variations or even physical specifications [20]. Regarding the basic taxonomic classification, there are nine divisions which are based on the pigments. The major groups are *Chlorophyceae* (green algae), *Phaeophyceae* (brown algae), *Pyrrophyceae* (dinoflagellates), *Chrysophyceae* (golden-brown algae), *Bacillariophyceae* (diatoms) and *Rhodophyceae* (red algae) [21, 22].

Even if the classification depends on the taxonomic specifications, in applied biotechnology the term "microalgae" usually covers all prokaryotic and eukaryotic microscopic algae [23]. This simplification is accepted by the researchers and serves a convenient ground especially in the area of bio-products and biofuels [3, 23 - 25].

From the biofuels point of view, algae can be used for numerous fuels, having microalgae higher production potential than macroalgae. The main reason is that macroalgae are used indirectly for fuel production. Macroalgae biomass is the source of fuel through various biological or chemical processes, while microalgae can be used both directly and indirectly. Microalgae can produce fuel alternatives like ethanol or hydrogen by its metabolic reactions in addition to its biomass-based fuels through various processes (Fig. **2**) similar to macroalgae [27].

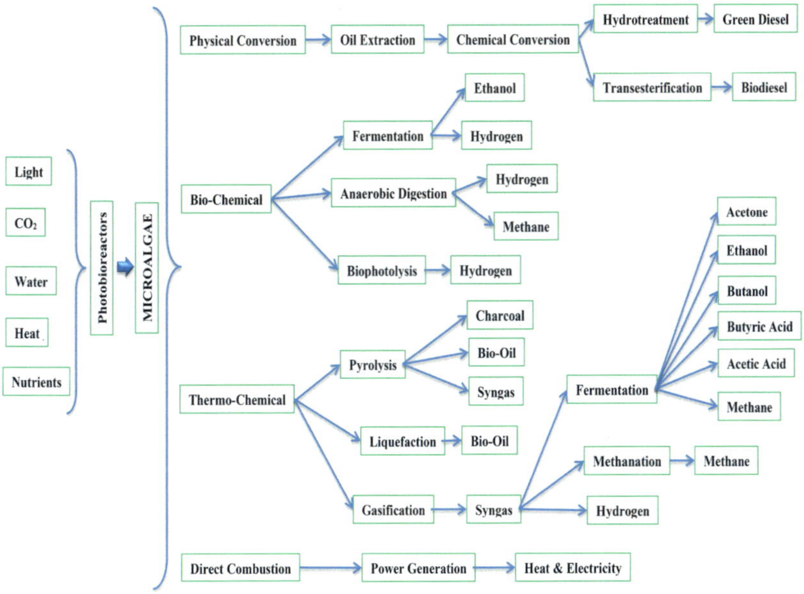

Fig. (2). Microalgae as a source for different biofuels the light to fuel concept: Modified from [7, 8, 26].

Compared to macroalgae (mainly focused in food industry), microalgae got more attention in the biofuels industry and this fact makes them consubstantiate to a good fuel source. With the progress in methods and changes in industrial priorities, biohydrogen shines out of the other alternatives like microalgal biodiesel, bioethanol or biogas. The main reason is that biohydrogen can be produced directly by the microalgae cells and with its unique properties like non-toxicity, high energy density, carbon free combustion and flexibility to be applied in economy with various production processes is a good source for sustainable future [17, 28, 29].

Even if the competition in the energy sector triggers, it is not easy to shift the direction rapidly, keeping in mind, the need of the transformation time for integrated workforce. However, from the society to policy makers' designation of new principles to build a better future that should focus on ethics, justice, solidarity, sustainability and stewardship, equity and human rights can be reached by alternative sources like microalgae. Microalgae as a valuable platform for biofuels without interfering potential food crops and limited clean water by integrating with different production routes and processes serve a flexibility to build a bio-economy that is eco-friendly, sustainable, safe and fits with the ethical values [3, 30].

With regard to the importance of biofuels and differences between macro and microalgae, the target of this chapter is to highlight the biohydrogen production from microalgae covering the current situation and possible future projections considering the theory, technology and economy, considering the major approaches in biohydrogen production: (i) direct (microalgae specific biohydrogen production by their metabolic reactions) (see Table **1**); and (ii) indirect (through the utilization of algal biomass by other microorganisms or processes similar to other biologic substrates like lignocellulosic stocks - see Table **2**). This review will focus on the more specific approach which is the direct production.

MICROALGAE: WHEN AND WHY

Microalgae come to the stage of the human civilization with the first observations from nature, long before they were scientifically defined. For example, algal blooms suddenly appearing in the lakes, which consumed by the birds was probably a real scene that first humans saw which trigger their curiosity. Human curiosity leading the trial and error mechanism resulted in the consumption of microalgae by ancient tribes around the globe where microalgae can be seen naturally like Africa or America [117, 118].

With the modern attempts in science pioneered by researchers like Martinus Beijernik, Otto H. Warburg, P. Dangeard or John S. Burlew, microalgae leap to a higher level considering cultivation and products [119 - 121]. Biofuel research performed in this period and crowned by the US Aquatic Species Program aimed to screen potential species for transportation fuels [122].

Today, the microalgae story (started in nature transferred to the laboratory and then to the industry) focused on the energy solutions with concentrated studies and attempts for realization. Considering the microalgae as a photosynthetic factory for fuel added to the direct production potential like in the hydrogen case, each part of microalgae is acting as a raw material for a fuel. For example, microalgal lipids can be used in biodiesel, green diesel or even jet fuel production. On the other hand, hydrocarbons or isoprenoids can be the source for the gasoline. The carbohydrates can be used for the bioethanol and biogas productions are a good sign of their potential as a valuable source [17, 123].

Table 1. Microalgal biohydrogen production, a brief survey of literature.

Microalgae	Cultivation	Process	PBR type	Culture Medium	Hydrogen Production Procedure	Time	H_2 Production	Ref.
Aphanothece halophytica	Batch-Photomixotrophic	Lab	Vial (10 mL)	BG-11+salt solution (containing various amounts of $NaCO_3$, $MgSO_4$, NaCl, Fe^{3+}, Ni^{2+},sugars)	*Under argon flushing *Under different light intensities	7-21 days	Up to 13.8 µmol mg $Chla^{-1}$ h^{-1} with N-free medium containing 0.4 µM Fe^{3+}	[31]
Anabaena variabilis (CCAP 1403/4B)	Continuous-Photoautotrophic	Lab	Stirred (300 mL)	Nitrogen free Allen-Arnon	*Nitrogen free Allen-Arnon medium *Under vacuum for 15 minutes *Increased light intensity for 5 hours at the 6th day of cultivation	35 days	12-14 mL g^{-1}_{DW} h^{-1}	[32]
Anabaena variabilis (ATCC 29413)	Continuous- Photoautotrophic	Lab	Vial (14 mL)	Allen-Arnon (containing Na_2MoO_4 or Na_3VO_4 or neither)	Under Argon flushing	80-100 h	Up to 5 nmol µg $Chla^{-1}$ h^{-1}at pH levels of 7-9 with Na_3VO_4 added cultures	[33]
Anabaena sp. PCC 7120 (3 hydrogenase mutants from PCC 7120)	Batch-Photoautotrophic	Lab	Sealed polystyrene cuvettes (1 cm light path, 4.7 ml capacity)	BG 11	*Nitrogen free BG 11 medium *Under anaerobic conditions *Various light intensities	32-40 h	Up to; 0.8-1 mmol per cuvette (4.7 ml)	[34]
Anabaena sp. PCC 7120 and its mutant AMC 414	Continuous-Photoautotrophic	Outdoor	Tubular-Coiled (4.35 L)	BG 11	*Nitrogen free BG 11 medium *Under argon atmosphere	7 days	14.9 ml $h^{-1}L^{-1}_{PBR}$ (373 mL total) with the mutant	[35]
Anabaena variabilis (ATCC 29413)	Batch-Photoautotrophic	Lab	Panel (500 ml capacity)	BG 11	*Nitrogen free BG 11 medium *Under anaerobic conditions *Increased light intensity	50 h	Up to; 40-50 mL	[36]
Anabaena variabilis ATCC 29413	Batch- Photoautotrophic-Photomixotrophic	Lab	Serum bottle (164 mL)	$BG11_0$ (nitrogen free)	* Two stage production (all N free) in anaerobic hydrogen phase, continuous illumination 40 mL culture * $BG11_0$ or $BG11_0$ mixed with various concentrations of molybdenum or 10 mM of glucose * Effect of cell density and CO_2 at the head space	>200 h	* 44 µmol H_2 mg^{-1} chl_a h^{-1} at cells density of 110 mg L^{-1}, 5% carbon dioxide headspace concentration, and molybdenum concentration of 1.6 mM * 49 µmol H_2 mg^{-1} chl_a h^{-1} with 10 mM glucose	[37]
Anabaena variabilis ATCC 29413	Batch- Photoautotrophic Immobilized	Lab	Glass column with Hydrophilic cuprammonium rayon hollow fibers (15 mL)	Allen-Arnon	* Under argon flushing * Effect of ammonium contamination in the medium * Continuous illumination under anaerobic environment	4 months	* Maximum in 3 months: 67 mL L^{-1} h^{-1} * Average in 3 months: 57 mL L^{-1} h^{-1}	[38]
Anabaena variabilis ATCC 29413	Batch-Photoautotrophic	Lab	Serum bottles (164 mL)	$BG11_0$	* Cells transferred to $BG11_0$ (5% CO2 flushed) anaerobic environment * Effect light intensity (35-350 µE $m^{-2}s^{-1}$) tested * Effect of light color (white, red, green, blue) tested	80-120 h	* Max 250 µmol H_2 at 80 h with blue light under 70 µE $m^{-2}s^{-1}$ * Max 275 µmol H_2 at 120 h with blue light under 350 µE $m^{-2}s^{-1}$	[39]
Chlamydomonas reinhardtii (C238)	Batch- Photomixotrophic- Mixed-culture	Lab	Flat glass bottles (1 L)	Modified Bristol Medium	*Microalgae was mixed cultivated with *Rhodospirillum rubrum* NCIB 8255 *Transferred to tubes (10 mL) for dark tests *Transferred to membrane reactor (15 mL) for L/D cycles (12:12) test	12 h (dark) 156 h (L/D)	Max 2.39 µmol mg^{-1} DW 12 h^{-1} (total mass base) during dark Max ~3 µmol mg^{-1} DW 12 h^{-1} with L/D cycle Both at 8:2 population ratio	[40]

(Table 1) contd.....

Microalgae	Cultivation	Process	PBR type	Culture Medium	Hydrogen Production Procedure	Time	H_2 Production	Ref.
Chlamydomonas reinhardtii (C137 mt+)	Batch-Photomixotrophic	Lab	Stirred glass bottles(1.2 L)	TAP	TAP-S medium	80 h	140 mL	[41]
Chlamydomonas reinhardtii (CC124)	Batch-Photomixotrophic, Synchronous and asynchronous growth,	Lab	Stirred glass bottles (1.2 L)	TAP	TAP-S medium (transferred by centrifugation)	140 h	*102 mL (with 4 hour synchronized cultures) *86 mL (with unsynchronized cultures)	[42]
Chlamydomonas reinhardtii (Dang 137C mt+)	Batch-Photomixotrophic	Lab	Stirred glass bottles (500 mL)	TAP	*TAP-S medium transfer by centrifugation *TAP-S medium inoculation by TAP culture (10% inoculation)	Up to 140 h	175 mL L^{-1} (with centrifuged cultures, under 20-40 $\mu E\ m^{-2}s^{-1}$ average light intensity)	[43]
Chlamydomonas reinhardtii (CC124)	Continuous-Photomixotrophic	Lab	Stirred glass bottles (1050 mL)	TAP-S (90 µmol sulfate added)	*TAP-S medium, two stage chemostat with aerobic stage and anaerobic stage	4000 hours	Up to; 0.58 mL $h^{-1}L^{-1}_{PBR}$	[44]
Chlamydomonas reinhardtii (Dang 137C mt+ and a nonmotile mutant CC 1036 pf18 mt+)	Batch-Photomixotrophic-Immobilized	Lab	Panel (160 mL)	TAP (0.46 mM sulphate replete)	*TAP-S medium *TAP medium with limiting sulphate (10-20 mM) *Anaerobic conditions	23 days	45 mL day^{-1} (380 mL total)	[45]
Chlamydomonas reinhardtii (Dang 137C mt+)	Batch-Photoautotrophic, Mixotrophic and Heterotrophic	Lab	Flat glass bottles (1.5 L volume)	HS (CO_2 bubbling); TAP with or without CO_2 bubbling	*Sulfur free medium *Anaerobic	60-80 h	* 1.1±0.4 mmol L^{-1} (~2.8 ml $(hL)^{-1}$) in photoautotrophic cultures, *4.5±1.6 mmol L^{-1} (~4.0 ml $(hL)^{-1}$) in photoheterotrophic cultures, *0.9±0.8 mmol L^{-1} (~6.9 ml $(hL)^{-1}$)in photomixotrophic cultures	[46]
Chlamydomonas reinhardtii (Dang 137C mt+)	Batch-Photoautotrophic	Lab	Flat glass bottles (1.5 L)	High Salt	*HS sulfur free medium *Argon purging during the first 24 hours of deprivation	100 h	*71±3 mL L^{-1} (175µE $m^{-2}s^{-1}$ light intensity, pH 7.7) *52±2 mL L^{-1} (420 µE $m^{-2}s^{-1}$ light intensity, pH 7.4)	[47]
Chlamydomonas reinhardtii (CC124)	Semi continuous-Photomixotrophic	Lab	Stirred tank (2.5 L)	TAP	*TAP-S medium *Various dilutions *Anaerobic conditions	127 days	1108 mL	[48]
Chlamydomonas reinhardtii (CC124)	Batch-Photomixotrophic-Immobilized	Lab	Vials (75 mL)	TAP	*TA-S-P (no sulphate, no phosphate) medium *Alginate entrapped harvested cells flushed with or without argon *Anaerobic conditions	160-180 h	0.30±0.02 mol m^{-2} (12.5 µmol $mg^{-1}_{Chl}\ h^{-1}$)	[49]
Chlamydomonas reinhardtii (WT and D1- mutants)	Batch- Photomixotrophic	Lab	Flat glass bottles (1.1 L)	TAP	*TAP-S medium transfer by centrifugation	Up to 240 h	*0.47 ± 0.11 mL L^{-1} h^{-1} with WT *1.54 ± 0.31 mL L^{-1} h^{-1} with D240 *2.60 ± 0.18 mL L^{-1} h^{-1} with D2390-40	[50]
Chlamydomonas reinhardtii CC124	Batch- Photomixotrophic	Lab	Flat glass bottles (1.1 L)	TAP and Olive Oil Mill Wastewater (OMWW)	*TAP-S medium transfer by centrifugation *Effect of OMWW concentration on hydrogen	120 h	*1.03 ± 0.18 mL L^{-1} h^{-1} with TAP grown culture *1.29 ± 0.14 mL L^{-1} h^{-1} with TAP-OMWW grown culture	[51]
Chlamydomonas reinhardtii (Dang 137C mt+)	Batch- Photomixotrophic	Lab	Stirred tank (0.5 L) and Flat glass bottles (0.55 L)	TAP	*TAP-S, TA-P or TA-S-P *TA-P also tested in flat PBRs (sealed after 80 h, initial Chl concentration diluted to ~1.5 mg L^{-1})	140 h (with tank) 300 h (with flat PBR)	Up to 80 mL L^{-1}	[52]
Chlamydomonas reinhardtii (CC124)	Batch-Photomixotrophic	Lab	Glass bottles (325 mL)	TAP	*TAP-S or TAP-N *DCMU mixed Ethanol addition *Dark anaerobic fermentation	192 h	*3.95 µmol x 10^6 $cells^{-1}$ h^{-1} with TAP-S *3.4 µmol x 10^6 $cells^{-1}$ h^{-1} with TAP-N	[53]

(Table 1) contd.....

Microalgae	Cultivation	Process	PBR type	Culture Medium	Hydrogen Production Procedure	Time	H_2 Production	Ref.
Chlamydomonas reinhardtii CC124	Batch- Photomixotrophic	Outdoor	Tubular (50 L)	TAP	*TAP-S medium transfer by centrifugation *Effect of outdoor solar light on the outdoor acclimated cultures	75 h	*No hydrogen produced with cultures directly transferred from laboratory *930 mL H_2 under solar light with acclimated cultures *930 ± 100 mL under artificial light	[54]
Chlamydomonas reinhardtii (CC124)	Batch-Photomixotrophic	Lab	Stirred tank (1, 2.5,5 L)	TAP	*TAP-S medium transfer by centrifugation *Various mixing time *Various light energy *Scale-up	192 h	Max with 1.22 kJ s-1m-3 light energy and 2.5 min mixing time: *1L PBR: 1.53±0.05 mL L^{-1} h^{-1} *2.5L PBR: 1.32±0.05 mL L^{-1} h^{-1} *5L PBR: 1.02±0.05 mL L^{-1} h^{-1}	[55]
Chlamydomonas reinhardtii (CC124)	Batch- Photomixotrophic	Lab	Tubular (110 L)	TAP	*TAP-S medium transfer by centrifugation *Modified with silica nanoparticle to enhance scattering	48 h	3121.5±178.9 mL (0.6 mL L^{-1} h^{-1})	[56]
Chlamydomonas reinhardtii Stm6Glc4 mutant	Batch- Photomixotrophic	Lab	Stirred tank (2 L)	TAP	*TAP medium with different sulfur concentrations selected (5-70 mg L^{-1}) *11 mgL^{-1} initial sulfur concentration applied for PBR experiments *At 120 h culture sealed and hydrogen phase started *Various light intensities were tested (80-721 μE $m^{-2}s^{-1}$)	660 h	650 mL g^{-1} dry biomass with 280 μE $m^{-2}s^{-1}$ intensity	[57]
Chlamydomonas reinhardtii (CC849 and Iba mutants)	Batch- Photoheterotrophic- Co-culture	Lab	Glass bottles (60 mL)	TAP	*TAP grown *Chlamydomonas* and YEM grown *Bradyrhizobium japonicum* was washed and transferred to TAP-S for co-culture *Various inoculation ratios tested	21 days	Max 300 μmol with 849+*B. japonicum* co-culture (100:1)	[58]
Chlamydomonas reinhardtii (CC400)	Batch- Photomixotrophic	Lab	Glass bottles (100 mL)	TAP	*TAP-S medium transfer by centrifugation * Wild type and fnr-RNAi mutants tested	120 h	Up to 500 μL mg $Chla^{-1}$ h^{-1} with mutants	[59]
Chlamydomonas reinhardtii (CC849)	Batch- Photoheterotrophic- Co-culture	Lab	Glass bottles (60 mL)	TAP	*TAP grown *Chlamydomonas* and YEM grown bacteria L2, L3, L4 (Genus: *Stenotrophomonas*, *Microbacterium*, *Pseudomonas*) were washed and transferred to TAP-S for co-culture * Various inoculation ratios tested	13 days	Max 61.5 μmol with 849+L4 co-culture (80:1)	[60]
Chlamydomonas reinhardtii (rbcS-T60-3 mt^- mutants; Y67A,Y68A,Y72A)	Batch- heterotrophic	Lab	Shake flasks (80 mL)	TAP	*Cells in dark harvested and transferred to S deplete and S replete medium	120 h	*Maximum 709.24 μmol L-1 with Y67A mutant under S-deplete medium *Maximum 433.23 μmol L-1 with Y67A mutant under S-replete medium	[61]
Chlamydomonas reinhardtii (CC124 mt^-) and 4 mutants (CC4348,CC2890,CC1354,CC3723)	Batch- Photoheterotrophic	Lab	Flat glass bottles (125 mL)	TAP	*Cells suspended in TAP medium under dark anaerobic conditions for 3 h with nitrogen flushing *Different lights were tested (White, Red690nm, no light) *Different light-dark cycles tested (15:15 h:h) *TAP-S cultures tested for comparison	18-48	Up to 8.81 mL^3 g^{-1} DW with CC3723 mutant	[62]
Chlamydomonas reinhardtii CC124	Batch- Photoheterotrophic	Lab	Glass bottles (130 mL)	TAP	*Cells flushed with nitrogen *CCCP (15 μmol L^{-1}) added	12 h	Up to 20 mmol mg^{-1} Chl with CCCP treated cultures	[63]
Chlamydomonas reinhardtii CC124	Batch- Photomixotrophic	Lab	Tubular and Flat panel (5 L)	TAP	*TAP-S medium transfer by centrifugation *Mixing time and light intensity as the comparison factors	120 h	*1.3 ± 0.05 mL L^{-1} h^{-1} with panel type PBR *1.05 ± 0.05 mL L^{-1} h^{-1} with tubular type PBR	[64]
Chlamydomonas reinhardtii CC125	Batch- Photomixotrophic	Lab	Glass bottles (1 L)	TAP	* TAP grown or TAP+ leachate (16%) cultures (12:12 cycles) transferred to anaerobic TAP-S or TAP+DCMU medium * Effect of leachate addition * NADPH fluorescence correlation to hydrogen production	10 days	* Up to 4×10^{-3} mL H_2 chl_a^{-1} with TAP grown cultures transferred to TAP-S * Up to 6×10^{-3} mL H_2 chl_a^{-1} with TAP+leachate grown cultures transferred to TAP-S	[65]

(Table 1) contd.....

Microalgae	Cultivation	Process	PBR type	Culture Medium	Hydrogen Production Procedure	Time	H_2 Production	Ref.
Chlamydomonas reinhardtii cell wall–deficient PCS CC-406 and two LHCBM9 knockdown transformants (kd-L9.1 and kd-L9.2)	Batch- Photomixotrophic	Lab	Vessel (3 mL)	TAP	* TAP transferred to TAP without sulfur (-S), nitrogen (-N), or phosphorus (-P) * Continuous illumination	< 250 h	* Up to 6 mL H_2 10^{-9} cell with PCS * Up to 1.5 mL H_2 10^{-9} cell with kd-L9.1 * Up to 1.5 mL H_2 10^{-9} cell with kd-L9.2	[66]
Chlamydomonas reinhardtii CC849 (cell Wall deficient) and its transgenic alga lba	Batch- Photomixotrophic	Lab	Bottle (40 mL)	TAP (+spectinomycin for the transgenic)	* TAP grown transferred to TAP-S * Continuous illumination	5-6 days	* 1171 µmol mg^{-1} chl with lba alga * 415 µmol mg^{-1} chl with CC849	[67]
Chlamydomonas reinhardtii CC-124 mt$^-$ and CC-4147 FUD7 mt$^+$	Batch- Photoheterotrophic to Photomixotrophic Immobilized	Lab	Vials (75 mL)	TAP	* TAP-S immobilized (1cm × 6 cm films) transferred to TAP-P (+S), TAP-S or TAP-S-P * Effect of 10^{-5} M DCMU dissolved ethanol (0.1%) addition (20, 44, 70 h) * Under argon	60-170 h	* > 0.1 mol m^{-2} with CC124 at TAP-P-S at ~85 h * < 0.06 mol m^{-2} with CC124 at TAP-P-S at 70 h (DCMU added at 46 h) * < 0.15 mol m^{-2} with CC124 at TAP-P-S reached at 168 h (no DCMU) * < 0.12 mol m^{-2} with CC124 at TAP-P-S reached at 168 h (DCMU added at 70 h) * < 0.09 mol m^{-2} with CC124 at TAP-P-S reached at 168 h (DCMU added at 44 h) * < 0.06 mol m^{-2} with CC124 at TAP-P-S reached at 168 h (DCMU added at 20 h) * No meaningful production in CC124 TAP-P+S * No meaningful production in CC-4147 TAP-P-S	[68]
Chlamydomonas reinhardtii CC124 (WT and D1- mutants)	Batch- Photomixotrophic	Lab	Flat glass bottles (1.1 L)	TAP	*TAP-S medium transfer by centrifugation *Effect of D1 mutations on hydrogen production	Up to 280 h	Mean production rates between: *0.3 ± 0.11 mL L^{-1} h^{-1} with WT *5.15 ± 0.11 mL L^{-1} h^{-1} with A250L	[69]
Chlamydomonas reinhardtii CC124	Batch- Photomixotrophic	Lab	Flat glass bottles (1.1 L)	TAP	* TAP-S transferred two stage protocol under continuous illumination * CAT, APOX and GPOX activities investigated during hydrogen production	144 h	* Maximum at 120 h: ~115 mL L^{-1}	[70]
Chlamydomonas moewusii FACHB-266, *Chlorella pyrenoidosa* FACHB-1222, *Scnedesmus obliquus* UTEX 393, *Chlorella fusca* UTEX 292 and *Chlamydomonas reinhardtii* CC124	Batch- Photomixotrophic	Lab	Bottles (150 mL)	TAP	* Cells nitrogen flushed and dark adapted in TAP * Cultured under continuous light * Cultures tested for oxygen tolerance (0-2%) * Further hydrogen tests with selected two	100-300 h	* All species reach around: 3-40 µmol H_2 mg^{-1} chl $_a$ h^{-1} oxygen free * Top 2 is *C. moewusi*:40.0±5.4 µmol H_2 mg^{-1} chl $_a$ h^{-1} and *C. reinhardtii*: 34.7 ± 1.7 µmol H_2 mg^{-1} chl $_a$ h^{-1} * Max total *C. reinhardtii*: 73.56 ± 2.81 mL L^{-1} at 60 h * Max total *C. moewusi*: 127.05±9.09 mL L^{-1} at 61 h	[71]
Chlamydomonas reinhardtii CC1010	Batch- Photoautotrophic	Lab	Flasks (500 mL)	TAP-C (no acetate)	* Autotrophic two-stage process based on S-deprivation (TAP-CS) and light-transition for efficient H2 production by *C. reinhardtii* in * Either CCM-uninduced cells or CCM-induced cells	144-160 h	* Max with CCM-uninduced cells: 58 ± 3 mL L^{-1} at 25 µE m^{-2}s^{-1} * Max with CCM-induced cells: 118 ± 4 mL L^{-1} at 50 µE m^{-2}s^{-1}	[72]
Chlamydomonas reinhardtii C541	Batch- Photomixotrophic	Lab	Beaker (100 mL)	TAP	* Cells transferred to TAP-S with the beaker * Beaker head closed put inside a desiccator filled with 0.8 atm nitrogen under continuous illumination * Sustainable production just by the addition of TAP medium and giving 2h for recovery is investigated	14 days	Max 100-150 ppm H_2	[73]

(Table 1) contd.....

Microalgae	Cultivation	Process	PBR type	Culture Medium	Hydrogen Production Procedure	Time	H_2 Production	Ref.
Chlamydomonas reinhardtii WT137C	Batch-Photoheterotrophically Immobilized	Lab	Vials (75 mL)	TAP	* Cells transferred to TAP or TAP-S * Alginate immobilized (1 cm×5cm films) * Effect of initial pH, light intensity (dark for control) and light color (blue, green, red, white) investigated	17-20 days	*Max 200 µmol H_2 m^{-2} at 80 µmol photons $m^{-2}s^{-1}$ * Max 213 mmol H_2 m^{-2} at pH 6.5 * Max 6 mmol H_2 m^{-2} with red light (690 nm) at 40 µmol photons $m^{-2}s^{-1}$	[74]
C. reinhardtii CC 849 transgenic algal strain (hemHc-lbac), CC124, CC503 And *Bradyrhizobium japonicum*	Batch- Photoheterotrophic and Heterotrophic under dark Co-culture (algae+bacteria)	Lab	Bottle (40 mL)	TAP (algae) YEM (bacteria)	*Aerobic cultures of microalgae were transferred to TAP-S with bacteria * Dark incubation and then continuous illumination	17-21 days	* Max 280 µmol H_2 mg^{-1} Chl with transgenic algae co-culture (80:1 mix algae-bacteria) under 40 µmolE $m^{-2}s^{-1}$ * Max 280 µmol H_2 mg^{-1} Chl with CC124 co-culture (80:1 mix algae-bacteria) under 200 µmolE $m^{-2}s^{-1}$ * Max 280 µmol H_2 mg^{-1} Chl with CC503 co-culture (20:1 mix algae-bacteria) under 200 µmolE $m^{-2}s^{-1}$	[75]
Chlorella ptotothecoides	Batch- Photoheterotrophic	Lab	Stirred vessels (650 mL)	TAP (with various NH_4Cl concentrations; 0.35-7 mM)	*Harvested, washed and transferred to TAP-S or TAP-N medium under continuous illumination	140 h	Max 233.7 mL L^{-1} with TAP-S cultures (0.35 mM NH_4Cl)	[76]
Chlorella protothecoides	Batch- Photomixotrophic	Lab	Glass tubes (20 mL)	TAP standard (7 mM ammonium) or TAP modified (0.7 mM ammonium)	*TAP cultures (14:10 cycle) transferred to TAP-S (sealed under continuous illumination) * TAP cultures (14:10 cycle) transferred to TAP-S (sealed under continuous illumination) * Effect of 10 µM DCMU added at 24 h to TAP modified	100-120 h	* 1.56 ml L^{-1} in TAP * 1.69 ml L^{-1} in TAP-S * 110.8 ml L^{-1} in TAP modified * 140.4 ml L^{-1} in TAP modified-S * ~ 30 ml L^{-1} in TAP modified-S+DCMU * ~ 20 ml L^{-1} in TAP modified +DCMU	[77]
Chlorella pyrenoidosa C-101	Batch-Photoautotrophic	Lab	Bubble column (650 mL)	Modified Bristol	Nitrogen flushing under dark for " >20" hours	60-65 h	$6.9x10^{-2}$ m^3 kg^{-1} cell	[78]
Chlorella pyrenoidosa	Batch- Photoautotrophic	Lab	Bottles (75 mL)	TAP (2.0 g L^{-1} sodium acetate)	* Aerobic cultures transferred to TAP-S or TCP ($NaHCO_3$ (10 mM) modified TAP to replace acetate) * Continuous illumination purged with nitrogen with DCMU addition (10 µM)	8 days	Max mL L^{-1} with TCP added DCMU at 9 h after illumination start	[79]
Chlorella vulgaris MSU 01	Batch-Photomixotrophic	Lab	Stirred (500 mL)	Modified BG11 and MJ	*Anaerobic conditions *Corn stalk as carbon source in the medium	6-7 days	26 mL	[80]
Chlorella vulgaris NIER-10003	Batch-Photomixotrophic-Immobilized	Lab	Glass bottle (1 L)	MA	*Harvested, washed and transferred to MA-S *Nitrogen flushed *Various carbon sources, their concentrations and medium pH were tested *Optical fibers tested	40-120 h	Maximum volume of 1315 mL L^{-1} with pH = 8, Sucrose = 5 mgL^{-1} Maximum rate of 24 mL L^{-1} h^{-1} was with fructose	[81]
Chlorella vulgaris YSL01 and YSL16	Batch- Photoautotrophic	Lab	Bottles (500 mL)	$EDTANa_2$-lacking bold basal medium (BBM)	* Cultures headspace with N_2, CO_2 (5,10,15%), O_2 (5,10,15%) gas mixtures under continuous illumination	5 weeks	* ~ 0.6 mL L^{-1} maximum at N_2, CO_2, O_2 mixture (70,15,15%) with YSL01 * < 0.4 mL L^{-1} maximum at N_2, CO_2, O_2 mixture (70,15,15%) with YSL16 * < 2 mL L^{-1} maximum at N_2, CO_2, O_2 mixture (80,10,5%) with YSL01 * < 1.5 mL L^{-1} maximum at N_2, CO_2, O_2 mixture (80,10,5%) with YSL16	[82]

(Table 1) contd.....

Microalgae	Cultivation	Process	PBR type	Culture Medium	Hydrogen Production Procedure	Time	H_2 Production	Ref.
Chlorella sp. IOAC707S	Batch- Photomixotrophic	Lab	Vials (45 mL)	TAP/NaCl and TAP-SeaWater	* Aerobic cultures transferred to various medium with anaerobic environment under continuous illumination * Chla+b concentration (0.15-20 mg L^{-1}) investigated for selection (~1 mg L^{-1} selected for experiments) * Effect of CO_2 (10%) addition * TA-P/NaCl, TA-P/SW, TA-P/NaCl+CO_2, TA-P/SW+CO_2	400-700 h	Maximum * TA-P/NaCl ~25 mL L^{-1} (initial Chl: 0.8 mg L^{-1}) * TA-P/NaCl: ~25 mL L^{-1} (> 500h) * TA-P/NaCl+CO_2: ~40 mL L^{-1} (< 500h) * TA-P/SW: ~40 mL L^{-1} (< 400h) * TA-P/SW+CO_2: ~35 mL L^{-1} (< 400h)	[83]
Cyanothece sp. 51142	Continuous vs Batch Photoautotrophic	Lab	Flat glass bottles (1 L) Series with Tubular vessel (3 L)	ASP2 (with the addition of 1.5 g NaNO3 L^{-1} and 10 mM of 99% purity glycerol)	*Two types of PBRs are connected one tubular for culture the other flat for hydrogen production under continuous illumination * Batch and Continuous process comparison	750 h	* Single stage batch: 272 * Multi-stage batch: 864 mL * Continuous: 1750 mL	[84]
Lyngbya sp.	Batch- Photomixotrophic	Lab	Vessel (150 mL)	Basal	* Cultures transferred to hydrogen production medium (benzoate rich) flushed with argon and sealed for anaerobic phase under continuous illumination	~100 h	Max rate 17.05 μmol H_2 g chl a^{-1} h^{-1} at 64 h	[85]
Microcystis flos-aquae	Batch-Photomixotrophic	Lab	Glass bottle (60 mL)	BG 11	*Sealed samples (30 mL) were placed under dark or light	26 days	6.11 μmol mg $Chla^{-1}$ h^{-1}	[86]
Nostoc PCC 7120	Batch-Photoautotrophic	Lab	Flat glass vessel (5.04 L)	$BG11_0$ (nitrogen free)	Under different aerobic and anaerobic conditions with a starting OD_{750} in all experiments at inoculation is 0.1	217 h	The highest buildup of hydrogen gas observed was 6.89% H_2 (100% argon phase) of the total PBR system with a maximal production of 4.85 mL L^{-1} h^{-1}	[87]
Platymonas subcordiformis	Batch-Photoautotrophic	Lab	Serum bottle (295 mL)	Sea Water Medium	*Sulfur deprive medium *Incubation under dark *Anaerobic conditions *Continuous illumination afterwards	50 h	11720 nL h^{-1} (with Seawater-S medium)	[88]
Platymonas subcordiformis	Batch-Photoautotrophic	Lab	Torus (1.5 L)	Defined mineral	*Dark incubation *Anaerobic nitrogen atmosphere *15 μM CCCP addition *Continuous illumination afterwards	4-6 days	7.20 mL h^{-1} (236.6±7.0 mL)	[89]
Platymonas helgolandica var. *tsingtaoensis*	Batch-Photoautotrophic	Lab	Jars (130 mL)	F2	*Anaerobic *Illuminated -S medium *Dark CCCP/DCMU or both added medium	25 h	*0.002 mmol L^{-1} with –S medium *0.160 mmol L^{-1} with CCCP medium *0.014 mmol L^{-1} with DCMU medium *0.290 mmol L^{-1} with CCCP+DCMU medium	[90]
Spirulina platensis (NIES-46)	Batch-Photoautotrophic	Lab	Erlenmeyer flasks (60 mL)	SOT	*Nitrogen free medium *Under dark *Anaerobic conditions	20 h	2 μmol mg_{DW}^{-1}	[91]
Synechocystis sp. PCC 6803 and its mutant NDH-1 complex deficient M55	Batch-Photoautotrophic-Encapsulated	Lab	Vials (2 mL)	BG 11	*Nitrogen atmosphere *Under cycled light and dark exposure	5 days	0.005-0.045 mM	[92]
Tetraspora sp. CU2551	Batch-Photomixotrophic	Lab	Vials	TAP (with, or without, 0.5 mM β-mercaptoethanol for 24 h)	*Sulfur and Nitrogen deprived medium *Argon flushed anaerobic atmosphere	> 24 h	17.3-61.7 μmol mg^{-1} chl_a h^{-1}	[93]
Volvox carteri f. nagariensis EVE	Batch-Photoautotrophic	Lab	Serum vials (10 mL)	Standard Volvox Medium	* Cultures incubated with 20 mM sodium bicarbonate for 72 hours * Transferred to anaerobic media (degassed SVM supplemented with 0.16 mM $MgCl_2$ rather than 0.16 mM $MgSO_4$ to exclude sulfate from the medium) under dark	45 h	3-4.5 μmol H_2 mg^{-1} chl	[94]

(Table 1) contd.....

Microalgae	Cultivation	Process	PBR type	Culture Medium	Hydrogen Production Procedure	Time	H_2 Production	Ref.
Isolates from Baltic Sea and Finnish lakes (*Calothrix* 336/3, *Calothrix* XPORK5E, *Nostoc* XHIID A6, *Calothrix* BECID 33, *Nodularia* TRO31, *Nodularia* AV33, *Nostoc* BECID 19, *Anabaena XSPORK* 7B, *Calothrix* XSPORK 36C and *Calothrix* XSPORK 11A) and *Anabaena* PCC 7120, *Anabaena* PCC 7120 the ΔhupL and ΔhupL/ΔhoxH mutant	Batch- Photoautotrophic	Lab	Vials (20 mL)	Z8× (± salt) or BG11 (according to the specie)	* Z8× or BG 11 cultures transferred to Argon flushed sealed vials under continuous illumination	24 h	* ~ 5 µmol H_2 mg^{-1} chl_a h^{-1} with *Anabaena* PCC 7120 * > 15 µmol H_2 mg^{-1} chl_a h^{-1} with *Anabaena* PCC 7120 ΔhupL/ΔhoxH * < 5 µmol H_2 mg^{-1} chl_a h^{-1} with *Anabaena XSPORK* 7B * ~ 20 µmol H_2 mg^{-1} chl_a h^{-1} with *Calothrix* 336/3 * ~ 5 µmol H_2 mg^{-1} chl_a h^{-1} with *Calothrix* BECID 33 * < 15 µmol H_2 mg^{-1} chl_a h^{-1} with *Calothrix* XPORK5E * < 5 µmol H_2 mg^{-1} chl_a h^{-1} with *Calothrix* XSPORK 36C * < 5 µmol H_2 mg^{-1} chl_a h^{-1} with *Calothrix* XSPORK 11A * > 5 µmol H_2 mg^{-1} chl_a h^{-1} with *Nodularia* AV33 * < 10 µmol H_2 mg^{-1} chl_a h^{-1} with *Nodularia* TRO31 * < 5 µmol H_2 mg^{-1} chl_a h^{-1} with *Nostoc* BECID 19 * < 15 µmol H_2 mg^{-1} chl_a h^{-1} with *Nostoc* XHIID A6	[95]

Table 2. Integrated biohydrogen processes with a special emphasis on microalgal biomass as a substrate for fermentation.

Microalgae	System	Max Hydrogen Production	Ref.
Anabaena sp. PCC7120	Fermentation of hydrogen produced residual algal biomass by *Enterobacter aerogenes*	0.0114 kg kg^{-1}biomass	[96]
Anabaena sp. PCC 7120	Processing of physico-chemically treated algal biomass by an enriched thermophilic mixed culture collected from the anaerobic digester	6.9 mL g^{-1} h^{-1}	[97]
Arthrospira (Spirulina) platensis, Nannochloropsis sp. and Dunaliella tertilecta	Anaerobic fermentation of thermal pretreated algal biomass by immobilized *Clostridium acetobutylicum*	8.5 mmol L-1 day^{-1}	[98]
Arthrospira (Spirulina) platensis	Fermentation of pretreated (boiling, bead milling, ultrasonication, enzymatic hydrolysis) wet biomass by anaerobic activated sludge	92 mL g^{-1}DW	[99]

(Table 2) contd.....

Microalgae	System	Max Hydrogen Production	Ref.
Arthrospira platensis and *Laminaria digitata*	Fermentation of algal biomass pre-treated with 2.5% dilute H_2SO_4 by anaerobic fermentative bacteria originated from anaerobic sludge from a farm digester	70-75 mL g^{-1} VS	[100]
Arthrospira maxima	Enzymatic hydrolysis and fermentation of algal biomass by activated sludge	78.7 mL g^{-1}	[101]
Chlorella vulgaris ESP6	Dark fermentation of acid or alkaline/enzyme pretreated algal biomass hydrolysate by *Clostridium butyricum* CGS5	246 mL L^{-1} h^{-1}	[102]
Chlorella vulgaris- INETI 58, *Scenedesmus obliguus* ACOI 204/07, Consortium isolated from the wastewater (including different strains such as *Chlorella*, *Scenedesmus*, *Chaetophora* and *Navicula*)	Processing of algal biomass through dark fermentation by a strain of the bacteria *Enterobacter aerogenes* ATCC 13048 Sputum	58.6 mL g^{-1} VS with *S. obliquus* biomass	[103]
Chlorella sorokiniana	Dark fermentation of HCl-heat pretreated algal biomass by *Enterobacter cloacae* IIT-BT08	148 mL L^{-1} h^{-1}	[104]
Chlorella sp.	Simultaneous hydrolysis and fermentation of algal biomass (*Chlorella* sp. ESP-6 and commercial *Chlorella* powder)by sewage sludge consortia	1.01±0.09 mmol g^{-1}powder	[105]
Chlorella vulgaris and *Dunaliella tertiolecta*	Fermentation of algal biomass by anaerobic activated sludge	12.6 mL $g^{-1}{}_{VS}$ from *D. tertiolecta*	[106]
Thalassiosira weissflogii	Dark fermentation of algal biomass with the thermophilic bacterium *Thermotoga neapolitana*	18.1±5 mL L^{-1} h^{-1}	[107]
Chlamydomonas reinhardtii (IAM C-238), *Chlorella pyrenoidosa* (IAM C-212) and *Dunaliella tertiolecta* (ATCC 30909)	Fermentation of algal biomass by *Lactobacillus amylovorous* (ATCC 33620) and *Rhodobacter sphaeroides* RV, *Rhodobacter capsulata* (ATCC 11166), *Rhodospirillum rubrum* (ATCC 11170), *Rhodovullum sulfidophilus* (ATCC 35886), *Rhodobium marinum* (ATCC 35675)	8 mol mol^{-1} starch-glucose equivalent from *C. reinhardtii* biomass with *R. marinum*	[108]

(Table 2) contd.....

Microalgae	System	Max Hydrogen Production	Ref.
Chlamydomonas reinhardtii (IAM C-238) and *Dunaliella tertiolecta* (ATCC 30909)	Fermentation of algal biomass by *Lactobacillus amylovorous* and *Rhodobium marinum* (A-501) mixed culture	2.47±0.21 mmol $h^{-1}L^{-1}$ with *D. tertiolecta* biomass	[109]
Chlamydomonas reinhardtii (UTEX 90)	Fermentation of algal biomass by *Clostridium butyricum* to hydrogen and organic acids followed by the photodissimilation of organic acids to hydrogen by *Rhodobacter sphaeroides*	8.30 mol mol^{-1} starch-glucose equivalent algal biomass	[110]
Chlamydomonas (MGA 161)	Fermentation of algal biomass by *Rhodovulum sulfidophilum* (W-1S)	0.1875 mmol L^{-1} h^{-1}	[111]
Chlamydomonas reinhardtii (UTEX 90)	Fermentation of heat-HCl and enzymatic pre-treated microalgal biomass by hyperthermotolerant eubacterium *Thermotoga neopolitana*	227±9.4 mL L^{-1} h^{-1}	[112]
Nannochloropsis sp.	Dark fermentation of algal biomass extraction residues by *Enterobacter aerogenes*	60.6 mL g^{-1}_{DW}	[113]
Scnedesmus sp.	Fermentation of algal biomass by anaerobic digested sludge	3.84 mL h^{-1} (19.53 mL $g^{-1}_{Volatile\ solids}$)	[114]
Scenedesmus obliguus	Fermentation of algal biomass by *Clostridium butyricum*	2.9±0.3 mol $mol^{-1}_{algal\ sugars}$	[96]
Scenedesmus obliguus (YSW15)	Fermentation of algal biomass by anaerobic digester consortia	120.3 mL L^{-1} h_{-1}	[115]
Scenedesmus obliguus	Fermentation of algal biomass by bacteria (Enterobacter aerogenes ATCC 13048 and Clostridium butyricum DSM 10702)	113.1 mL g^{-1}_{VS}*C. butyricum*	[103]
Spirulina platensis, *Anabaena variabilis* and *Chlorella* sp.	Processing of algal biomass by an integrated system of methanogenic culture, *Rhodobacter capsulatus* and *Thermohydrogenium kirishi*	1080 mL L^{-1} day^{-1} with *Chlorella* sp. Biomass by *R. capsulatus*	[116]

The focus on microalgae for the hydrogen production followed a similar route as the other fuel alternatives like ethanol or biogas. Jackson and Ellms who reported the biohydrogen production in the cyanobacterium *Anabaena* in 1896 followed by Gaffron and Rubin who reported in 1940's the hydrogen potential of green microalgae *Scenedesmus obliguus* triggered the research on the microalgal hydrogen production [121, 124, 125]. The investigations on biohydrogen after Gaffron and Rubin shifted the attention to the metabolism of the microalgae. The first attempts after their findings highlighted the importance of anaerobic conditions for the production. These clues were used in the studies of Wykoff *et al.* that showed the effect of sulfur deprivation on PSII blocking and later, in 2000, Melis *et al.* realized the photobiohydrogen production through a two stage protocol which separated the aerobic phases and anaerobic phases by using the key effect of sulfur deprived conditions and first time in history microalgal biohydrogen production reached reasonable productivities [41, 126].

Besides being a carbon neutral source like other biomass sources, the microalgal potential for biohydrogen production is a key advantage. First of all, microalgae will not compete with the food crops in two major junctions. Microalgae will not need any fertile soil (also no risk for land clearing as an indirect effect) or clean water to be cultured which are actually the main criticism for the oil plants which need these valuable sources that should be shifted for food. In addition, microalgae are not a basic edible source like terrestrial corn or soybean so no stress on the society or economy with the extreme sensitivity on scarce food and feed sources [4, 24, 127]. Other advantages can be listed to strengthen the usage of microalgae like: (i) CO_2 sequestration; (ii) efficient use of solar energy; (iii) yearlong harvesting; (iv) culture in non-arable or non-arid lands; (v) culture with an inoculation; (vi) rapid growth; (vii) use of various waters from waste to salty without underestimating the importance of microalgal fuel realization related with commercialization strategies, fuel prices, realistic vision supported by objective research and attention of society [3, 24, 128 - 130].

MICROALGAE: A PHOTOSYNTHETIC GREEN FACTORY

Starting with the classic sentence "*microalgae are photosynthetic green factories which can utilize solar energy and CO_2 in order to produce their own food in other words carbohydrates*" will give the clues about the nature of microalgae but to understand the whole mechanism is quite a challenge for the scientists for years. Comprehensive reviews by the leading scientist in the plant biology area actually highlight the importance of this reaction as a milestone in life [118, 131 - 133].

Basically photosynthesis is a complex machinery of routes that can interactively respond with the environment. This process can be separated into two stages: (i) the light dependent; and (ii) the light independent. Light dependent stage occurs in the thylakoid membranes that contain light harvesting antenna complex (and light-harvesting-like proteins which are low mass proteins in most photosynthetic eukaryotes). These complexes contain three transmembrane α-helices and bind chlorophyll a, b or c, carotenoids and lipids. According to the reactions that they catalyse, photosystem I (PSI) is defined as a plastocyanin-ferredoxin oxidoreductase, photosystem II (PSII) as a water-plastoquinone oxidoreductase, cytochrome b_6f as a plastoquinone-plastocyanin oxidoreductase and F-ATPase as a proton-motive force driven ATP synthase [134 - 136].

Photosystems have reaction centers constructed with a pair of chlorophyll molecules: P700 (PSI) and P680 (PSII) where specific wavelength of light can be harvested. These molecules also have the ability for oxidation and reduction reactions. The systems also comprise antenna chlorophylls, and other pigments like carotenoids, phycobilins that can harvest light quanta and deliver the excitation to the chlorophyll pairs. Excitation energy that is routed from the light harvesting antenna to these centers triggers the charge separation as the ignition of photosynthetic electron transport. Photosystem proteins are conserved in the primary sequence, mechanism and supramolecular architecture but very diverse in the antenna systems [136 - 139]. The light reactions are the source for the NADPH and ATP to fix inorganic carbon during the Calvin-Benson Cycle of dark stage. These reactions are not dependent directly to light, but use the products of the light reactions. Carbon fixation by using the enzyme RuBisCo produces three carbon sugars as an intermediate product, which is utilized for the production of final carbohydrate [128, 140, 141].

With regard to the major differences between the cell organizations of eukaryotic and prokaryotic algae, photosynthesis occurs in thylakoids membranes of the chloroplasts or in the prokaryotic thylakoid membranes. Eukaryotic chloroplasts comprise various compartments for photosynthetic proteins. PSI is located in the stroma lamellae, PSII is located in the grana, the ATP synthase is located the stroma lamellae, and the cytochrome b_6f complex is found in grana and grana margins [142]. Even if eukaryotic and prokaryotic PSI and PSII reaction centers are nearly identical regarding their subunit composition, their integration with light harvesting complexes is different. Cyanobacterial light antenna is made by phycobiliproteins, which builds the main light harvesters phycobilisome that bound extrinsically to the stromal surface of the PSII core compared to the specialized complexes in eukaryotes that intrinsically binds the harvesting antenna with chlorophyll a and chlorophyll b [143, 144]. However, the basic photosynthetic reactions are similar in both groups of microscopic algae.

After the introduction of the major life support mechanism of microalgae the following step is to understand the specific processes related to biohydrogen production.

MICROALGAL HYDROGEN PRODUCTION CONSIDERING BOTH PROKARYOTES AND EUKARYOTES

Hydrogen is an important source of reductant in photosynthesis and is a key player for the life of microalgae [140, 145]. Considering the differences between the eukaryotic and prokaryotic microscopic algae, the hydrogen production routes will depend on the cellular organization. Giving the highest priority to the hydrogenase enzymes, their diversity and sensitivity to oxygen are the keys in the hydrogen production [117, 146, 147].

Starting with cyanobacteria, the hydrogen production in these prokaryotic microscopic algae is catalyzed mainly by bidirectional and uptake [NiFe] hydrogenases. Specifically, cyanobacteria use [NiFe] type uptake hydrogenase for energy conservation and hydrogen oxidation, [NiFe] type bidirectional hydrogenase for energy conservation and maintaining intracellular redox balance and nitrogenase used for hydrogen production and nitrogen fixation [148 - 150]. Uptake hydrogenases are not wanted because they oxidize the produced hydrogen. So suppression of *hup* genes (*hup*S and *hup*L) which are encoding these hydrogenases is a key in the sustainable hydrogen production from cyanobacteria [151]. Some species (especially the heterocystous cyanobacteria) can also use nitrogenase enzymes in hydrogen production. Nitrogenase catalyzed reactions need energy (16 ATP for a mole of N_2) to produce hydrogen as a by-product during the fixation of dinitrogen to ammonium. Thus, they do not contribute to an efficient process but compared to the hydrogenases they are less sensitive to oxygen [152 - 156]. Nitrogenase enzymes contain metallic active sites like molybdenum, vanadium or iron in their prosthetic groups, with the most dominant as the molybdenum-iron (MoFe) nitrogenases consisted of dinitrogenase (MoFe protein) and dinitrogenase reductase (Fe protein) [37, 157, 158].

Biohydrogen production in cyanobacteria can follow three different routes. The first one is the direct biophotolysis of water by using PS II and PS I *via* light based electron transfer from water to ferredoxin. Water will be processed to oxygen and electrons in PSII and then the formed electrons will be reduced to ferredoxin which can be converted to hydrogen by hydrogenases. The reduced ferredoxin can transfer the electrons to NAD or NADP, which can be reoxidized to hydrogen [154, 156]. The second route (an anaerobic light independent one) depends on the utilization of the photosynthetic storage molecules through light independent reactions in such a way that NADPH from polysaccharides or lipids is the main electron donor, which are transferred to plastoquinone pool (PQ-pool) for the hydrogen production [159].

Non-heterocystous cyanobacteria could not prevent the effects of oxygen produced by the vegetative cells. However, they can still fix nitrogen and produce hydrogen by the on-off control of the photosynthesis with specialized mechanism, related to the special storage granules (cyanophycin) and nitrogenase comprising group of cells [160]. In addition, some unicellular cyanobacteria can produce hydrogen with the indirect separation of the oxygenic processes under light and dark cycles of the day. Oxygen from photosynthesis continued by the nitrogen fixation in the day followed by the oxygen consuming respiration ended up in the production of biohydrogen at night.

Cyanobacteria with the nitrogen fixation ability use a photofermentative third pathway that is actually the integration of the first two in the nitrogen fixing units named as "heterocyts". Heterocysts are the sights of indirect biophotolysis where the spatial separation of oxygenic photosynthesis and hypoxic nitrogen fixation occur due to their specialized structure of oxygen impermeable glycolipid layer with the lack of oxygen evolving PSII other than their ability of high respiration rates which helps to utilize the produced oxygen easily [161]. NADP from the aerobic reactions in the vegetative cell which is related with the electrons and H^+ transported to the PQ-pool, is reduced through PSI alone inside the anoxygenic environment of the heterocysts. During this process uptake hydrogenase [158] and bidirectional hydrogenase are inactivated forcing the saccharides to supply the electrons for nitrogenase reactions [162 - 165].

Considering the eukaryotic microalgae, similar mechanism of hydrogen production with a special diversification due to the taxonomy can be also observed. In eukaryotic microalgae [FeFe] hydrogenases (also called "iron only" hydrogenases) are responsible for the hydrogen production process, where they play a role in both energy conservation and fermentation regarding their ability to transfer electrons from H_2 or to balance cells overall redox potential [148 - 150, 166 - 168]. Besides their sensitivity to oxygen, [FeFe] hydrogenases display higher activity compared to the [NiFe] hydrogenases with a high turnover potential up to 10^6. s^{-1} [162, 169 - 172]. They both have common architecture in mature enzyme structure with CO and CN^- ligands, bimetallic active site and iron atom existence in the active site mosaic, but they have different three dimensional structure [166, 173 - 175]. Their monomeric structure is relatively simpler in shape when compared to [NiFe] hydrogenases and they have two iron atoms in their active site [155, 176]. Active cluster of these enzymes only contains cysteine as organic residues. The metal center is ligated *via* sulfur bridges which hold the structure as a whole [173]. Microalgal hydrogenases are 48 kDa in molecular weight with 350 amino acids and are very small compared to bacterial hydrogenases [169, 177]. In order to have an active enzyme for hydrogen production, the accessory maturation proteins that are regulated by *hyd*A1 and *hyd*A2 genes are required [178]. Basically, these genes are present in oxygen free environment. *Hyd*A1 controls the activity of the hydrogenase enzyme in the light but *hyd*A2 is still not clear [179]. As an important feature, [FeFe] hydrogenases have also the ability to reach a thermodynamically favorable hydrogen production through electron bifurcation which is found in anaerobic prokaryotes [180].

Similar to cyanobacteria, eukaryotic microalgae can also follow various routes, two of which are light dependent and one is light independent, in hydrogen production related with the dissipation of excess electrons of the electron transport chain when the main electron acceptors CO_2 and O_2 are absent or the other processes like Calvin Cycle, photorespiration and Mehler reactions used as sinks are blocked [140, 181]. In all the routes, reduced ferredoxin is the source of electrons to the hydrogenase through the irreversible reaction of 2 H^+ to H_2. The key in these routes is the transfer of electrons to hydrogen evolving enzymes, whether supplied directly from water splitting or indirectly from organic materials such as starch [182, 183].

First route is the direct biophotolysis of water into hydrogen and oxygen where the water splitting PSII and ferredoxin reducing PSI acts cooperatively. Second route is the indirect biophotolysis where the electrons (from the catabolism of endogenous substrates through glycolysis pathway and citric acid cycle) are transferred to the linear electron transport chain utilized only by a PSI active hydrogen evolution pathway. In this route, the reductant from catabolic reactions are transferred through the enzyme NADP plastoquinone oxidoreductase directly to the plastoquinone pool, by passing PSII [180, 184].

A third route (other than these light dependent routes) is the dark anoxic fermentation of the decarboxylated pyruvate from glycolysis by pyruvate ferredoxin oxidoreductase. During this process, microalgae try to protect their basal metabolism generating fermentative end products like formate, acetate, ethanol, and hydrogen [69, 161, 181, 183, 185].

Keeping in mind the mechanism of the microscopic algae hydrogen production the interrelation of the environment and the cells plays the key role in the efficiency. The specific requirements for the hydrogen production need to highlight the production systems in other words photobioreactors. Hydrogen production from microalgae actually hides the key issue for a sustainable process: the dependence on microalgae. One should primarily produce microalgae effectively in order to produce the secondary product hydrogen.

MICROALGAL HYDROGEN PRODUCTION SYSTEMS: THE PHOTOBIOREACTORS

With the accelerated studies after the first attempts in Stanford, Essen and Tokyo, which are reviewed in the book of Burlew in 1953, the microalgae cultivation becomes a hot topic in food and feed industries as well as the space studies for their gas exchange capabilities with a rising interest in the production systems [186].

Microalgae need a special living environment like every other organisms with a special emphasis on the environment interrelated parameters like light, temperature, nutrients, pH and even population limits in order to survive [187 - 189]. Production systems are the special units that can support the microalgae by providing the optimum conditions with regard to these parameters. With a key relation in the interaction with the atmosphere, these systems can be classified into two types: the open systems that have direct contact with the atmosphere and closed systems that are isolated from the atmosphere. In a more specific statement, closed systems or technically photobioreactors are more sophisticated culture systems that prevent the direct exchange of gases or contaminants between the culture and the atmosphere [23, 121, 190].

Remembering the observations from nature, open systems are actually the simple copies of the lakes and lagoons that house the microalgae and they are widely spread around the globe long before the photobioreactor technology catches the attention even if the earlier cultivation studies tried to be held in the closed systems inside the laboratories. The key of the open systems is their economic feasibility with minimum expenditure by focusing on the more resistant species like *Spirulina*, *Dunaliella* or *Chlorella* which are actually highly selective species that can survive in nature under open environmental conditions [188, 191]. Considering the sub classes open systems can be divided into two main groups unmixed and mixed systems in which the dry biomass productivities can be around 0.1 and 0.5 g L^{-1} considering the dynamic outdoor conditions [186].

Unmixed systems are usually natural lagoons or constructed shallow (up to 50 cm deep) water pools without any mechanic mixing units, direct mimicking of nature where aeration and mixing is accomplished by wind and convection. Today one of the major *Dunaliella* production plans owned by BASF produces β-carotene with a productivity of about 1 g $m^{-2}d^{-1}$ in the natural lagoons of Hutt and Whyalla in Australia is a good example of the strength of these primitive systems in commercial scale [120].

Unmixed systems are quite limited with the natural lagoons or water ponds where naturally occurring microalgae serve the advantage for feasibility. Especially the land limitations and the need of better mixing to obtain a homogeneous environment regarding light and nutrients increased the use of mixed open ponds having an average surface to volume ratio around 10 m^{-1}. Considering the classification, two main designs related with the geometry can be seen: raceways and circular ponds [23, 186, 192]. Raceway ponds (can also be named as high rate algal ponds, which are developed in 1950's especially for wastewater treatment plants) are the most employed designs in the industry of microalgae cultivation since 1960s. They are around 30 cm deep, rectangular ponds with semicircular ends having a paddle wheel type mixer, which is mounted in the flow channels that are defined by separator walls. With a direct interaction by the changes in the outdoor conditions, annually dry biomass production can reach up to 22 g $m^{-2}d^{-1}$ [192 - 194]. Some of the conventional examples like the Earthrise Farms' raceways are covering a facility area about 440.000 m^2 with 5000 m^2 raceways each and some examples like in the Cyanotech have 2900 m^2 raceways in the facilities [192, 195, 196].

Circular ponds the other important design, circular ponds are less popular but they can be scaled up to 50 m diameter and 50.000 m^2 in area, which are mainly used in Japan, Taiwan and Indonesia. These ponds can have a depth of 30-70 cm and are basically mixed with a center pivoted turning blade [120, 197 - 199]. Some productions like in the case of *Oscillatoria* can reach a productivity of about 15 g $m^{-2}d^{-1}$ in dry weight basis. Today circular ponds using various impeller designs like grid plates or pitched blades are becoming more recognized especially in industrial production like South China [200].

Another design of open ponds (a more sophisticated one) designed to sustain high light utilization and turbulent mixing regime is the cascade (also called thin layer cascade or inclined) open systems. Cascade systems were developed in Trebon, Czech Republic in 1960's have the advantage to reach higher culture densities (up to 35 g L^{-1}) to the micro depths of less than 1 cm and having a higher surface area to volume ratios (100 m^{-1}). The design principle is based on the inclined solar stage where the cultures are interacting with light and air while flowing down and collected in the main storage tank where the main volume filled in. These systems are expensive compared to the other open pond designs but they can have chance in the less illuminated areas like in the central Europe [119, 120, 186, 197, 201].

Considering the hydrogen production where the need of a better controlled environment, open systems are not enough to satisfy the needs even if they can be helpful as a support system for sustainable biomass or taking a role in a biorefinery approach to fulfill the economic expectations. Thus, closed systems (or technically termed as photobioreactors - PBRs) should be highlighted. PBRs are systems where the culture is partially isolated from the surrounding environment. They can be constructed outdoors (or in green houses) to use solar energy but completely isolated from the atmosphere (no direct exchange of gas or contaminants with the outdoors).

Compared to open ponds, PBRs have a wide spectrum of designs, which gave them the ability to culture different species other than the highly selective ones. This is an important advantage to investigate more species that can be useful in microalgae industry especially considering the energy in the search of high producers [23, 197, 202]. PBRs have some key features that serve as an advantage when compared to open systems like the possibility of full control of the culture environment, better light and gas utilization, prevention of contamination risks, flexible for heterotrophic, mixotrophic or photoautotrophic cultivations with different modes like continuous, batch or semi-continuous to investigate various process parameters with proper modifications, all leading to a high productivities with a standard product [121, 203]. However, the cost of these bioreactors is the main disadvantage. Even if the economy is the main drawback, the chance to reach high productivities serves the motivation for the use of PBRs.

PBR as a system has two key features: the illumination stage (solar receiver or solar stage) and the culture reservoir. The high volume illumination stage is the area of light harvesting where the intensity levels play a vital role in optimum efficiency or risk of inhibition. Culture reservoir comprises a smaller volume that is usually light limited, houses degassing, control, mixing and circulation systems. Depending on the process these units can be separated or integrated leading to the major designs: flat panels (flat plates), tubular and fermenter types (tanks and columns) [23, 121].

Flat panel PBRs are vertical or inclined rectangular vessels, which are widely employed in laboratory and outdoors. They present a high surface to volume ratio resulting in a better light utilization and the fluid can be mixed hydraulically, pneumatically or mechanically in order to sustain a homogeneous culture environment with low shear stress [23, 204]. The height and width dimensions are usually 1.5 m x 2.5 m with an optical path of 0.1 m (light path can vary between 1 cm to 30 cm) [138, 203]. The modular characteristic makes these designs able to be oriented relative to the light (sun or artificial) for a better light harvesting. This ability also helps to control the outdoor cultivation regarding the risks due to high light intensity or temperature [205, 206]. From plastic materials to costly glasses, various materials can be used in the construction of the flat panels without a complicated construction process give them an economic feasibility and competitiveness for diverse applications from biofuels to high value chemicals with their high performance in terms of efficiency and productivity [207].

Tubular PBRs are other highly recognized designs with various geometries that are applicable for both laboratory and commercial scales. They are horizontal, vertical or even inclined oriented stack of transparent tubes with ≤ 0.1 m diameters (usually between 1 cm to 6 cm) attached by U-bends or manifolds, in which culture circulation and mixing accompanied by pumps or airlifts reaching linear velocities of 20 cm/s to 50 cm/s for the cultures [56, 138, 208]. They have surface to volume ratios of 80 m^{-1} to 100 m^{-1} as well as some very high ratios up to 2000 m^{-1} also reported especially in experimental designs. The tubular PBRs are well adapted for commercial facilities and some examples in Israel, Spain or Germany can reach up to 500 km of total tubing length with a special attention on the prevention for O_2 accumulation, light efficient orientation and high pumping power need especially considering the average energy consumption of around 2000 $W \cdot m^{-3}$ compared to 50 $W \cdot m^{-3}$ for bubble column and flat plate PBRs [23, 121, 187, 192, 209, 210].

Fermenter type PBRs can be classified into stirred tanks and aerated columns, which are basically the modified standard bioreactors used in bioprocesses usually by illumination systems for microalgae cultivation. Stirred tank PBRs are mechanically mixed, usually baffled, sparger or nozzle aerated, glass or steel tanks usually used in laboratory scales with a typical height to diameter ratio up to 3 for an adequate mixing. High disengagement zone separates the excess sparged gas and oxygen by photosynthesis from gassed liquid to gas phase. Relatively lower surface area to volume ratio and high shear stress due to mechanical agitation need to be considered especially in large scale applications. These PBRs can be illuminated internally or externally according to the design and application area and light source [3, 204, 205]. Airlift and bubble columns are the vertical PBRs with no moving mechanical parts (unless modified with internal mixers) having mass transfer coefficients about 0.006 s^{-1} and low power inputs for mixing by supplying aeration rates of 0.25 vvm [138, 205, 211]. Airlift PBRs can be classified into internal loop, external loop and divided column types by which the main key feature of forming circular patterns during mixing causing a flashing light effect on algal cells [207, 209]. Even if the circulation principle still lies in the density difference due to aerated liquid phase, the patterns in bubble columns are more chaotic without a defined trajectory like in airlifts and especially gas flow rates less than 60 m s^{-1} no circulation pattern is observed. Generally, the column PBRs have a height up to 2 m and diameter up to 50 cm to sustain enough circulation and light capture. The PBRs can be constructed by simple polyethylene bags or in more sophisticated designs using glass and steel constructions up to 500 L working capacity [3, 55, 64, 187, 205, 212].

Today the hydrogen production from microalgae still depends on the anoxic processes to trigger the microalgae shift their metabolism for hydrogen. Thus, open ponds can only be used (at least till the researches on oxygen tolerant productions are successful) for the biomass production that may later be transferred to anaerobic environment supplied in the PBRs for hydrogen production. When the focus is the hydrogen production, a special attention should be on the leakage and impermeability to prevent gas escape from the system without underestimating the need of high volume productions in other words the flexibility to scale-up.

Besides the leakage proof characteristics, construction materials should also be non-toxic against inhibition risk on the cells. In addition, they should be selected according to the needs of the designs in which some should be rigid and some should be flexible with a good ability to be attached and detached [117, 121, 209, 211]. In addition, produced hydrogen should be collected easily through gas tight ports for storage or direct utilization without increasing the head space biohydrogen partial pressure which decreases the productivity [202, 213]. Mixing should be enough to keep the homogeneous culture environment illuminated, keeping in mind the possible shear stresses and the wall growth depending on the strains, without disturbing anaerobic environment [211]. Some modifications with static mixers or light scattering particles are good examples to enhance the mixing and light efficiency at the same time without applying stress on the culture [210]. Even more sophisticated applications like Fresnel lenses, Himawari lenses, solar concentrators, optical fibers or light tubes were used for effective solar energy harvesting [121].

When considering outdoor productions, PBRs are well designed and equipped with the proper systems to control the temperature and light interactions due to the dynamic nature of the outdoor environment. High response rate temperature control systems can be mounted to prevent temperature fluctuations during the hydrogen production process. On the other hand, solar tracking systems and orientation modules can be constructed to harvest enough solar energy or protect from high intensities in order to sustain flexibility for different strains. PBRs should also be designed according to proper dimensions with a special emphasis on various engineering dimensions like surface to volume ratio or illuminated area ground surface area ratio [187, 207, 211, 214].

Scaling up is an overall process, which connects the laboratory and commercial scales taking into account all the steps from unit volume to total volume. The key factors for an efficient scale-up strategy covers various base points, which are well known for their importance in biotechnology design engineering like power input, $k_L a$, dissolved oxygen, surface to volume ratio, light supply coefficient, light dark cycle frequency, mixing time, Reynolds number, dissolved CO_2 concentration, blade tip velocity, optical path length or other dimensional similarity factors like height to width ratio [3]. Considering the basic PBR designs, the scale-up strategy usually focuses on the approach of modules. Rather than a dimensional increase in the single PBR (for example increasing the tube length or height of a flat panel) the increase in the number of modules (for example increasing the number of tubes or panels without changing their dimensions) are preferred keeping in mind the energy need. This approach can be a circumventing way to pass the construction and maintenance challenges such as durability, gas build-up, cleaning and sterilization, temperature control, uniform illumination, orientation, leakage, *etc*., which can be faced during dimensional increase of a PBR. This gives elasticity in operation against contamination and productivity losses with the ability to respond locally to a single module without disturbing all the system [187, 205]. Keeping in mind the basic design differences, each PBR type has their own specifications and flexibility for scale-up but for hydrogen production there are some major limits due to the need of anaerobic culture environment like the proper mixing, gas accumulation, head space pressure and light capture. Considering the types of PBRs, flat panels or tanks are easier for the adaptation to hydrogen production with their advantages like simple integration of a mixing system, easy design as modular units, easier implementation of light harvesting enhancers compared to long tubular PBRs [121].

MICROALGAL HYDROGEN PRODUCTION SUCCESS STORY: THE *CHLAMYDOMONAS* CASE

Considering the studies in photobiological hydrogen production, the major bottleneck is the sensitivity to oxygen due to the enzymes in charge. Strategies, such as applying PSII inhibitors (DCMU, SAL, DBMIB, C1-CCP), purging inert gas or using light-dark cycles, were tested in the previous years to overcome oxygen sensitivity. However, as they are not practical or even damaging to the cells, especially in case of inhibitors, they are not preferred for a sustainable production [3, 41, 214, 215].

The studies of Gaffron and Rubin on *Scendesmus obliquus* in 1940s showed that after a dark anaerobic adaptation period, these green microalgae can produce hydrogen in the dark at low rates and can even reach much higher rates under light [124, 216].

A two stage protocol of an indirect biophotolysis with *Chlamydomonas reinhardtii* (used to pass the oxygen sensitivity by separating temporarily the aerobic environment from hydrogen evolving anaerobic environment) is a success by providing a sustainable hydrogen production under light [41, 140]. Focusing on the microalgal hydrogen production by using two stage protocol, various applications (batch, semi-continuous, continuous or immobilized cultures) with different light regimes and light cycles are investigated for their potential [43, 46, 48, 217, 218].

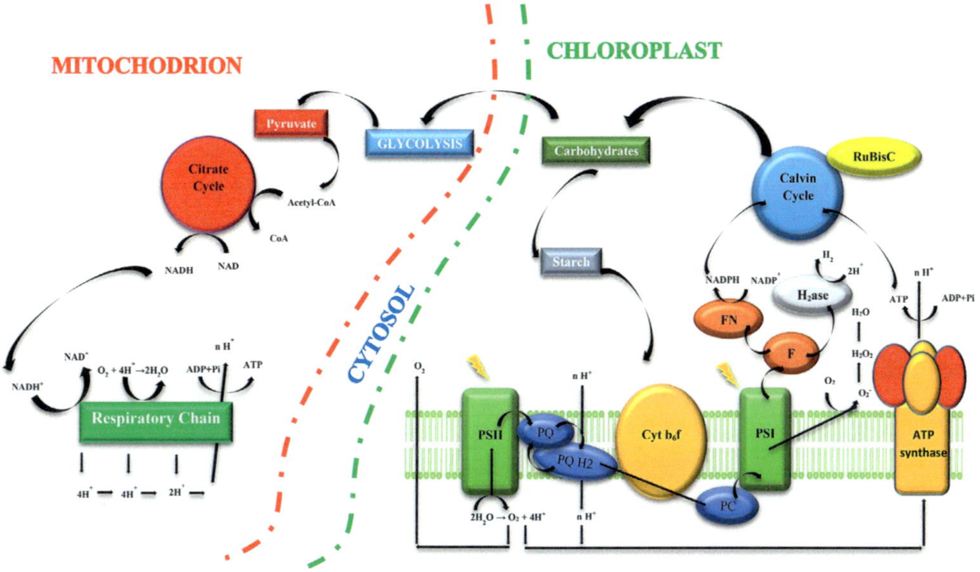

Fig. (3). Biohydrogen production routes in *Chlamydomonas reinhardtii*: Modified from [219, 220].

The biochemistry of hydrogen production (Fig. **3**) [219, 220] is an interrelated process between oxygenic photosynthesis, mitochondrial respiration, catabolism of endogenous substrate, and electron transport by the [Fe]-hydrogenase pathway. The key in this protocol, leading to a sustainable production process considering these integrated complex routes, is to block the PSII repair cycle under light by sulfur deprivation. This is based on the fact that sulfur a metabolic router, is essential for the amino acids (cysteine and methionine) of the D1 a 32 kD protein in the PSII reaction center that is responsible for the oxygen evolution [41, 150, 215, 221]. Basically this process is a two-step reversible process. The cells can be reacclimated in aerobic culture for biomass, all of which are under light that forces some metabolic shift in the cell to produce hydrogen which can be observed in several stages. First step is the standard photosynthetic aerobic cultivation, where tri acetate phosphate (TAP) medium is used under aerobic conditions (the biomass production stage). Following this step, the cells are transferred, usually in their mid-logarithmic phase, from the TAP medium to the sulfur deficient TAP-S medium (usually the biomass is separated and washed few times with TAP-S medium to wash out the standard TAP medium) during which the cells are now sealed for strict anaerobiosis (sometimes bubbled with nitrogen before closing the PBR to reach anaerobiosis faster but usually just sealing is preferred). The cells in this new medium and environment start to produce hydrogen [189, 222].

Considering the metabolic activities, the aerobic cultivation leads to photosynthetic reactions resulting in the CO_2 fixation (or in case of TAP medium acetate also acts as a supportive carbon source depending on the cultivation mode to be photoautotrophic, photoheterotrophic or photomixotrophic), O_2 evolution by the H_2O oxidation and the biomass increase. After the transfer to anaerobic environment, a lag time about a day in sealed cultures will be observed during which the residual PSII activity drops down the respiration limit of mitochondria triggering anaerobiosis, which can be monitored through the decrease in the fluorescence measurements. In the first hours of oxygen limited environment cells stop growth and start to breakdown proteins, especially RuBisCo, resulting an increase in storage compounds like starch and lipids (which are probably the electron sinks for the decrease of PSII electrons) and shift to hydrogen metabolism for survival [41, 150, 181]. Electrons needed during the hydrogen production are supplied from water splitting by the residual PSII activity and from the breaking down of the stored starch linked with bidirectional [FeFe] hydrogenase (which also has hydrogen uptake activity) related with PSI activity [140, 176, 181]. Actually the anaerobiosis by S deprivation is an energy dependent process, which needs a carbon source for respiration. The main source during the lag phase is acetate, but as the environment shifts anaerobic, acetate utilization stops and does not play a role in the hydrogen production showing the major role of acetate is to support cellular respiration and anaerobiosis establishment [41, 223].

After reaching the maximum levels within 3-4 days [41, 46, 55] hydrogen production starts to slow down and ceased even if some amount can be detected in longer terms up to 30 days [48]. This prolonged production is probably related with the minimum need of energy to survive. During the absence of a sink (in other words lack of O_2) for electrons from H_2O, electron transport in thylakoids would stop, blocking the photosynthetic phosphorylation. Under anaerobic conditions, oxidative phosphorylation in mitochondria of the green algae would also be inhibited, causing the cells to lose another ATP source. However, hydrogenase activity and H_2 gas production trigger a slow electron transport flow in the thylakoid membrane that is connected to electron transport in mitochondria, sustaining ATP for the cell just enough for survival under these stress conditions [41, 46, 224]. Considering the decrease in hydrogen production, the reason is related with the negative effect of sulfur deprivation causing the conversion of active Q_B reducing PSII centers into inactive Q_B non-reducing PSII centers addition to the accumulation of the inhibiting products like ethanol and formate under anaerobic conditions [44, 181].

Besides the temporal separation of oxygen machinery from hydrogen production, genetic modifications can be used to circumvent the oxygen stress. Basically, molecular approaches are focused on the photosynthesis machinery, hydrogenase enzymes, starch metabolism, D1 protein and sulfur metabolism.

Photosynthesis in microalgae has a primary role in the regulation of biohydrogen generation in terms of energy supply to hydrogen generating machinery. The photosynthetic activity directly regulates the number of the electron shuttled through the electron transport chain to the H^+ ions. The photosynthetic activity of the culture is a parameter of light harvested *via* light harvesting complexes (LHC) of PSI and PSII mechanism [225]. In addition, the chlorophyll antenna size is an issue to harvest the maximum light. When light harvesting capacity is up or down regulated *via* genetic manipulation, the photosynthetic activity is increased and hydrogen generation capacity is indirectly elevated as well [61]. Considering large antennas, which can have 600 chlorophyll molecules in the reaction centers, a minimum number of 132 chlorophylls can be enough for the cell which makes the antenna truncation a valuable approach [221]. Antenna size truncation generates a dilution effect and increases the photosynthetic efficiency up to 3 fold [221, 226] and a list of truncated chlorophyll antenna transformants *tla*1, *tla*X and *tla*-CW^+ of *Chlamydomonas* is reported with tla1 reaching higher biohydrogen productivities when immobilized [148, 214, 221]. Site directed mutagenesis, gene silencing, knock in or knock out mutations are designed to manipulate photosynthesis to trigger hydrogen production; however, it should also be noted that the light driven energy is not only utilized to generate hydrogen but the cells use energy for reproduction, growth and maintenance [61, 137, 187, 227]. Thus, hydrogen is only a small metabolic regulation as a response to anaerobiosis.

Hydrogenase enzymes are the main catalysts to convert free H^+ ions into hydrogen in gas form. The activity of the hydrogenase enzyme is dependent on the availability of oxygen within the culture chamber [228]. The reversible deactivation of hydrogenase enzyme occurs when oxygen is sensed and it travels through the micro gas channels of hydrogenase enzymes active site, H cluster. The competitive race in the micro channels and H-cluster is deactivated *via* oxygen transport [228]. To sustain an effective hydrogen production, the oxygen sensitivity problem should be overcome. To activate hydrogenase enzyme, a serial reactions are required in the name of gene expression. There is a process called "maturation", which expresses maturation proteins to form active enzyme. The responsible genes for hydrogenase enzyme is *Hyd*A1 and *Hyd*A2 [179, 229]; among them, *Hyd*A1 is the main gene responsible for enzyme expression. However, RNA silencing methods revealed that there is another gene Hyd3 required for the activation of the enzyme [230]. The enzyme activity of eukaryotic and prokaryotic enzymes and maturation processes differ due to the differences of the taxonomic diversity. However, the [FeFe] hydrogenases are the most active ones in catalytic units, so considering *Chlamydomonas reinhardtii* a eukaryotic hydrogen producing model the crystal structure of the [FeFe] hydrogenases are of importance to distinguish from the bacterial enzymes.

Hydrogen production requires electron transfer and since the cells cannot sustain oxygenic respiration metabolism the starch anabolism is re-organized [170]. With the signals of anaerobiosis, cells start to synthesize great amount of starch to sustain energy requiring steps for cell survival [231]. Starch is the main reserve to start the hydrogen production, mutation studies show that strains with no starch accumulation capabilities has a decreased level of hydrogen generation. One of the essentials in starch metabolism is that in nitrogen limiting culture conditions there is a completion of fatty acid synthesis pathway and starch accumulation. In that case, strain selection becomes a priority for hydrogen generation [53, 232]. Starch accumulating strains rather than fatty acids reported to generate more hydrogen. In genetic level of understanding, isoamylase gene controls the starch metabolism in the cells [233]. The overexpression of isoamylase gene also alters the starch accumulation in a positive way that cells become capable to synthesize hydrogen more effectively. The combinational effect of light irradiation to starch accumulation also changes the hydrogen levels. Tla1 mutant strains which are capable to grow under high light conditions [223, 226]. Cyclic electron flow around PSI active center, light harvesting capacity of the cells, RuBisCo enzyme activity and stress conditions mainly affect the hydrogen generation [61, 234]. A comprehensive understanding to hydrogen relation with starch metabolism can be altered *via* genetic manipulation. Since the pathways are well known, more energy required pathways like flagellar movement can be silenced, or fatty acid metabolism can be suppressed [235].

On the other hand, the D1 protein expression rate and PSII repair cycle is the main trigger point of hydrogen generation in microalgae cells [236]. Turnover rate of D1 protein can be tuned *via* classical mutagenesis approaches to understand the role in cell cycle and hydrogen generation as well. PSII and relation with D1 protein synthesis is directly related with the hydrogenase activation, since oxygen generation due to water pyrolysis occurs at the PSII reaction center. D1 protein encoding gene *psb*A is constantly expressed and D1 protein is continuously synthesized when photosynthesis is regularly sustained [237]. The evidence shows that *psb*A gene is composed of highly conserved sequences. The genetic structure of the *psb*A gene is composed of 4 introns [237]. The introns in the gene structure of *psb*A indicate that intron-less mutants have a 1.1 kb of length meanwhile introns were take a space of approximately 5.6 kb. The results also showed that the photosynthetic activity and growth of the IL-*Chlamydomonas* mutant strain was comparable with the wild type one [237, 238]. This study is a key one to understand the intron and exon relation and translation of these genetic parts into a mature D1 protein. D1 protein is also related with the sulfur metabolism [239]. As D1 protein contains sulfur dependent amino acids, lack of sulfated salts in the culture decrease the amount of D1 protein synthesized in the cells. Sulfate is transported inside the cell *via* sulfate transport genes (*Sul*P) encoded in the nucleus [240]. There are two sulfate transporter in the cell: one is at the cell membrane and the other is localized at the chloroplast membrane [241]. The sulfate transport system also alters the cells hydrogen generation capacity. If sulfate transport at the chloroplast membrane layer is down regulated, the D1 protein synthesis will decrease to a level so hydrogen generation may occur. Considering the key role of the D1 protein, strains of *Chlamydomonas reinhardtii* with different amino acid deletions have reached maximum daily productivities up to 7.1 mL $L^{-1}h^{-1}$ [50, 69]. The modifications are done in enzymatic basis both for maturation proteins or mature enzyme itself in order to achieve more tolerant enzyme active site to oxygen [148, 149]. Other mutants with low sulfate permease which is responsible for the transfer of the sulfate to chloroplasts will be a potential high producer candidate [140].

Targeted genome engineering is one of the latest tools to manipulate organism's genome [242]. The targeted genome engineering is accomplished *via* endonucleases to cleavage DNA at the specific regions of the genome [243]. Endonucleases create a double stranded breaks in the genome [243] Gene of interest is directed to the double stranded breaks and inserted into the genome. Targeted genome engineering tools show success with the engineering of bacteria, yeast, plant and animal genomes. There are 3 specific methods: (i) Zinc Finger Nucleases (ZFNs); (ii) Transcription Activator-Like Effector Nucleases (TALENs); and (iii) Clustered Regularly Interspaced Short Palindromic Repeats (CRISPR) [244]. Tools are designed to use cells own DNA repair metabolism. They are claimed to be easy and are able to control and monitor. They help in rapid selection of mutants, designing high-throughput analysis libraries and are able to deliver stable mutations without the requirement of specific selectivity design modules. ZFNs and TALENs require enzyme assistance to deliver DNA [243, 245, 246]. However, CRISPR uses guide RNA with the assistant of Cas9 protein and recognize PAM sequences in the genome where double stranded breaks are created and gene of interest is ligated [247, 248]. Targeted genome engineering tools and their application to the microalgae are limited yet promising. The important decision is rather use chloroplast genome or nuclear genome to engineer and edit. TALEN and ZFN are found to be effective for algal gene editing [246, 249]. However, it is suggested to optimize codons before editing due to the unknown gene silencing metabolism of the microalgae. There is a recent study showing the effectiveness on creating the knock in or knock out mutations in *Chlamydomonas reinhardtii via* CRISPR [248, 250 - 252].

The future of targeted genome engineering especially using CRISPR to re-organize the hydrogen production in microalgae is becoming the key topic. The ultimate goal is breeding a strain or strains to produce hydrogen under oxygen containing conditions and increasing the hydrogen production efficiency to evaluate microalgal hydrogen as a future promising energy source.

On the other hand, new genes from other organisms can help for the progress for a high producer like in the case of HUP1 hexose uptake protein from *Chlorella kessleri* introduced to *Chlamydomonas* giving the ability to assimilate glucose in dark heterotrophic environment [148], or using marine phage PsaJF subunit in *Synechocytis* enabled its transformant to accept electrons from alternative sources [253], or introducing the hydrogenase enzymes from *Clostridium pasteurianum* into *Synechococcus* sp. resulted in the active FeFe hydrogenase production are good examples of the progress [150].

MICROALGAL HYDROGEN PRODUCTION: KEYS FOR THE FUTURE

The future of microalgal hydrogen production will depend on the realistic approach towards the topic considering all the potential key factors and possibilities for progress. There is still more barriers to pass to reach high productivities. However, keeping in mind the vast number of species, the chance to catch a high producer is a good motivation for the researchers. Taking into account the theoretical estimations sustainable production of 20 g hydrogen per m^2 of culture area can be a possibility for the future. However, this amount could not be reached just by today's conventional processes a new approach integrating genetic engineering tools, omics and systems biology having the projection from a single cell to a large scale production with a special emphasis on the possible high production factors like reduced amount of chlorophyll per antenna, high photosynthetic efficiency, high respiration rates, high carbohydrate accumulation ability to survive under stress conditions and high xanthophyll cycle activity [41, 69, 182]. Basically, there are two main spots in the microalgal hydrogen realization: the cell and the production systems both of which are directly related to the economy.

Starting with the cell, for the ultimate target the search for high producer and resistant species will be a research platform for more enhancement through biochemical or molecular modifications focused on key factors like single step hydrogen production ability, growth rate, antenna truncation, D1 protein expression, enzyme maturation, photosynthetic activity, respiration mechanism, hydrogen related gene sequencing, starch metabolism, sources of electron sinks, thermal resistance, oxygen sensitivity, light adaptation flexibility or water dependence (survive in different sources like salty, brackish or waste) to reach a stable and robust nature. These factors will play a key role as the bridge from laboratory scale to commercial scale that will help to reach a competitive microalgal hydrogen production [3, 64, 69, 181, 192, 205, 206, 213, 254].

Considering the production systems, the future will be related with the new materials and innovative designs, which should be effective and cost efficient. According to the thermodynamic analysis two stage hydrogen production in PBRs can reach an efficiency between 1.8% and 4.1%, which is quite lower than the possible single step production that can reach a value between 2% and 5.6% [255]. On the other hand, in the conversion of solar energy to biohydrogen by direct biophotolysis, the losses due to the reflection and scattering, radiation outside photosynthetic active radiation region and non-photochemical reactions decrease the efficiency to a theoretical limit of 13±1%. However, taking into account the two step process like sulfur deprivation, addition of cell maintenance and down regulation of PSII steps will further decrease to only 1.0±0.5% so single stage production should be an important target for efficient production [121]. Hydrogen production will need a focus on the basic parameters inside the PBRs so the robust and leakage proof materials should be focused with proper production techniques for flexible application and easy construction keeping in mind the scale-up. Thinking the other major factors like mixing, light harvesting or culture environment control modified systems like: compact PBRs (with easier management and less land dependent), static or dynamic mixing systems (that can be applied to different scales or designs with minimal modifications), special illumination systems (like optical fibers, light scattering particles or lenses targeting more efficient light capture) are designed. Also energy efficient temperature control systems that are easy to apply to different designs is becoming a key factor for feasible processes [121]. On the other hand, flexible sterilization and cleaning systems (that can be effective to prevent contamination and fouling without underestimating the energy efficiency), highly efficient recycle and waste management systems (that are well integrated with the overall process), progress in hydrogen employment grids like direct usage (employing fuel cells) or storage in special systems (like storage; in high-pressure gas cylinders, as liquefied hydrogen by cryogenic applications, by adsorption on carbon structures, by absorption in host metals and metallic hydrides, in complex compounds-such as borohydrides and, by metals and complexes together with water) for later usage, supplying energy need from renewable sources (like wind, solar or biomass with a strong integration through all the process steps), well adapted upstream and downstream processes (to get benefit from all side products) will be very important for the future applications to be more competitive thinking in a biorefinery concept which totalize the process with all its subunits (like potential alternatives for nutrients, by-products especially other potential fuels (methane, ethanol, biodiesel) or high value chemicals (pigments, lipids, vitamins *etc.*) and even wastes to reach the light to fuel target) [121, 187, 218, 236, 256 - 258].

When the topic comes to the point of commercialization, all pathways lead to economy issues. With a special emphasis on a very dynamic and hard to penetrate market like energy the competition is simply wild with every progressive attempt will count as points for the lead.

Hydrogen (an energy carrier which is not bound to carbon like fossil fuels) has the highest energy content per unit weight compared to any known fuel (143 GJ $tonne^{-1}$) and the claim of biological hydrogen to be sustainable, environmentally friendly and feasible should be validated by the realistic numbers forecasting the market. The capital cost in biohydrogen production comprise the costs of subunits like PBRs (42%), hydrogen collection assembly (34%), microalgae feed assembly (4%), recycle and pumping systems (13%) and other costs like control systems, consumables (7%). According to today's analysis, the microalgal production should target a cost for both biomass ($100\$/m^2$) and for biofuel ($15\$/m^2$) production to be competitive in the retail price [121, 259, 260]. The economic perspective about microalgae actually root back to the conventional food, feed, cosmeceuticals and pharmaceuticals industries where the cost of production of the raw biomass changes from 1 to 7 \$/kg. More specific products like valuable chemicals can reach up to 1000 \$/kg or even higher like in the case of astaxanthin which can rise up to 10.000 \$/kg depending on the production system and downstream processes for the desired product [3]. The estimation for microalgal biohydrogen without considering the storage costs can be 1.38 \$/kg which is lower than the estimate of USDE and NREL having set a target to 2.99 \$/kg biohydrogen by using oxygen tolerant microalgae [150, 261]. These numbers may not be competitive for today compared to the fossil fuels but the sharp oscillations in the fossil fuel market especially due to the pressure of the chaos in the potential fields around the globe force the industry to search for the alternatives which will help the commercial viability of microalgal biohydrogen over the long term.

CONCLUSION

Admit or not the catalyzing effect of fossil era helped the humanity to leap to a higher stage of civilization but the change is too rapid to give the nature an adaptation time leading to a complex and out of control web of problems. Today the key word of sustainability should also comprise the word continuity which is the must for the new energy approaches and it is not easy and achievable to compensate the fossil energy dependency directly to renewable or biobased but humanity should start from somewhere. Defining this, somewhere, for the future energy of microalgal biohydrogen can play an important role if the enhancements on the key factors like productivity, alternative production methods, scale-up, molecular approaches, photosynthetic efficiency, cost or commercialization can be accomplished. With a realistic thinking, using microalgae as a source of energy can help as a support in today's energy grid but with the new progresses in science, they can be one of the strongest candidates to carry more responsibility in the future as a fossil fuel substitute.

CONFLICT OF INTEREST

The author declares no conflict of interest, financial or otherwise.

ACKNOWLEDGEMENTS

Declared none.

REFERENCES

[1] B. Hankamer, F. Lehr, J. Rupprecht, J.H. Mussgnug, C. Posten, and O. Kruse, "Photosynthetic biomass and H2 production by green algae: from bioengineering to bioreactor scale-up", *Physiol. Plant.*, vol. 131, no. 1, pp. 10-21, 2007.
[http://dx.doi.org/10.1111/j.1399-3054.2007.00924.x] [PMID: 18251920]

[2] R.H. Wijffels, O. Kruse, and K.J. Hellingwerf, "Potential of industrial biotechnology with cyanobacteria and eukaryotic microalgae", *Curr. Opin. Biotechnol.*, vol. 24, no. 3, pp. 405-413, 2013.
[http://dx.doi.org/10.1016/j.copbio.2013.04.004] [PMID: 23647970]

[3] N.L. Panwar, S.C. Kaushik, and S. Kothari, "Role of renewable energy sources in environmental protection: A review", *Renew. Sustain. Energy Rev.*, vol. 15, pp. 1513-1524, 2011.
[http://dx.doi.org/10.1016/j.rser.2010.11.037]

[4] A.K. Azad, M.G. Rasul, M.M. Khan, S.C. Sharma, and M.A. Hazrat, "Prospect of biofuels as an alternative transport fuel in Australia", *Renew. Sustain. Energy Rev.*, vol. 43, pp. 331-351, 2015.
[http://dx.doi.org/10.1016/j.rser.2014.11.047]

[5] B.S. Harish, M.J. Ramaiah, and K. Babu, "Bioengineering strategies on catalysis for the effective production of renewable and sustainable energy", *Renew. Sustain. Energy Rev.*, vol. 51, pp. 533-547, 2015.
[http://dx.doi.org/10.1016/j.rser.2015.06.030]

[6] M.A. Scaife, A. Merkx-Jacques, D.L. Woodhall, and R.E. Armenta, "Algal biofuels in Canada: Status and potential", *Renew. Sustain. Energy Rev.*, vol. 44, pp. 620-642, 2015.
[http://dx.doi.org/10.1016/j.rser.2014.12.024]

[7] A. Singh, P.S. Nigam, and J.D. Murphy, "Renewable fuels from algae: an answer to debatable land based fuels", *Bioresour. Technol.*, vol. 102, no. 1, pp. 10-16, 2011.
[http://dx.doi.org/10.1016/j.biortech.2010.06.032] [PMID: 20615690]

[8] S.K. Bardhan, S. Gupta, M.E. Gorman, and M.A. Haider, *Biorenewable chemicals : Feedstocks, technologies and the conflict with food production*, 2015.

[9] A. Cancela, S. Urréjola, A. Sánchez, R. Maceiras, and M. Rodrı, *Macroalgae : Raw material for biodiesel production*, 2011.

[10] S.G. Hays, and D.C. Ducat, "Engineering cyanobacteria as photosynthetic feedstock factories", *Photosynth. Res.*, pp. 1-11, 2014.
[PMID: 24526260]

[11] R.P. John, G.S. Anisha, K.M. Nampoothiri, and A. Pandey, "Micro and macroalgal biomass: a renewable source for bioethanol", *Bioresour. Technol.,* vol. 102, no. 1, pp. 186-193, 2011. [http://dx.doi.org/10.1016/j.biortech.2010.06.139] [PMID: 20663661]

[12] C.S. Goh, and K.T. Lee, *A visionary and conceptual macroalgae-based third-generation bioethanol (TGB) biorefinery in Sabah, Malaysia as an underlay for renewable and sustainable development*, 2010.

[13] H.P. Makkar, G. Tran, V. Heuzé, S. Giger-Reverdin, M. Lessire, F. Lebas, and P. Ankers, "Seaweeds for livestock diets: A review", *Anim. Feed Sci. Technol.,* vol. 212, pp. 1-17, 2016. [http://dx.doi.org/10.1016/j.anifeedsci.2015.09.018]

[14] H. Chen, D. Zhou, G. Luo, S. Zhang, and J. Chen, "Macroalgae for biofuels production: Progress and perspectives", *Renew. Sustain. Energy Rev.,* vol. 47, pp. 427-437, 2015. [http://dx.doi.org/10.1016/j.rser.2015.03.086]

[15] H.J. Bixler, and H. Porse, "A decade of change in the seaweed hydrocolloids industry", *J. Appl. Phycol.,* vol. 23, no. 3, pp. 321-335, 2011. [http://dx.doi.org/10.1007/s10811-010-9529-3]

[16] E. Forján, F. Navarro, M. Cuaresma, I. Vaquero, M.C. Ruíz-Domínguez, Ž. Gojkovic, M. Vázquez, M. Márquez, B. Mogedas, E. Bermejo, S. Girlich, M.J. Domínguez, C. Vílchez, J.M. Vega, and I. Garbayo, "Microalgae: Fast-Growth Sustainable Green Factories", *Crit. Rev. Environ. Sci. Technol.,* vol. 45, no. 16, pp. 1705-1755, 2015. [http://dx.doi.org/10.1080/10643389.2014.966426]

[17] A. Parmar, N. Kumar, A. Pandey, E. Gnansounou, and D. Madamwar, *Bioresource Technology Cyanobacteria and microalgae : A positive prospect for biofuels.* vol. Vol. 102. , 2011, pp. 10163-10172.

[18] M. Kumar, S. Kaushik, and P. Goswami, "Biomass and Bioenergy Cyanobacteria : A metabolic power house for harvesting solar energy to produce bio-electricity and biofuels", *Biomass Bioenergy,* vol. 90, pp. 187-201, 2016. [http://dx.doi.org/10.1016/j.biombioe.2016.03.043]

[19] R. Ramanan, B.H. Kim, D.H. Cho, H.M. Oh, and H.S. Kim, "Algae-bacteria interactions: Evolution, ecology and emerging applications", *Biotechnol. Adv.,* vol. 34, no. 1, pp. 14-29, 2016. [http://dx.doi.org/10.1016/j.biotechadv.2015.12.003] [PMID: 26657897]

[20] "P. Drews-, R. G. Colares, P. Machado, M. de Faria, A. Detoni, and V. Tavano, "Microalgae classification using semi-supervised and active learning based on Gaussian mixture models", *J. Braz. Comput. Soc.,* vol. 19, no. 4, pp. 411-422, 2013. [http://dx.doi.org/10.1007/s13173-013-0121-y]

[21] J.L. Harwood, and I.A. Guschina, "The versatility of algae and their lipid metabolism", *Biochimie,* vol. 91, no. 6, pp. 679-684, 2009. [http://dx.doi.org/10.1016/j.biochi.2008.11.004] [PMID: 19063932]

[22] S. Hemaiswarya, R. Raja, R. Ravikumar, and I.S. Carvalho, *Microalgae Taxonomy and Breeding*, 2013. [http://dx.doi.org/10.1079/9781845938857.0044]

[23] O. Pulz, "Photobioreactors: production systems for phototrophic microorganisms", *Appl. Microbiol. Biotechnol.,* vol. 57, no. 3, pp. 287-293, 2001. [http://dx.doi.org/10.1007/s002530100702] [PMID: 11759675]

[24] A. Singh, P.S. Nigam, and J.D. Murphy, "Mechanism and challenges in commercialisation of algal biofuels", *Bioresour. Technol.,* vol. 102, no. 1, pp. 26-34, 2011. [http://dx.doi.org/10.1016/j.biortech.2010.06.057] [PMID: 20609580]

[25] A. Bahadar, and M.B. Khan, "Progress in energy from microalgae", *RE:view,* vol. 27, pp. 128-148, 2013.

[26] S.S. Oncel, "Microalgae for a macroenergy world", *Renew. Sustain. Energy Rev.,* vol. 26, pp. 241-264, 2013.
[http://dx.doi.org/10.1016/j.rser.2013.05.059]

[27] V. Singh, D.K. Chaudhary, I. Mani, and P.K. Dhar, "Recent advances and challenges of the use of cyanobacteria towards the production of biofuels", *Renew. Sustain. Energy Rev.,* vol. 60, pp. 1-10, 2016.
[http://dx.doi.org/10.1016/j.rser.2016.01.099]

[28] A. Midilli, M. Ay, I. Dincer, and M.A. Rosen, "On hydrogen and hydrogen energy strategies: I: current status and needs", *Renew. Sustain. Energy Rev.,* vol. 9, no. 3, pp. 255-271, 2005.
[http://dx.doi.org/10.1016/j.rser.2004.05.003]

[29] A. Midilli, M. Ay, I. Dincer, and M.A. Rosen, "On hydrogen and hydrogen energy strategies II: future projections affecting global stability and unrest", *Renew. Sustain. Energy Rev.,* vol. 9, no. 3, pp. 273-287, 2005.
[http://dx.doi.org/10.1016/j.rser.2004.05.002]

[30] A. Biofuels, and L. Mcgraw, *The Ethics of Adoption and Development of Case Study 1 of the Adoption and Development of Energy Technologies.* State of the Art Review, 2009.

[31] S. Taikhao, S. Junyapoon, A. Incharoensakdi, and S. Phunpruch, "Factors affecting biohydrogen production by unicellular halotolerant cyanobacterium Aphanothece halophytica", *J. Appl. Phycol.,* vol. 25, no. 2, pp. 575-585, 2013.
[http://dx.doi.org/10.1007/s10811-012-9892-3]

[32] S.A. Markov, A.D. Thomas, M.J. Bazin, and D.O. Hall, "Photoproduction of hydrogen by cyanobacteria under partial vacuum in batch culture or in a photobioreactor", *Int. J. Hydrogen Energy,* vol. 22, no. 5, pp. 521-524, 1997.
[http://dx.doi.org/10.1016/S0360-3199(96)00134-6]

[33] A.S. Tsygankov, L.T. Serebryakova, D.A. Sveshnikov, K.K. Rao, I.N. Gogotov, and D.O. Hall, "Hydrogen photoproduction by three different nitrogenases in whole cells of Anabaena variabilis and the dependence on pH", *Int. J. Hydrogen Energy,* vol. 22, no. 9, pp. 859-867, 1997.
[http://dx.doi.org/10.1016/S0360-3199(96)00242-X]

[34] H. Masukawa, M. Mochimaru, and H. Sakurai, "Hydrogenases and photobiological hydrogen production utilizing nitrogenase system in cyanobacteria", *Int. J. Hydrogen Energy,* vol. 27, no. 11, pp. 1471-1474, 2002.
[http://dx.doi.org/10.1016/S0360-3199(02)00125-8]

[35] P. Lindblad, K. Christensson, P. Lindberg, A. Fedorov, F. Pinto, and A. Tsygankov, "Photoproduction of H2 by wildtype Anabaena PCC 7120 and a hydrogen uptake deficient mutant: from laboratory experiments to outdoor culture", *Int. J. Hydrogen Energy,* vol. 27, no. 11, pp. 1271-1281, 2002.
[http://dx.doi.org/10.1016/S0360-3199(02)00111-8]

[36] J.H. Yoon, J. Hae Shin, M-S. Kim, S. Jun Sim, and T.H. Park, "Evaluation of conversion efficiency of light to hydrogen energy by Anabaena variabilis", *Int. J. Hydrogen Energy,* vol. 31, no. 6, pp. 721-727, 2006.
[http://dx.doi.org/10.1016/j.ijhydene.2005.06.023]

[37] S.F. Salleh, A. Kamaruddin, M.H. Uzir, and A.R. Mohamed, "Effects of cell density, carbon dioxide and molybdenum concentration on biohydrogen production by Anabaena variabilis ATCC 29413", *Energy Convers. Manage.,* vol. 87, pp. 599-605, 2014.
[http://dx.doi.org/10.1016/j.enconman.2014.06.069]

[38] S.A. Markov, E.S. Protasov, V.A. Bybin, E.R. Eivazova, and D.I. Stom, "Using immobilized cyanobacteria and culture medium contaminated with ammonium for H2 production in a hollow-fiber photobioreactor", *Int. J. Hydrogen Energy,* vol. 40, no. 14, pp. 4752-4757, 2015.
[http://dx.doi.org/10.1016/j.ijhydene.2015.02.053]

[39] S.F. Salleh, A. Kamaruddin, M.H. Uzir, A.R. Mohamed, and A.H. Shamsuddin, "Light irradiance and spectral distribution effects on cyanobacterial hydrogen production", *IOP Conf. Ser. Earth Environ. Sci.*, vol. vol. 32, 2016p. 12046 [http://dx.doi.org/10.1088/1755-1315/32/1/012046]

[40] K. Miyamoto, S. Ohta, Y. Nawa, Y. Mori, and Y. Miura, "Hydrogen Production by a Mixed Culture of a Green Alga, Chlamydomonas reinhardtii and a Photosynthetic Bacterium, Rhodospirilum rubrum", *Agric. Biol. Chem.*, vol. 51, no. 5, pp. 1319-1324, 1987.

[41] A. Melis, L. Zhang, M. Forestier, M.L. Ghirardi, and M. Seibert, "Sustained photobiological hydrogen gas production upon reversible inactivation of oxygen evolution in the green alga Chlamydomonas reinhardtii", *Plant Physiol.*, vol. 122, no. 1, pp. 127-136, 2000. [http://dx.doi.org/10.1104/pp.122.1.127] [PMID: 10631256]

[42] A. Tsygankov, S. Kosourov, M. Seibert, and M.L. Ghirardi, "Hydrogen Production Under Continous Illumination by Sulfer-deprived, Synchronous Chlamydomonas reinhardtii Cultures", *Int. J. Hydrogen Energy*, vol. 27, pp. 1239-1244, 2002. [http://dx.doi.org/10.1016/S0360-3199(02)00108-8]

[43] T. Laurinavichene, I. Tolstygina, and A. Tsygankov, "The effect of light intensity on hydrogen production by sulfur-deprived Chlamydomonas reinhardtii", *J. Biotechnol.*, vol. 114, no. 1-2, pp. 143-151, 2004. [http://dx.doi.org/10.1016/j.jbiotec.2004.05.012] [PMID: 15464608]

[44] S. Kosourov, V. Makarova, A.S. Fedorov, A. Tsygankov, M. Seibert, and M.L. Ghirardi, "The effect of sulfur re-addition on H(2) photoproduction by sulfur-deprived green algae", *Photosynth. Res.*, vol. 85, no. 3, pp. 295-305, 2005. [http://dx.doi.org/10.1007/s11120-005-5105-0] [PMID: 16170632]

[45] T.V. Laurinavichene, A.S. Fedorov, M.L. Ghirardi, M. Seibert, and A.A. Tsygankov, "Demonstration of sustained hydrogen photoproduction by immobilized, sulfur-deprived Chlamydomonas reinhardtii cells", *Int. J. Hydrogen Energy*, vol. 31, no. 5, pp. 659-667, 2006. [http://dx.doi.org/10.1016/j.ijhydene.2005.05.002]

[46] S. Kosourov, E. Patrusheva, M.L. Ghirardi, M. Seibert, and A. Tsygankov, "A comparison of hydrogen photoproduction by sulfur-deprived Chlamydomonas reinhardtii under different growth conditions", *J. Biotechnol.*, vol. 128, no. 4, pp. 776-787, 2007. [http://dx.doi.org/10.1016/j.jbiotec.2006.12.025] [PMID: 17275940]

[47] I.V. Tolstygina, T.K. Antal, S.N. Kosourov, T.E. Krendeleva, A.B. Rubin, and A.A. Tsygankov, "Hydrogen production by photoautotrophic sulfur-deprived Chlamydomonas reinhardtii pre-grown and incubated under high light", *Biotechnol. Bioeng.*, vol. 102, no. 4, pp. 1055-1061, 2009. [http://dx.doi.org/10.1002/bit.22148] [PMID: 18985615]

[48] S. Oncel, and F. Vardar-Sukan, "Photo-bioproduction of hydrogen by Chlamydomonas reinhardtii using a semi-continuous process regime", *Int. J. Hydrogen Energy*, vol. 34, no. 18, pp. 7592-7602, 2009. [http://dx.doi.org/10.1016/j.ijhydene.2009.07.027]

[49] S.N. Kosourov, and M. Seibert, "Hydrogen photoproduction by nutrient-deprived Chlamydomonas reinhardtii cells immobilized within thin alginate films under aerobic and anaerobic conditions", *Biotechnol. Bioeng.*, vol. 102, no. 1, pp. 50-58, 2009. [http://dx.doi.org/10.1002/bit.22050] [PMID: 18823051]

[50] C. Faraloni, and G. Torzillo, "Phenotypic characterization and hydrogen production in chlamydomonas reinhardtii Qb-binding D1-protein mutants under sulfur starvation: Changes in chl fluorescence and pigment composition", *J. Phycol.*, vol. 46, no. 4, pp. 788-799, 2010. [http://dx.doi.org/10.1111/j.1529-8817.2010.00857.x]

[51] C. Faraloni, A. Ena, C. Pintucci, and G. Torzillo, "Enhanced hydrogen production by means of sulfur-deprived Chlamydomonas reinhardtii cultures grown in pretreated olive mill wastewater", *Int. J.*

Hydrogen Energy, vol. 36, no. 10, pp. 5920-5931, 2011. [http://dx.doi.org/10.1016/j.ijhydene.2011.02.007]

[52] K.A. Batyrova, A.A. Tsygankov, and S.N. Kosourov, "Sustained hydrogen photoproduction by phosphorus-deprived Chlamydomonas reinhardtii cultures", *Int. J. Hydrogen Energy,* vol. 37, no. 10, pp. 8834-8839, 2012. [http://dx.doi.org/10.1016/j.ijhydene.2012.01.068]

[53] G. Philipps, T. Happe, and A. Hemschemeier, "Nitrogen deprivation results in photosynthetic hydrogen production in Chlamydomonas reinhardtii", *Planta,* vol. 235, no. 4, pp. 729-745, 2012. [http://dx.doi.org/10.1007/s00425-011-1537-2] [PMID: 22020754]

[54] A. Scoma, L. Giannelli, C. Faraloni, and G. Torzillo, "Outdoor H2 production in a 50-L tubular photobioreactor by means of a sulfur-deprived culture of the microalga Chlamydomonas reinhardtii", *J. Biotechnol.,* vol. 157, no. 4, pp. 620-627, 2012. [http://dx.doi.org/10.1016/j.jbiotec.2011.06.040] [PMID: 21771618]

[55] S. Oncel, and M. Sabankay, "Microalgal biohydrogen production considering light energy and mixing time as the two key features for scale-up", *Bioresour. Technol.,* vol. 121, pp. 228-234, 2012. [http://dx.doi.org/10.1016/j.biortech.2012.06.079] [PMID: 22858490]

[56] L. Giannelli, and G. Torzillo, "Hydrogen production with the microalga Chlamydomonas reinhardtii grown in a compact tubular photobioreactor immersed in a scattering light nanoparticle suspension", *Int. J. Hydrogen Energy,* vol. 37, no. 22, pp. 16951-16961, 2012. [http://dx.doi.org/10.1016/j.ijhydene.2012.08.103]

[57] F. Lehr, M. Morweiser, R. Rosello Sastre, O. Kruse, and C. Posten, "Process development for hydrogen production with Chlamydomonas reinhardtii based on growth and product formation kinetics", *J. Biotechnol.,* vol. 162, no. 1, pp. 89-96, 2012. [http://dx.doi.org/10.1016/j.jbiotec.2012.06.002] [PMID: 22750091]

[58] S. Wu, L. Xu, R. Wang, X. Liu, and Q. Wang, "A high yield mutant of Chlamydomonas reinhardtii for photoproduction of hydrogen", *Int. J. Hydrogen Energy,* vol. 36, no. 21, pp. 14134-14140, 2011. [http://dx.doi.org/10.1016/j.ijhydene.2011.05.001]

[59] X. Sun, Y. Cao, H. Xu, Y. Liu, J. Sun, D. Qiao, and Y. Cao, "Effect of nitrogen-starvation, light intensity and iron on triacylglyceride/carbohydrate production and fatty acid profile of Neochloris oleoabundans HK-129 by a two-stage process", *Bioresour. Technol.,* vol. 155, pp. 204-212, 2014. [http://dx.doi.org/10.1016/j.biortech.2013.12.109] [PMID: 24457305]

[60] X. Li, S. Huang, J. Yu, Q. Wang, and S. Wu, "Improvement of hydrogen production of Chlamydomonas reinhardtii by co-cultivation with isolated bacteria", *Int. J. Hydrogen Energy,* vol. 38, no. 25, pp. 10779-10787, 2013. [http://dx.doi.org/10.1016/j.ijhydene.2013.02.102]

[61] T.S. Pinto, F.X. Malcata, J.D. Arrabaça, J.M. Silva, R.J. Spreitzer, and M.G. Esquível, "Rubisco mutants of Chlamydomonas reinhardtii enhance photosynthetic hydrogen production", *Appl. Microbiol. Biotechnol.,* vol. 97, no. 12, pp. 5635-5643, 2013. [http://dx.doi.org/10.1007/s00253-013-4920-z] [PMID: 23649352]

[62] T. Hoshino, D.J. Johnson, M. Scholz, and J.L. Cuello, "Effects of implementing PSI-light on hydrogen production via biophotolysis in Chlamydomonas reinhardtii mutant strains", *Biomass Bioenergy,* vol. 59, pp. 243-252, 2013. [http://dx.doi.org/10.1016/j.biombioe.2013.09.004]

[63] D. Yang, Y. Zhang, D.K. Barupal, X. Fan, R. Gustafson, R. Guo, and O. Fiehn, "Metabolomics of photobiological hydrogen production induced by CCCP in Chlamydomonas reinhardtii", *Int. J. Hydrogen Energy,* vol. 39, no. 1, pp. 150-158, 2014. [http://dx.doi.org/10.1016/j.ijhydene.2013.09.116]

[64] S. Oncel, and A. Kose, "Comparison of tubular and panel type photobioreactors for biohydrogen production utilizing Chlamydomonas reinhardtii considering mixing time and light intensity",

Bioresour. Technol., vol. 151, pp. 265-270, 2014.
[http://dx.doi.org/10.1016/j.biortech.2013.10.076] [PMID: 24246482]

[65] S. White, A. Anandraj, and C. Trois, "NADPH fluorescence as an indicator of hydrogen production in the green algae Chlamydomonas reinhardtii", *Int. J. Hydrogen Energy,* vol. 39, no. 4, pp. 1640-1647, 2014.
[http://dx.doi.org/10.1016/j.ijhydene.2013.11.016]

[66] S. Grewe, M. Ballottari, M. Alcocer, C. D'Andrea, O. Blifernez-Klassen, B. Hankamer, J.H. Mussgnug, R. Bassi, and O. Kruse, "Light-Harvesting Complex Protein LHCBM9 Is Critical for Photosystem II Activity and Hydrogen Production in Chlamydomonas reinhardtii", *Plant Cell,* vol. 26, no. 4, pp. 1598-1611, 2014.
[http://dx.doi.org/10.1105/tpc.114.124198] [PMID: 24706511]

[67] L. Xu, Q. Wang, S. Wu, and D. Li, "Improvement of hydrogen yield of lba-transgenic Chlamydomonas reinhardtii caused by increasing respiration and impairing photosynthesis", *Int. J. Hydrogen Energy,* vol. 39, no. 25, pp. 13347-13352, 2014.
[http://dx.doi.org/10.1016/j.ijhydene.2014.04.178]

[68] T.K. Antal, D.N. Matorin, G.P. Kukarskikh, M.D. Lambreva, E. Tyystjärvi, T.E. Krendeleva, A.A. Tsygankov, and A.B. Rubin, "Pathways of hydrogen photoproduction by immobilized Chlamydomonas reinhardtii cells deprived of sulfur", *Int. J. Hydrogen Energy,* vol. 39, no. 32, pp. 18194-18203, 2014.
[http://dx.doi.org/10.1016/j.ijhydene.2014.08.135]

[69] G. Torzillo, A. Scoma, C. Faraloni, and L. Giannelli, "Advances in the biotechnology of hydrogen production with the microalga Chlamydomonas reinhardtii", *Crit. Rev. Biotechnol.,* vol. 8551, no. March, pp. 1-12, 2014.
[PMID: 24754449]

[70] M.E. Sáenz, K. Bišová, E. Touloupakis, C. Faraloni, W.D. Di Marzio, and G. Torzillo, "Evidences of oxidative stress during hydrogen photoproduction in sulfur-deprived cultures of Chlamydomonas reinhardtii", *Int. J. Hydrogen Energy,* vol. 40, no. 33, pp. 10410-10417, 2015.
[http://dx.doi.org/10.1016/j.ijhydene.2015.06.124]

[71] X. Fan, H. Wang, R. Guo, D. Yang, Y. Zhang, X. Yuan, Y. Qiu, Z. Yang, and X. Zhao, "Comparative study of the oxygen tolerance of Chlorella pyrenoidosa and Chlamydomonas reinhardtii CC124 in photobiological hydrogen production", *Algal Res.,* vol. 16, pp. 240-244, 2016.
[http://dx.doi.org/10.1016/j.algal.2016.03.025]

[72] M.E. Hong, Y.S. Shin, B.W. Kim, and S.J. Sim, "Autotrophic hydrogen photoproduction by operation of carbon-concentrating mechanism in Chlamydomonas reinhardtii under sulfur deprivation condition", *J. Biotechnol.,* vol. 221, pp. 55-61, 2016.
[http://dx.doi.org/10.1016/j.jbiotec.2016.01.023] [PMID: 26812657]

[73] T. Yagi, K. Yamashita, N. Okada, T. Isono, D. Momose, S. Mineki, and E. Tokunaga, "Hydrogen photoproduction in green algae Chlamydomonas reinhardtii sustainable over 2 weeks with the original cell culture without supply of fresh cells nor exchange of the whole culture medium", *J. Plant Res.,* vol. 129, no. 4, pp. 771-779, 2016.
[http://dx.doi.org/10.1007/s10265-016-0825-0] [PMID: 27083446]

[74] T.K. Antal, G.P. Kukarskikh, A.A. Volgusheva, T.E. Krendeleva, E. Tyystjärvi, and A.B. Rubin, "Hydrogen photoproduction by immobilized S-deprived Chlamydomonas reinhardtii: Effect of light intensity and spectrum, and initial medium pH", *Algal Res.,* vol. 17, pp. 38-45, 2016.
[http://dx.doi.org/10.1016/j.algal.2016.04.009]

[75] L. Xu, D. Li, Q. Wang, and S. Wu, "Improved hydrogen production and biomass through the co-cultivation of Chlamydomonas reinhardtii and Bradyrhizobium japonicum", *Int. J. Hydrogen Energy,* vol. 41, no. 22, pp. 9276-9283, 2016.
[http://dx.doi.org/10.1016/j.ijhydene.2016.04.009]

[76] M. He, L. Li, L. Zhang, and J. Liu, "The enhancement of hydrogen photoproduction in Chlorella protothecoides exposed to nitrogen limitation and sulfur deprivation", *Int. J. Hydrogen Energy,* vol. 37, no. 22, pp. 16903-16915, 2012. [http://dx.doi.org/10.1016/j.ijhydene.2012.08.121]

[77] L. Zhang, M. He, and J. Liu, "The enhancement mechanism of hydrogen photoproduction in Chlorella protothecoides under nitrogen limitation and sulfur deprivation", *Int. J. Hydrogen Energy,* vol. 39, no. 17, pp. 8969-8976, 2014. [http://dx.doi.org/10.1016/j.ijhydene.2014.04.045]

[78] E. Kojima, and B. Lin, "Effect of partial shading on photoproduction of hydrogen by Chlorella", *J. Biosci. Bioeng.,* vol. 97, no. 5, pp. 317-321, 2004. [http://dx.doi.org/10.1016/S1389-1723(04)70212-1] [PMID: 16233636]

[79] J.Z. Liu, Y.M. Ge, S.Y. Xia, J.Y. Sun, and J. Mu, "Photoautotrophic hydrogen production by Chlorella pyrenoidosa without sulfur-deprivation", *Int. J. Hydrogen Energy,* vol. 41, no. 20, pp. 8427-8432, 2016. [http://dx.doi.org/10.1016/j.ijhydene.2016.03.191]

[80] K. Bala Amutha, and A.G. Murugesan, "Biological hydrogen production by the algal biomass Chlorella vulgaris MSU 01 strain isolated from pond sediment", *Bioresour. Technol.,* vol. 102, no. 1, pp. 194-199, 2011. [http://dx.doi.org/10.1016/j.biortech.2010.06.008] [PMID: 20620045]

[81] N. Rashid, K. Lee, and Q. Mahmood, "Bio-hydrogen production by Chlorella vulgaris under diverse photoperiods", *Bioresour. Technol.,* vol. 102, no. 2, pp. 2101-2104, 2011. [http://dx.doi.org/10.1016/j.biortech.2010.08.032] [PMID: 20826084]

[82] J-H. Hwang, H-C. Kim, J-A. Choi, R.A. Abou-Shanab, B.A. Dempsey, J.M. Regan, J.R. Kim, H. Song, I.H. Nam, S.N. Kim, W. Lee, D. Park, Y. Kim, J. Choi, M.K. Ji, W. Jung, and B.H. Jeon, "Photoautotrophic hydrogen production by eukaryotic microalgae under aerobic conditions", *Nat. Commun.,* vol. 5, p. 3234, 2014. [http://dx.doi.org/10.1038/ncomms4234] [PMID: 24492668]

[83] K. Batyrova, A. Gavrisheva, E. Ivanova, J. Liu, and A. Tsygankov, "Sustainable hydrogen photoproduction by phosphorus-deprived marine green microalgae Chlorella sp", *Int. J. Mol. Sci.,* vol. 16, no. 2, pp. 2705-2716, 2015. [http://dx.doi.org/10.3390/ijms16022705] [PMID: 25629229]

[84] P. Dechatiwongse, G. Maitland, and K. Hellgardt, "Demonstration of a two-stage aerobic/anaerobic chemostat for the enhanced production of hydrogen and biomass from unicellular nitrogen-fixing cyanobacterium", *Algal Res.,* vol. 10, pp. 189-201, 2015. [http://dx.doi.org/10.1016/j.algal.2015.05.004]

[85] X.Y. Shi, and H.Q. Yu, "Simultaneous metabolism of benzoate and photobiological hydrogen production by Lyngbya sp", *Renew. Energy,* vol. 95, pp. 474-477, 2016. [http://dx.doi.org/10.1016/j.renene.2016.04.051]

[86] L. Wei, X. Li, J. Yi, Z. Yang, Q. Wang, and W. Ma, "A simple approach for the efficient production of hydrogen from Taihu Lake Microcystis spp. blooms", *Bioresour. Technol.,* vol. 139, pp. 136-140, 2013. [http://dx.doi.org/10.1016/j.biortech.2013.04.026] [PMID: 23648763]

[87] M. Nyberg, T. Heidorn, and P. Lindblad, "Hydrogen production by the engineered cyanobacterial strain Nostoc PCC 7120 ΔhupW examined in a flat panel photobioreactor system", *J. Biotechnol.,* vol. 215, pp. 35-43, 2015. [http://dx.doi.org/10.1016/j.jbiotec.2015.08.028] [PMID: 26325196]

[88] Y. Guan, M. Deng, X. Yu, and W. Zhang, "Two-stage photo-biological production of hydrogen by marine green alga Platymonas subcordiformis", *Biochem. Eng. J.,* vol. 19, no. 1, pp. 69-73, 2004. [http://dx.doi.org/10.1016/j.bej.2003.10.006]

[89] C-F. Ji, J. Legrand, J. Pruvost, Z-A. Chen, and W. Zhang, "Characterization of hydrogen production by Platymonas Subcordiformis in torus photobioreactor", *Int. J. Hydrogen Energy,* vol. 35, no. 13, pp. 7200-7205, 2010.
[http://dx.doi.org/10.1016/j.ijhydene.2010.02.085]

[90] Y. Zhang, X. Fan, Z. Yang, H. Wang, D. Yang, and R. Guo, "Characterization of H2 photoproduction by a new marine green alga, Platymonas helgolandica var. tsingtaoensis", *Appl. Energy,* vol. 92, pp. 38-43, 2012.
[http://dx.doi.org/10.1016/j.apenergy.2011.09.044]

[91] K. Aoyama, I. Uemura, J. Miyake, and Y. Asada, "Fermentative metabolism to produce hydrogen gas and organic compounds in a cyanobacterium, Spirulina platensis", *J. Ferment. Bioeng.,* vol. 83, no. 1, pp. 17-20, 1997.
[http://dx.doi.org/10.1016/S0922-338X(97)87320-5]

[92] D.J. Dickson, C.J. Page, and R.L. Ely, "Photobiological hydrogen production from Synechocystis sp. PCC 6803 encapsulated in silica sol-gel", *Int. J. Hydrogen Energy,* vol. 34, no. 1, pp. 204-215, 2009.
[http://dx.doi.org/10.1016/j.ijhydene.2008.10.021]

[93] C. Maneeruttanarungroj, P. Lindblad, and A. Incharoensakdi, "A newly isolated green alga, Tetraspora sp. CU2551, from Thailand with efficient hydrogen production", *Int. J. Hydrogen Energy,* vol. 35, no. 24, pp. 13193-13199, 2010.
[http://dx.doi.org/10.1016/j.ijhydene.2010.08.096]

[94] A.J. Cornish, R. Green, K. Gärtner, S. Mason, and E.L. Hegg, "Characterization of hydrogen metabolism in the multicellular green alga Volvox carteri", *PLoS One,* vol. 10, no. 4, p. e0125324, 2015.
[http://dx.doi.org/10.1371/journal.pone.0125324] [PMID: 25927230]

[95] H. Leino, and S. Shunmugam, "J. Isoj??rvi, P. Oliveira, P. Mulo, L. Saari, N. Battchikova, K. Sivonen, P. Lindblad, E. M. Aro, and Y. Allahverdiyeva, "Characterization of ten H2 producing cyanobacteria isolated from the Baltic Sea and Finnish lakes", *Int. J. Hydrogen Energy,* vol. 39, no. 17, pp. 8983-8991, 2014.
[http://dx.doi.org/10.1016/j.ijhydene.2014.03.171]

[96] A.F. Ferreira, J. Ortigueira, L. Alves, L. Gouveia, P. Moura, and C.M. Silva, "Energy requirement and CO2 emissions of bioH2 production from microalgal biomass", *Biomass Bioenergy,* vol. 49, pp. 249-259, 2013.
[http://dx.doi.org/10.1016/j.biombioe.2012.12.033]

[97] B.K. Nayak, S. Roy, and D. Das, "Biohydrogen production from algal biomass (Anabaena sp. PCC 7120) cultivated in airlift photobioreactor", *Int. J. Hydrogen Energy,* vol. 39, no. 14, pp. 7553-7560, 2014.
[http://dx.doi.org/10.1016/j.ijhydene.2013.07.120]

[98] E.N. Efremenko, A.B. Nikolskaya, I.V. Lyagin, O.V. Senko, T.A. Makhlis, N.A. Stepanov, O.V. Maslova, F. Mamedova, and S.D. Varfolomeev, "Production of biofuels from pretreated microalgae biomass by anaerobic fermentation with immobilized Clostridium acetobutylicum cells", *Bioresour. Technol.,* vol. 114, pp. 342-348, 2012.
[http://dx.doi.org/10.1016/j.biortech.2012.03.049] [PMID: 22483558]

[99] J. Cheng, A. Xia, W. Song, H. Su, J. Zhou, and K. Cen, "Comparison between heterofermentation and autofermentation in hydrogen production from Arthrospira (Spirulina) platensis wet biomass", *Int. J. Hydrogen Energy,* vol. 37, no. 8, pp. 6536-6544, 2012.
[http://dx.doi.org/10.1016/j.ijhydene.2012.01.025]

[100] A. Xia, A. Jacob, M.R. Tabassum, C. Herrmann, and J.D. Murphy, "Production of hydrogen, ethanol and volatile fatty acids through co-fermentation of macro- and micro-algae", *Bioresour. Technol.,* vol. 205, pp. 118-125, 2016.
[http://dx.doi.org/10.1016/j.biortech.2016.01.025] [PMID: 26820925]

[101] J. Cheng, M. Zhang, W. Song, A. Xia, J. Zhou, and K. Cen, "Cogeneration of hydrogen and methane from Arthrospira maxima biomass with bacteria domestication and enzymatic hydrolysis", *Int. J. Hydrogen Energy,* vol. 36, no. 2, pp. 1474-1481, 2011. [http://dx.doi.org/10.1016/j.ijhydene.2010.11.009]

[102] C.H. Liu, C.Y. Chang, Q. Liao, X. Zhu, and J.S. Chang, "Photoheterotrophic growth of Chlorella vulgaris ESP6 on organic acids from dark hydrogen fermentation effluents", *Bioresour. Technol.,* vol. 145, pp. 331-336, 2013. [http://dx.doi.org/10.1016/j.biortech.2012.12.111] [PMID: 23305898]

[103] A.P. Batista, L. Ambrosano, S. Graça, C. Sousa, P.A. Marques, B. Ribeiro, E.P. Botrel, P. Castro Neto, and L. Gouveia, "Combining urban wastewater treatment with biohydrogen production--an integrated microalgae-based approach", *Bioresour. Technol.,* vol. 184, pp. 230-235, 2015. [http://dx.doi.org/10.1016/j.biortech.2014.10.064] [PMID: 25453433]

[104] K. Kumar, S. Roy, and D. Das, "Continuous mode of carbon dioxide sequestration by C. sorokiniana and subsequent use of its biomass for hydrogen production by E. cloacae IIT-BT 08", *Bioresour. Technol.,* vol. 145, pp. 116-122, 2013. [http://dx.doi.org/10.1016/j.biortech.2013.01.137] [PMID: 23453984]

[105] K-L. Ho, D-J. Lee, A. Su, and J-S. Chang, "Biohydrogen from lignocellulosic feedstock via one-step process", *Int. J. Hydrogen Energy,* vol. 37, no. 20, pp. 15569-15574, 2012. [http://dx.doi.org/10.1016/j.ijhydene.2012.01.137]

[106] A-M. Lakaniemi, O.H. Tuovinen, and J.A. Puhakka, "Anaerobic conversion of microalgal biomass to sustainable energy carriers--a review", *Bioresour. Technol.,* vol. 135, pp. 222-231, 2013. [http://dx.doi.org/10.1016/j.biortech.2012.08.096] [PMID: 23021960]

[107] L. Dipasquale, G. d'Ippolito, C. Gallo, F.M. Vella, A. Gambacorta, G. Picariello, and A. Fontana, "Hydrogen production by the thermophilic eubacterium Thermotoga neapolitana from storage polysaccharides of the CO2-fixing diatom Thalassiosira weissflogii", *Int. J. Hydrogen Energy,* vol. 37, no. 17, pp. 12250-12257, 2012. [http://dx.doi.org/10.1016/j.ijhydene.2012.05.160]

[108] A. Ike, N. Toda, K. Hirata, and K. Miyamoto, "Hydrogen photoproduction from CO2-fixing microalgal biomass: Application of lactic acid fermentation by Lactobacillus amylovorus", *J. Ferment. Bioeng.,* vol. 84, no. 5, pp. 428-433, 1997. [http://dx.doi.org/10.1016/S0922-338X(97)82003-X]

[109] H. Kawaguchi, K. Hashimoto, K. Hirata, and K. Miyamoto, "H2 production from algal biomass by a mixed culture of Rhodobium marinum A-501 and Lactobacillus amylovorus", *J. Biosci. Bioeng.,* vol. 91, no. 3, pp. 277-282, 2001. [http://dx.doi.org/10.1016/S1389-1723(01)80134-1] [PMID: 16232989]

[110] M-S. Kim, J-S. Baek, Y-S. Yun, S. Jun Sim, S. Park, and S-C. Kim, "Hydrogen production from Chlamydomonas reinhardtii biomass using a two-step conversion process: Anaerobic conversion and photosynthetic fermentation", *Int. J. Hydrogen Energy,* vol. 31, no. 6, pp. 812-816, 2006. [http://dx.doi.org/10.1016/j.ijhydene.2005.06.009]

[111] Y. Miura, T. Akano, K. Fukatsu, H. Miyasaka, T. Mizoguchi, K. Yagi, I. Maeda, Y. Ikuta, and H. Matsumoto, "Stably sustained hydrogen production by biophotolysis in natural day/night cycle", *Energy Convers. Manage.,* vol. 38, pp. S533-S537, 1997. [http://dx.doi.org/10.1016/S0196-8904(96)00323-8]

[112] T-A. Nguyen, K-R. Kim, M-T. Nguyen, M.S. Kim, D. Kim, and S.J. Sim, "Enhancement of fermentative hydrogen production from green algal biomass of Thermotoga neapolitana by various pretreatment methods", *Int. J. Hydrogen Energy,* vol. 35, no. 23, pp. 13035-13040, 2010. [http://dx.doi.org/10.1016/j.ijhydene.2010.04.062]

[113] B.P. Nobre, F. Villalobos, B.E. Barragán, A.C. Oliveira, A.P. Batista, P.A. Marques, R.L. Mendes, H. Sovová, A.F. Palavra, and L. Gouveia, "A biorefinery from Nannochloropsis sp. microalga--extraction

of oils and pigments. Production of biohydrogen from the leftover biomass", *Bioresour. Technol.*, vol. 135, pp. 128-136, 2013.
[http://dx.doi.org/10.1016/j.biortech.2012.11.084] [PMID: 23265815]

[114] Z. Yang, R. Guo, X. Xu, X. Fan, and S. Luo, "Fermentative hydrogen production from lipid-extracted microalgal biomass residues", *Appl. Energy,* vol. 88, no. 10, pp. 3468-3472, 2011.
[http://dx.doi.org/10.1016/j.apenergy.2010.09.009]

[115] J.A. Choi, J.H. Hwang, B.A. Dempsey, R.A. Abou-Shanab, B. Min, H. Song, D.S. Lee, J.R. Kim, Y. Cho, S. Hong, and B.H. Jeon, "Enhancement of fermentative bioenergy (ethanol/hydrogen) production using ultrasonication of Scenedesmus obliquus YSW15 cultivated in swine wastewater effluent", *Energy Environ. Sci.,* vol. 4, pp. 3513-3520, 2011.
[http://dx.doi.org/10.1039/c1ee01068a]

[116] V. Teplyakov, L. Gassanova, E. Sostina, E. Slepova, M. Modigell, and A. Netrusov, "Lab-scale bioreactor integrated with active membrane system for hydrogen production: experience and prospects", *Int. J. Hydrogen Energy,* vol. 27, no. 11, pp. 1149-1155, 2002.
[http://dx.doi.org/10.1016/S0360-3199(02)00093-9]

[117] K. Skjånes, C. Rebours, and P. Lindblad, *Potential for green microalgae to produce hydrogen, pharmaceuticals and other high value products in a combined process*, 2012.

[118] S.S. Oncel, A. Kose, F. Vardar, and G. Torzillo, *Handbook of Marine Microalgae.* Elsevier, 2015.

[119] M. Sergejevová, V. Jirka, J. Masojídek, Š. Papác, J. Kunc, O. Verbovikova, J. Kopecký, D. Štys, G. Torzillo, and J. Korec, "“A closed solar photobioreactor for cultivation of microalgae under supra-high irradiance : basic design and performance,” no", *Burlew,* vol. 1953, pp. 239-248, 2003.

[120] M. a. Borowitzka, *Algae for Biofuels and Energy Developments in Applied Phycology 5,* 2013.

[121] S.S. Oncel, *Handbook of Marine Microalgae.* Elsevier, 2015.

[122] J. Sheehan, *A Look Back at the U . S . Department of Energy ' s Aquatic Species Program : Biodiesel from Algae.*

[123] C. Yang, R. Nie, J. Fu, Z. Hou, and X. Lu, "Production of aviation fuel via catalytic hydrothermal decarboxylation of fatty acids in microalgae oil", *Bioresour. Technol.,* vol. 146, pp. 569-573, 2013.
[http://dx.doi.org/10.1016/j.biortech.2013.07.131] [PMID: 23973977]

[124] O. Reaction, I.N. Algae, and B.Y. Gaffron, "OXYHYDROGEN REACTION IN ALGAE (From the Department of Chemistry, The University of Chicago, Chicago)", *J. Gen. Physiol.,* 1942.

[125] J.R. Benemann, "Feasibility analysis of photobiological hydrogen production", *Int. J. Hydrogen Energy,* vol. 22, no. 10-11, pp. 979-987, 1997.
[http://dx.doi.org/10.1016/S0360-3199(96)00189-9]

[126] D.D. Wykoff, J.P. Davies, A. Melis, and A.R. Grossman, "The regulation of photosynthetic electron transport during nutrient deprivation in Chlamydomonas reinhardtii", *Plant Physiol.,* vol. 117, no. 1, pp. 129-139, 1998.
[http://dx.doi.org/10.1104/pp.117.1.129] [PMID: 9576782]

[127] D. Rajagopal, S.E. Sexton, D. Roland-Holst, and D. Zilberman, "Challenge of biofuel: filling the tank without emptying the stomach?", *Environ. Res. Lett.,* vol. 2, no. 4, p. 44004, 2007.
[http://dx.doi.org/10.1088/1748-9326/2/4/044004]

[128] D.A. Walker, "Biofuels, facts, fantasy, and feasibility", *J. Appl. Phycol.,* vol. 21, no. 5, pp. 509-517, 2009.
[http://dx.doi.org/10.1007/s10811-009-9446-5]

[129] J. Singh, and S. Gu, *Commercialization potential of microalgae for biofuels production.* vol. Vol. 14. , 2010, pp. 2596-2610.

[130] S. Berteotti, M. Ballottari, and R. Bassi, *Increased biomass productivity in green algae by tuning non-photochemical quenching.*, 2016.

[131] E.S. Johnston, and J.E. Myers, *Andjack.*, 1939, pp. 473-492.

[132] B.M. Calvin, J.A. Bassham, and A.A. Benson, *PHOTOSYNTHESISl,* vol. 2, pp. 215-228, 1952.

[133] B. G. Hoch, PHOTOSYNTHESISl,2,3, vol. 700, no. Rg 6692, 1960.

[134] "the Thermodynamic Efficiency (Quantum Demand) and Dynamics of Photosynthetic Growth", *New Phytol.,* vol. 102, no. 1, pp. 3-37, 1986. [http://dx.doi.org/10.1111/j.1469-8137.1986.tb00794.x]

[135] S. Eberhard, G. Finazzi, and F-A. Wollman, "The dynamics of photosynthesis", *Annu. Rev. Genet.,* vol. 42, no. 1, pp. 463-515, 2008. [http://dx.doi.org/10.1146/annurev.genet.42.110807.091452] [PMID: 18983262]

[136] N. Nelson, *Photosystems and global effects of oxygenic photosynthesis*, 2011.

[137] B. Drop, M. Webber-Birungi, S.K. Yadav, A. Filipowicz-Szymanska, F. Fusetti, E.J. Boekema, and R. Croce, "Light-harvesting complex II (LHCII) and its supramolecular organization in Chlamydomonas reinhardtii", *Biochim. Biophys. Acta,* vol. 1837, no. 1, pp. 63-72, 2014. [http://dx.doi.org/10.1016/j.bbabio.2013.07.012] [PMID: 23933017]

[138] J. Beckmann, F. Lehr, G. Finazzi, B. Hankamer, C. Posten, L. Wobbe, and O. Kruse, "Improvement of light to biomass conversion by de-regulation of light-harvesting protein translation in Chlamydomonas reinhardtii", *J. Biotechnol.,* vol. 142, no. 1, pp. 70-77, 2009. [http://dx.doi.org/10.1016/j.jbiotec.2009.02.015] [PMID: 19480949]

[139] L. Wobbe, R. Bassi, and O. Kruse, "Multi-Level Light Capture Control in Plants and Green Algae", *Trends Plant Sci.,* vol. 21, no. 1, pp. 55-68, 2016. [http://dx.doi.org/10.1016/j.tplants.2015.10.004] [PMID: 26545578]

[140] T.K. Antal, T.E. Krendeleva, and A.B. Rubin, "Acclimation of green algae to sulfur deficiency: underlying mechanisms and application for hydrogen production", *Appl. Microbiol. Biotechnol.,* vol. 89, no. 1, pp. 3-15, 2011. [http://dx.doi.org/10.1007/s00253-010-2879-6] [PMID: 20878321]

[141] M.S. Khan, "Engineering photorespiration in chloroplasts: a novel strategy for increasing biomass production", *Trends Biotechnol.,* vol. 25, no. 10, pp. 437-440, 2007. [http://dx.doi.org/10.1016/j.tibtech.2007.08.007] [PMID: 17870196]

[142] S.I. Allakhverdiev, V.D. Kreslavski, V. Thavasi, S.K. Zharmukhamedov, V.V. Klimov, T. Nagata, H. Nishihara, and S. Ramakrishna, "Hydrogen photoproduction by use of photosynthetic organisms and biomimetic systems", *Photochem. Photobiol. Sci.,* vol. 8, no. 2, pp. 148-156, 2009. [http://dx.doi.org/10.1039/B814932A] [PMID: 19247505]

[143] N. Nelson, and C.F. Yocum, "Structure and function of photosystems I and II", *Annu. Rev. Plant Biol.,* vol. 57, no. 1, pp. 521-565, 2006. [http://dx.doi.org/10.1146/annurev.arplant.57.032905.105350] [PMID: 16669773]

[144] Govindjee, and D. Shevela, "Adventures with cyanobacteria: a personal perspective", *Front. Plant Sci.,* vol. 2, no. July, p. 28, 2011. [PMID: 22645530]

[145] N. Rashid, M.S. Rehman, S. Memon, Z. Ur Rahman, K. Lee, and J.I. Han, "Current status, barriers and developments in biohydrogen production by microalgae", *Renew. Sustain. Energy Rev.,* vol. 22, pp. 571-579, 2013. [http://dx.doi.org/10.1016/j.rser.2013.01.051]

[146] A. Melis, "Green alga hydrogen production: progress, challenges and prospects", *Int. J. Hydrogen Energy,* vol. 27, no. 11-12, pp. 1217-1228, 2002. [http://dx.doi.org/10.1016/S0360-3199(02)00110-6]

[147] M.M. Mendes-Pinto, M.F. Raposo, J. Bowen, A.J. Young, and R. Morais, "Evaluation of different cell disruption processes on encysted cells of Haematococcus pluvialis: Effects on astaxanthin recovery

and implications for bio-availability", *J. Appl. Phycol.,* vol. 13, no. 1, pp. 19-24, 2001. [http://dx.doi.org/10.1023/A:1008183429747]

[148] J. Mathews, and G. Wang, "Metabolic pathway engineering for enhanced biohydrogen production", *Int. J. Hydrogen Energy,* vol. 34, no. 17, pp. 7404-7416, 2009. [http://dx.doi.org/10.1016/j.ijhydene.2009.05.078]

[149] S. Yang, M.T. Guarnieri, S. Smolinski, M. Ghirardi, and P.T. Pienkos, "De novo transcriptomic analysis of hydrogen production in the green alga Chlamydomonas moewusii through RNA-Seq", *Biotechnol. Biofuels,* vol. 6, no. 1, p. 118, 2013. [http://dx.doi.org/10.1186/1754-6834-6-118] [PMID: 23971877]

[150] H. Sakurai, H. Masukawa, M. Kitashima, and K. Inoue, "Photobiological hydrogen production: Bioenergetics and challenges for its practical application", *J. Photochem. Photobiol. Photochem. Rev.,* vol. 17, pp. 1-25, 2013. [http://dx.doi.org/10.1016/j.jphotochemrev.2013.05.001]

[151] W. Khetkorn, W. Baebprasert, P. Lindblad, and A. Incharoensakdi, "Redirecting the electron flow towards the nitrogenase and bidirectional Hox-hydrogenase by using specific inhibitors results in enhanced H2 production in the cyanobacterium Anabaena siamensis TISTR 8012", *Bioresour. Technol.,* vol. 118, pp. 265-271, 2012. [http://dx.doi.org/10.1016/j.biortech.2012.05.052] [PMID: 22705533]

[152] D.C. Rees, and J.B. Howard, "Nitrogenase: standing at the crossroads", *Curr. Opin. Chem. Biol.,* vol. 4, no. 5, pp. 559-566, 2000. [http://dx.doi.org/10.1016/S1367-5931(00)00132-0] [PMID: 11006545]

[153] Y. Zhao, S.M. Bian, H.N. Zhou, and J.F. Huang, "Diversity of nitrogenase systems in diazotrophs", *J. Integr. Plant Biol.,* vol. 48, no. 7, pp. 745-755, 2006. [http://dx.doi.org/10.1111/j.1744-7909.2006.00271.x]

[154] D. Carrieri, K. Wawrousek, C. Eckert, J. Yu, and P.C. Maness, "The role of the bidirectional hydrogenase in cyanobacteria", *Bioresour. Technol.,* vol. 102, no. 18, pp. 8368-8377, 2011. [http://dx.doi.org/10.1016/j.biortech.2011.03.103] [PMID: 21514820]

[155] Y.K. Oh, S.M. Raj, G.Y. Jung, and S. Park, "Current status of the metabolic engineering of microorganisms for biohydrogen production", *Bioresour. Technol.,* vol. 102, no. 18, pp. 8357-8367, 2011. [http://dx.doi.org/10.1016/j.biortech.2011.04.054] [PMID: 21733680]

[156] D-H. Kim, and M-S. Kim, "Hydrogenases for biological hydrogen production", *Bioresour. Technol.,* vol. 102, no. 18, pp. 8423-8431, 2011. [http://dx.doi.org/10.1016/j.biortech.2011.02.113] [PMID: 21435869]

[157] P. Tamagnini, E. Leitão, P. Oliveira, D. Ferreira, F. Pinto, D.J. Harris, T. Heidorn, and P. Lindblad, "Cyanobacterial hydrogenases: diversity, regulation and applications", *FEMS Microbiol. Rev.,* vol. 31, no. 6, pp. 692-720, 2007. [http://dx.doi.org/10.1111/j.1574-6976.2007.00085.x] [PMID: 17903205]

[158] P. Tamagnini, R. Axelsson, P. Lindberg, F. Oxelfelt, R. Wünschiers, and P. Lindblad, *Hydrogenases and hydrogen metabolism of cyanobacteria.*, 2002.

[159] M.L. Ghirardi, L. Zhang, J.W. Lee, T. Flynn, M. Seibert, E. Greenbaum, and A. Melis, "Microalgae: a green source of renewable H(2)", *Trends Biotechnol.,* vol. 18, no. 12, pp. 506-511, 2000. [http://dx.doi.org/10.1016/S0167-7799(00)01511-0] [PMID: 11102662]

[160] M. Olaizola, "Commercial development of microalgal biotechnology: from the test tube to the marketplace", *Biomol. Eng.,* vol. 20, no. 4-6, pp. 459-466, 2003. [http://dx.doi.org/10.1016/S1389-0344(03)00076-5] [PMID: 12919832]

[161] M.L. Ghirardi, A. Dubini, J. Yu, and P-C. Maness, "Photobiological hydrogen-producing systems", *Chem. Soc. Rev.,* vol. 38, no. 1, pp. 52-61, 2009.

[http://dx.doi.org/10.1039/B718939G] [PMID: 19088964]

[162] O. Kruse, J. Rupprecht, K.P. Bader, S. Thomas-Hall, P.M. Schenk, G. Finazzi, and B. Hankamer, "Improved photobiological H2 production in engineered green algal cells", *J. Biol. Chem.,* vol. 280, no. 40, pp. 34170-34177, 2005. [http://dx.doi.org/10.1074/jbc.M503840200] [PMID: 16100118]

[163] P.C. Hallenbeck, "Microbial technologies in advanced biofuels production", *Microb. Technol. Adv. Biofuels Prod.,* vol. 9781461412, pp. 1-272, 2012.

[164] D. Madamwar, N. Garg, and V. Shah, "Cyanobacterial hydrogen production", *World J. Microbiol. Biotechnol.,* vol. 16, no. 8-9, pp. 757-767, 2000. [http://dx.doi.org/10.1023/A:1008919200103]

[165] G. Kufryk, "Advances in utilizing cyanobacteria for hydrogen production", *Adv. Microbiol.,* vol. 3, no. 6, pp. 60-68, 2013. [http://dx.doi.org/10.4236/aim.2013.36A008]

[166] W. Lubitz, H. Ogata, O. Rüdiger, and E. Reijerse, "Hydrogenases", *Chem. Rev.,* vol. 114, no. 8, pp. 4081-4148, 2014. [http://dx.doi.org/10.1021/cr4005814] [PMID: 24655035]

[167] F. Leroux, P.P. Liebgott, L. Cournac, P. Richaud, A. Kpebe, B. Burlat, B. Guigliarelli, and P. Bertrand, "C. L??ger, M. Rousset, and S. Dementin, "Is engineering O2-tolerant hydrogenases just a matter of reproducing the active sites of the naturally occurring O2-resistant enzymes?", *Int. J. Hydrogen Energy,* vol. 35, no. 19, pp. 10770-10777, 2010. [http://dx.doi.org/10.1016/j.ijhydene.2010.02.071]

[168] T. Yagi, and Y. Higuchi, ""Studies on hydrogenase," Proc. Jpn. Acad", *Ser. B,* vol. 89, pp. 16-33, 2013.

[169] J. Meyer, "[FeFe] hydrogenases and their evolution: a genomic perspective", *Cell. Mol. Life Sci.,* vol. 64, no. 9, pp. 1063-1084, 2007. [http://dx.doi.org/10.1007/s00018-007-6477-4] [PMID: 17353991]

[170] M. Horch, L. Lauterbach, O. Lenz, P. Hildebrandt, and I. Zebger, "NAD(H)-coupled hydrogen cycling - structure-function relationships of bidirectional [NiFe] hydrogenases", *FEBS Lett.,* vol. 586, no. 5, pp. 545-556, 2012. [http://dx.doi.org/10.1016/j.febslet.2011.10.010] [PMID: 22056977]

[171] H.S. Shafaat, O. Rüdiger, H. Ogata, and W. Lubitz, *[NiFe] hydrogenases: A common active site for hydrogen metabolism under diverse conditions*, 2013.

[172] J.W. Peters, G.J. Schut, E.S. Boyd, D.W. Mulder, E.M. Shepard, J.B. Broderick, P.W. King, and M.W. Adams, "[FeFe]- and [NiFe]-hydrogenase diversity, mechanism, and maturation", *Biochim. Biophys. Acta,* vol. 1853, no. 6, pp. 1350-1369, 2015. [http://dx.doi.org/10.1016/j.bbamcr.2014.11.021] [PMID: 25461840]

[173] D.M. Heinekey, "Hydrogenase enzymes: Recent structural studies and active site models", *J. Organomet. Chem.,* vol. 694, no. 17, pp. 2671-2680, 2009. [http://dx.doi.org/10.1016/j.jorganchem.2009.03.047]

[174] M.J. Corr, and J.A. Murphy, "Evolution in the understanding of [Fe]-hydrogenase", *Chem. Soc. Rev.,* vol. 40, no. 5, pp. 2279-2292, 2011. [http://dx.doi.org/10.1039/c0cs00150c] [PMID: 21365080]

[175] D.W. Mulder, E.M. Shepard, J.E. Meuser, N. Joshi, P.W. King, M.C. Posewitz, J.B. Broderick, and J.W. Peters, "Insights into [FeFe]-hydrogenase structure, mechanism, and maturation", *Structure,* vol. 19, no. 8, pp. 1038-1052, 2011. [http://dx.doi.org/10.1016/j.str.2011.06.008] [PMID: 21827941]

[176] J.W. Peters, and R.K. Szilagyi, "Exploring new frontiers of nitrogenase structure and mechanism", *Curr. Opin. Chem. Biol.,* vol. 10, no. 2, pp. 101-108, 2006.

[http://dx.doi.org/10.1016/j.cbpa.2006.02.019] [PMID: 16510305]

[177] L. Zhang, T. Happe, and A. Melis, "Biochemical and morphological characterization of sulfur-deprived and H2-producing Chlamydomonas reinhardtii (green alga)", *Planta,* vol. 214, no. 4, pp. 552-561, 2002.
[http://dx.doi.org/10.1007/s004250100660] [PMID: 11925039]

[178] E.M. Shepard, F. Mus, J.N. Betz, A.S. Byer, B.R. Duffus, J.W. Peters, and J.B. Broderick, "[FeFe]-hydrogenase maturation", *Biochemistry,* vol. 53, no. 25, pp. 4090-4104, 2014.
[http://dx.doi.org/10.1021/bi500210x] [PMID: 24878200]

[179] J.E. Meuser, S. D'Adamo, R.E. Jinkerson, F. Mus, W. Yang, M.L. Ghirardi, M. Seibert, A.R. Grossman, and M.C. Posewitz, "Genetic disruption of both Chlamydomonas reinhardtii [FeFe]-hydrogenases: Insight into the role of HYDA2 in H2 production", *Biochem. Biophys. Res. Commun.,* vol. 417, no. 2, pp. 704-709, 2012.
[http://dx.doi.org/10.1016/j.bbrc.2011.12.002] [PMID: 22177948]

[180] A. Atteia, R. Van Lis, A.G. Tielens, and W.F. Martin, *Anaerobic energy metabolism in unicellular photosynthetic eukaryotes*, 2013.

[181] B. Ghysels, and F. Franck, "Hydrogen photo-evolution upon S deprivation stepwise: an illustration of microalgal photosynthetic and metabolic flexibility and a step stone for future biotechnological methods of renewable H(2) production", *Photosynth. Res.,* vol. 106, no. 1-2, pp. 145-154, 2010.
[http://dx.doi.org/10.1007/s11120-010-9582-4] [PMID: 20658193]

[182] O. Kruse, and B. Hankamer, "Microalgal hydrogen production", *Curr. Opin. Biotechnol.,* vol. 21, no. 3, pp. 238-243, 2010.
[http://dx.doi.org/10.1016/j.copbio.2010.03.012] [PMID: 20399635]

[183] K. Srirangan, M.E. Pyne, and C. Perry Chou, "Biochemical and genetic engineering strategies to enhance hydrogen production in photosynthetic algae and cyanobacteria", *Bioresour. Technol.,* vol. 102, no. 18, pp. 8589-8604, 2011.
[http://dx.doi.org/10.1016/j.biortech.2011.03.087] [PMID: 21514821]

[184] J. Weber, F. Krujatz, and G. Hilpmann, "S. Gr??tzner, J. Herrmann, S. Thierfelder, G. Bienert, R. Illing, K. Helbig, A. Hurtado, G. Cuniberti, M. Mertig, R. Lange, E. G??nther, J. Opitz, W. Lippmann, T. Bley, and N. Haufe, "Biotechnological hydrogen production by photosynthesis", *Eng. Life Sci.,* vol. 14, no. 6, pp. 592-606, 2014.
[http://dx.doi.org/10.1002/elsc.201400056]

[185] L.L. Beer, E.S. Boyd, J.W. Peters, and M.C. Posewitz, "Engineering algae for biohydrogen and biofuel production", *Curr. Opin. Biotechnol.,* vol. 20, no. 3, pp. 264-271, 2009.
[http://dx.doi.org/10.1016/j.copbio.2009.06.002] [PMID: 19560336]

[186] M.a. Borowitzka, "Commercial production of microalgae: ponds, tanks, tubes and fermenters", *J. Biotechnol.,* vol. 70, no. 1-3, pp. 313-321, 1999.
[http://dx.doi.org/10.1016/S0168-1656(99)00083-8]

[187] F.G. Fernandez, J.M. Sevilla, and E.M. Grima, "Photobioreactors for the production of microalgae", *Rev. Environ. Sci. Biotechnol.,* vol. 12, pp. 131-151, 2013.
[http://dx.doi.org/10.1007/s11157-012-9307-6]

[188] M.O. Pulz, and W. Gross, *Valuable products from biotechnology of microalgae.*, 2004, pp. 635-648.

[189] S.S. Oncel, A. Kose, C. Faraloni, E. Imamoglu, M. Elibol, G. Torzillo, and F. Vardar Sukan, "Biohydrogen production from model microalgae Chlamydomonas reinhardtii: A simulation of environmental conditions for outdoor experiments", *Int. J. Hydrogen Energy,* 2015.
[http://dx.doi.org/10.1016/j.ijhydene.2014.12.121]

[190] J.U. Grobbelaar, "Factors governing algal growth in photobioreactors: The 'open' versus 'closed' debate", *J. Appl. Phycol.,* vol. 21, no. 5, pp. 489-492, 2009.
[http://dx.doi.org/10.1007/s10811-008-9365-x]

[191] O. Jorquera, A. Kiperstok, E.A. Sales, M. Embiruçu, and M.L. Ghirardi, "Comparative energy life-cycle analyses of microalgal biomass production in open ponds and photobioreactors", *Bioresour. Technol.,* vol. 101, no. 4, pp. 1406-1413, 2010. [http://dx.doi.org/10.1016/j.biortech.2009.09.038] [PMID: 19800784]

[192] Y. Chisti, "Biodiesel from microalgae", *Biotechnol. Adv.,* vol. 25, no. 3, pp. 294-306, 2007. [http://dx.doi.org/10.1016/j.biotechadv.2007.02.001] [PMID: 17350212]

[193] P.G. Stephenson, C.M. Moore, M.J. Terry, M.V. Zubkov, and T.S. Bibby, "Improving photosynthesis for algal biofuels: toward a green revolution", *Trends Biotechnol.,* vol. 29, no. 12, pp. 615-623, 2011. [http://dx.doi.org/10.1016/j.tibtech.2011.06.005] [PMID: 21775004]

[194] M.A. Borowitzka, "Microalgae for aquaculture: Opportunites and constraints", *J. Appl. Phycol.,* vol. 9, no. 5, pp. 393-401, 1997. [http://dx.doi.org/10.1023/A:1007921728300]

[195] A. Saeid, and K. Chojnacka, "Chemical Engineering Research and Design Toward production of microalgae in photobioreactors under temperate climate", *Chem. Eng. Res. Des.,* vol. 93, no. June, pp. 377-391, 2014.

[196] W. Farooq, M. Moon, B. Ryu, W.I. Suh, A. Shrivastav, M.S. Park, S.K. Mishra, and J-W. Yang, "Effect of harvesting methods on the reusability of water for cultivation of Chlorella vulgaris, its lipid productivity and biodiesel quality", *Algal Res.,* vol. 8, pp. 1-7, 2015. [http://dx.doi.org/10.1016/j.algal.2014.12.007]

[197] T.L. Walker, S. Purton, D.K. Becker, and C. Collet, "Microalgae as bioreactors", *Plant Cell Rep.,* vol. 24, no. 11, pp. 629-641, 2005. [http://dx.doi.org/10.1007/s00299-005-0004-6] [PMID: 16136314]

[198] Y.K. Lee, "Microalgal mass culture systems and methods: Their limitation and potential", *J. Appl. Phycol.,* vol. 13, no. 4, pp. 307-315, 2001. [http://dx.doi.org/10.1023/A:1017560006941]

[199] M.F. de Jesus Raposo, R.M. de Morais, and A.M. de Morais, "Health applications of bioactive compounds from marine microalgae", *Life Sci.,* vol. 93, no. 15, pp. 479-486, 2013. [http://dx.doi.org/10.1016/j.lfs.2013.08.002] [PMID: 23994664]

[200] C. Meng, J. Huang, C. Ye, W. Cheng, J. Chen, and Y. Li, "Comparing the performances of circular ponds with different impellers by CFD simulation and microalgae culture experiments", *Bioprocess Biosyst. Eng.,* vol. 38, no. 7, pp. 1347-1363, 2015. [http://dx.doi.org/10.1007/s00449-015-1376-9] [PMID: 25680396]

[201] J. Doucha, and K. Lívanský, "Outdoor open thin-layer microalgal photobioreactor: Potential productivity", *J. Appl. Phycol.,* vol. 21, no. 1, pp. 111-117, 2009. [http://dx.doi.org/10.1007/s10811-008-9336-2]

[202] A. Kose, and S.S. Oncel, *Biohydrogen production from engineered microalgae Chlamydomonas reinhardtii*, 2014.

[203] F.G. Acién, J.M. Fernández, J.J. Magán, and E. Molina, "Production cost of a real microalgae production plant and strategies to reduce it", *Biotechnol. Adv.,* vol. 30, no. 6, pp. 1344-1353, 2012. [http://dx.doi.org/10.1016/j.biotechadv.2012.02.005] [PMID: 22361647]

[204] C.U. Ugwu, H. Aoyagi, and H. Uchiyama, "Photobioreactors for mass cultivation of algae", *Bioresour. Technol.,* vol. 99, no. 10, pp. 4021-4028, 2008. [http://dx.doi.org/10.1016/j.biortech.2007.01.046] [PMID: 17379512]

[205] L. Xu, P.J. Weathers, X.R. Xiong, and C.Z. Liu, "Microalgal bioreactors: Challenges and opportunities", *Eng. Life Sci.,* vol. 9, no. 3, pp. 178-189, 2009. [http://dx.doi.org/10.1002/elsc.200800111]

[206] N.H. Norsker, M.J. Barbosa, M.H. Vermuë, and R.H. Wijffels, "Microalgal production--a close look at

the economics", *Biotechnol. Adv.,* vol. 29, no. 1, pp. 24-27, 2011. [http://dx.doi.org/10.1016/j.biotechadv.2010.08.005] [PMID: 20728528]

[207] G. Olivieri, P. Salatino, and A. Marzocchella, "Advances in photobioreactors for intensive microalgal production: Configurations, operating strategies and applications", *J. Chem. Technol. Biotechnol.,* vol. 89, no. 2, pp. 178-195, 2014. [http://dx.doi.org/10.1002/jctb.4218]

[208] E. Molina Grima, E-H. Belarbi, F.G. Acién Fernández, A. Robles Medina, and Y. Chisti, "Recovery of microalgal biomass and metabolites: process options and economics", *Biotechnol. Adv.,* vol. 20, no. 7-8, pp. 491-515, 2003. [http://dx.doi.org/10.1016/S0734-9750(02)00050-2] [PMID: 14550018]

[209] K. Kumar, C.N. Dasgupta, B. Nayak, P. Lindblad, and D. Das, "Development of suitable photobioreactors for CO2 sequestration addressing global warming using green algae and cyanobacteria", *Bioresour. Technol.,* vol. 102, no. 8, pp. 4945-4953, 2011. [http://dx.doi.org/10.1016/j.biortech.2011.01.054] [PMID: 21334885]

[210] B. Wang, C.Q. Lan, and M. Horsman, *Closed photobioreactors for production of microalgal biomasses.* vol. Vol. 30. , 2012, pp. 904-912.

[211] C.N. Dasgupta, J. Jose Gilbert, P. Lindblad, T. Heidorn, S.A. Borgvang, K. Skjanes, and D. Das, "Recent trends on the development of photobiological processes and photobioreactors for the improvement of hydrogen production", *Int. J. Hydrogen Energy,* vol. 35, no. 19, pp. 10218-10238, 2010. [http://dx.doi.org/10.1016/j.ijhydene.2010.06.029]

[212] R.N. Singh, and S. Sharma, "Development of suitable photobioreactor for algae production -", *RE:view,* vol. 16, pp. 2347-2353, 2012.

[213] A. Hemschemeier, A. Melis, and T. Happe, "Analytical approaches to photobiological hydrogen production in unicellular green algae", *Photosynth. Res.,* vol. 102, no. 2-3, pp. 523-540, 2009. [http://dx.doi.org/10.1007/s11120-009-9415-5] [PMID: 19291418]

[214] G. Torzillo, and M. Seibert, *Hydrogen Production by Chlamydomonas reinhardtii.* Microalgal Cult. Appl. Phycol, 2013.

[215] A. Melis, and T. Happe, "Hydrogen production. Green algae as a source of energy", *Plant Physiol.,* vol. 127, no. 3, pp. 740-748, 2001. [http://dx.doi.org/10.1104/pp.010498] [PMID: 11706159]

[216] Y. Asada, and J. Miyake, "Photobiological hydrogen production", *J. Biosci. Bioeng.,* vol. 88, no. 1, pp. 1-6, 1999. [http://dx.doi.org/10.1016/S1389-1723(99)80166-2] [PMID: 16232564]

[217] T.K. Antal, T.E. Krendeleva, T.V. Laurinavichene, V.V. Makarova, M.L. Ghirardi, A.B. Rubin, A.A. Tsygankov, and M. Seibert, *The dependence of algal H2 production on Photosystem II and O2 consumption activities in sulfur-deprived Chlamydomonas reinhardtii cells*, 2003.

[218] S.S. Oncel, E. Imamoglu, E. Gunerken, and F.V. Sukan, "Comparison of different cultivation modes and light intensities using mono-cultures and co-cultures of Haematococcus pluvialis and Chlorella zofingiensis", *J. Chem. Technol. Biotechnol.,* vol. 86, no. 3, pp. 414-420, 2011. [http://dx.doi.org/10.1002/jctb.2532]

[219] A. Dubini, and M.L. Ghirardi, *Engineering photosynthetic organisms for the production of biohydrogen.*, 2015, pp. 241-253.

[220] O. Kruse, J. Rupprecht, J.H. Mussgnug, G.C. Dismukes, and B. Hankamer, "Photosynthesis: a blueprint for solar energy capture and biohydrogen production technologies", *Photochem. Photobiol. Sci.,* vol. 4, no. 12, pp. 957-970, 2005. [http://dx.doi.org/10.1039/b506923h] [PMID: 16307108]

[221] A. Melis, "Solar energy conversion efficiencies in photosynthesis: Minimizing the chlorophyll

antennae to maximize efficiency", *Plant Sci.,* vol. 177, no. 4, pp. 272-280, 2009. [http://dx.doi.org/10.1016/j.plantsci.2009.06.005]

[222] S.S. Oncel, A. Kose, C. Faraloni, E. Imamoglu, M. Elibol, G. Torzillo, and F.V. Sukan, "Biohydrogen production using mutant strains of Chlamydomonas reinhardtii: The effects of light intensity and illumination patterns", *Biochem. Eng. J.,* vol. 92, pp. 47-52, 2014. [http://dx.doi.org/10.1016/j.bej.2014.06.019]

[223] S.D. Tetali, M. Mitra, and A. Melis, "Development of the light-harvesting chlorophyll antenna in the green alga Chlamydomonas reinhardtii is regulated by the novel Tla1 gene", *Planta,* vol. 225, no. 4, pp. 813-829, 2007. [http://dx.doi.org/10.1007/s00425-006-0392-z] [PMID: 16977454]

[224] A. Melis, *Photosynthetic H 2 metabolism in Chlamydomonas reinhardtii (unicellular green algae)*, 2007.

[225] J.H. Mussgnug, S. Thomas-Hall, J. Rupprecht, A. Foo, V. Klassen, A. McDowall, P.M. Schenk, O. Kruse, and B. Hankamer, "Engineering photosynthetic light capture: impacts on improved solar energy to biomass conversion", *Plant Biotechnol. J.,* vol. 5, no. 6, pp. 802-814, 2007. [http://dx.doi.org/10.1111/j.1467-7652.2007.00285.x] [PMID: 17764518]

[226] J.E. Polle, S-D. Kanakagiri, and A. Melis, "tla1, a DNA insertional transformant of the green alga Chlamydomonas reinhardtii with a truncated light-harvesting chlorophyll antenna size", *Planta,* vol. 217, no. 1, pp. 49-59, 2003. [PMID: 12721848]

[227] M. Oey, I.L. Ross, E. Stephens, J. Steinbeck, J. Wolf, K.A. Radzun, J. Kügler, A.K. Ringsmuth, O. Kruse, and B. Hankamer, "RNAi knock-down of LHCBM1, 2 and 3 increases photosynthetic H2 production efficiency of the green alga Chlamydomonas reinhardtii", *PLoS One,* vol. 8, no. 4, p. e61375, 2013. [http://dx.doi.org/10.1371/journal.pone.0061375] [PMID: 23613840]

[228] K.D. Swanson, M.W. Ratzloff, D.W. Mulder, J.H. Artz, S. Ghose, A. Hoffman, S. White, O.A. Zadvornyy, J.B. Broderick, B. Bothner, P.W. King, and J.W. Peters, "[FeFe]-hydrogenase oxygen inactivation is initiated at the H cluster 2Fe subcluster", *J. Am. Chem. Soc.,* vol. 137, no. 5, pp. 1809-1816, 2015. [http://dx.doi.org/10.1021/ja510169s] [PMID: 25579778]

[229] K. Sybirna, P. Ezanno, C. Baffert, C. Léger, and H. Bottin, "Arginine171 of Chlamydomonas reinhardtii [Fe-Fe] hydrogenase HydA1 plays a crucial role in electron transfer to its catalytic center", *Int. J. Hydrogen Energy,* vol. 38, no. 7, pp. 2998-3002, 2013. [http://dx.doi.org/10.1016/j.ijhydene.2012.12.078]

[230] J.E. Godman, A. Molnár, D.C. Baulcombe, and J. Balk, "RNA silencing of hydrogenase(-like) genes and investigation of their physiological roles in the green alga Chlamydomonas reinhardtii", *Biochem. J.,* vol. 431, no. 3, pp. 345-351, 2010. [http://dx.doi.org/10.1042/BJ20100932] [PMID: 20726841]

[231] V. Chochois, D. Dauvillée, A. Beyly, D. Tolleter, S. Cuiné, H. Timpano, S. Ball, L. Cournac, and G. Peltier, "Hydrogen production in Chlamydomonas: photosystem II-dependent and -independent pathways differ in their requirement for starch metabolism", *Plant Physiol.,* vol. 151, no. 2, pp. 631-640, 2009. [http://dx.doi.org/10.1104/pp.109.144576] [PMID: 19700559]

[232] J. Msanne, D. Xu, A.R. Konda, J.A. Casas-Mollano, T. Awada, E.B. Cahoon, and H. Cerutti, "Metabolic and gene expression changes triggered by nitrogen deprivation in the photoautotrophically grown microalgae Chlamydomonas reinhardtii and Coccomyxa sp. C-169", *Phytochemistry,* vol. 75, pp. 50-59, 2012. [http://dx.doi.org/10.1016/j.phytochem.2011.12.007] [PMID: 22226037]

[233] M.C. Posewitz, S.L. Smolinski, S. Kanakagiri, A. Melis, M. Seibert, and M.L. Ghirardi, "Hydrogen

photoproduction is attenuated by disruption of an isoamylase gene in Chlamydomonas reinhardtii", *Plant Cell,* vol. 16, no. 8, pp. 2151-2163, 2004.
[http://dx.doi.org/10.1105/tpc.104.021972] [PMID: 15269330]

[234] A. Scoma, L. Bertin, C. Pintucci, S. Raddi, and F. Fava, "Inhibition of photosystem 2 in starch-enriched Chlamydomonas reinhardtii cells prevents the efficient induction of H 2 production in sulfur-depleted cultures", *Int. J. Hydrogen Energy,* vol. 37, no. 14, pp. 10604-10610, 2012.
[http://dx.doi.org/10.1016/j.ijhydene.2012.04.046]

[235] B.S. Hamilton, K. Nakamura, and D.A. Roncari, "Accumulation of starch in Chlamydomonas reinhardtii flagellar mutants", *Biochem. Cell Biol.,* vol. 70, no. 3-4, pp. 255-258, 1992.
[http://dx.doi.org/10.1139/o92-039] [PMID: 1515125]

[236] A. Scoma, D. Krawietz, C. Faraloni, L. Giannelli, T. Happe, and G. Torzillo, "Sustained H2 production in a Chlamydomonas reinhardtii D1 protein mutant", *J. Biotechnol.,* vol. 157, no. 4, pp. 613-619, 2012.
[http://dx.doi.org/10.1016/j.jbiotec.2011.06.019] [PMID: 21723340]

[237] U. Johanningmeier, and S. Heiss, "Construction of a Chlamydomonas reinhardtii mutant with an intronless psbA gene", *Plant Mol. Biol.,* vol. 22, no. 1, pp. 91-99, 1993.
[http://dx.doi.org/10.1007/BF00038998] [PMID: 8499620]

[238] P. Mulo, I. Sakurai, and E-M. Aro, *Strategies for psbA gene expression in cyanobacteria, green algae and higher plants: From transcription to PSII repair*, 2012.

[239] M. Bochenek, G.J. Etherington, A. Koprivova, S.T. Mugford, T.G. Bell, G. Malin, and S. Kopriva, "Transcriptome analysis of the sulfate deficiency response in the marine microalga Emiliania huxleyi", *New Phytol.,* vol. 199, no. 3, pp. 650-662, 2013.
[http://dx.doi.org/10.1111/nph.12303] [PMID: 23692606]

[240] A. Melis, and H.C. Chen, "Chloroplast sulfate transport in green algae--genes, proteins and effects", *Photosynth. Res.,* vol. 86, no. 3, pp. 299-307, 2005.
[http://dx.doi.org/10.1007/s11120-005-7382-z] [PMID: 16307303]

[241] H.C. Chen, A.J. Newton, and A. Melis, "Role of SulP, a nuclear-encoded chloroplast sulfate permease, in sulfate transport and H2 evolution in Chlamydomonas reinhardtii", *Photosynth. Res.,* vol. 84, no. 1-3, pp. 289-296, 2005.
[http://dx.doi.org/10.1007/s11120-004-7157-y] [PMID: 16049788]

[242] K. Pauwels, *Engineered nucleases for targeted genome editing and gene regulation.*.

[243] H. Kim, and J.S. Kim, "A guide to genome engineering with programmable nucleases", *Nat. Rev. Genet.,* vol. 15, no. 5, pp. 321-334, 2014.
[http://dx.doi.org/10.1038/nrg3686] [PMID: 24690881]

[244] T. Gaj, *ZFN, TALEN and CRISPR/Cas based methods for genome engineering*, 2014.

[245] J.N. Abelson, M.I. Simon, A.M. Pyle, S.P. Colowick, and N.O. Kaplan, *The Use of CRISPR/cas9, ZFNs, TALENs in Generating Site Specific Genome Alterations*, vol. 4. 2014.

[246] I. Sizova, A. Greiner, M. Awasthi, S. Kateriya, and P. Hegemann, "Nuclear gene targeting in Chlamydomonas using engineered zinc-finger nucleases", *Plant J.,* vol. 73, no. 5, pp. 873-882, 2013.
[http://dx.doi.org/10.1111/tpj.12066] [PMID: 23137232]

[247] S.M. Schaeffer, and P.A. Nakata, "CRISPR/Cas9-mediated genome editing and gene replacement in plants: Transitioning from lab to field", *Plant Sci.,* vol. 240, pp. 130-142, 2015.
[http://dx.doi.org/10.1016/j.plantsci.2015.09.011] [PMID: 26475194]

[248] L. Yao, I. Cengic, J. Anfelt, and E.P. Hudson, "Multiple Gene Repression in Cyanobacteria Using CRISPRi", *ACS Synth. Biol.,* vol. 5, no. 3, pp. 207-212, 2016.
[http://dx.doi.org/10.1021/acssynbio.5b00264] [PMID: 26689101]

[249] P.D. Weyman, K. Beeri, S.C. Lefebvre, J. Rivera, J.K. McCarthy, A.L. Heuberger, G. Peers, A.E.

Allen, and C.L. Dupont, "Inactivation of Phaeodactylum tricornutum urease gene using transcription activator-like effector nuclease-based targeted mutagenesis", *Plant Biotechnol. J.,* vol. 13, no. 4, pp. 460-470, 2015.
[http://dx.doi.org/10.1111/pbi.12254] [PMID: 25302562]

[250] S-E. Shin, J-M. Lim, H.G. Koh, E.K. Kim, N.K. Kang, S. Jeon, S. Kwon, W-S. Shin, B. Lee, K. Hwangbo, J. Kim, S.H. Ye, J-Y. Yun, H. Seo, H-M. Oh, K-J. Kim, J-S. Kim, W-J. Jeong, Y.K. Chang, and B.R. Jeong, "CRISPR/Cas9-induced knockout and knock-in mutations in Chlamydomonas reinhardtii", *Sci. Rep.,* vol. 6, no. April, p. 27810, 2016.
[http://dx.doi.org/10.1038/srep27810] [PMID: 27291619]

[251] W. Jiang, A.J. Brueggeman, K.M. Horken, T.M. Plucinak, and D.P. Weeks, "Successful transient expression of Cas9 and single guide RNA genes in Chlamydomonas reinhardtii", *Eukaryot. Cell,* vol. 13, no. 11, pp. 1465-1469, 2014.
[http://dx.doi.org/10.1128/EC.00213-14] [PMID: 25239977]

[252] K. Baek, D.H. Kim, J. Jeong, S.J. Sim, A. Melis, J-S. Kim, E. Jin, and S. Bae, "DNA-free two-gene knockout in Chlamydomonas reinhardtii via CRISPR-Cas9 ribonucleoproteins", *Sci. Rep.,* vol. 6, p. 30620, 2016.
[http://dx.doi.org/10.1038/srep30620] [PMID: 27466170]

[253] Y. Mazor, H. Toporik, and N. Nelson, *Temperature-sensitive PSII and promiscuous PSI as a possible solution for sustainable photosynthetic hydrogen production*, 2012.

[254] J.C. Ortiz-Marquez, M. Do Nascimento, J.P. Zehr, and L. Curatti, "Genetic engineering of multispecies microbial cell factories as an alternative for bioenergy production", *Trends Biotechnol.,* vol. 31, no. 9, pp. 521-529, 2013.
[http://dx.doi.org/10.1016/j.tibtech.2013.05.006] [PMID: 23791304]

[255] T. Holtermann, and R. Madlener, "Assessment of the technological development and economic potential of photobioreactors", *Appl. Energy,* vol. 88, no. 5, pp. 1906-1919, 2011.
[http://dx.doi.org/10.1016/j.apenergy.2010.11.018]

[256] M. L. Ghirardi, P. W. King, M. C. Posewitz, P. C. Maness, A. Fedorov, K. Kim, J. Cohen, K. Schulten, and M. Seibert, *Approaches to developing biological H(2)-photoproducing organisms and processes, .*

[257] W. Chen, P. Zhou, M. Zhang, Y. Zhu, X. Wang, X. Luo, Z. Bao, and L. Yu, "Transcriptome analysis reveals that up-regulation of the fatty acid synthase gene promotes the accumulation of docosahexaenoic acid in Schizochytrium sp. S056 when glycerol is used", *Algal Res.,* vol. 15, pp. 83-92, 2016.
[http://dx.doi.org/10.1016/j.algal.2016.02.007]

[258] S.S. Oncel, A. Kose, and C. Faraloni, *Handbook of Marine Microalgae.* Elsevier, 2015.

[259] X. Zeng, M.K. Danquah, X.D. Chen, and Y. Lu, "Microalgae bioengineering: From CO2 fixation to biofuel production", *Renew. Sustain. Energy Rev.,* vol. 15, no. 6, pp. 3252-3260, 2011.
[http://dx.doi.org/10.1016/j.rser.2011.04.014]

[260] S. Venkata Mohan, G.N. Nikhil, P. Chiranjeevi, C. Nagendranatha Reddy, M.V. Rohit, A.N. Kumar, and O. Sarkar, "Waste biorefinery models towards sustainable circular bioeconomy: Critical review and future perspectives", *Bioresour. Technol.,* vol. 215, pp. 2-12, 2016.
[http://dx.doi.org/10.1016/j.biortech.2016.03.130] [PMID: 27068056]

[261] B.D. James, G.N. Baum, J. Perez, and K.N. Baum, *Technoeconomic Boundary Analysis of Biological Pathways to Hydrogen Production, .*
[http://dx.doi.org/10.2172/1218404]

CHAPTER 6

Algal Production Platforms

Alessandro Marco Lizzul[1,*] and **Aitor Lekuona-Amundarain**[2]

[1] *Department of Environmental Engineering, University College London, Gower Street, WC1E 6BT, London, United Kingdom*

[2] *Tecnalia Research and Innovation, Leonardo Da Vinci 11, E-01510, Miñano (Araba), Spain*

Abstract: The design and construction of any algal photobioreactor is a complex multi-parametric problem, with successful examples requiring the consideration of many different factors. In practical terms, this means that most feasible cultivation platforms require a compromise between biotic, abiotic and economic factors. As a result, most of the successful designs are found commercially prompt to maximise favourable parameter ranges within practical cost limits. The end result is the construction and subsequent operation of the photobioreactor within a multi-parametric 'sweet-spot', which will vary dependent on the production process and end product. Such cost considerations are particularly important for the production of lower value commodities such as biofuels, which are yet to achieve economical production at scale. Considering these aforementioned constraints; 60 years of photobioreactor design and production have resulted in several general blue-prints for larger scale algal production platforms. Despite this fact, most of the photobioreactors in successful operation, today, can be categorised as having characteristics that are based on a relatively limited repertoire of basic designs and construction materials. The most common separation is between open and closed systems which differ fundamentally in their levels of parameter control and cost base. Further configurational commonalities can be seen in many examples and are most notable in their external appearances. The most common photobioreactor systems are horizontal tubular, vertical column, panel or plate and pond based systems.

Keywords: Photobioreactor, Biofuel, Algae, Bubble column, Flat plate, Airlift, Tubular, Serpentine, Manifold, Horizontal, Vertical, Turbulent, Light path, Mass transfer.

INTRODUCTION

Many photobioreactor design considerations are relatively well established within the literature [1 - 5], and can be simplified into several major categories. For the purposes of this chapter, we will consider phototrophic or mixotrophic growth, as

* **Corresponding author Alessandro Marco Lizzul:** Department of Environmental Engineering, University College London, Gower Street, WC1E 6BT, London, United Kingdom; Tel: +44 1684 312980; E-mail: marcolizzul@variconaqua.com

José Carlos Magalhães Pires (Ed.)

heterotrophic cultivation can be undertaken in the conventional non illuminated fermenter designs [6]. For phototrophic growth the parameter of greatest importance is a suitable surface for exposure to a source of irradiance, as without adequate light, the photosynthetic processes cannot occur. The next most important factor is to ensure suitable provision for the containment of the culture and addition of the nutrients necessary for algal growth. In practical terms, this means that the designs should include entry points for the introduction of an appropriate carbon source; which is usually bubbled into the culture as carbon dioxide or added as dissolved bicarbonate. Within the cultivation medium, suitable temperature control is also a very important factor for the maintenance of optimal growth rates and must be kept within set constraints according to strain preferences. Likewise, the ability to remove inhibitory waste products from the process is imperative, especially when growing the culture to higher densities. This is particularly true for dissolved oxygen levels, which at high levels, are known to inhibit photosynthesis [3]. A final factor of considerable importance is to ensure that these biotic and abiotic factors interface successfully; this is most commonly achieved by high rates of mass transfer under well mixed (or turbulent) conditions within the reactor.

IMPORTANT CONSIDERATIONS

Lighting

Access to sufficient light is imperative for phototrophic algal cultivation, and its maximisation within biological and economic constraints is desirable at all times. Algal cultivation can be undertaken using either natural or artificial light; providing that they have sufficient quanta at the correct wavelengths. Traditional open ponds often rely solely on natural sunlight and growth within these systems is subject to seasonal and diurnal cycles [7]. Reliance on natural light has obvious drawbacks, not least because a sizable proportion of 24 hour day has insufficient light, suggesting that photosynthesis will stop and respiration will become the predominant metabolic activity within the culture. Another important factor to consider when growing algae under natural conditions is that many strains respond differently to diurnal and seasonal cycles. This has the combined effect of altering the final biomass composition, and as a result can change the yield of metabolites necessary for biofuel production [8]. Given these considerations, algal production under natural lighting is better suited to latitudes closer to the equator, where irradiation is more consistent and of a better quality. Despite these inherent problems the one notable benefit of natural light is the significant reduction in energy costs imparted onto the cultivation process when used. This simple fact means that to date a vast majority of commercially grown algae are produced with natural lighting [8, 9], and it is likely that biofuels will be produced in this way.

Artificially illuminated systems have the benefit of offering a consistent light source, but at considerable energetic costs. Conventional systems for lighting include fluorescent bulbs, which are often deployed at laboratory scale due to a favourable wavelength profile and a reasonable power consumption (see Fig. **1A**). However, these lights often lack the power output and penetration required for larger and more densely growing cultures. For these bigger systems, popular choices for irradiance include metal halide lights (see Fig. **1B**), as they have a spectral output that closely matches natural light, making them ideal for photosynthetic processes. Recent developments in light emitting diode (LED) technology have also made them an attractive alternative to conventional lighting systems, as they can be manufactured to specific wavelengths at reasonable cost. This means that the cultivator has the potential to tailor the incident wavelengths to suit the strain or process requirements. For example, red LEDs with wavelengths between 620-700 nm and blue LEDs with wavelengths between 455-492 nm can be incorporated in varying proportions (see Fig. **1C**). The result is that wavelengths, not used during photosynthesis, can be excluded, leading to an increase in quantum efficiency [10]. Relationships between light intensity and penetration can be used to predict the productivity and final yield of algal biomass produced in the system with a suitable example found in the work of [11].

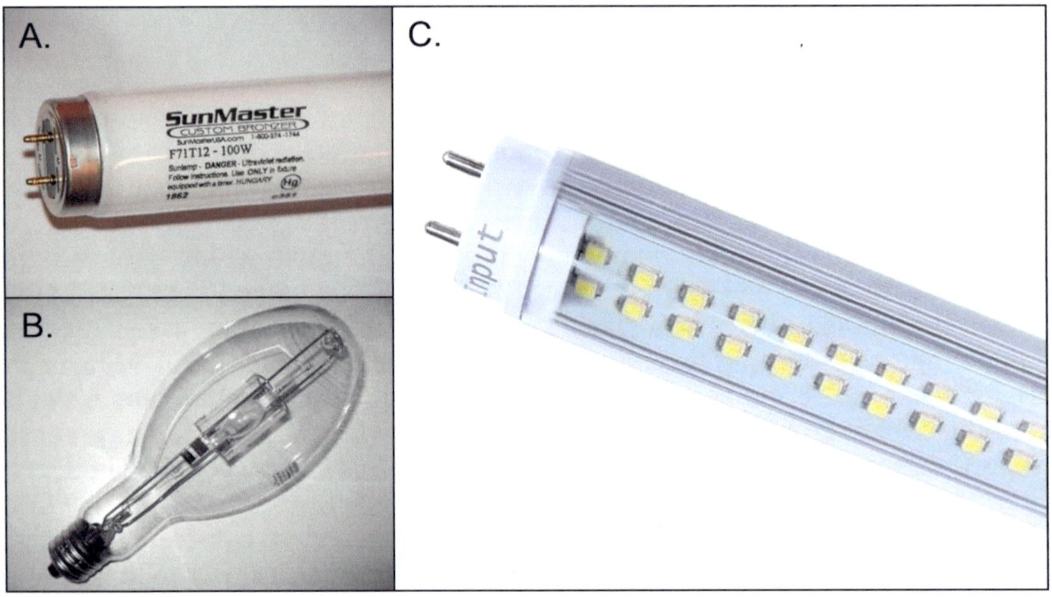

Fig. (1). Examples of common lighting arrangements.
Photograph A shows a fluorescent light tube [12]. Photograph B shows a Metal Halide lamp [13]. Photograph C shows an LED lighting block [14].

Mixing and Mass Transfer

Adequate mixing is required in all bioprocesses to maintain homogenous culture conditions within a heterogonous mixture [15]. In the case of algal cultivation, this is particularly important so as to prevent cells settling out of culture which would decrease the overall process efficiency. Adequate mixing helps to ensure access to nutrients, photons of light and to increase gaseous oxygen and carbon dioxide exchange. The mixing within typical bioprocesses can be induced *via* convective or intensive methods, although within most algal applications intensive and direct fluid displacement is preferred for the aforementioned reasons. This means that fully turbulent reactor operation is preferable, and a simple overall characterisation of system turbulence can be achieved by ascertaining the dimensionless Reynolds number [15]. Characteristic Reynolds values differ dependent on the system geometry but values greater than 4,000-10,000 describe a fully turbulent regime and would indicate that the cells are well mixed. Mass transfer within reactor systems can be characterized by the mass transfer coefficient (k_L) and further theory and approaches to measure it can be found in [15]. The displacement required for turbulent fluid mixing and high mass transfer can be undertaken in one of many ways. Conventional systems include: impeller or paddlewheel agitation, direct liquid displacement *via* pumping, or airlift systems [9].

Impellers and paddle wheels can be used to move an aqueous dispersion around predictable circulatory patterns by introducing kinetic energy into the fluid. Impellers tend to be used within smaller laboratory scale systems, or within classical enclosed fermenters (see Fig. **2A**). This method of mixing has the benefit of transferring a large amount of energy into the fluid, but can also have some negative attributes, such as high energy consumption and the creation of considerable shearing effects [15]. Furthermore, it has been found that cultivating photosynthetic algae with impellers on a larger scale is challenging due to the poor light penetration in bigger fermentation vessels [16]. This means that large scale impeller mixed systems have been deployed solely for heterotrophic cultivation of algae, with Solazyme producing algal oil for biofuel applications in this way in the U.S [17]. Paddlewheel systems differ from impellers in that they are only partially submerged and positioned in a horizontal plane to the culture; they also run at a much lower cycling speed (see Fig. **2B**). This mode of mixing is preferred for larger scale algal production, due to its lower running costs and comparative ease of maintenance [18]. However, the low liquid velocities induced by this type of mixing result in an algal culture that is sub-optimally mixed, and dead zones can form within certain areas of the pond. Some of these problems can be overcome to an extent by introducing baffles and maintaining certain depth to width and length relationships [19, 20].

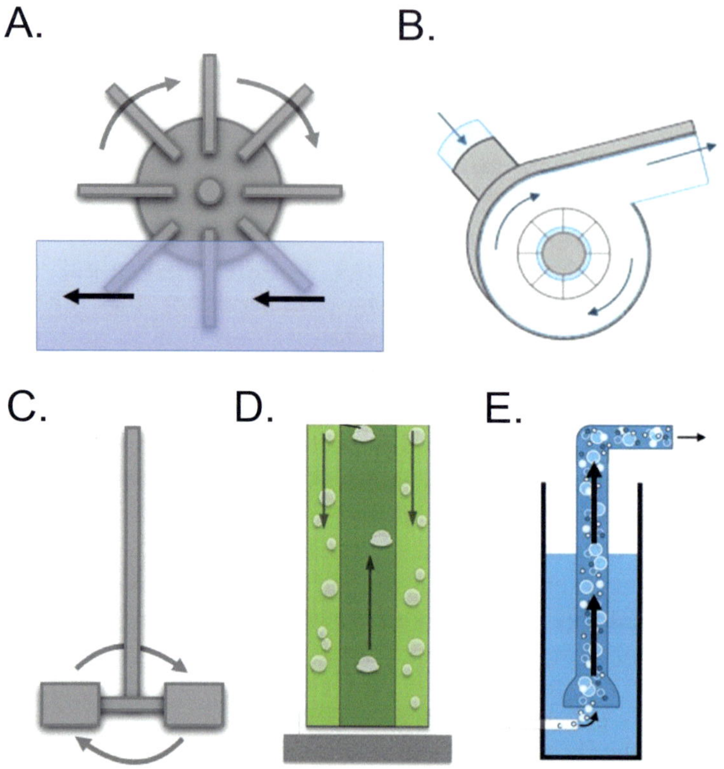

Fig. (2). Diagrams illustrating different mixing methods. (**A**) Paddlewheel, used to create directional fluid mixing in open ponds. (**B**) Velocity pump, used in tubular systems, showing the inlet and outlet. (**C**) Rushton impeller, deployed to create radial mixing within stirred tank reactors. (**D**) Bubble column reactor, turbulence is created by direct bubbling of the culture. (**E**) Airlift reactor where air injection creates directional liquid circulation.

Other prominent mixing methods deployed within closed systems include liquid or air pumps. These conventional pumps can be grouped into three main categories, based on how they create the actual mixing. The first type of pump is the impulse or airlift pump, which creates a density difference in the fluid circulation pathway, thereby forcing liquid circulation to occur. These reactors are often described as having a bubble column or airlift configuration, see Fig. (**2C**) and (**D**), and are discussed further in this chapter and at length in the work of [21]. Another prominent category of system pumping is by positive displacement, which generates movement by trapping a fixed volume of fluid and moving it through into a discharge pipe. Pumping in this way is achieved by creating a driving motion through reciprocating or rotary motion. The final major category of pumps includes velocity driven motors. These operate by adding kinetic energy into the fluid, increasing the pressure and flow rate around a set pathway (see Fig. **2E**). These types of pump form a broad category, which include rotodynamic and centrifugal motors. Centrifugal pumps are frequently used to move liquids through piping systems and are therefore a popular choice for horizontal tubular reactors [22]. These pumps operate by allowing fluid to enter the impeller along or near to the rotating axis. The liquid is then accelerated by the impeller flowing outward in either radial or axial directions where it enters a diffuser or volute chamber from which it can exit towards the downstream piping. Centrifugal systems can find particular applications where large discharge through smaller heads is required.

Control Systems and Construction Materials

Larger scale algal applications have a requirement for varying degrees of process detection, control and automation. This allows for maintenance of optimal parameters and conditions, whilst minimising personnel workload and operational costs. The actual core components of the control system takes the form of a series of sensor and control relays that can measure, relay and then adjust parameters to a set point. More complex systems allow for the ability of dynamic control, including turbidostat culture where the culture density can be maintained. Several common parameters are most often measured and controlled during algal cultivation. These include light intensity, temperature, pH, dissolved oxygen and cell density. For more specialised applications other more specific probes can be incorporated, including probes that trace nitrogen species levels. Once values for these variables are detected by the sensors they are converted into a readable electrical signal that is passed onto a computer. In the case of a control loop, the readings can be used to trigger the adjustment of conditions back to the set point (see Fig. (**3**)).

Fig. (3). AlgaeConnect Control System and pH Probe. The probes can interface with control systems such as AlgaeConnect produced by AlgaeLab Systems [23, 24].

The decision on the materials used for photobioreactor construction is often driven by a trade-off in lower upfront expenditure versus longer lasting and more expensive building materials. Modern trends in reactor design include an increasing focus on the environmental impact of construction materials, achieved *via* detailed life-cycle assessment [25]. In many conventional photobioreactors and open ponds construction is often undertaken from cheap, durable and readily available materials. In the case of raceway pond systems this requires the levelling and compacting of the ground upon which it will be built, over which concrete or plastic under-lining is spread to contain the culture [5, 19]. In the case of conventional photobioreactors, the choice of construction material is often decided by the requirements for transparent materials with high optical clarity, such as durable polymers like acrylic or polycarbonate [5]. Glass is also being increasingly deployed, and has many benefits in comparison to plastics, especially in terms of overall longevity. Other less optically important parts of the reactor can be constructed from cheaper materials such as polyvinyl chloride (PVC) [9].

CULTIVATION SYSTEMS

Pond Based Systems

The initial design for high yield open algal ponds can be traced back to work initiated in the latter part of the 1940s and early 1950s [26]. Such systems can also be described as ‘high rate algal ponds’ (HRAPs) within the literature. To date open ponds have remained the most widespread and historically successful of large scale production systems. There are several common variations in the design of an open system, with most examples taking the form of concrete or plastic lined channels that form a raceway loop system [27]. Most ponds tend to be fairly shallow, so as to allow for maximal solar penetration of the culture, usually within the range of 0.1-0.3 metres in depth [28]. Mixing within such systems is usually achieved by paddle wheel agitation and nutrients can be added either in batch or continuously. Although most ponds are mixed with paddle wheels there have also been some attempts to design airlift ponds, however, the findings from these studies have shown that they compare unfavourably with conventional paddle wheel systems [9]. Carbon dioxide is often introduced into the pond *via* submerged aerators located within sumps [19], although adequate gaseous retention and distribution within the medium can be hard to maintain due to the large contact areas involved. Harvesting methods are dependent on the desired product and can occur in either continuous or batch unit operations downstream. As the name suggests, most ponds are open to the elements, although some coverings have been employed in smaller pilot type projects or inoculation ponds [27]. Typical arrangements are shown in Fig. (**4**).

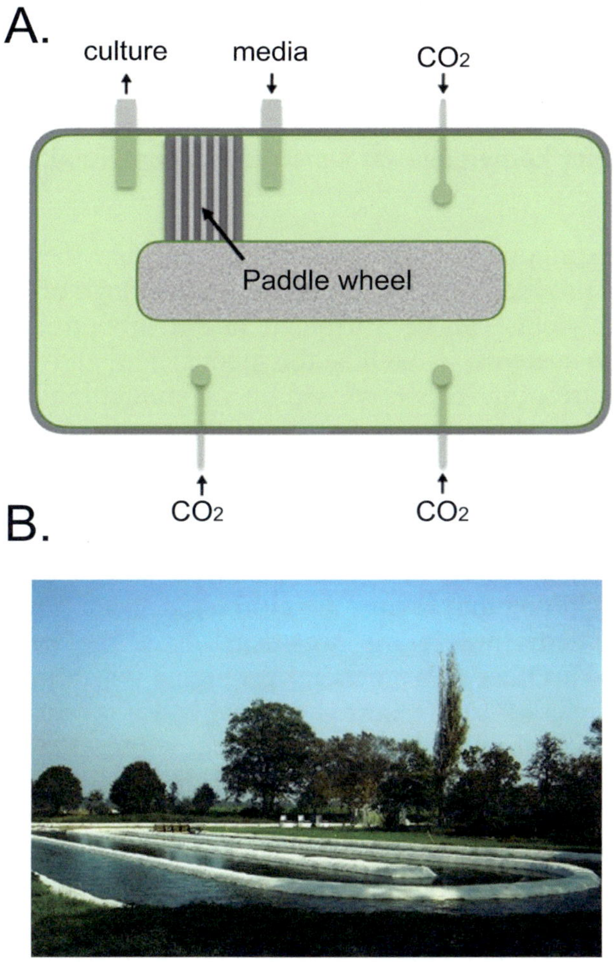

Fig. (4). Schematic and photograph showing the typical arrangement of a raceway pond. (A) Shows an aerial view, indicating how the pond is mixed and sparged with CO_2 (modified from [29]). (B) Shows a raceway pond operated by Ingrepro in the Netherlands [30].

There are several process benefits that can be realised by the adoption of an open pond system, which helps to explain why it remains the preferred production system for many algal applications. The single most important factor is the low energy inputs required to maintain a sufficient level of mixing in comparison to other production systems [31]. This makes most production processes more economical within open pond systems. Other benefits include cheap and simple constituent parts, as well as easy access to the whole system. This means that fouling can be cleaned relatively easily, whilst the lack of internalised parts also makes the systems relatively flexible should operational changes be required.

There are also some biotic benefits to the relatively low liquid velocity within open ponds. Namely the reduction in the shear levels encountered within the system, which can allow for the cultivation of more sensitive algal species. Another important factor that should not be understated is that raceway ponds have been in existence for a number of years, and there is already a considerable body of operational knowledge on successful commercial operating procedures [28].

Limitations of open pond designs include the fact that they have relatively low levels of biomass productivity, with averages in the range of 0.18-0.25 g L^{-1} d^{-1} [3, 32]. These lower yields can be attributed to the suboptimal mixing conditions found within such systems, as well as the inconsistent and variable light profiles found through the reactor. This is caused by a combination of low liquid velocity and culture depth which means that much of the sunlight is blocked from all but the uppermost part of the pond, slowing down the average growth rate, especially in denser cultures [33]. Other issues with the use of open ponds concern their exposed nature, making contamination with competing organisms or predator species particularly problematic. These problems can be partially overcome by using growth conditions that favour the cultivated strain. For example the use of extremophilic or extremotolerant organisms is a preferred option to avoid contamination [34]. Other more general problems with open ponds include the relatively large areas of land required to establish a production facility. Due to their large size, open ponds are also particularly vulnerable to changes in abiotic factors such as temperature and light fluctuations. This makes most open ponds unsuitable above certain latitudes. Finally, from an environmental perspective there are also considerable concerns regarding the evaporative losses that can occur within open pond systems [29].

In terms of biomass production, open pond systems are particularly suited to the low or middle value range of biomass products, including biofuels [9] and feed production. They have also found considerable usage in the production of some higher value pigments [35], providing that contamination issues can be mitigated. Several companies produce algae in this way, including Cyanotech, who cultivate *Haematococcus pluvialis* to produce astaxanthin in a two stage system, with the first vegetative stage within closed photobioreactors and the second maturation stage within HRAP systems [26]. Other companies such as Seambiotic produce algae in open ponds grown from flue gas and wastewater, which may be better suited to biofuel type applications. Likewise, the potential use of algae in wastewater treatment infrastructure using HRAP type systems, particularly within warmer climates could be a potential avenue for cost reduction [36, 37]. Given the cost implications of algal production it is likely that any serious attempt to generate large quantities of lower value products such as algal biomass for biofuel will require the large scale deployment of HRAP type systems due to their comparably favourable operational costs.

Plate Based Systems

In theory flat panel reactors are the most efficient enclosed photobioreactor system due to their large surface area to volume ratio which results in very little culture induced shading. This means that these systems display a particularly high conversion efficiency of incident sunlight [38]. These favourable characteristics can result in some of the highest reported daily levels of biomass production, with some averages reaching 2.4 g L^{-1} d^{-1} [3, 32]. Additionally, a panelled array can easily be tilted towards the sun to maximise solar penetration [39], see Fig. (**5A**). Dependent upon configuration, plate reactors can benefit from lower levels of dissolved oxygen build-up than many other types of closed reactor. This is due to the relatively short circulatory path, meaning the culture is able to rapidly de-gas. Operational costs of panelled systems are comparable to tubular or column reactors, and are heavily dependent on the selected mixing mode and intensity. Indeed, the mixing within plate systems tends not to follow constrained directional patterns, and could be described as generally very turbulent, in a spectrum between bubble column and airlift flow patterns, which depend on the presence of internal structures such as baffles or channels. Like other closed photobioreactors, external contamination can be kept to a minimum, with an ability to be sterilised more effectively than many other configurations [38].

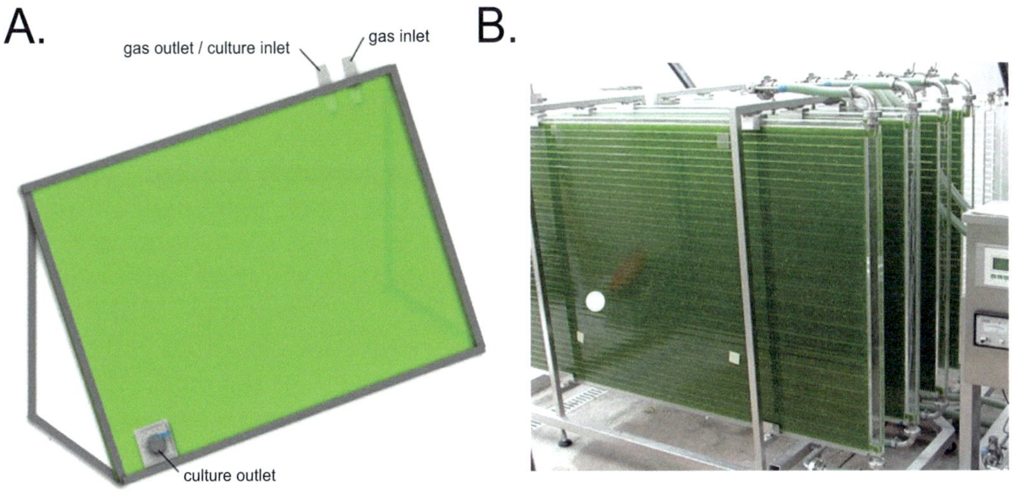

Fig. (5). Diagram shows the potential arrangements of flat panelled reactors. (A) Gives an indication of the many different tilt angles that are possible to optimise solar penetration. (B) Shows an array of glass plated panelled reactors manufactured by IGV in Germany [40].

One potential drawback of flat panelled systems is the practical scalability of the reactors, as some level of compromise has to be struck between solar collecting surface area, culture depth and areal considerations, see Fig. (**5**). This problem can be overcome to some extent by connecting a modular array of panels in tandem, although the resultant areal footprint would be likely to be bigger than an equivalent volume of other types closed reactors. Other physical drawbacks are associated with temperature control within the system, which can be more energy intensive than other closed systems due to the large surface area to volume ratio. The large surface areas and short light paths within plate reactors can also create problems with photo-damage if the light levels are too high; although this damaging effect can be lessened by using turbidostat based cultivation techniques. As with all bioreactors it is also likely that some degree of wall growth and fouling is unavoidable, especially at the meeting of straight edges, where the liquid velocity is lower. The large surface area to volume ratio also means that the levels of hydrodynamic stress placed on the algae can be higher than in some other systems [32]. Another notable point is that the rapid degassing seen in flat panels could be seen as advantageous in most applications, as it prevents the build-up of excess dissolved oxygen. However, the converse is true if gas use efficiency is sought, as gas retention is considerably lower in these systems.

Examples of larger scale plate or panel PBR arrays (Fig. **5B**) are less commonly deployed than other closed systems, meaning that to date there are relatively few examples of commercial scale algal production using these configurations. This is most probably due to numerous manufacturing and cost related issues that still need to be overcome before they become widely utilised. For example, there have been reports of difficulty with manufacturing at scale, and problems associated with sealing and warping of the materials during temperature stress. As process problems are overcome it is likely that flat panelled systems will increasingly find similar applications to those employed by tubular reactors, although with some distinct process advantages in terms of degassing, and solar penetration. At smaller scale the combination of large surface area and short light path make panelled systems a favourable option, and systems such as Infors Labfors reactor have found considerable use in growth modelling and screening. In the shorter term it is likely that higher value indoor algal cultivation may find considerable use for flat panels, especially if good productivity levels are required. Algenol are one example, using large modular bag based panels to produce bioethanol.

Horizontal Tubular Systems

Tubular photobioreactors encompass a broad range of designs, and can be arranged either horizontally or vertically, in serpentine or manifold arrangements. They can also be laid flat on the ground or positioned stacked above one another.

Most traditional commercial and research designs tend to favour a single serpentine or horizontal manifold arrangement, with the tubes stacked vertically in order to maximise areal productivity. Prominent examples within the literature include a ground based, flat horizontal tubular reactor in Cadiz, Spain [1] or the vertically stacked manifold system at the Ben Gurion University, Israel [41]. Mixing within these tubular reactors is conventionally achieved with a variety of pumping systems, including centrifugal or diaphragm pumps, as well as airlift driven systems. Popular velocity based pumping systems include centrifugal pumps, in which a rotating impeller increases the pressure and flow rate of a fluid. Examples of manifold type systems can be seen in Fig. (**6A** and **6B**), whilst serpentine configurations can be found in Fig. (**6C** and **6D**).

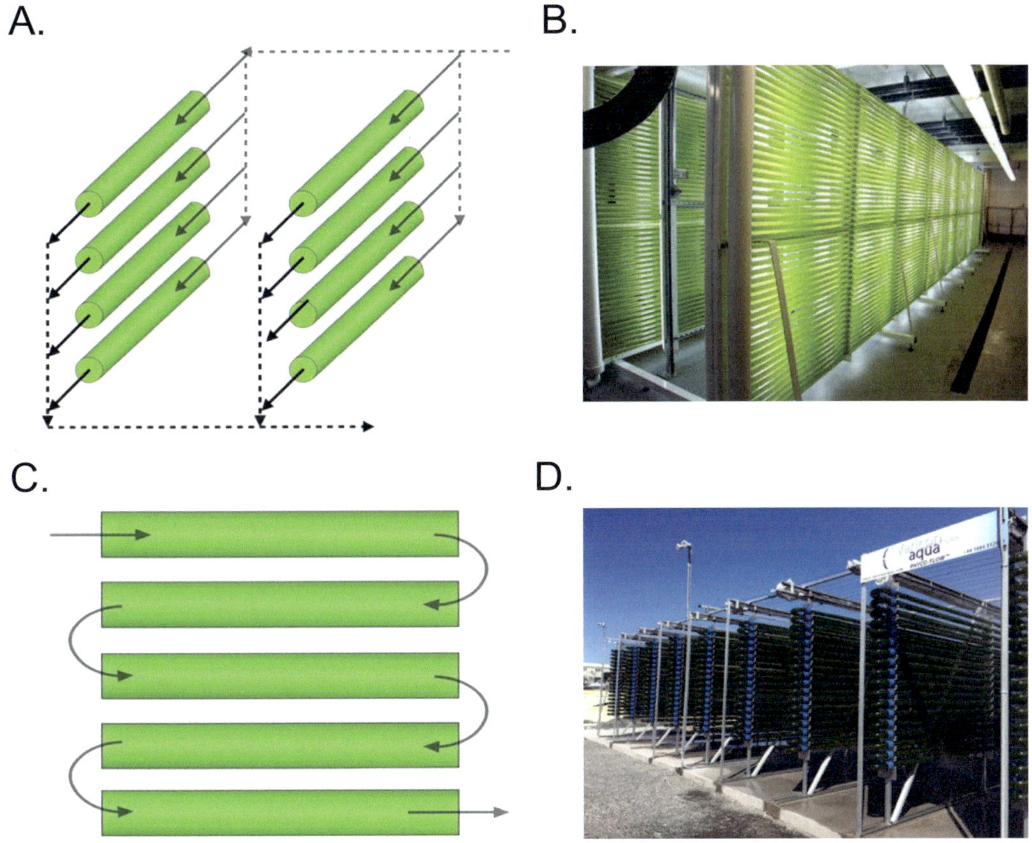

Fig. (6). Schematic showing the potential arrangements of tubular reactors. (A) Illustration of a manifold system, showing how the manifolds split the main flow, modified from [29]. (B) The BioFence™ produced by Varicon Aqua operates using a manifold configuration [43]. (C) Illustration of a serpentine arrangement, modified from [42]. (D) Varicon Aqua's Phyco-Flow™ which is a prominent example of a serpentine system. Photographs courtesy of Varicon Aqua Solutions, UK.

Tubular systems have numerous comparative advantages over open pond systems stemming from their enclosed nature. Most importantly they have been shown to achieve higher and more consistent levels of areal productivity than can be obtained within open ponds [42]. This is because horizontal tubular systems have a greater proportion of illuminated surface area than open ponds, with tubes often in the region of 0.03 – 0.1 m in diameter. An enclosed system means that there is better control over both biotic and abiotic factors during cultivation, making it easier to control contamination, whilst also allowing for a wider repertoire of strains to be cultivated. Another factor that improves productivity within tubular systems is the greater level of turbulence that is achievable within them [26], caused by the higher liquid velocities. In combination with better light penetration, this allows for far higher algal growth rates than open ponds, with average reports of biomass productivities ranging from 0.25-1.22 g L^{-1} d^{-1} [3, 32]. Tubular configurations also have the comparative benefit of lower levels of water loss in than in open pond systems.

Despite the many process benefits that can be conferred by utilising a tubular configuration there are also some limitations. One of the major issues with tubular reactors is the fact that the capital and operational costs are considerably higher than that of open ponds [44]; this is due in part to the relative cost of building materials as well as the considerable energy requirements for operation. As with all closed systems it is also common to find some degree of wall growth and fouling. This can increase the costs associated with the cleaning and maintenance of such a system, and can become a particular challenge with larger and more elaborate photobioreactors. To overcome this problem, several companies have devised ingenious methods of keeping the reactor walls clean, a prominent example being the use of Bio-Beads™ in Varicon Aqua's BioFence™. In manufacturing terms the scale up of tubular reactors can be a comparatively straight forward process; however larger deployments can suffer from a variety of biotic and abiotic problems. One major issue with large scale tubular systems is that considerable gradients of dissolved oxygen, carbon dioxide and pH can develop across the system, causing inhibition or underperformance of algal growth [45]. These considerations can place a practical constraint on the run length of many tubular reactors, meaning that very large scale systems require interruptions in the run length for degassing. It is also worth noting that many of these tubular systems are inappropriate for the more fragile strains of algae, due to the relatively high liquid velocities causing considerable shear within the culture [29]. Another important consideration when scaling a tubular system is the multiplication of pumping energy requirements to overcome the frictional forces of multiple tubes and bends.

Many horizontal tubular reactors are already on the market, and may be purchased in the form of specialised equipment from a variety of suppliers. One prominent example would be Varicon Aqua's BioFence™ and Phyco-Flow™ systems which between them have been deployed in over 80 locations worldwide for a variety of uses, including both research and industrial applications. Both systems have found widespread use in the production of higher value nutraceutical and cosmetic compounds. With one example being the deployment of the BioFence™ for the production of cyanobacterial metabolites for use in cosmetic products by Blue Lagoon in Iceland. Other examples of commercial production in tubular systems can be found in Ketura, Israel where *Haematococcus pluvialis* is grown for the production of astaxanthin in a custom built manifold system [41]. Despite these successes current data indicates that around 10 times more algal biomass is produced in open pond systems than in closed reactors [42]. This highlights the fact that it is somewhat unlikely that closed tubular reactors will be able to produce algal biomass for lower value applications such as biofuels on a large scale. Instead it is more likely that closed photobioreactors will integrate with open pond systems either as high rate inoculation platforms or within two stage production processes [46]. Moving forward it is also likely that tubular photobioreactors will continue to find uses in the production of higher grade biochemicals, where quality control is of the utmost importance and where the use of genetically modified microalgae may be required [4].

Bubble Columns

Column reactors have found widespread use within conventional fermentation processes; mainly because the technology can be readily transferred to the growth of many different micro-organisms (see Fig. **7**). The mixing regime within column reactors is created *via* air displacement, which is often cheaper than conventional liquid pumping. In fact one of the major benefits of a bubble column is that the turbulence created by aeration can be greater than equivalent power inputs with other types of mixing. This aeration results in high levels of mass transfer and maintains low levels of dissolved oxygen [47]. Shear stress is also lower than many velocity driven systems. Reports from the literature would indicate that productivity can reach an average of 0.24-0.5 g L^{-1} d^{-1} [3, 32]. Whilst other advantages of column reactors include the fact that they are fairly compact in terms of planar space, as well as being relatively simple to construct in comparison to the larger horizontal tubular systems. Bubble columns can be made from a variety of materials including PVC and polythene which can dramatically lower the capital expenditure in many larger scale operations, whilst the absence of internal parts translates to lower overall maintenance costs.

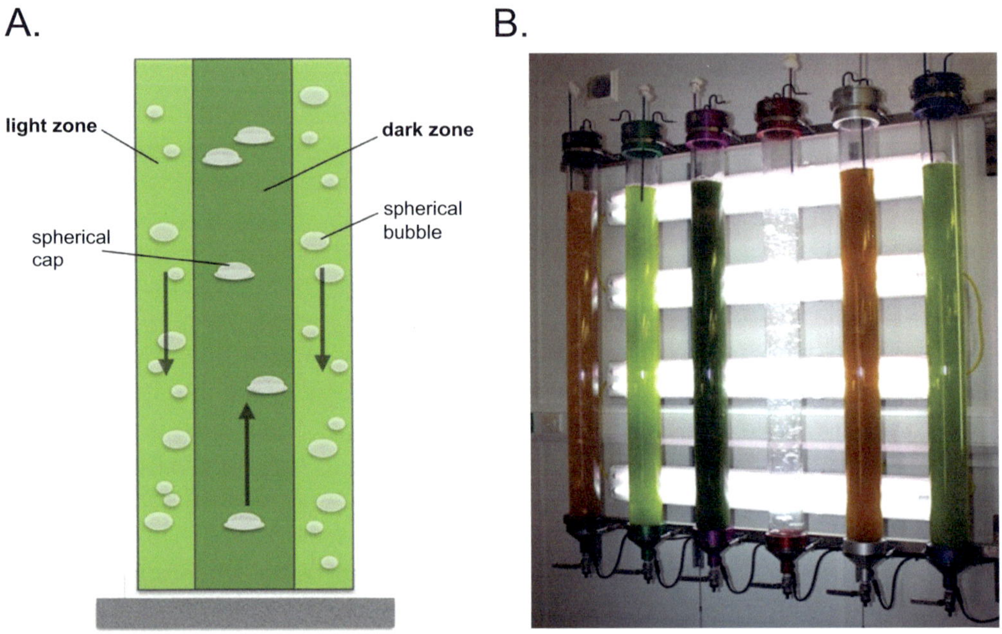

Fig. 7. Diagram shows a simplified schematic of a bubble column. (a) The diagram indicates the relative sections of light and dark zonation within the bubble column. Diagram modified from [48]. It also shows how air bubbles can deform as they rise through the column. (b) A battery of bubble columns [49].

Bubble columns share some of the drawbacks of other tubular systems in terms of construction materials and scale-up. However, there are also some design specific problems; the foremost being the relatively small illumination surface to volume ratio. This is especially the case when compared to other types of tubular reactor or flat panels which conventionally have shorter geometric light paths. These larger column diameters can lead to sub-optimal productivity caused by the large dark zone within the reactor (see Fig. **7A**). Another prominent consideration is that although bubble columns compare well in terms of cost with other tubular photobioreactor designs, they are still considerably more expensive than open ponds. Other important considerations for bubble column deployment include practical issues around scale-up; this is because by their very nature, they are essentially individual systems. This means that scale up can only be undertaken by increasing the number of individual units. Such a scenario means that each column is an individual photobioreactor, which results in increased costs both in terms of equipment and labour.

Bubble columns are a popular choice for both laboratory and intermediate sized systems where the relative ease of scaling by number using repeating units has led them to be used in both low and high value production processes. They find particular prominence within the aquaculture industry for the production of algae as feed, or as inoculation vessels for larger photobioreactors. To date several commercial reactors are on offer within the marketplace, including a UK spin-out from Plymouth Marine Laboratory, see Fig. (**7**) as well as the Phyco-Bubble system sold by Varicon Aqua. However, the relative ease of manufacturing bubble columns means that they can be fabricated with limited resources and specialist equipment, resulting in a great variety of in-house designs and configurations. Prominent designs of novel and modular column reactors can now be found within the literature, and address some of the issues regarding scale-up of bubble columns; this includes vertical systems built by AlgEternal and systems deployed by the University of Texas.

Airlift Reactors

Airlift photobioreactors encompass a broad family of pneumatic gas-liquid contacting devices, which act to create circulatory motion within a defined geometry. This type of flow regime differs from that found within a bubble column, being characterised instead by cyclical flow patterns [50]. The circulatory patterns within airlift photobioreactors are a function of the geometry and velocity within the system, and are created through interconnecting channels designed specifically for this purpose. The channels are often described as riser and downcomer sections, corresponding to the direction of the liquid travelling within them. The actual motion within the reactor is created by the injection of a mixing gas into the reactor from the bottom of the riser section. The hold-up of the mixing gas within the riser creates a density difference between the riser and downcomer sections. As the gas leaves the fluid by disengaging at the de-gassing zone of the riser, the denser fluid of the riser moves down through the downcomer, and the circulatory motion continues [51].

Airlift reactors can be categorised as having either an internal or external loop configuration. Internal loop reactors separate their riser and downcomer, either with a draft tube or a split-cylinder arrangement. External loop airlifts have a physically separated riser and downcomer, taking the appearance of two separate interconnected tubes. Bubble size within airlift reactors is usually in the order of 0.5–10 mm. Fig. (**8**) illustrates some of the more common airlift reactor configurations, which can impart distinct benefits over other types of reactor design. These include relatively high gas and mass transfer, uniform turbulent mixing, low hydrodynamic stress and ease of control, particularly with regards to liquid velocity [51]. They also display lower levels of dissolved oxygen in the culture medium in comparison to horizontal tubular reactors, due to short liquid circulation loops and rapid de-gassing. Many of these characteristics are of particular benefit to certain microbial processes, including algal cultivation.

Productivity within airlift reactor types has been reported to reach as high as 0.5-1.5 g L^{-1} d^{-1} [1], despite this fact, they have still not reached widespread commercial deployment.

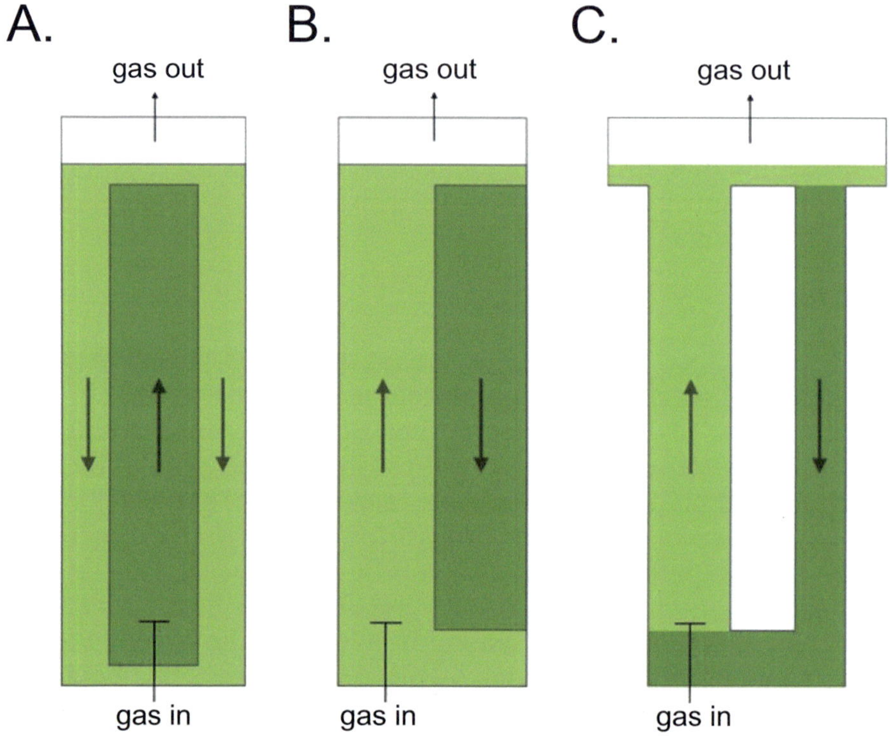

Fig. (8). Different configurations of airlift bioreactors. (A) Concentric draught-tube internal-loop; (B) Split-cylinder internal-loop; (C) Vertical draught-tube [51].

Notwithstanding the low levels of commercial uptake, there are numerous examples of airlift photobioreactors in deployment across the globe. One prominent reactor within the literature is the horizontal serpentine system deployed in Almeria, Spain. The study of this system has provided considerable guidance for subsequent airlift photobioreactor design [1, 52]. Another prominent academic example includes the Massachusetts Institute of Technology (MIT) photobioreactor, which was deployed to investigate the potential of flue gas scrubbing with algae. This system takes the form of individual 30 litre triangular modules arranged in an elegant array formation; see Fig. (**9A** and **B**) [53]. Other airlift configurations include vertical planar designs, an example of which can be seen in Fig. (**9C**) showing a 10 litre Phyco-Lift produced by Varicon Aqua (UK).

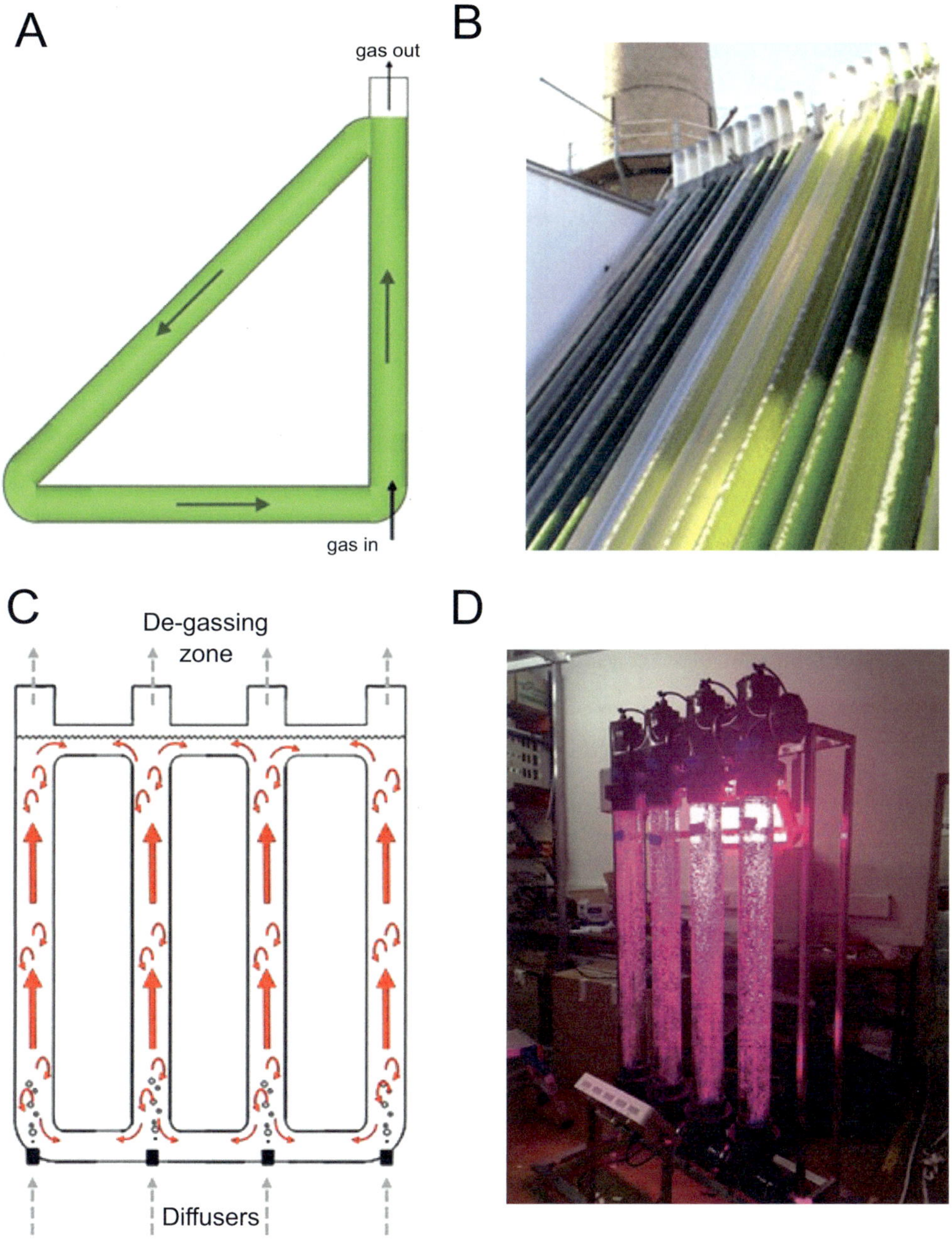

Fig. (9). Diagram shows some potential arrangements of airlift reactors. (A) and (B) show how an airlift system can be created by directing the flow of the airlift, images courtesy of [53] at MIT. (C) and (D) shows the Phyco-Lift designed and built by Varicon Aqua Solutions in collaboration with University College London [54].

PHOTOBIOREACTOR DESIGN SUMMARY

Table **1** outlines some of the most common photobioreactor design considerations, and acts as a good basis for the decision making process. This includes a summation of the relative advantages and disadvantages of attributes associated with the most common reactor designs.

Table 1. Common photobioreactor design considerations. One * represents 'low' and *** represents 'high'. Modified from the work of [3, 5, 55].**

		Open Ponds	Flat Panel	Tubular	Bubble Column	Airlift
Biotic	Productivity/Yield	*	*****	***	***	***
	Shear stress	*	*****	****	**	***
	Fouling	*	*****	***	***	***
	Contamination	*****	*	**	**	**
Abiotic	Scale-up	*****	*	****	**	***
	Mixing	*	*****	****	*****	****
	Light penetration	*	*****	****	***	****
Operational	Material cost	*	*****	***	***	***
	Energy cost	*	*****	***	****	****
	Ease of operation	*	****	****	**	**
	Maintenance	*	*****	****	***	***

PHOTOBIOREACTOR USE FOR BIOFUELS

The best designs for biofuel production would have to remain relatively cheap to construct, whilst being easy to operate and maintain. Key criteria include scalability (volumetric size between 100 m^3 and 10,000 m^3), without compromising the key design parameters. Manufacturing costs should be kept to a minimum; which can be achieved in part through fabrication using standardised parts and procedures. Considering the stringent cost requirements for the production of biofuels, it is an inevitable conclusion that a strong candidate for reactor selection would be an open raceway pond. This design has generally been favoured by industry, due to low operational cost and ease of maintenance [55], meaning that there are a considerable number of successful operational examples to act as guidance for scale-up [28]. However, several problems remain with pond use within certain contexts. These include poor light penetration, poor mass transfer, insufficient mixing and gas hold-up, lack of temperature control, large areal footprint and high risk of contamination [3, 55]. In combination these factors can lead to relatively poor areal yields and studies have shown better productivities in tubular systems than open ponds when grown on the same nutrients [56].

CONFLICT OF INTEREST

The authors declare no conflict of interest, financial or otherwise.

ACKNOWLEDGEMENTS

We would like to thank the support of University College London and Varicon Aqua (UK) in providing material for this chapter.

REFERENCES

[1] F. G. Acién Fernández, J. M. Fernández Sevilla, J. A. Sánchez Pérez, E. Molina Grima, and Y. Chisti, *Airlift-driven external-loop tubular photobioreactors for outdoor production of microalgae: assessment of design and performance,* . [http://dx.doi.org/10.1016/S0009-2509(00)00521-2]

[2] J.C. Weissman, R.P. Goebel, and J.R. Benemann, "Photobioreactor design: Mixing, carbon utilization, and oxygen accumulation", *Biotechnol. Bioeng.,* vol. 31, no. 4, pp. 336-344, 1988. [http://dx.doi.org/10.1002/bit.260310409] [PMID: 18584613]

[3] C.U. Ugwu, H. Aoyagi, and H. Uchiyama, "Photobioreactors for mass cultivation of algae", *Bioresour. Technol.,* vol. 99, no. 10, pp. 4021-4028, 2008. [http://dx.doi.org/10.1016/j.biortech.2007.01.046] [PMID: 17379512]

[4] O. Pulz, "Photobioreactors: production systems for phototrophic microorganisms", *Appl. Microbiol. Biotechnol.,* vol. 57, no. 3, pp. 287-293, 2001. [http://dx.doi.org/10.1007/s002530100702] [PMID: 11759675]

[5] M.R. Tredici, *Mass production of microalgae: photobioreactors*, 2004.

[6] M. C. Cerón-García, M. D. Macías-Sánchez, A. Sánchez-Mirón, F. García-Camacho, and E. Molina-Grima, *A process for biodiesel production involving the heterotrophic fermentation of Chlorella protothecoides with glycerol as the carbon source,* . [http://dx.doi.org/10.1016/j.apenergy.2012.09.054]

[7] J.B. Park, R.J. Craggs, and A.N. Shilton, "Wastewater treatment high rate algal ponds for biofuel production", *Bioresour. Technol.,* vol. 102, no. 1, pp. 35-42, 2011. [http://dx.doi.org/10.1016/j.biortech.2010.06.158] [PMID: 20674341]

[8] C.-Y. Chen, K.-L. Yeh, R. Aisyah, D.-J. Lee, and J.-S. Chang, *Cultivation, photobioreactor design and harvesting of microalgae for biodiesel production: A critical review,* . [http://dx.doi.org/10.1016/j.biortech.2010.06.159]

[9] D. Chaumont, "Biotechnology of algal biomass production: a review of systems for outdoor mass culture", *J. Appl. Phycol.,* vol. 5, no. 6, pp. 593-604, 1993. [http://dx.doi.org/10.1007/BF02184638]

[10] C.G. Lee, and B.Ø. Palsson, "Photoacclimation of Chlorella vulgaris to Red Light from Light-Emitting Diodes Leads to Autospore Release Following Each Cellular Division", *Biotechnol. Prog.,* vol. 12, no. 2, pp. 249-256, 1996. [http://dx.doi.org/10.1021/bp950084t]

[11] E.M. Grima, F.G. Camacho, J.A. Pérez, J.M. Sevilla, F.G. Fernández, and A.C. Gómez, "A mathematical model of microalgal growth in light-limited chemostat culture", *J. Chem. Technol. Biotechnol.,* vol. 61, no. 2, pp. 167-173, 1994. [http://dx.doi.org/10.1002/jctb.280610212]

[12] C. Taube, "variouc fluorescent lamps," ed. www.wikipedia.com, 2005.

[13] Gerben, "metal halide lamp," ed. www.wikipedia.com, 2006.

[14] M. Capdevila, "Fluo-45W LED-17W," ed. www.wikipedia.com, 2012.

[15] P.M. Doran, *Bioprocess Engineering Principles.* Elsevier Academic Press, 1995, p. 439.

[16] R. N. Singh, and S. Sharma, "Development of suitable photobioreactor for algae production – A review", *Renewable and Sustainable Energy Reviews,* vol. 16, no. 4, pp. 2353-5, 2012. [http://dx.doi.org/10.1016/j.rser.2012.01.026]

[17] S. Franklin, A. Somanchi, K. Espina, G. Rudenko, and P. Chua, "Renewable chemical production from novel fatty acid feedstocks," ed: Google Patents, 2012.

[18] K. L. Terry, and L. P. Raymond, *System design for the autotrophic production of microalgae,* . [http://dx.doi.org/10.1016/0141-0229(85)90148-6]

[19] J.C. Weissman, and R. Goebel, *Design and analysis of microalgal open pond systems for the purpose of producing fuels: a subcontract report.* Solar Energy Research Inst.: Golden, CO, USA, 1987. [http://dx.doi.org/10.2172/6546458]

[20] H. Hadiyanto, S. Elmore, T. Van Gerven, and A. Stankiewicz, *Hydrodynamic evaluations in high rate algae pond (HRAP) design,* . [http://dx.doi.org/10.1016/j.cej.2012.12.015]

[21] Y. Chisti, *Airlift Bioreactors.* Elsevier Applied Science, 1989, p. 353.

[22] J. McDonald, "Varicon Engineering and Projects", A. M. Lizzul, Ed., ed. Malvern: Varicon Aqua Solutions, 2013.

[23] C. Lee, "AlgaeConnect system", 2012. http://algaelabsystems.com/algaeconnect/

[24] C. Lee, "pH probe", 2012. http://algaelabsystems.com/algaeconnect/

[25] K. Soratana, and A.E. Landis, "Evaluating industrial symbiosis and algae cultivation from a life cycle perspective", *Bioresour. Technol.,* vol. 102, no. 13, pp. 6892-6901, 2011. [http://dx.doi.org/10.1016/j.biortech.2011.04.018] [PMID: 21543224]

[26] M.A. Borowitzka, "Commercial production of microalgae: ponds, tanks, tubes and fermenters", *J. Biotechnol.,* vol. 70, no. 1-3, pp. 313-321, 1999. [http://dx.doi.org/10.1016/S0168-1656(99)00083-8]

[27] C. Jiménez, B.R. Cossío, D. Labella, and F. Xavier Niell, "The Feasibility of industrial production of Spirulina (Arthrospira) in Southern Spain", *Aquaculture,* vol. 217, no. 1-4, pp. 179-190, 2003. [http://dx.doi.org/10.1016/S0044-8486(02)00118-7]

[28] W.J. Oswald, "Ponds in the twenty-first century", *Water Sci. Technol.,* vol. 31, no. 12, pp. 1-8, 1995. [http://dx.doi.org/10.1016/0273-1223(95)00487-8]

[29] Y. Chisti, "Biodiesel from microalgae", *Biotechnol. Adv.,* vol. 25, no. 3, pp. 294-306, 2007. [http://dx.doi.org/10.1016/j.biotechadv.2007.02.001] [PMID: 17350212]

[30] JanB46, "Raceway pond used for the cultivation of microalgae. The water is kept in constant motion with a powered paddle wheel.", 2011. https://en.wikipedia.org/wiki/Raceway_pond#/media/File: Microalgenkwekerij_te_Heure_bij_Borculo.jpg

[31] A.L. Stephenson, E. Kazamia, J.S. Dennis, C.J. Howe, S.A. Scott, and A.G. Smith, "Life-Cycle Assessment of Potential Algal Biodiesel Production in the United Kingdom: A Comparison of Raceways and Air-Lift Tubular Bioreactors", *Energy Fuels,* vol. 24, no. 7, pp. 4062-4077, 2010. [http://dx.doi.org/10.1021/ef1003123]

[32] L. Brennan, and P. Owende, "Biofuels from microalgae--A review of technologies for production, processing, and extractions of biofuels and co-products", *Renew. Sustain. Energy Rev.*, vol. 14, no. 2, pp. 557-577, 2010. [http://dx.doi.org/10.1016/j.rser.2009.10.009]

[33] A.E. Rabe, and R.J. Benoit, "Mean light intensity—a useful concept in correlating growth rates of dense cultures of microalgae", *Biotechnol. Bioeng.*, vol. 4, no. 4, pp. 377-390, 1962. [http://dx.doi.org/10.1002/bit.260040404]

[34] P. Schenk, "Second Generation Biofuels: High-Efficiency Microalgae for Biodiesel Production", *BioEnergy Res.*, vol. 1, no. 1, pp. 20-43, 2008. [http://dx.doi.org/10.1007/s12155-008-9008-8]

[35] M. Borowitzka, "Algal biotechnology products and processes — matching science and economics", *J. Appl. Phycol.*, vol. 4, no. 3, pp. 267-279, 1992. [http://dx.doi.org/10.1007/BF02161212]

[36] J. Sheehan, T. Dunahay, J. Benemann, and P. Roessler, "Look Back at the U.S. Department of Energy's Aquatic Species Program: Biodiesel from Algae; Close-Out Report," in "Other Information: PBD: 1 Jul 1998," NREL/TP-580-24190; TRN: US200420%%272 United States10.2172/ 15003040TRN: US200420%%272Tue Feb 12 21:48:39 EST 2008NREL; RN04066091English, 1998, Available: http://www.osti.gov/energycitations/servlets/purl/15003040-tW7nZs/native/.

[37] W. Oswald, *Micro-algae and waste-water treatment.* Micro-algal Biotechnology, 1988, pp. 305-328.

[38] Q. Hu, H. Guterman, and A. Richmond, "A flat inclined modular photobioreactor for outdoor mass cultivation of photoautotrophs", *Biotechnol. Bioeng.*, vol. 51, no. 1, pp. 51-60, 1996. [http://dx.doi.org/10.1002/(SICI)1097-0290(19960705)51:1<51::AID-BIT6>3.0.CO;2-#] [PMID: 18627087]

[39] A. Richmond, *Handbook of Microalgal Culture: Biotechnology and Applied Phycology.* Wiley-Blackwell, 2003, p. 584. [http://dx.doi.org/10.1002/9780470995280]

[40] I. Biotech, "Photobioreactor PBR 500 P IGV Biotech", P. P. P. I. Biotech, Ed., ed. wikipedia, 2003.

[41] A. Richmond, S. Boussiba, A. Vonshak, and R. Kopel, "A new tubular reactor for mass production of microalgae outdoors", *J. Appl. Phycol.*, vol. 5, no. 3, pp. 327-332, 1993. [http://dx.doi.org/10.1007/BF02186235]

[42] C. Posten, "Design principles of photo-bioreactors for cultivation of microalgae", *Eng. Life Sci.*, vol. 9, no. 3, pp. 165-177, 2009. [http://dx.doi.org/10.1002/elsc.200900003]

[43] H. C. Greenwell, L. M. L. Laurens, R. J. Shields, R. W. Lovitt, and K. J. Flynn, *Placing microalgae on the biofuels priority list: a review of the technological challenges, .* [http://dx.doi.org/10.1098/rsif.2009.0322]

[44] O. Jorquera, A. Kiperstok, E.A. Sales, M. Embiruçu, and M.L. Ghirardi, "Comparative energy life-cycle analyses of microalgal biomass production in open ponds and photobioreactors", *Bioresour. Technol.*, vol. 101, no. 4, pp. 1406-1413, 2010. [http://dx.doi.org/10.1016/j.biortech.2009.09.038] [PMID: 19800784]

[45] T. Mazzuca Sobczuk, F. García Camacho, F. Camacho Rubio, F.G. Acién Fernández, and E. Molina Grima, "Carbon dioxide uptake efficiency by outdoor microalgal cultures in tubular airlift photobioreactors", *Biotechnol. Bioeng.*, vol. 67, no. 4, pp. 465-475, 2000. [http://dx.doi.org/10.1002/(SICI)1097-0290(20000220)67:4<465::AID-BIT10>3.0.CO;2-9] [PMID: 10620762]

[46] L. Rodolfi, G. Chini Zittelli, N. Bassi, G. Padovani, N. Biondi, G. Bonini, and M.R. Tredici, "Microalgae for oil: strain selection, induction of lipid synthesis and outdoor mass cultivation in a low-cost photobioreactor", *Biotechnol. Bioeng.*, vol. 102, no. 1, pp. 100-112, 2009.

[http://dx.doi.org/10.1002/bit.22033] [PMID: 18683258]

[47] A. S. Mirón, F. G. Camacho, A. C. Gómez, E. M. Grima, and Y. Chisti, "Bubble-column and airlift photobioreactors for algal culture," 2000.

[48] A. Sánchez Mirón, A. Contreras Gómez, F. García Camacho, E. Molina Grima, and Y. Chisti, "Comparative evaluation of compact photobioreactors for large-scale monoculture of microalgae", *J. Biotechnol.,* vol. 70, no. 1-3, pp. 249-270, 1999. [http://dx.doi.org/10.1016/S0168-1656(99)00079-6]

[49] M. Allen, "Bubble column," ed. http://bubble-columns.com/: bubble-columns.com, 2013.

[50] "Godbole S, Deckwer W-D, "Design parameters estimations for bubble column reactors", *AIChE J.,* vol. 28, no. 3, pp. 353-379, 1982. [http://dx.doi.org/10.1002/aic.690280302]

[51] M.Y. Chisti, *Airlift Bioreactors.* Elsevier Applied Science, 1989, p. 353.

[52] F.G. Acién, J.M. Fernández, J.J. Magán, and E. Molina, "Production cost of a real microalgae production plant and strategies to reduce it", *Biotechnol. Adv.,* vol. 30, no. 6, pp. 1344-1353, 2012. [http://dx.doi.org/10.1016/j.biotechadv.2012.02.005] [PMID: 22361647]

[53] G. Vunjak-Novakovic, Y. Kim, X. Wu, I. Berzin, and J.C. Merchuk, "Air-Lift Bioreactors for Algal Growth on Flue Gas: Mathematical Modeling and Pilot-Plant Studies", *Ind. Eng. Chem. Res.,* vol. 44, no. 16, pp. 6154-6163, 2005. [http://dx.doi.org/10.1021/ie049099z]

[54] A.M. Lizzul, *Integrated Production of Algal Biomass.* Engineering Doctorate, Civil, Environmental and Geometic Engineering, University College London, United Kingdom, 2015.

[55] M. Borowitzka, "Commercial production of microalgae: ponds, tanks, tubes and fermenters", *J. Biotechnol.,* vol. 70, no. 1-3, pp. 313-321, 1999. [http://dx.doi.org/10.1016/S0168-1656(99)00083-8]

[56] Z. Arbib, J. Ruiz, P. Álvarez-Díaz, C. Garrido-Pérez, J. Barragan, and J.A. Perales, "Long term outdoor operation of a tubular airlift pilot photobioreactor and a high rate algal pond as tertiary treatment of urban wastewater", *Ecol. Eng.,* vol. 52, pp. 143-153, 2013. [http://dx.doi.org/10.1016/j.ecoleng.2012.12.089]

CHAPTER 7

Use of Flue Gas as Carbon Source

Jorge Alberto V. Costa[1], **Adriano S. A. Henrard**[2], **Luiza Moraes**[1], **Etiele G. Morais**[1], **Igor S. Gonçalves**[2] and **Michele G. Morais**[2*]

[1] *Laboratory of Biochemical Engineering, College of Chemistry and Food Engineering, Federal University of Rio Grande, P.O. Box 474, Av. Itália km 8, 96203-900 Rio Grande, RS, Brazil*

[2] *Laboratory of Microbiology and Biochemistry, College of Chemistry and Food Engineering, Federal University of Rio Grande, P.O. Box 474, Av. Itália km 8, 96203-900 Rio Grande, RS, Brazil*

Abstract: Carbon dioxide (CO_2) concentration in atmosphere has increased since the beginning of industrial age. CO_2 biofixation by microalgae is a technology based on the use of solar energy through photosynthesis to capture and use this carbon source, mainly from industrial combustion flue gases. This CO_2 mitigation technology enables microalgal cultivation with an alternative source of carbon, decreasing the emissions of industries (such as power generation) and making processes more sustainable and environmentally friendly. The chapter aimed to approach the implementation of using flue gas as a carbon source in microalgal cultures, as well as the metabolism of CO_2 biofixation and the associated environmental issues.

Keywords: Biofixation, Biomass, Bioproducts, Carbon dioxide, Cultivation, Flue gases, Global warming, Microalgae, Photosynthesis.

INTRODUCTION

The increase in world population density, the burning of fossil fuels and deforestation on a large-scale have contributed to the increase of greenhouse gases (GHGs) concentrations in atmosphere, contributing to the global warming [1 - 3]. This temperature rise on the planet caused by anthropic activities has become one of the most worrying environmental problems in the world [4].

Continued use of fossil fuels has become unsustainable. This resource is limited and its combustion for power generation results in excessive emission of GHGs into the atmosphere. Among them, CO_2 is considered the biggest enhancer of greenhouse effect. Strategies have been explored to reduce CO_2 emissions into the

* **Corresponding author Michele G. de Morais:** Laboratory of Microbiology and Biochemistry, College of Chemistry and Food Engineering, Federal University of Rio Grande, P.O. Box 474, Av. Itália km 8, 96203-900 Rio Grande, RS, Brazil; Tel: +55 53 3233 6977; E-mail: michele.morais@pq.cnpq.br

José Carlos Magalhães Pires (Ed.)

atmosphere, and microalgal culture is considered an alternative. Several studies employed microalgae in fixing flue gases emitted by municipal waste incinerators [5] and thermal power plants [6 - 8].

The biological fixation of carbon dioxide has been considered a promising approach compared to other methods such as carbon capture and storage technology [9]. Microalgae have received attention in recent years due to high cell growth rates and higher photosynthetic efficiency when compared to terrestrial plants [10]. These microorganisms have capability to convert CO_2 into oxygen and biomass, accumulating high-value biocompounds, with applicability in the most different knowledge areas [11]. In this way, the chapter aimed to present the use of flue gas as a carbon source in microalgal cultivation, as well as the metabolism of CO_2 biofixation and the associated environmental issues.

GLOBAL WARMING AND GREENHOUSE GASES

Global warming and the excessive emission of GHGs into the atmosphere have caused great concern worldwide. In this way, solutions to sustainable development have become the focus of discussions and international actions. Deforestation, industrial production increase and intensive use of fossil fuels have contributed to the raise of CO_2 concentration in the atmosphere [12]. According to IPCC (Intergovernmental Panel on Climate Change) [13], the annual emissions of GHGs increased 10 Gt_{CO2eq} between 2000 and 2010, derived directly from the supply energy sectors (47%), industries (30%), transport (11%) and constructions (3%). The global energy matrix is still based on non-renewable sources, such as oil, coal and natural gas. These sources represent about 80% of energy consumption in the world [14].

In fossil fuels burning (like oil, natural gas and coal), flue gases are released, containing up to 142 compounds. Among these; are oxygen (O_2), water steam (H_2O), nitrogen (N_2), carbon dioxide (CO_2), nitrogen oxides (NO_X), sulfur oxides (SO_X), carbon monoxide (CO), hydrocarbons from incomplete combustion (CxHy), acids of halogens and particulates [15].

Environmental pollution, flora and fauna degradation and climate change are associated with oil companies activities, which produce comfort and promoted social well-being in the 20th century. Modern society still enjoys the facilities and benefits provided by oil refineries such as the use of high energy fuels, plastics and chemicals production. However, the society charges the negative environmental cost caused in the last century by excessive CO_2 emissions, making it a challenge for 21st century [14].

Phenomena such as the warming of the atmosphere and oceans, melting of Antarctica and Greenland glaciers, rise in sea levels and more frequent and long heat waves are threatening the planet [16 - 19]. These phenomena can lead to harmful effects on human life and biodiversity, increasing the number of diseases and deaths related to heat. The world economy is also affected by the destruction of cultures and flood of coastal cities at sea level [16].

MEASURES FOR GREENHOUSE GASES REDUCTION

In the last two decades, several CO_2 mitigation strategies have been investigated by research addressing physical, chemical and biological processes [20]. Strategies include carbon capture and storage (CCS) that is considered a short-term solution to reduce GHG emissions. The CCS technology comprises three steps: capture, transport and storage of CO_2 [21]. Associated technologies are chemical absorption [22] and adsorption [23], membrane separation [24], chemical looping combustion [25], underground storage.

Among several CO_2 mitigation strategies, biological fixation with microalgal cultures has been highlighted, considering the economical value obtained with the production of high-value products from biomass. The microalgae application is widely accepted in practice, one of the best strategies outlined by bioengineering [26]. The use of microalgae in CO_2 capture has the advantages of high growth rate compared to terrestrial plants [27, 28] and low impact on food demand [29]. These microorganisms have the ability and specificity in CO_2 mitigation [30] and biomass with high added value for application in nutrition [7], pharmaceutical products [31], fertilizers [32] and biofuels [33].

Studies concerning CO_2 mitigation by microalgae have explored the potential of using flue gas as CO_2 source. Radmann *et al.* [34] and Duarte *et al.* [8] use simulated flue gas of coal in the cultivation of different species of microalgae. Using real flue gas from coal burn, Costa *et al.* [6] observed an increase in *Spirulina* biomass productivity. Rosa *et al.* [35] proposed the integration of chemical and biological fixation. Chemical absorbent monoethanolamine (MEA) increased the concentration of carbon dissolved in the liquid medium contributing to *Spirulina* sp. LEB 18 growth and to macromolecules production in biomass.

Besides CO_2, flue gases also contains other compounds, such as SO_2 and NO, which in high concentrations may have toxic effects on the cells. The addition of flue gas (containing CO_2, SO_2, NO and ashes) to *Chlorella fusca* LEB 111 culture was reported by Duarte *et al.* [8] and Vaz *et al.* [7]. The culture growth and CO_2 biofixation was evaluated and microalgal biomass composition was characterized. Duarte *et al.* [8] have found more efficient CO_2 fixing ($50.0 \pm 0.8\%$, w w^{-1}) using 10% of CO_2, 200 ppm of SO_2 and NO and 40 ppm of ashes. The microalgal biomass composition was similar in all cultures, around 19.7% (w w^{-1}) of carbohydrates, 15.5% (w w^{-1}) of lipids and 50.2% (w w^{-1}) of protein.

Combining strategies through microalgae cultivation, CO_2 mitigation and bio-products production make the process economically viable and environmentally sustainable. One of the possible strategy may be enhanced by CO_2 fixation of flue gases from industries [2] and power plants [6].

CYCLIC MICROALGAE PRODUCTION PROCESS

In a cyclic microalgal production process, the system returns to the same starting point after a series of transformations, where the intermediates are microalgal cultivation, biomass production, one or more bioproducts obtaining, in the effluent generation and use the same as a source of nutrients for cultivation (Fig. **1**). This process aims to make it more sustainable and profitable microalgal cultivation for obtaining bioproducts, reducing effluents emissions and cultivation costs using these wastes as a nutritional source. The major components of industrial wastewaters are also those from microalgal culture medium, carbon and nitrogen, which, together, represent more than 50% of the compounds used in cultivations.

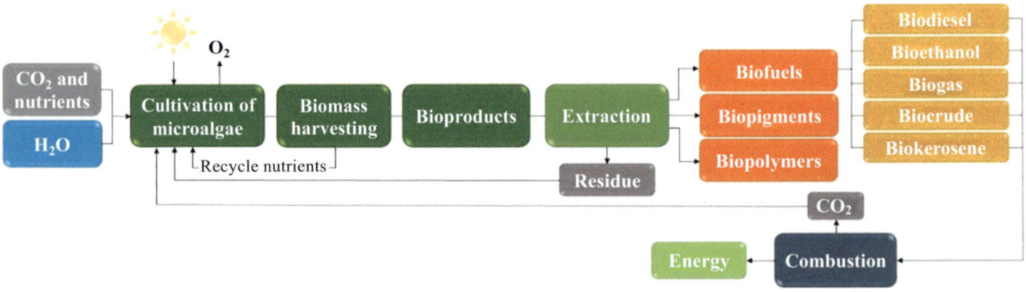

Fig. (1). Schematic diagram of the microalgae production cyclic process.

Among the products obtained by microalgae biofuels, biodiesel, bioethanol, bio-kerosene, bio-oil and bio-methane are the most studied. The production and application of microalgal biofuels are limited because of the production costs involved in the process [36]. Thus, the implementation of microalgae cyclic production systems beyond biorefinery (where more than one bioproduct is obtained from biomass) can be the pathway to make biofuels production economically feasible [37]. Although the emission of gases such as CO_2 by biofuels is reduced when compared to fossil fuels, this emitted amount can return to microalgal culture to be used as carbon source [38] (Table **1**).

Borges *et al.* [39] have applied the effluent from *Spirulina* sp. LEB 18 biomass anaerobic digestion as a substrate for this strain cultivation. The addition of this effluent (composed by carbon, nitrogen and phosphorus) to culture medium did not change the productivity of microalgae or maximum specific growth rate. Morais [31] have studied the cultivation of different microalgae with medium containing glycerine, evaluating the potential for biodiesel production. The authors have found efficiency up to 50% of biodiesel in batch of *Spirulina* sp. LEB 18 using 0.01 mol L^{-1} of glycerine (0,17 $g.L^{-1}$ of biodiesel in an assay which yielded 0.33 $g.L^{-1}$ of lipid).

Table 1. Bioproducts of microalgae cultivation, generated effluent and potential use effluent as a nutritional source in cyclical process.

Bioproducts	Effluent	Nutritional Source	Reference
Biofuels	CO_2	Inorganic Carbon	Morais and Costa [38]
Biodiesel	Glycerol	Inorganic Carbon	Morais [31]
Biomethane	Anaerobic Digestion Waste	Organic Carbon and Nitrogen	Borges *et al.* [39]

POTENTIAL USE OF FLUE GASES IN MICROALGAE CULTIVATION

The emissions of flue gases from power plants account for more than 7% of total CO_2 emissions in the world [40]. Among the flue gases, it stands out the CO_2 produced from natural gas combustion (5 to 6% of CO_2) and coal combustion (10 to 15% of CO_2) [15, 41]. To reduce the CO_2 emissions, several processes have been studied and applied. In this context, the CO_2 biofixation by microalgae has been considered a promising method for post-combustion processes, CO_2 capture and storage *in situ*. Microalgae have a high capacity to fix and store carbon through photosynthesis, converting CO_2 and water into organic compounds, without consuming additional energy and without providing secondary pollution [42].

Among the nutritional factors for the cultivation of microalgae, carbon is considered one of the essential elements for growth, accounting for about 50% of the cellular matter [35, 39, 43]. The carbon can be provided from various sources, such as atmospheric CO_2 from flue gases and chemically fixed in the form of soluble carbonates [44, 45]. Since the CO_2 concentration atmospheric is low (0.04-0.06%), this may limit the growth of microalgae, necessitating the addition of carbon as soluble carbonates or compressed CO_2, which results in increased production costs. According to Molina Grima [46], the compressed CO_2 supply can represent up to 41% of the total production costs of microalgal biomass. Flue gases contain 6-15% of CO_2 and are considered a promising source of carbon for the microalgae cultivation and a potentially more efficient route for CO_2 fixation.

In addition, the use of flue gas from industrial units (power plants) contributes to the mitigation of this GHG emissions to the atmosphere [47]. According to Van Den Hende *et al.* [15], the direct use of flue gases becomes an economically and environmentally better option compared to the compressed CO_2, especially when microalgae cultivation is located near the industrial facilities.

Several studies have been conducted to evaluate the use of flue gas as a source of nutrients for the cultivation of microalgae, confirming promising results regarding the growth and sequestration of CO_2 [6 - 8, 34, 48]. Application of microalgal cultivation in industrial grade using flue gas is still little known. Among the examples, there is the *Seambiotic* Ltd, an Israeli company that patented the method of cultivation of photosynthetic organisms with flue gases from fossil fuels power plant [49]. *Cyanotech Corporation*, located in Hawaii, patented a system that uses a microalgae production plant integrated with the generation of electricity. The CO_2 produced by burning fossil fuels for power generation is recovered and used as a nutrient for cultivation [50].

In Brazil, the Candiota Thermoelectric complex property of Eletrobrás CGTEE (Thermal Generation of Electricity Company) uses coal as the primary fuel and its maximum capacity of power generation is 796 MW. The flue gas emitted by the Presidente Médici Thermal Power Plant (UTPM) contains 10-12% (v v^{-1}) of CO_2, 680 ppm of NOx, 1700 ppm of SOx and 265 ppm of ash (Phase C - UTPM) (data provided by CGTEE). In this context, in order to reduce emissions of these gases, an agreement was signed in 2005 between the Laboratory of Biochemical Engineering (LEB)/Federal University of Rio Grande (FURG) with Eletrobrás CGTEE for the development of research in the field of CO_2 biofixation by microalgae. From this agreement, a pilot plant with 6,000 m^2, containing 70 m^2 of laboratories (Fig. **2a**), two raceway type bioreactors of 18 m^3 each (Fig. **2b**) and a lower volume bioreactor (1 m^3) was built in UTPM for CO_2 biofixation by microalgae.

a **b**

Fig. (2). **a**) Pilot Plant for CO_2 biofixation by microalgae and **b**) Raceway bioreactor with volume of 18 m^3, located in Presidente Médici Thermal Power Plant, Candiota-RS, Brazil.

With the project, several studies have been developed both in laboratory scale and pilot scale aiming to facilitate and maximize CO_2 fixation. Research has focused on: (i) strain isolation [34, 52]; (ii) CO_2 fixation in different photobioreactors and CO_2 concentrations [38, 53, 54]; (iii) mass transfer study with different diffuser configurations [55]; (iv) use of real and simulated flue gas [6 - 8]; (v) integration of chemical and biological fixation [35, 56]; (vi) nutrients recycling [53]; and (vii) evaluation of biomass composition [8, 35, 55 - 58].

FACTORS INFLUENCING CO_2 FIXATION FROM FLUE GAS BY MICROALGAE

Several studies have been performed not only to enable the use of microalgae in the capture of CO_2 from flue gases, but also to evaluate the influence of the high CO_2 concentration, the presence of other gases and particulate matter [15, 34, 48, 51, 59]. CO_2 fixation by microalgae biomass and productivity depend on the applied cultivation conditions. In addition to the species of microalgae, the factors that influence CO_2 fixation usually include physicochemical (*e.g.*, CO_2 concentration, pH, toxic pollutants in flue gas, temperature and light) and hydrodynamic parameters (*e.g.*, flow rate, mixing and mass transfer) [42].

Microalgae Strains

The selection of suitable strains for CO_2 fixation has a significant effect on the competitiveness of costs and effectiveness of the process. To achieve high CO_2 fixation rates, desirable features include: (i) high cell growth rate; (ii) high CO_2 utilization efficiency; (iii) tolerance to flue gases constituents; (iv) ability to produce bioproducts of interest; (v) ease of harvest; (vi) tolerance to high temperatures of the liquid medium; and (vii) capability of being cultivated with wastewater [60].

According to Ono and Cuello [61], the use of species that can tolerate high CO_2 concentrations, contamination of flue gases, and high temperatures can minimize pre-gas treatment costs. Morais and Costa [52] and Radmann *et al.* [34] showed that the native species isolated near thermal power plants are the best options for CO_2 capture from the gaseous effluents of these industries. Morais and Costa [52] verified that the strains *Chlorella kessleri* LEB 15 and *Scenedesmus obliquus* LEB 22 isolated from lakes near the UTPM thermal power plant showed cell growth when exposed to CO_2 concentrations (6%, 12% and 18% v v^{-1}), indicating the potential of native strains in the fixation of CO_2. Growing *Chlorella fusca* LEB 111 isolated from wastewater stabilization ponds near the UTPM thermal power plant, Vaz *et al.* [7] found that the strain showed tolerance to the flue gas from coal and ashes, besides achieving high CO_2 fixation rate.

CO_2 Concentration in Flue Gas

Several research studies focused on CO_2 fixation by microalgae evaluate the tolerance of the microalgae against high CO_2 concentrations. Studies have shown the positive effect of using pure CO_2, real and simulated flue gases as a carbon source [7, 8, 15, 38, 48, 52]. Sung *et al.* [62] evaluated the effect of CO_2 (10%, 30%, 50% and 70%), pH, temperature and light intensity in cultivations of *Chlorella* sp. KR-1. The authors have found that the highest growth rate was found with the injection of 10% of CO_2. The microalgae have also proved to be tolerant to pH and temperature changes, allowing to conclude that the KR-1 strain has the potential for large-scale cultivation to fix CO_2 from flue gas.

According to Costa *et al.* [6], the use of the coal combustion flue gas promoted a 35% increase in *Spirulina* sp. LEB 18 biomass productivity and reduction of 24% of CO_2 from this gaseous stream. Another study applying flue gases as carbon source was performed by Chiu *et al.* [48]. These authors have found that the growth rate of *Chlorella* sp. MTF-7 was approximately 1.5, 2.5 and 2.0 times higher than those achieved by *Chlorella* sp. WT with 2%, 10% and 25% CO_2, respectively.

pH

The medium pH is considered an important factor, which significantly affects the microalgal growth [63], nutrients solubility and bioavailability [64]. The transport of substances through the cytoplasmic membrane, activity of intracellular and extracellular enzymes and electron transport in photosynthesis and osmotic potential of the cytoplasm can be influenced by modifying the pH [65, 66].

During microalgal cultivation, a complex relationship between pH and CO_2 concentration occurs due to chemical equilibrium between the carbon species in the liquid medium ($CO_{2\ (aq)} \leftrightarrow H_2CO_3 \leftrightarrow HCO_3^- \leftrightarrow CO_3^{2-}$) [67]. The addition of flue gases in the cultivation of microalgae can reduce the pH of medium, not only due to CO_2 but also due to dissolution of NOx and SOx. Moreover, the injection of these gases in the cultivation can cause growth inhibition [59]. Maeda *et al.* [68] have observed a reduction in the pH in the cultivation of *Chlorella* sp. T-1 with the addition of high concentrations of SOx and NOx. The authors have found a decrease to levels below pH 3 when were added 15% CO_2, 50 to 80 ppm of SO_X and 150-240 ppm NOx. However, the T-1 strain showed tolerance to reduced pH and cell growth was observed when high concentration of CO_2 was added (50% to 100%). During the cultivation of *Chlorella pyrenoidosa* SJTU-2, Tang *et al.* [11] reported a reduction from pH 8.7 to 5.2 with increasing CO_2 concentration (0.03% to 50% v v^{-1}).

NO_X, SO_X and Particulate Materials

The direct utilization of the flue gas reduces the cost of pre-treatment, but requires extreme cultivation conditions for microalgae, such as high concentrations of CO_2 (10-15%) and the presence of growth inhibiting compounds, such as NOx, SOx [69, 70] particulate materials, among others [7, 8].

Flue gases exhibit different species of NOx, with nitric oxide (NO) and nitrogen dioxide (NO_2) being the main ones [71]. The addition of NOx in the cultivation of microalgae can have toxic effects [15]. However, some research show that NOx present in low concentrations can be used by some species of microalgae as a source of nitrogen [72]. The tolerance of the microalgae to NOx depends on factors such as cell density, concentration and flow of the gas, the cultivation agitation, bioreactor and microalgal species [42].

Unlike the NOx, sulfur oxides, including sulfur dioxide (SO_2) stand out for having high solubility in aqueous medium and are found in higher concentrations in the flue gas [15]. The main effect of SO_2 verified in the cultivation of microalgae is the reduction in pH due to the release of H^+ ions by the hydrolysis of the SO_2, which can result in decreased CO_2 fixation efficiency by the microalgae [73]. Several authors report the toxic effects of NOx and SOx in the cultivation of microalgae. Maeda *et al.* [68] have proved the toxic effect suffered by the microalgae *Chlorella* when exposed to high NOx (>150 ppm) and SOx (>50 ppm) concentrations. Lee *et al.* [74] have observed that the alkalization of the medium with calcium carbonate can reduce the toxicity of SO_2, while maintaining the pH above 6 to prevent the formation of sulfites.

Some microalgae species may show tolerance to high levels of the SOx and NOx present in flue gases. Duarte *et al.* [8] found that the addition of 60 ppm SO_2, 100 ppm NO and 40 ppm of ash in the cultivation of *Chlorella fusca* LEB 111 promoted results similar to those observed in the control cultivation, without the addition of industrial waste. Furthermore, the authors observed that the microalgae were able to tolerate SO_2 and NO concentrations of up to 400 ppm. According to Kao *et al.* [75], removal efficiencies of 80% for NOx and 93% for SO_2 were found when *Chlorella* sp. MTF-15 was cultivated with power plant flue gas (24% of CO_2, 25-30 ppm of NOx - NO and NO_2 - and 15-20 ppm of SO_2).

Particulate materials from combustion processes can vary with emission sources and they are sorted in aerosols, fly ash and soot [15, 76]. The effect of flue gas ashes in the cultivation of *Spirulina* sp. LEB 18 and *Chlorella fusca* LEB 111 was evaluated by Vaz *et al.* [7]. The authors have observed that the addition of 40 ppm of ash from the burning coal of UTPM (Candiota, Brazil) in the culture medium did not cause inhibition of growth of microalgae; likewise, the use of this particulate material may behave as a source of trace elements for the metabolism of microalgae. Similarly, Duarte *et al.* [8] have observed that the addition of simulated flue gas and ash in the same concentration (40 ppm) in the cultivation of *Chlorella fusca* LEB 111 did not affect cell growth and CO_2 fixation efficiency by the microalgae.

Temperature and Light

Photosynthesis is a process composed of a number of temperature-dependent physicochemical reactions [42, 73]. Thus, temperature of cultivation is a determinant factor in CO_2 fixation by microalgae and in biomass production. At low temperatures, the reduction of ribulose bisphosphate carboxylase-oxygenase (Rubisco) enzymatic activity can occur, leading to reduced photosynthetic rate. However, at high temperatures, inhibition of the metabolic rate of microalgae is verified along with the decrease of CO_2 solubility in the liquid medium [42].

Flue gases from power plants can reach high temperatures (120°C), making the use of thermophilic microalgae an interesting form of enabling the CO_2 sequestration from flue gases [9]. Besides being used for CO_2 sequestration, microalgae can contribute to reducing cooling costs of the exhaust gases and produce biomass rich in secondary metabolites [21]. In studies with *Thermosynechococcus* sp. TCL-1, at 50°C and CO_2 injection of 10%, Hsueh *et al.* [77] have found the specific growth rate of 2.7 d^{-1} and a biomass with a high concentration of carbohydrates (60.6% w w^{-1}). Because of the thermophilic characteristic, the TCL-1 strain demonstrated potential for application in cultivation using the flue gases.

Light is considered one of the most important factors in the CO_2 fixation process and biomass production. In microalgal cultivation using the CO_2 from flue gases, the light source is generally divided into solar (applied both in open and closed systems) and artificial (mainly applied in closed cultivation). Generally, the effect of the light in microalgal cultivation is evaluated for intensity and dark/light cycles [42]. Ho *et al.* [78] have found the highest specific growth rates (1.3 d^{-1}) and CO_2 fixation rate (853.7 mg L^{-1} d^{-1}) for *Scenedesmus obliquus* CNW-N, when it was exposed to a light intensity of 180 μmol.m^{-2}·s^{-1} and 2.5% of CO_2. In their studies with four strains of microalgae, Gonçalves *et al.* [79] observed that the application of higher light intensity (180 μE m^{-2} s^{-1}) and light cycle (24:0) resulted in higher rates of growth and CO_2 fixation.

Rosa *et al.* [80] have evaluated the growth of microalgae *Spirulina* sp. LEB 18 and *Chlorella kessleri* under controlled and uncontrolled light and temperature conditions. The authors have observed that in uncontrolled conditions *Spirulina* had a maximum cell concentration (X_{max} = 4.95 g L^{-1}) when cultivated with 6% of

CO_2, while for the *Chlorella* cultivation, the X_{max} values were obtained under concentrations of 12% and 18% of CO_2.

Mass Transfer in Bioreactors

The CO_2 mass transfer to the liquid medium and the mixing conditions are critical parameters in the CO_2 fixation by microalgae in photobioreactors [81]. The CO_2 mass transfer process for microalgae cultivation occurs through of the three-phase: mass transfer of CO_2 for the culture medium, CO_2 in the culture medium to inside of microalgae cell, and inside of the microalgae, carbon metabolic conversion at microalgae macromolecules [42]. Factors such as flow rate, bubbles size, gas retention time in the liquid medium, gas-liquid contact area, CO_2 concentration may influence the CO_2 mass transfer in microalgae cultivation [55, 82 - 84].

The mass transfer coefficient (k_La) is a parameter commonly used to evaluate the photobioreactor performance. The volumetric mass transfer coefficient depends on factors such as stirring rate, diffuser configuration, surfactants/antifoams agents and temperature [85]. The k_La parameter characterizes the mass transfer capacity of CO_2 in the bioreactor, and thus determines whether a reactor is able to sustain a given cell growth rate. As regards the transfer of CO_2, this parameter is critical in the design stages, in the scale-up and in the operation of a cultivation system for the production of biomass [86].

Moraes *et al.* [55] have examined the effect of different diffusers configurations and CO_2 flow rates in vertical tubular photobioreactors on mass transfer and CO_2 fixation by *Spirulina* sp. LEB 18. The authors have found that the use of a smaller flow injection and porous diffusers can promote lower CO_2 k_La values for the liquid medium, however, under these conditions higher CO_2 fixation efficiencies are observed for the microalgae.

Bioreactor Application in CO_2 Fixation by Microalgae

Bioreactors can be classified into closed (photobioreactors) or open systems [59]. The main advantage of open systems is the relative ease of construction and operation compared to most closed systems [85]. However, they usually have a disadvantage of being not possible the culture parameters control, such as the availability of light, stirring, temperature, pH and nutrient content [60]. When used in CO_2 fixation, these systems can result in losses of CO_2 to atmosphere due to low gas residence time in the liquid medium (lower liquid columns offered by these systems) [87].

To enable the use of open bioreactors in the CO_2 fixation process, research is being carried out in laboratory and pilot scales. Duarte-Santos *et al.* [88] have evaluated the CO_2 molar fraction and gas flow rate for control of pH in microalgae cultivation in a raceway equipped with sump in pilot scale. In this study, it was found that the use of low CO_2 molar fraction (2-6%) or gas flow rate (75-100 L min^{-1}) resulted in carbon efficiency in the sump was higher than 95% and 85% of the injected carbon being converted into biomass.

Several studies have reported that photobioreactors (closed systems) have better results when used in CO_2 fixation process by microalgae [9, 89]. The photobioreactors can have several configurations such as: vertical and horizontal tubular, bubble columns, air-lift, flat plates, helical type, stirred tank, hybrid type [9] and membrane photobioreactor [90]. This bioreactor configurations have advantages over open systems, such as high photosynthetic efficiency, high CO_2 fixation efficiency by microalgae, high biomass productivity, better parameters control of cultivation and larger CO_2 mass transfer efficiency [45, 73, 89].

Morais and Costa [38] have evaluated different bioreactors configurations in fixing CO_2 in the cultivation of four strains of microalgae. The maximum CO_2 fixation were observed in the *Spirulina* sp. LEB 18 cultivation in vertical tubular photobioreactors with volumes of 2 L (9.3%) and 4 L (9.2%) with injection of 6% of CO_2. Chai and Zhao [81] used a bubble tank-type stirrer to enhance the CO_2 removal efficiency and improve the photobioreactor mixing conditions in the cultivation of *Chlorococcum* sp. The authors found that the use of this system promoted 94% removal of CO_2 from the gas stream enriched with 10% of CO_2.

Radmann *et al.* [34] have found the greatest results of maximum daily CO_2 fixation (14.9% and 13.4%) by *Spirulina* sp. LEB 18 and *Chlorella vulgaris* LEB 106, respectively, in the cultivations with vertical tubular photobioreactors in series with simulated coal flue gas injection (12% of CO_2, 100 ppm of NO and 60 ppm of SO_2). The authors concluded that these microalgae have higher CO_2 fixation capacity in medium containing SO_2 and NO, and they can be cultivated in thermal power plants with coal flue gas, contributing to the reduction of global warming.

From studies using a photobioreactor with a polytetrafluoroethylene membrane system, Chai *et al.* [91] have reported high concentrations of CO_2 dissolved in the liquid medium, resulting in high transfer rates of CO_2 in the cultivation of microalgae *Chlorococcum* sp. At high concentrations of dissolved carbon and longer retention time in this medium, higher fixation rates and lower losses of CO_2 due to exhaustion are observed.

CO_2 BIOFIXATION METABOLISM

The biological fixation of CO_2 can be attributed to fixation made by the higher plants and the microalgae through photosynthesis. Although they are very similar mechanisms, the fixation by microalgae is considered more efficient, because these microorganisms grow rapidly and the CO_2 fixing efficiency is 10 to 50 times higher than plants [92, 93]. In microalgae biofixation process, CO_2 can be captured directly from the emission source, while higher plants capture this gas from atmosphere. Furthermore, the organic material produced through photosynthesis can be processed into several products such as biofuels, food additives, and health and beauty products. This represents additional benefits of CO_2 capture process by microalgae [58].

The CO_2 biofixation mechanism by microalgae is based on skill of these microorganisms have to perform photosynthesis. In photosynthesis, plants and photosynthetic microorganisms convert solar energy into chemical energy through the following reaction: $6CO_2 + 6H_2O$ + energy solar $\rightarrow C_6H_{12}O_6 + 6O_2$. In this reaction, the light intensity and the quality of radiation plays a vital role in the cell growth rate [94].

Photosynthesis occurs in two stages: (i) light phase, which happens only when the cells are illuminated; and (i) carbon fixation reactions, also known as the dark phase, which occurs both in presence and absence of light. In the centre of the photosynthetic apparatus there are two main photoactive complexes, photosystem I (PSI) and photosystem II (PSII), which absorb sunlight to the electron transport chain through an excited chlorophyll dimer (Fig. **3**) [95].

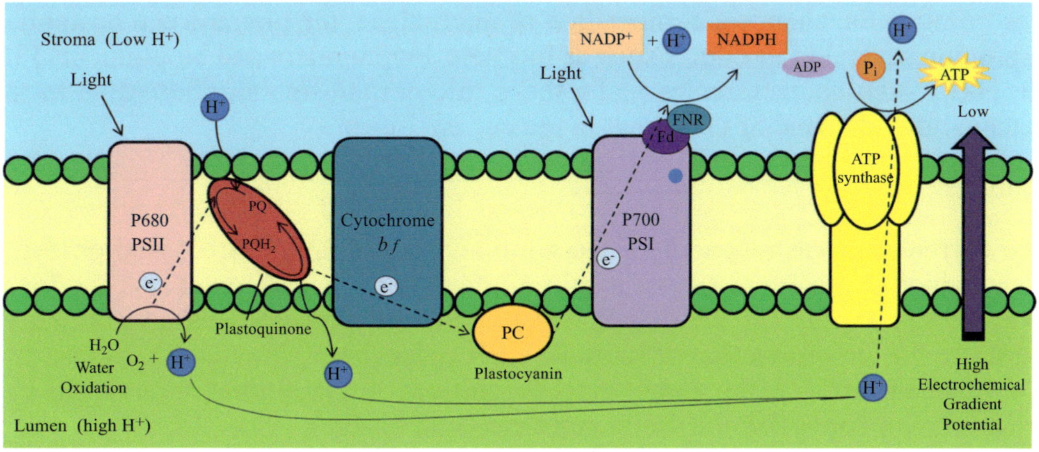

Fig. (3). Schematic view of the photosynthetic electron transport system in the thylakoid membrane.

Photosynthesis starts in PSII complex and then transfers electrons through the excitation of P680 chlorophyll dimer to PSI. Thus, PSI transfers the received electrons from PSII *via* chlorophyll dimer P700, which is oxidized by the antenna complex reducing NADPH [96] and ferredoxin, which is subsequently used in CO_2 fixation reactions through the Calvin cycle (dark reaction) [95, 97]. In the Calvin cycle, three ATP molecules and two NADPH molecules are consumed for every CO_2 molecule converted to carbohydrate according to the reaction: $3CO_2$ + 9ATP + 6NADPH + water → glyceraldehyde 3-fosfato + 8Pi + $6NADP^+$.

The enzymes D-ribulose 1,5-bisphosphate carboxylase-oxygenase (Rubisco) and phosphorylasekinase are essential to the Calvin cycle, as they are responsible for CO_2 sequestration and catalyze the regeneration of the CO_2 acceptor. In the central reaction of carbon fixation, an inorganic carbon is converted to organic carbon. The available CO_2 combines with ribulose 1,5-bisphosphate and water to yield two molecules of 3-phosphoglycerate compound. The fixation reaction is catalyzed by the stromal chloroplast ribulose bisphosphate carboxylase [98].

BIOPRODUCTS FROM MICROALGAL BIOMASS GROWN WITH FLUE GAS

Microalgae are presented as an ideal model for use of the CO_2 from the combustion of thermal power plants and industrial activities, generating microalgal biomass that is potentially applicable for the production of high value-added bioproducts. The CO_2 biofixation by microalgae has proven to be an efficient and economical method, mainly due to the photosynthetic skill of these microorganisms to use this gas as a source of nutrient.

According to the chemical composition of microalgae, the biomass can be applied to provide several bioproducts such as biofuels, biopolymers and biopigments [36, 99, 100]. Such characteristics make these microorganisms an alternative to the sustainability demand of the planet in the current times.

Biofuels

The current patterns of energy generation and consumption are based on fossil fuels, which cause emissions of local pollutants, GHGs and put at risk the long-term energy supply on the planet. Microalgae are presented as an alternative substrate for the production of biofuels such as biodiesel, bioethanol, biogas and biohydrogen [101 - 105]. Depending on cultivation conditions, microalgae can produce compounds that the easily converted into biofuels [99].

Among the different obtainable biofuel derived from microalgal biomass, biodiesel is highlighted due to their chemical characteristics similar to diesel oil

[36]. For biodiesel production, high concentrations of saturated fatty acids are desired in microalgal biomass. Depending on the genus and species, microalgae produce different types of lipids [106] and their fatty acid profiles can be influenced by the CO_2 concentration injected in the cultures [58]. After extraction of lipids for biodiesel production, the residual biomass obtained can be used as a substrate to produce bioethanol and biogas [107].

Ethanol is a biofuel obtained by the fermentation of carbohydrates-rich biomass, after hydrolysis of starch and cellulose [108]. Several carbohydrates are present in microalgae, such as cellulose, xylose, galactose, arabinose, glucose and mannose, and they can be used as substrates for bioethanol production [109, 110]. Depending on the cultivation conditions, the microalgae biomass can provide high concentrations of carbon compounds, and these compounds can be used directly in the fermentation or after pre-treatment and used for ethanol production [36]. Microalgae cultivation has lower environmental impact, not competing for space with food cultures, unlike biofuels produced from other raw materials [111 - 113].

Biogas is the product of anaerobic digestion of organic matter mainly consisting of methane and CO_2 in the presence of methanogenic bacteria [107]. Substrates normally used for anaerobic digestion involve domestic and industrial effluents [114, 115], swine and bovine manure [116, 117] and algal biomass [104, 118, 119]. Microalgae present proteins, carbohydrates and lipids in their composition that can be digested to produce methane, which, by burning, can generate carbon-free electricity [107]. When compared to other biofuels, the biogas production is operationally simple and eliminates drying and chemical pre-treatment of the raw material. Moreover, the CO_2 produced during the anaerobic process can be consumed as a nutrient in the microalgal cultivation, thus increasing the biogas purity [120] and the residual biomass from the anaerobic digester can go through reprocessing for use as a fertilizer and/or animal feed [118].

Another promising biofuel that can be produced by using microalgal cultivation for industrial flue gases is biohydrogen [99]. Hydrogen is an alternative fuel compared to conventional fossil; it releases energy in combustion without pollutants [36]. Bioprocesses for biological hydrogen production can be water biophotolysis by algae; hydrogen production in the absence of light during the acidogenic phase in anaerobic digestion of organic matter; photo-fermentative hydrogen production [121, 122].

Microalgae stand out in this proceeding, because they can be cultivation in closed systems, allowing the capture of hydrogen. The photobiological hydrogen production can be increased according to the concentration of carbon in the biomass [123]. The biomass obtained from hydrogen production process can be burned for energy generation due to the calorific value of these microorganisms being greater than some coals [36].

Biopigments

Biopigments can be found in several microorganisms, such as microalgae and cyanobacteria. Microalgae are presented as an excellent biological resource for obtaining a range of pigments such as chlorophyll, carotenoids and phycobiliproteins [124]. These pigments have several health benefits such as: vitamin precursors, antioxidants, strengtheners of the immune system and anti-inflammatory agents [125].

Chlorophylls are green pigments, hydrophobic, soluble in lipids found in algae, higher plants and cyanobacteria [126]. The green colour present in algae is due to the presence of chlorophylls *a*, *b*, c_1, c_2 and *d*, which absorb wavelength corresponding to its absorption spectrum (450 - 675 nm) and reflect the wavelengths of its colour. Its structure is composed of a chain of phytol and a porphyrin ring and appear dissolved in the lipid membrane of chloroplasts, while at the centre of the porphyrin ring structure there is a magnesium atom carrying the electrons to where the photosynthetic reactions occur [127]. Several factors can influence chlorophyll content in cyanobacteria biomass, culture medium composition, cultivation time and light intensity are some of them [128]. The nitrogen (N) added in the cultivation favours the accumulation of chlorophylls (*a* and *b*) in microalgal biomass, because chlorophyll is N rich, and it is easily accessible, whereas high luminosity can reduce chlorophyll *a* content in biomass [129, 130]. Exposure to the electromagnetic fields is another promising technique that can stimulate chlorophyll accumulation in microalgae [124]. Chlorophyll can be used as a natural colorant and antioxidant in foods, and as an antimutagenic agent [36].

Carotenoids are pigments widely distributed in nature and found in plants, algae and some fungi and bacteria species. Chemically, carotenoid compounds differ only in a few characteristics, which give rise to different structures, to the class of hydrocarbons (carotenes) and their oxygenated derivatives (xanthophylls) [131]. There are over 600 types of carotenoids that perform several functions in various organisms, including microalgae. Among the biopigments present in algae and cyanobacteria, β-carotene, astaxanthin, lutein, canthaxanthin, violaxanthin, neoxantina, zeaxanthin and myxoxanthophyll can be found. The green algae are presented as a source of lutein, whereas cyanobacteria can be exploited for the production of zeaxanthin and myxoxanthophyll and β-carotene is found in all classes and orders of microalgae [132].

Carotenoids are responsible for the yellow, orange and red colours of many foods such as fruits, vegetables, fish and crustaceans. For these colorants to be synthesized only by plants, algae and some microorganisms, they must be supplied in the diet of animals often converted to other carotenoids and embedded within the muscle tissue [133]. In addition, they have antioxidant properties, and are precursors of vitamin A, work in the prevention of cancer, ulcer inhibitors, heart attacks, food colouring, cosmetic additives and health food [133 - 135].

Phycobiliproteins are water-soluble proteins, components of photosynthetic light collector antenna complex, are bright coloured, highly fluorescent, present in cyanobacteria, red algae and cryptophytes. These proteins are divided into two major groups based on their colours, phycoerythrin (red) and phycocyanin (blue). The phycocyanins include C-phycocyanin, R-phycocyanin and allophycocyanin [136], with phycocyanin being the main pigment found in microalgae cells, while allophycocyanin is present in less quantity [137]. C-phycocyanin is strongly fluorescent, and is blue in colour. Together with the chlorophyll, which has green a colour, they are responsible for the blue-green staining of cyanobacteria [138]. It can be used as biopigment in food and cosmetics, as well a promising source for the treatment of cancer and support for tissue engineering [139]. Among the existing cyanobacteria, *Spirulina* receives more attention, since C-phycocyanin may be present by up to 20% dry weight of total proteins of the microorganism [140]. *Spirulina* is an excellent CO_2 biofixation agent, using this carbon source for its own growth and producing simultaneous C-phycocyanin, valuable product with excellent benefits [70].

Biopolymers

Polyhydroxybutyrate (PHB) is a biodegradable plastic with similar properties to conventional oil-based plastics, an is synthesized by a variety of organisms through intracellular material storage. PHB belongs to polyhydroxyalkanoates (PHA) group, which are hydroxyalkanoic acid polyesters which can be obtained from renewable carbon sources and organic waste, being produced by various bacteria and cyanobacteria as a reserve material as in the from of intracellular granules [141, 142]. These microorganisms can be considered an alternative system for biopolymer production, due to their photoautotrophic nature [143].

PHBs are a class of polymers formed by monomeric units which can have different numbers of carbon atoms, in which the most studied is the poly(3-hydroxybutyrate) (P(3HB)) and poly(3-hydroxybutyrate-co-3-hydroxyvalerate) (P(3HB-co-3HV)). The monomeric unit depends on the composition or the side chain of the radical R and the value of n (Fig. **4**).

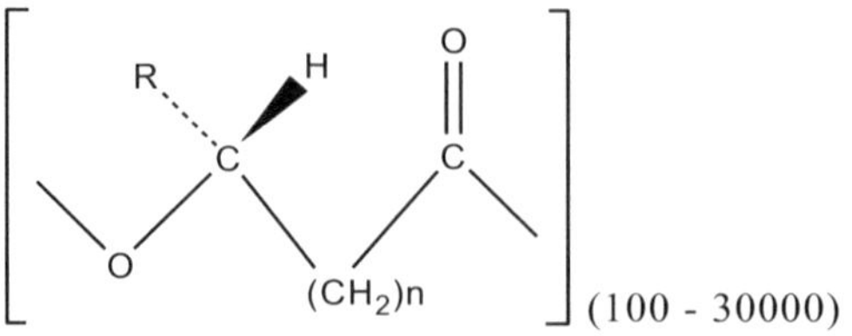

n=1	R=H	Poly(3-hydroxypropienate)	P(3HP)
	$R=CH_3$	Poly(3-hydroxybutyrate)	P(3HB)
	$R=C_2H_5$	Poly(3-hydroxyvalerate)	P(3HV)
	$R=C_3H_7$	Poly(3-hydroxyhexanoate)	P(3HX)
n=2	R=H	Poly(4-hydroxybutyrate)	P(4HB)
n=3	R=H	Poly(5-hydroxyvalerate)	P(5HV)

Fig. (4). Structure of polyhydroxyalkanoates.

PHB biosynthesis in the microalgae cell is directly related to the physicochemical conditions and nutrition of the microorganism. Cultivation submitted to limitation of nutrients such as phosphorus and nitrogen, carbon excess as CO_2, stress conditions, temperature, pH, control of metabolic and flux and different modes of cultivation may increase the amount of PHB in the cells [47, 144, 145]. Carbon sources such as glycerine obtained as by-products of biodiesel production, and CO_2 from industries and power plants can be used as raw material in cell growth. The use of industrial wastes has become increasingly promising, offering a sustainable alternative, in addition to closing the carbon cycle [146, 147].

Mallick *et al.* [148] have studied the PHB-HV production in *Nostoc muscorum* using BG-11 medium with propionate addition, achieving 28.2% of biopolymers. Martins *et al.* [149] obtained a PHB yield of 44.2% when *Spirulina* sp. LEB 18 was cultivated in a modified Zarrouk medium containing 8.4 g L^{-1} of sodium bicarbonate and 0.25 g L^{-1} of sodium nitrate. Bhati and Mallick [150] cultivated *Nostoc muscorum* Agarth cyanobacteria in different carbon sources like glucose, acetate, fructose, propionate and valerate for (P(3HB-co-3HV)) production, obtaining biopolymer yields of 28% e 26% using 0.4% of acetate and 0.4% of glucose, respectively. However, on adding 0.4% of valerate and 0.4% of acetate, with nitrogen restriction, the biopolymer yield reached 60%. Kaewbai-ngam *et al.* [151] evaluated the polydroxybutyrate accumulation in *Calothrix scytonemicola* TISTR 8095 using CO_2 and observed a PHB accumulation of 25.4% (w w^{-1}) in biomass, presenting PHB productivity equivalent to the CO_2 consumption of 729.2 mg L^{-1}.

CONCLUDING REMARKS

Concerning what was presented, the CO_2 biofixation process by microalgae from flue gas contributes to the reduction of GHGs in atmosphere. The advantages of using these microorganisms for this purpose bring together the reduction the production costs of a biomass rich in high-value compounds with the viability of the operation of large industries. Furthermore, several microalgae species and cultivation conditions can be used to increase of the biofixation process efficiency and with thus produce biomass with applicability in the obtention of several bioproducts such as biofuels, biopigments and biopolymers.

CONFLICT OF INTEREST

The authors declare no conflict of interest, financial or otherwise.

ACKNOWLEDGEMENTS

The authors would like thanking the Brazilian development agencies: CAPES (Coordination for the Improvement of Higher Education Personnel), R & D Program ANEEL-Eletrobrás CGTEE (National Agency of Electric Energy-Company of Thermal Generation of Electric Power), CNPq (National Council of Technological and Scientific Development) and MCTI (Ministry of Science, Technology and Innovation).

REFERENCES

[1] K. Benis, and P. Ferrão, "Potential mitigation of the environmental impacts of food systems through urban and peri-urban agriculture (UPA) e a life cycle assessment approach", *J. Clean. Prod.,* vol. 140, pp. 784-795, 2017.
[http://dx.doi.org/10.1016/j.jclepro.2016.05.176]

[2] B. Wang, Y. Li, N. Wu, and C.Q. Lan, "CO(2) bio-mitigation using microalgae", *Appl. Microbiol. Biotechnol.,* vol. 79, no. 5, pp. 707-718, 2008.
[http://dx.doi.org/10.1007/s00253-008-1518-y] [PMID: 18483734]

[3] E.M. Kondili, and J.K. Kaldellis, "Biofuel implementation in East Europe: Current status and future prospects", *Renew. Sustain. Energy Rev.,* vol. 11, pp. 2137-2151, 2007.
[http://dx.doi.org/10.1016/j.rser.2006.05.001]

[4] J. Houghton, "Global warming", *Rep. Prog. Phys.,* vol. 68, pp. 1343-1403, 2005.
[http://dx.doi.org/10.1088/0034-4885/68/6/R02]

[5] I. Douskova, J. Doucha, K. Livansky, J. Machat, P. Novak, D. Umysova, V. Zachleder, and M. Vitova, "Simultaneous flue gas bioremediation and reduction of microalgal biomass production costs", *Appl. Microbiol. Biotechnol.,* vol. 82, no. 1, pp. 179-185, 2009.
[http://dx.doi.org/10.1007/s00253-008-1811-9] [PMID: 19096837]

[6] J.A.V. Costa, M.G. Morais, E.M. Radmann, F.B. Santana, F.V. Camerini, M.R.A.Z. Souza, A.A. Henrard, A.P.C. Rosa, and L.R.F. Brusch, "Biofixation of carbon dioxide from coal station flue gas using *Spirulina* sp. LEB 18 and *Scenedesmus obliquus* LEB 22", *Afr. J. Microbiol. Res.,* vol. 9, no. 44, pp. 2202-2208, 2015.
[http://dx.doi.org/10.5897/AJMR2015.7640]

[7] B.S. Vaz, J.A.V. Costa, and M.G. Morais, "CO_2 Biofixation by the Cyanobacterium *Spirulina* sp. LEB 18 and the Green Alga *Chlorella fusca* LEB 111 Grown Using Gas Effluents and Solid Residues of Thermoelectric Origin", *Appl. Biochem. Biotechnol.,* vol. 178, no. 2, pp. 418-429, 2016. [http://dx.doi.org/10.1007/s12010-015-1876-8] [PMID: 26453033]

[8] J.H. Duarte, L.S. Fanka, and J.A.V. Costa, "Utilization of simulated flue gas containing CO_2, SO_2, NO and ash for *Chlorella fusca* cultivation", *Bioresour. Technol.,* vol. 214, pp. 159-165, 2016. [http://dx.doi.org/10.1016/j.biortech.2016.04.078] [PMID: 27132223]

[9] K. Kumar, C.N. Dasgupta, B. Nayak, P. Lindblad, and D. Das, "Development of suitable photobioreactors for CO_2 sequestration addressing global warming using green algae and cyanobacteria", *Bioresour. Technol.,* vol. 102, no. 8, pp. 4945-4953, 2011. [http://dx.doi.org/10.1016/j.biortech.2011.01.054] [PMID: 21334885]

[10] M. Anjos, B.D. Fernandes, A.A. Vicente, J.A. Teixeira, and G. Dragone, "Optimization of CO_2 bio-mitigation by *Chlorella vulgaris*", *Bioresour. Technol.,* vol. 139, pp. 149-154, 2013. [http://dx.doi.org/10.1016/j.biortech.2013.04.032] [PMID: 23648764]

[11] D. Tang, W. Han, P. Li, X. Miao, and J. Zhong, "CO_2 biofixation and fatty acid composition of *Scenedesmus obliquus* and *Chlorella pyrenoidosa* in response to different CO_2 levels", *Bioresour. Technol.,* vol. 102, no. 3, pp. 3071-3076, 2011. [http://dx.doi.org/10.1016/j.biortech.2010.10.047] [PMID: 21041075]

[12] I. P. O. C. CHANGE, *Report IPCC/ONU - New Climate Scenarios - initiative Ecolatina 1 (in portuguese),* 2007.

[13] *SPM : Summary for Policymakers Contents.* Mitig. Clim. Chang. Summ. Policymakers, 2014.

[14] C.J.A. Mota, and R.S. Monteiro, "Quim. Nova", *Química e Sustentabilidade Novas Front. em Biocombustíveis,* vol. 36, no. 10, pp. 1483-1490, 2013.

[15] S. Van Den Hende, H. Vervaeren, and N. Boon, "Flue gas compounds and microalgae: (Bio-)chemical interactions leading to biotechnological opportunities", *Biotechnol. Adv.,* vol. 30, no. 6, pp. 1405-1424, 2012. [http://dx.doi.org/10.1016/j.biotechadv.2012.02.015] [PMID: 22425735]

[16] D. Moreira, and J.C.M. Pires, "Atmospheric CO_2 capture by algae: Negative carbon dioxide emission path", *Bioresour. Technol.,* vol. 215, pp. 371-379, 2016. [http://dx.doi.org/10.1016/j.biortech.2016.03.060] [PMID: 27005790]

[17] T.P. Dawson, S.T. Jackson, J.I. House, I.C. Prentice, and G.M. Mace, "Beyond predictions: biodiversity conservation in a changing climate", *Science,* vol. 332, no. 6025, pp. 53-58, 2011. [http://dx.doi.org/10.1126/science.1200303] [PMID: 21454781]

[18] M. Meinshausen, N. Meinshausen, W. Hare, S.C.B. Raper, K. Frieler, R. Knutti, D.J. Frame, and M.R. Allen, "Greenhouse-gas emission targets for limiting global warming to 2 °C", *Nature,* vol. 458, no. 7242, pp. 1158-1162, 2009. [http://dx.doi.org/10.1038/nature08017] [PMID: 19407799]

[19] J. Rockström, W. Steffen, K. Noone, A. Persson, F.S. Chapin III, E.F. Lambin, T.M. Lenton, M. Scheffer, C. Folke, H.J. Schellnhuber, B. Nykvist, C.A. de Wit, T. Hughes, S. van der Leeuw, H. Rodhe, S. Sörlin, P.K. Snyder, R. Costanza, U. Svedin, M. Falkenmark, L. Karlberg, R.W. Corell, V.J. Fabry, J. Hansen, B. Walker, D. Liverman, K. Richardson, P. Crutzen, and J.A. Foley, "A safe operating space for humanity", *Nature,* vol. 461, no. 7263, pp. 472-475, 2009. [http://dx.doi.org/10.1038/461472a] [PMID: 19779433]

[20] Y. Peng, B. Zhao, and L. Li, "Advance in Post-Combustion CO_2 Capture with Alkaline Solution: A Brief Review", *Energy Procedia,* vol. 14, pp. 1515-1522, 2012. [http://dx.doi.org/10.1016/j.egypro.2011.12.1126]

[21] U.B. Singh, and A.S. Ahluwalia, "Microalgae: a promising tool for carbon sequestration", *Mitig. Adapt. Strategies Glob. Change,* vol. 18, no. 1, pp. 73-95, 2013.

[http://dx.doi.org/10.1007/s11027-012-9393-3]

[22] S-W. Park, B-S. Choi, and J-W. Lee, "Chemical absorption of carbon dioxide with triethanolamine in non-aqueous solutions", *Korean J. Chem. Eng.*, vol. 23, pp. 138-143, 2006. [http://dx.doi.org/10.1007/BF02705705]

[23] A.L. Chaffee, G.P. Knowles, Z. Liang, J. Zhang, P. Xiao, and P.A. Webley, "CO_2 capture by adsorption: Materials and process development", *Int. J. Greenh. Gas Control,* vol. 1, pp. 11-18, 2007. [http://dx.doi.org/10.1016/S1750-5836(07)00031-X]

[24] T.E. Rufford, S. Smart, G.C.Y. Watson, B.F. Graham, J. Boxall, J.C. Diniz, and E.F. May, "The removal of CO_2 and N_2 from natural gas: A review of conventional and emerging process technologies", *J. Petrol. Sci. Eng.,* vol. 94-95, pp. 123-154, 2012. [http://dx.doi.org/10.1016/j.petrol.2012.06.016]

[25] X. Huang, X. Wang, M. Fan, Y. Wang, H. Adidharma, K.A.M. Gasem, and Maciej Radosz, "A cost-effective approach to reducing carbon deposition and resulting deactivation of oxygen carriers for improvement of energy efficiency and CO2 capture during methane chemical-looping combustion", *Appl. Energy,* vol. 193, pp. 381-392, 2017. [http://dx.doi.org/10.1016/j.apenergy.2017.02.059]

[26] D.D Tsai, R. Ramaraj, and P.H. Chen, "Carbon dioxide bio-fixation by algae of high rate pond on natural water medium", *Ecol. Eng.,* vol. 92, pp. 106-110, 2016. [http://dx.doi.org/10.1016/j.ecoleng.2016.03.021]

[27] T. Minowa, S. Yokoyama, M. Kishimoto, and T. Okakurat, "Oil production from algal cells of *Dunaliella tertiolecta* by direct thermochemical liquefaction", In: *Butterworth Heinemann* vol. 74. , 1995, no. 12, pp. 1735-173.

[28] A. Demirbas, "Production of Biodiesel from Algae Oils", *Energy Sources,* vol. 168, p. 13163, 2009.

[29] P.M. Schenk, S.R. Thomas-hall, E. Stephens, U.C. Marx, J.H. Mussgnug, C. Posten, O. Kruse, and B. Hankamer, "Second Generation Biofuels: High-Efficiency Microalgae for Biodiesel Production", *BioEnergy Res.,* vol. 1, pp. 20-43, 2008. [http://dx.doi.org/10.1007/s12155-008-9008-8]

[30] J-S. Lee, and J-P. Lee, "Review of Advances in Biological CO_2 Mitigation Technology", *Biotechnol. Bioprocess Eng.; BBE,* vol. 8, pp. 354-359, 2003. [http://dx.doi.org/10.1007/BF02949279]

[31] E.G. Morais, Cultivo de microalgas utilizando efluentes da indústria bioenergética e avaliação da conversão em biodiesel, 2015.

[32] S.C. Wuang, M.C. Khin, P.Q. Chua, and Y.D. Luo, "Use of *Spirulina* biomass produced from treatment of aquaculture wastewater as agricultural fertilizers", *Algal Res.,* vol. 15, pp. 59-64, 2016. [http://dx.doi.org/10.1016/j.algal.2016.02.009]

[33] J. Milano, H. Chyuan, H.H. Masjuki, W.T. Chong, and M. Kee, "Microalgae biofuels as an alternative to fossil fuel for power generation", *Renew. Sustain. Energy Rev.,* vol. 58, pp. 180-197, 2016. [http://dx.doi.org/10.1016/j.rser.2015.12.150]

[34] E.M. Radmann, F.V. Camerini, T.D. Santos, and J.A.V. Costa, "Isolation and application of SO_X and NO_X resistant microalgae in Biofixation of CO_2 from thermoelectricity plants", *Energy Convers. Manage.,* vol. 52, pp. 3132-3136, 2011. [http://dx.doi.org/10.1016/j.enconman.2011.04.021]

[35] G.M. Rosa, L. Moraes, B.B. Cardias, M.R.A.Z. Souza, and J.A.V. Costa, "Chemical absorption and CO_2 biofixation via the cultivation of Spirulina in semicontinuous mode with nutrient recycle", *Bioresour. Technol.,* vol. 192, pp. 321-327, 2015. [http://dx.doi.org/10.1016/j.biortech.2015.05.020] [PMID: 26051496]

[36] A. Pandey, D.-J. Lee, Y. Chisti, and C. R. Soccol, *Biofuels from Algae,* 2014.

[37] C.K. da Silva, *Produção biotecnológica de biopolímeros aplicando processo cíclico de reutilização de resíduo da extração de poli-hidroxibutirato (PHB)*, 2016.

[38] M.G. Morais, and J.A.V. Costa, "Biofixation of carbon dioxide by *Spirulina* sp. and *Scenedesmus obliquus* cultivated in a three-stage serial tubular photobioreactor", *J. Biotechnol.*, vol. 129, no. 3, pp. 439-445, 2007.
[http://dx.doi.org/10.1016/j.jbiotec.2007.01.009] [PMID: 17320994]

[39] J.A. Borges, G.M. Rosa, L.H. Meza, A.A. Henrard, M.R.A.Z. Souza, and J.A.V. Costa, "*Spirulina* sp. LEB-18 culture using effluent from the anaerobic digestion", *Braz. J. Chem. Eng.*, vol. 30, no. 2, pp. 277-287, 2013.
[http://dx.doi.org/10.1590/S0104-66322013000200006]

[40] K.L. Kadam, "Power plant flue gas as a source of CO_2 for microalgae cultivation: economic impact of different process options", *Energy Convers. Manage.*, vol. 38, pp. 505-510, 1997.
[http://dx.doi.org/10.1016/S0196-8904(96)00318-4]

[41] *"National Algal Biofuels Technology Roadmap," no. May.* U.S. Department of Energy, Office of Energy Efficiency and Renewable Energy, 2010.

[42] B. Zhao, and Y. Su, "Process effect of microalgal-carbon dioxide fixation and biomass production: A review", *Renew. Sustain. Energy Rev.*, vol. 31, pp. 121-132, 2014.
[http://dx.doi.org/10.1016/j.rser.2013.11.054]

[43] A. Richmond, *Handbook of microalgae culture: Biotechnology and applied phycology,* vol. 296, no. 2, 2004.

[44] E.W. Becker, *Microalgae: Biotechnology and Microbiology.* Cambridge University Press, 1994.

[45] A. Kumar, S. Ergas, X. Yuan, A. Sahu, Q. Zhang, J. Dewulf, F.X. Malcata, and H. van Langenhove, "Enhanced $CO(_2)$ fixation and biofuel production *via* microalgae: recent developments and future directions", *Trends Biotechnol.*, vol. 28, no. 7, pp. 371-380, 2010.
[http://dx.doi.org/10.1016/j.tibtech.2010.04.004] [PMID: 20541270]

[46] E. Molina Grima, E.H. Belarbi, F.G. Acién Fernández, A. Robles Medina, and Y. Chisti, "Recovery of microalgal biomass and metabolites: process options and economics", *Biotechnol. Adv.*, vol. 20, no. 7-8, pp. 491-515, 2003.
[http://dx.doi.org/10.1016/S0734-9750(02)00050-2] [PMID: 14550018]

[47] T.M. Mata, A.A. Martins, and N.S. Caetano, "Microalgae for biodiesel production and other applications: A review", *Renew. Sustain. Energy Rev.*, vol. 14, no. 1, pp. 217-232, 2010.
[http://dx.doi.org/10.1016/j.rser.2009.07.020]

[48] S.Y. Chiu, C.Y. Kao, T.T. Huang, C.J. Lin, S.C. Ong, C.D. Chen, J.S. Chang, and C.S. Lin, "Microalgal biomass production and on-site bioremediation of carbon dioxide, nitrogen oxide and sulfur dioxide from flue gas using *Chlorella* sp. cultures", *Bioresour. Technol.*, vol. 102, no. 19, pp. 9135-9142, 2011.
[http://dx.doi.org/10.1016/j.biortech.2011.06.091] [PMID: 21802285]

[49] H. Weiss, Method for growing photosynthetic organisms - United States patent application US20080220486 A1., 2008.

[50] G. Jensen, K-K. Hi, H.R. Eric, and N. Pnnceton, *Integrated Microalgae Production And Electricity Cogeneration.* U.S. Patent 5,659,977, August 26, 1997.

[51] G.V. Swarnalatha, N.S. Hegde, V.S. Chauhan, and R. Sarada, "The effect of carbon dioxide rich environment on carbonic anhydrase activity, growth and metabolite production in indigenous freshwater microalgae", *Algal Res.*, vol. 9, pp. 151-159, 2015.
[http://dx.doi.org/10.1016/j.algal.2015.02.014]

[52] M.G. Morais, and J.A.V. Costa, "Isolation and selection of microalgae from coal fired thermoelectric power plant for biofixation of carbon dioxide", *Energy Convers. Manage.*, vol. 48, no. 7, pp. 2169-

2173, 2007.
[http://dx.doi.org/10.1016/j.enconman.2006.12.011]

[53] M.G. Morais, C.K. Silva, A.A. Henrard, and J.A.V. Costa, "Carbon dioxide mitigation by microalga in a vertical tubular reactor with recycling of the culture medium", *Afr. J. Microbiol. Res.*, vol. 9, no. 33, pp. 1935-1940, 2015.
[http://dx.doi.org/10.5897/AJMR2015.7632]

[54] M.G. Morais, and J.A.V. Costa, "Carbon dioxide fixation by *Chlorella kessleri, C. vulgaris, Scenedesmus obliquus* and *Spirulina* sp. cultivated in flasks and vertical tubular photobioreactors", *Biotechnol. Lett.*, vol. 29, no. 9, pp. 1349-1352, 2007.
[http://dx.doi.org/10.1007/s10529-007-9394-6] [PMID: 17503002]

[55] L. Moraes, G.M. Rosa, B.B. Cardias, L. Oliveira, and J.A.V. Costa, "Microalgal biotechnology for greenhouse gas control: Carbon dioxide fixation by *Spirulina* sp. at different diffusers", *Ecol. Eng.*, vol. 91, pp. 426-431, 2016.
[http://dx.doi.org/10.1016/j.ecoleng.2016.02.035]

[56] G.M. Rosa, L. Moraes, M.R.A.Z. Souza, and J.A.V. Costa, "Spirulina cultivation with a CO_2 absorbent: Influence on growth parameters and macromolecule production", *Bioresour. Technol.*, vol. 200, pp. 528-534, 2016.
[http://dx.doi.org/10.1016/j.biortech.2015.10.025] [PMID: 26524251]

[57] E.M. Radmann, and J.A.V. Costa, "Lipid content and fatty acids composition variation of microalgae exposed to CO_2, SO_2 and NO", *Quim. Nova,* vol. 31, no. 7, pp. 1609-1612, 2008.
[http://dx.doi.org/10.1590/S0100-40422008000700002]

[58] M.G. Morais, and J.A.V. Costa, "Fatty acids profile of microalgae cultivated with carbondioxide", *Sci.Agrotechnology,* vol. 32, no. 4, pp. 1245-1251, 2008.

[59] J.C.M. Pires, F.G. Martins, and M. Simões, "Carbondioxide capture from flue gases using microalgae: Engineering aspects and biorefinery concept", *Renew. Sustain. Energy Rev.,* vol. 16, no. 5, pp. 3043-3053, 2012.
[http://dx.doi.org/10.1016/j.rser.2012.02.055]

[60] L. Brennan, and P. Owende, "Biofuels from microalgae-A review of technologies for production, processing, and extractions of biofuels and co-products", *Renew. Sustain. Energy Rev.,* vol. 14, no. 2, pp. 557-577, 2010.
[http://dx.doi.org/10.1016/j.rser.2009.10.009]

[61] E. Ono, and J.L. Cuello, "Carbon Dioxide Mitigation using Thermophilic Cyanobacteria", *Biosyst. Eng.,* vol. 96, no. 1, pp. 129-134, 2007.
[http://dx.doi.org/10.1016/j.biosystemseng.2006.09.010]

[62] K. Sung, J. Lee, C. Shin, S. Park, and M-J. Choi, "CO_2 fixation by *Chlorella* sp. KR-1 and its cultural characteristics", *Bioresour. Technol.,* vol. 68, pp. 269-273, 1999.
[http://dx.doi.org/10.1016/S0960-8524(98)00152-7]

[63] V. Sankar, D.K. Daniel, and A. Krastanov, "Carbon Dioxide Fixation by *Chlorella minutissima* Batch Cultures in a Stirred Tank Bioreactor", *Biotechnol. Biotechnol. Equip.,* vol. 25, no. 3, pp. 2468-2476, 2011.
[http://dx.doi.org/10.5504/BBEQ.2011.0058]

[64] W. Stumn, and J. Morgan, *Aquatic Chemistry, 2nd*. Wiley: New York, 1981.

[65] A. Walsby, "Cell-water and cell-solute relations", In: *The biology of cyanobacteria,* N. Carr, B. Whitton, Eds., Blackwell Science Publications: Oxford, 1982, pp. 237-262.

[66] C. Poza-Carrión, E. Fernández-Valiente, F.F. Piñas, and F. Leganés, "Acclimation of photosynthetic pigments and photosynthesis of the cyanobacterium *Nostoc* sp. strain UAM206 to combined fluctuations of irradiance, pH, and inorganic carbon availability", *J. Plant Phiysiology,* vol. 158, pp. 1455-1461, 2001.

[http://dx.doi.org/10.1078/0176-1617-00555]

[67] A. Richmond, and J.U. Grobbelaar, "Factors affecting the output rate of *Spirulina platensis* with reference to mass cultivation", *Biomass,* vol. 10, no. 4, pp. 253-264, 1986. [http://dx.doi.org/10.1016/0144-4565(86)90002-8]

[68] K. Maeda, M. Owada, N. Kimura, K. Omata, and I. Karube, "CO_2 fixation from the flue gas on coal-fired thermal power plant by microalgae", *Energy Convers. Manage.,* vol. 36, no. 6, pp. 717-720, 1995. [http://dx.doi.org/10.1016/0196-8904(95)00105-M]

[69] J.S. Lee, D.K. Kim, J.P. Lee, S.C. Park, J.H. Koh, H.S. Cho, and S.W. Kim, "Effects of SO_2 and NO on growth of *Chlorella* sp. KR-1", *Bioresour. Technol.,* vol. 82, no. 1, pp. 1-4, 2002. [http://dx.doi.org/10.1016/S0960-8524(01)00158-4] [PMID: 11848373]

[70] C.Y. Chen, P.C. Kao, C.J. Tsai, D.J. Lee, and J.S. Chang, "Engineering strategies for simultaneous enhancement of C-phycocyanin production and CO_2 fixation with *Spirulina platensis*", *Bioresour. Technol.,* vol. 145, pp. 307-312, 2013. [http://dx.doi.org/10.1016/j.biortech.2013.01.054] [PMID: 23664178]

[71] K. Skalska, J.S. Miller, and S. Ledakowicz, "Trends in $NO_{(x)}$ abatement: a review", *Sci. Total Environ.,* vol. 408, no. 19, pp. 3976-3989, 2010. [http://dx.doi.org/10.1016/j.scitotenv.2010.06.001] [PMID: 20580060]

[72] H. Nagase, K. Yoshihara, K. Eguchi, Y. Okamoto, S. Murasaki, R. Yamashita, K. Hirata, and K. Miyamoto, "Uptake pathway and continuous removal of nitric oxide from flue gas using microalgae", *Biochem. Eng. J.,* vol. 7, pp. 241-246, 2001.

[73] W.Y. Cheah, P.L. Show, J-S. Chang, T.C. Ling, and J.C. Juan, "Biosequestration of atmospheric CO_2 and flue gas-containing CO_2 by microalgae", *Bioresour. Technol.,* vol. 184, pp. 190-201, 2015. [PMID: 25497054]

[74] J.N. Lee, J.S. Lee, C.S. Shin, S.C. Park, and S.W. Kim, "Methods to enhance tolerances of *Chlorella* KR-1 to toxic compounds in flue gas", *Appl. Biochem. Biotechnol.,* vol. 84, pp. 329-342, 2000. [http://dx.doi.org/10.1385/ABAB:84-86:1-9:329] [PMID: 10849800]

[75] C.Y. Kao, S.Y. Chiu, T.T. Huang, L. Dai, L.K. Hsu, and C.S. Lin, "Ability of a mutant strain of the microalga *Chlorella* sp. to capture carbon dioxide for biogas upgrading", *Appl. Energy,* vol. 93, pp. 176-183, 2012. [http://dx.doi.org/10.1016/j.apenergy.2011.12.082]

[76] G. Merker, C. Schwarz, G. Stiesch, and F. Otto, *Simulating Combustion.* 1st ed. Springer-Verlag Berlin Heidelberg, 2006.

[77] H.T. Hsueh, W.J. Li, H.H. Chen, and H. Chu, "Carbon bio-fixation by photosynthesis of *Thermosynechococcus* sp. CL-1 and *Nannochloropsis oculta*", *J. Photochem. Photobiol. B,* vol. 95, no. 1, pp. 33-39, 2009. [http://dx.doi.org/10.1016/j.jphotobiol.2008.11.010] [PMID: 19167907]

[78] S.H. Ho, C.Y. Chen, and J.S. Chang, "Effect of light intensity and nitrogen starvation on CO_2 fixation and lipid/carbohydrate production of an indigenous microalga *Scenedesmus obliquus* CNW-N", *Bioresour. Technol.,* vol. 113, pp. 244-252, 2012. [http://dx.doi.org/10.1016/j.biortech.2011.11.133] [PMID: 22209130]

[79] A.L. Gonçalves, M. Simões, and J.C.M. Pires, "The effect of light supply on microalgal growth, CO2 uptake and nutrient removal from wastewater", *Energy Convers. Manage.,* vol. 85, pp. 530-536, 2014. [http://dx.doi.org/10.1016/j.enconman.2014.05.085]

[80] A.P. da Rosa, L.F. Carvalho, L. Goldbeck, and J.A.V. Costa, "Carbon dioxide fixation by microalgae cultivated in open bioreactors", *Energy Convers. Manage.,* vol. 52, no. 8-9, pp. 3071-3073, 2011. [http://dx.doi.org/10.1016/j.enconman.2011.01.008]

[81] X. Chai, and X. Zhao, "Enhanced removal of carbon dioxide and alleviation of dissolved oxygen

accumulation in photobioreactor with bubble tank", *Bioresour. Technol.,* vol. 116, pp. 360-365, 2012. [http://dx.doi.org/10.1016/j.biortech.2012.03.105] [PMID: 22531167]

[82] L. Fan, Y. Zhang, L. Cheng, L. Zhang, D. Tang, and H. Chen, "Optimization of carbon dioxide fixation by *Chlorella vulgaris* cultivated in a membrane-photobioreactor", *Chem. Eng. Technol.,* vol. 30, no. 8, pp. 1094-1099, 2007. [http://dx.doi.org/10.1002/ceat.200700141]

[83] L.H. Fan, Y.T. Zhang, L. Zhang, and H.L. Chen, "Evaluation of a membrane-sparged helical tubular photobioreactor for carbon dioxide biofixation by *Chlorella vulgaris*", *J. Membr. Sci.,* vol. 325, no. 1, pp. 336-345, 2008. [http://dx.doi.org/10.1016/j.memsci.2008.07.044]

[84] H.J. Ryu, K.K. Oh, and Y.S. Kim, "Optimization of the influential factors for the improvement of CO_2 utilization efficiency and CO_2 mass transfer rate", *J. Ind. Eng. Chem.,* vol. 15, pp. 471-475, 2009. [http://dx.doi.org/10.1016/j.jiec.2008.12.012]

[85] C.U. Ugwu, H. Aoyagi, and H. Uchiyama, "Photobioreactors for mass cultivation of algae", *Bioresour. Technol.,* vol. 99, no. 10, pp. 4021-4028, 2008. [http://dx.doi.org/10.1016/j.biortech.2007.01.046] [PMID: 17379512]

[86] P. Talbot, M.P. Gortares, R.W. Lencki, and J. de la Noüe, "Absorption of CO_2 in algal mass culture systems: a different characterization approach", *Biotechnol. Bioeng.,* vol. 37, no. 9, pp. 834-842, 1991. [http://dx.doi.org/10.1002/bit.260370907] [PMID: 18600683]

[87] A.P. Carvalho, L.A. Meireles, and F.X. Malcata, "Microalgal reactors: a review of enclosed system designs and performances", *Biotechnol. Prog.,* vol. 22, no. 6, pp. 1490-1506, 2006. [http://dx.doi.org/10.1002/bp060065r] [PMID: 17137294]

[88] T. Duarte-Santos, J.L. Mendoza-Martín, F.G. Acién Fernández, E. Molina, J.A. Vieira-Costa, and S. Heaven, "Optimization of carbon dioxide supply in raceway reactors: Influence of carbon dioxide molar fraction and gas flow rate", *Bioresour. Technol.,* vol. 212, pp. 72-81, 2016. [http://dx.doi.org/10.1016/j.biortech.2016.04.023] [PMID: 27085148]

[89] S.H. Ho, C.Y. Chen, D.J. Lee, and J.S. Chang, "Perspectives on microalgal CO_2-emission mitigation systems--a review", *Biotechnol. Adv.,* vol. 29, no. 2, pp. 189-198, 2011. [http://dx.doi.org/10.1016/j.biotechadv.2010.11.001] [PMID: 21094248]

[90] M.K. Lam, K.T. Lee, and A.R. Mohamed, "Current status and challenges on microalgae-based carbon capture", *Int. J. Greenh. Gas Control,* vol. 10, pp. 456-469, 2012. [http://dx.doi.org/10.1016/j.ijggc.2012.07.010]

[91] X. Chai, Z. Zhao, and W. Baoying, "Biofixation of carbon dioxide by *Chlorococcum* sp. in a photobioreactor with polytetrafluoroethene membrane sparger", *Afr. J. Biotechnol.,* vol. 11, no. 29, pp. 7445-7453, 2012.

[92] A. Demirbas, "Biodiesel from oilgae, biofixation of carbon dioxide by microalgae: A solution to pollution problems", *Appl. Energy,* vol. 88, no. 2, pp. 3541-3547, 2011. [http://dx.doi.org/10.1016/j.apenergy.2010.12.050]

[93] S-H. Ho, A. Kondo, T. Hasunuma, and J-S. Chang, "Engineering strategies for improving the CO_2 fixation and carbohydrate productivity of *Scenedesmus obliquus* CNW-N used for bioethanol fermentation", *Bioresour. Technol.,* vol. 143, pp. 163-171, 2013. [http://dx.doi.org/10.1016/j.biortech.2013.05.043] [PMID: 23792755]

[94] P.H. Ravelonandro, D.H. Ratianarivo, C. Joannis-Cassan, A. Isambert, and M. Raherimandimby, "Influence of light quality and intensity in the cultivation of *Spirulina platensis* from Toliara (Madagascar) in a closed system", *J. Chem. Technol. Biotechnol.,* vol. 83, no. 2, pp. 842-848, 2008. [http://dx.doi.org/10.1002/jctb.1878]

[95] T.M. Iverson, "Evolution and unique bioenergetic mechanisms in oxygenic photosynthesis", *Curr. Opin. Chem. Biol.,* vol. 10, no. 2, pp. 91-100, 2006.

[http://dx.doi.org/10.1016/j.cbpa.2006.02.013] [PMID: 16504567]

[96] J. Červený, I. Sˇetlı'k, M. Trtı'lek, and L. Nedbal, "Photobioreactor for cultivation and real-time, *in-situ* measurement of O_2 and CO_2 exchange rates, growth dynamics, and of chlorophyll fluorescence emission of photoautotrophic microorganisms", *Eng. Life Sci.*, vol. 9, pp. 247-253, 2009. [http://dx.doi.org/10.1002/elsc.200800123]

[97] M. Calvin, "Forty years of photosynthesis and related activities", *Photosynth. Res.*, vol. 21, 1989.

[98] M. Giordano, J. Beardall, and J.A. Raven, "CO_2 concentrating mechanisms in algae: mechanisms, environmental modulation, and evolution", *Annu. Rev. Plant Biol.*, vol. 56, pp. 99-131, 2005. [http://dx.doi.org/10.1146/annurev.arplant.56.032604.144052] [PMID: 15862091]

[99] J.A.V. Costa, and M.G. Morais, "The role of biochemical engineering in the production of biofuels from microalgae", *Bioresour. Technol.*, vol. 102, no. 1, pp. 2-9, 2011. [http://dx.doi.org/10.1016/j.biortech.2010.06.014] [PMID: 20580548]

[100] V.C. Coelho, C.K. da Silva, A.L. Terra, J.A.V. Costa, and M.G. Morais, "Polyhydroxybutyrate production by *Spirulina* sp. LEB 18 grown under different nutrient concentrations", *Afr. J. Microbiol. Res.*, vol. 9, no. 24, pp. 1586-1594, 2015. [http://dx.doi.org/10.5897/AJMR2015.7530]

[101] M.K. Lam, and K.T. Lee, "Effect of carbon source towards the growth of *Chlorella vulgaris* for CO_2 bio-mitigation and biodiesel production", *Int. J. Greenh. Gas Control,* vol. 14, pp. 169-176, 2013. [http://dx.doi.org/10.1016/j.ijggc.2013.01.016]

[102] M.C. Cerón-García, M.D. Macías-Sánchez, A. Sánchez-Mirón, F. García-Camacho, and E. Molina-Grima, "A process for biodiesel production involving the heterotrophic fermentation of *Chlorella* protothecoides with glycerol as the carbon source", *Appl. Energy,* vol. 103, pp. 341-349, 2013. [http://dx.doi.org/10.1016/j.apenergy.2012.09.054]

[103] H. Guo, M. Daroch, L. Liu, G. Qiu, S. Geng, and G. Wang, "Biochemical features and bioethanol production of microalgae from coastal waters of Pearl River Delta", *Bioresour. Technol.*, vol. 127, pp. 422-428, 2013. [http://dx.doi.org/10.1016/j.biortech.2012.10.006] [PMID: 23138065]

[104] C. Zamalloa, N. Boon, and W. Verstraete, "Anaerobic digestibility of *Scenedesmus obliquus* and *Phaeodactylum tricornutum* under mesophilic and thermophilic conditions", *Appl. Energy,* vol. 92, pp. 733-738, 2012. [http://dx.doi.org/10.1016/j.apenergy.2011.08.017]

[105] N. Rashid, K. Lee, and Q. Mahmood, "Bio-hydrogen production by *Chlorella vulgaris* under diverse photoperiods", *Bioresour. Technol.*, vol. 102, no. 2, pp. 2101-2104, 2011. [http://dx.doi.org/10.1016/j.biortech.2010.08.032] [PMID: 20826084]

[106] I.A. Guschina, and J.L. Harwood, "Lipids and lipid metabolism in eukaryotic algae", *Prog. Lipid Res.*, vol. 45, no. 2, pp. 160-186, 2006. [http://dx.doi.org/10.1016/j.plipres.2006.01.001] [PMID: 16492482]

[107] R. Harun, M. Davidson, M. Doyle, R. Gopiraj, M. Danquah, and G. Forde, "Technoeconomic analysis of an integrated microalgae photobioreactor, biodiesel and biogas production facility", *Biomass Bioenergy,* vol. 35, pp. 741-747, 2011. [http://dx.doi.org/10.1016/j.biombioe.2010.10.007]

[108] A.S. Cardoso, G.E.G. Vieira, and A.K. Marques, "O uso de microalgas para a obtenção de biocombustíveis", *Rev. Bras. Biocienc.,* vol. 9, no. 4, pp. 542-549, 2011. [In portuguese].

[109] C.E. de F. Silva, and A. Bertucco, "Bioethanol from microalgae and cyanobacteria: A review and technological outlook", *Process Biochem.,* vol. 51, pp. 1833-1842, 2016.

[110] K.B. Möllers, D. Cannella, H. Jørgensen, and N-U. Frigaard, "Cyanobacterial biomass as carbohydrate and nutrient feedstock for bioethanol production by yeast fermentation", *Biotechnol. Biofuels,* vol. 7, no. 7, p. 64, 2014.

[http://dx.doi.org/10.1186/1754-6834-7-64] [PMID: 24739806]

[111] Y. Chisti, "Biodiesel from microalgae", *Biotechnol. Adv.,* vol. 25, no. 3, pp. 294-306, 2007. [http://dx.doi.org/10.1016/j.biotechadv.2007.02.001] [PMID: 17350212]

[112] R. Harun, K. Danquah, and G.M. Forde, "Microalgal biomass as a fermentation feedstock for bioethanol production", *J. Chem. Technol. Biotechnol.,* vol. 85, pp. 199-203, 2010.

[113] L.D. Zhu, E. Hiltunen, E. Antila, J.J. Zhong, Z.H. Yuan, and Z.M. Wang, "Microalgal biofuels: Flexible bioenergies for sustainable development", *Renew. Sustain. Energy Rev.,* vol. 30, pp. 1035-1046, 2014. [http://dx.doi.org/10.1016/j.rser.2013.11.003]

[114] H.W. Kim, J.Y. Nam, and H.S. Shin, "A comparison study on the high-rate co-digestion of sewage sludge and food waste using a temperature-phased anaerobic sequencing batch reactor system", *Bioresour. Technol.,* vol. 102, no. 15, pp. 7272-7279, 2011. [http://dx.doi.org/10.1016/j.biortech.2011.04.088] [PMID: 21600764]

[115] R.U. Rani, S.A. Kumar, S. Kaliappan, I.T. Yeom, and J.R. Banu, "Low temperature thermo-chemical pretreatment of dairy waste activated sludge for anaerobic digestion process", *Bioresour. Technol.,* vol. 103, no. 1, pp. 415-424, 2012. [http://dx.doi.org/10.1016/j.biortech.2011.09.124] [PMID: 22047655]

[116] R. Alvarez, and G. Lide, "Low temperature anaerobic digestion of mixtures of llama, cow and sheep manure for improved methane production", *Biomass Bioenergy,* vol. 33, no. 3, pp. 527-533, 2009. [http://dx.doi.org/10.1016/j.biombioe.2008.08.012]

[117] T. Amon, B. Amon, V. Kryvoruchko, W. Zollitsch, K. Mayer, and L. Gruber, "Biogas production from maize and dairy cattle manure - Influence of biomass composition on the methane yield of biomass composition on the methane yield", *Agric. Ecosyst. Environ.,* vol. 118, pp. 173-182, 2007. [http://dx.doi.org/10.1016/j.agee.2006.05.007]

[118] B. Sialve, N. Bernet, and O. Bernard, "Anaerobic digestion of microalgae as a necessary step to make microalgal biodiesel sustainable", *Biotechnol. Adv.,* vol. 27, no. 4, pp. 409-416, 2009. [http://dx.doi.org/10.1016/j.biotechadv.2009.03.001] [PMID: 19289163]

[119] J.C. Costa, P.R. Gonçalves, A. Nobre, and M.M. Alves, "Biomethanation potential of macroalgae *Ulva* spp. and *Gracilaria* spp. and in co-digestion with waste activated sludge", *Bioresour. Technol.,* vol. 114, pp. 320-326, 2012. [http://dx.doi.org/10.1016/j.biortech.2012.03.011] [PMID: 22459959]

[120] C. Kao, S. Chiu, T. Huang, L. Dai, G. Wang, C. Tseng, C. Chen, and C. Lin, "A mutant strain of microalga *Chlorella* sp. for the carbon dioxide capture from biogas", *Biomass Bioenergy,* vol. 36, pp. 132-140, 2012. [http://dx.doi.org/10.1016/j.biombioe.2011.10.046]

[121] I.K. Kapdan, and F. Kargi, "Bio-hydrogen Production from Waste Materials", *Enzyme Microb. Technol.,* vol. 38, pp. 569-582, 2006. [http://dx.doi.org/10.1016/j.enzmictec.2005.09.015]

[122] F. Romagnoli, D. Blumberga, and I. Pilicka, "Life cycle assessment of biohydrogen production in photosynthetic processes", *Int. J. Hydrogen Energy,* vol. 36, pp. 7866-7871, 2011. [http://dx.doi.org/10.1016/j.ijhydene.2011.02.004]

[123] J.R. Benemann, "Feasibility analysis of photobiological hydrogen production", *Int. J. Hydrogen Energy,* vol. 22, pp. 979-987, 1997. [http://dx.doi.org/10.1016/S0360-3199(96)00189-9]

[124] J.R. Benavente-Valdés, C. Aguilar, J.C. Contreras-Esquivel, A. Méndez-Zavala, and J. Montañez, "Strategies to enhance the production of photosynthetic pigments and lipids in chlorophyceae species", *Biotechnol. Rep. (Amst.),* vol. 10, pp. 117-125, 2016. [http://dx.doi.org/10.1016/j.btre.2016.04.001] [PMID: 28352532]

[125] R. Pangestuti, and S. Kim, "Biological activities and health benefit effects of natural pigments derived from marine algae", *J. Funct. Foods,* vol. 3, no. 4, pp. 255-266, 2011. [http://dx.doi.org/10.1016/j.jff.2011.07.001]

[126] R.S. Rasmussen, and M.T. Morrissey, "Marine biotechnology for production of food ingredients", *Adv. Food Nutr. Res,* vol. 52, 2007. [http://dx.doi.org/10.1016/S1043-4526(06)52005-4]

[127] A.P. Carvalho, S.O. Silva, J.M. Baptista, and F.X. Malcata, "Light requirements in microalgal photobioreactors: an overview of biophotonic aspects", *Appl. Microbiol. Biotechnol.,* vol. 89, no. 5, pp. 1275-1288, 2011. [http://dx.doi.org/10.1007/s00253-010-3047-8] [PMID: 21181149]

[128] C. de O. Rangel-Yagui, E.D.G. Danesi, J.C.M. Carvalho, and S. Sato, "Chlorophyll production from *Spirulina platensis*: cultivation with urea addition by fed-batch process", *Bioresour. Technol.,* vol. 92, no. 2, pp. 133-141, 2004. [http://dx.doi.org/10.1016/j.biortech.2003.09.002] [PMID: 14693445]

[129] V. Ördög, W.A. Stirk, P. Bálint, J. Van Staden, and C. Lovász, "Changes in lipid, protein and pigment concentrations in nitrogen-stressed *Chlorella minutissima* cultures", *J. Appl. Phycol.,* vol. 24, pp. 907-914, 2012. [http://dx.doi.org/10.1007/s10811-011-9711-2]

[130] J. Seyfabadi, Z. Ramezanpour, and Z.A. Khoeyi, "Protein, fatty acid, and pigment content of *Chlorella vulgaris* under different light regimes", *J. Appl. Phycol.,* vol. 23, pp. 721-726, 2011. [http://dx.doi.org/10.1007/s10811-010-9569-8]

[131] R. Esteban, J.F. Moran, J.M. Becerril, and J.I. García-Plazaola, "Versatility of carotenoids: An integrated view on diversity, evolution, functional roles and environmental interactions", *Environ. Exp. Bot.,* vol. 119, pp. 63-75, 2015. [http://dx.doi.org/10.1016/j.envexpbot.2015.04.009]

[132] C. Paliwal, T. Ghosh, B. George, I. Pancha, R. Maurya, K. Chokshi, A. Ghosh, and S. Mishra, "Microalgal carotenoids: Potential nutraceutical compounds with chemotaxonomic importance", *Algal Res.,* vol. 15, pp. 24-31, 2016. [http://dx.doi.org/10.1016/j.algal.2016.01.017]

[133] B. Sivathanu, and S. Palaniswamy, "Purification and characterization of carotenoids from green algae *Chlorococcum humicola* by HPLC-NMR and LC-MS-APCI", *Biomed. Prev. Nutr.,* vol. 2, no. 4, pp. 276-282, 2012. [http://dx.doi.org/10.1016/j.bionut.2012.04.006]

[134] Y. Cheng, and C. Yang, "Using strain *Rhodotorula mucilaginosa* to produce carotenoids using food wastes", *J. Taiwan Inst. Chem. Eng.,* vol. 61, pp. 270-275, 2016. [http://dx.doi.org/10.1016/j.jtice.2015.12.027]

[135] P. Spolaore, C. Joannis-Cassan, E. Duran, and A. Isambert, "Commercial applications of microalgae", *J. Biosci. Bioeng.,* vol. 101, no. 2, pp. 87-96, 2006. [http://dx.doi.org/10.1263/jbb.101.87] [PMID: 16569602]

[136] M. Kuddus, P. Singh, G. Thomas, and A. Al-Hazimi, "Recent developments in production and biotechnological applications of C-phycocyanin", *BioMed Res. Int.,* vol. 2013, 2013. [http://dx.doi.org/10.1155/2013/742859] [PMID: 24063013]

[137] C.C. Moraes, J.F.M. Burkert, and S.J. Kalil, "C-phycocyanin extraction process for large-scale use", *J. Food Biochem.,* vol. 34, pp. 133-148, 2010. [http://dx.doi.org/10.1111/j.1745-4514.2009.00317.x]

[138] M. Gantar, D. Simović, S. Djilas, W.W. Gonzalez, and J. Miksovska, "Isolation, characterization and antioxidative activity of C-phycocyanin from *Limnothrix* sp. strain 37-2-1", *J. Biotechnol.,* vol. 159, no. 1-2, pp. 21-26, 2012.

[http://dx.doi.org/10.1016/j.jbiotec.2012.02.004] [PMID: 22353597]

[139] F.S. Figueira, J.G. Gettens, J.A.V. Costa, M.G. Morais, C.C. Moraes, and S.J. Kalil, "Production of Nanofibers Containing the Bioactive Compound C-Phycocyanin", *J. Nanosci. Nanotechnol.,* vol. 16, no. 1, pp. 944-949, 2016.
[http://dx.doi.org/10.1166/jnn.2016.10906] [PMID: 27398551]

[140] A. Vonshak, *Spirulina Platensis Arthrospira: Physiology.* Cell-Biology And Biotechnology, 2002.

[141] B. Laycock, P. Halley, S. Pratt, A. Werker, and P. Lant, "The chemomechanical properties of microbial polyhydroxyalkanoates", *Prog. Polym. Sci.,* vol. 38, no. 3-4, pp. 536-583, 2013.
[http://dx.doi.org/10.1016/j.progpolymsci.2012.06.003]

[142] B. Wang, S. Pugh, D.R. Nielsen, W. Zhang, and D.R. Meldrum, "Engineering cyanobacteria for photosynthetic production of 3-hydroxybutyrate directly from CO_2", *Metab. Eng.,* vol. 16, pp. 68-77, 2013.
[http://dx.doi.org/10.1016/j.ymben.2013.01.001] [PMID: 23333586]

[143] S. Balaji, K. Gopi, and B. Muthuvelan, "A review on production of poly-β-hydroxybutyrates from cyanobacteria for the production of bioplastics", *Algal Res.,* vol. 2, no. 2, pp. 278-285, 2013.
[http://dx.doi.org/10.1016/j.algal.2013.03.002]

[144] M.I. Campos, T.V. Figueiredo, L.S. Sousa, and J.I. Druzian, "The influence of crude glycerin and nitrogen concentrations on the production of PHA by *Cupriavidus necator* using a response surface methodology and its characterizations", *Ind. Crops Prod.,* vol. 52, pp. 338-346, 2014.
[http://dx.doi.org/10.1016/j.indcrop.2013.11.008]

[145] A. Shrivastav, S.K. Mishra, and S. Mishra, "Polyhydroxyalkanoate (PHA) synthesis by *Spirulina subsalsa* from Gujarat coast of India", *Int. J. Biol. Macromol.,* vol. 46, no. 2, pp. 255-260, 2010.
[http://dx.doi.org/10.1016/j.ijbiomac.2010.01.001] [PMID: 20060853]

[146] R.D. Ashby, D.K. Solaiman, G.D. Strahan, C. Zhu, R.C. Tappel, and C.T. Nomura, "Glycerine and levulinic acid: renewable co-substrates for the fermentative synthesis of short-chain poly(hydroxyalkanoate) biopolymers", *Bioresour. Technol.,* vol. 118, pp. 272-280, 2012.
[http://dx.doi.org/10.1016/j.biortech.2012.05.092] [PMID: 22705534]

[147] B. Drosg, I. Fritz, F. Gattermayr, and L. Silvestrini, "Photo-autotrophic Production of Poly(hydroxyalkanoates) in Cyanobacteria", *Chem. Biochem. Eng. Q.,* vol. 29, no. 2, pp. 145-156, 2015.
[http://dx.doi.org/10.15255/CABEQ.2014.2254]

[148] N. Mallick, L. Sharma, and A. Kumar Singh, "Poly-β-hydroxybutyrate accumulation in *Nostoc muscorum*: effects of metabolic inhibitors", *J. Plant Physiol.,* vol. 164, no. 3, pp. 312-317, 2007.
[http://dx.doi.org/10.1016/j.jplph.2006.01.012] [PMID: 16542751]

[149] R.G. Martins, I.S. Gonçalves, M.G. Morais, and J.A.V. Costa, "Bioprocess Engineering Aspects of Biopolymer Production by the Cyanobacterium *Spirulina* Strain LEB 18", *Int. J. Polym. Sci.,* vol. 2014, pp. 1-7, 2014.
[http://dx.doi.org/10.1155/2014/895237]

[150] R. Bhati, and N. Mallick, "Production and characterization of poly(3-hydroxybutyrate-c--3-hydroxyvalerate) co-polymer by a N_2-fixing cyanobacterium, *Nostoc muscorum* Agardh", *J. Chem. Technol. Biotechnol.,* vol. 87, no. 4, pp. 505-512, 2012.
[http://dx.doi.org/10.1002/jctb.2737]

[151] A. Kaewbai-Ngam, A. Incharoensakdi, and T. Monshupanee, "Increased accumulation of polyhydroxybutyrate in divergent cyanobacteria under nutrient-deprived photoautotrophy: An efficient conversion of solar energy and carbon dioxide to polyhydroxybutyrate by *Calothrix scytonemicola* TISTR 8095", *Bioresour. Technol.,* vol. 212, pp. 342-347, 2016.
[http://dx.doi.org/10.1016/j.biortech.2016.04.035] [PMID: 27130227]

CHAPTER 8

Harvesting, Thickening and Dewatering Processes

Dries Vandamme*

KU Leuven Campus Kulak, Laboratory for Aquatic Biology, E. Sabbelaan 53, 8500 Kortrijk, Belgium

Abstract: Microalgae harvesting is an essential process for the production of almost all types of microalgae products. However, these harvesting, thickening and dewatering processes have a significant cost, despite extensive research efforts for cost reduction. Microalgal harvesting is especially challenging because these unicellular cells are small and have a similar density to water, and because the biomass concentration in culture is relatively low. The microalgal biomass (0.05% w/w) needs to be concentrated to a paste with 15-25% water content. This process is usually conducted in a multi-step scheme by primary concentration, thickening and dewatering of the biomass using a combination of various technologies. The most important implementation criteria are to avoid any biomass contamination, minimize any influence on biomass quality, and to allow water recycling to reduce the water footprint. This chapter gives an overview of several primary harvesting, thickening and dewatering methods with a focus on their advantages for implementation into a microalgae harvesting strategy.

Keywords: Harvesting, Recovery, Dewatering, Primary concentration, Coagulation, Filtration, Membrane, Centrifugation, Separation, Sedimentation, Flotation, Skimming, Thickening, Microalgae, Biomass.

INTRODUCTION

Microalgae have a high areal productivity and do not require fertile land for cultivation. Therefore, they are considered a promising new source for biomass that could be used in the biobased products for food, feed, chemicals or fuel markets [1]. The production of microalgal biomass is currently limited to 10-20 thousand tons dry matter per year, which is mainly used in high-value products such as pigments, nutritional supplements or pharmaceuticals [2 - 4]. The large-scale commercial production of microalgae is generally hampered by several factors including high-cost harvesting [5 - 8]. Realizing a significant cost

* **Corresponding author Dries Vandamme:** KU Leuven Campus Kulak, Laboratory for Aquatic Biology, E. Sabbelaan 53, 8500 Kortrijk, Belgium; Tel: +32 56 24 6041; Fax: +32 56 24 6999; E-mail: dries.vandamme@kuleuven.be

José Carlos Magalhães Pires (Ed.)

reduction in this process will increase the feasibility of microalgae production for commodity products such as biofuels [9, 10].

The challenges of microalgal harvesting are associated with several factors, namely (a) the nature of the microalgal cells, such as their size (2-20 μm), cell density, surface charge and cell morphology, (b) low cell-concentration in large-scale cultures systems, and (c) high capital investment cost, especially for marine microalgae. These varying biological, physico-chemical and practical parameters make it challenging to design a harvesting process that is suitable for most microalgal species with a wide range of production processes. This chapter gives an overview of the general requirements for effective microalgae harvesting, and of the advantages and disadvantages of the currently available harvesting technologies.

GENERAL REQUIREMENTS FOR EFFECTIVE MICROALGAE HARVESTING

The effectiveness of any solid-liquid separation process can be quantified in terms of the separation efficiency (SE) and the concentration factor (CF) [11]. For harvesting microalgae, the SE is defined as the ratio of mass of the microalgal cells recovered in the final product to the total mass of the microalgal cells in the initial culture (eq. 1). Thus, harvesting processes with a high SE will capture most of the microalgal cells from the feed, resulting in an effective separation of the cells from their cultivation medium. The concentration factor (CF) is the ratio of the concentration of the microalgae biomass in the final products to the initial concentration in the culture (eq. 2). The CF provides information related to the relative biomass concentration after a particular harvesting step and can be used to calculate the total water removal achieved in that process. Generally, one should aim for a harvesting process that can achieve a high SE and CF with minimal energy inputs. Thus, both SE and CF are of equal importance to assess the effectiveness of a harvesting process for microalgae [12].

$$\text{Separation efficiency (SE)} = \frac{\text{mass of microalgae cells recovered}}{\text{initial mass of microalgae cells}} \quad \textbf{(1)}$$

$$\text{Concentration factor (CF)} = \frac{microalgae\ concentration\ in\ final\ product}{initial\ concentration\ of\ microalgae\ cells} \quad \textbf{(2)}$$

Microalgae are a heterogenous group of micro organisms from different evolutionary lines of eukaryotes. Consequently, there is a large variation in the physico-chemical properties across different species and strains. Microalgal cells can differ greatly in size, shape, cell rigidity and cell surface properties (*e.g.* charge, functionality or hydrophobicity) [13 - 16]. Secondly, microalgae excrete extracellular algal organic matter (AOM) in their medium, which affects microalgal harvesting processes, for example by inhibiting flocculation or causing fouling during filtration [17 - 19]. As a result, it may be necessary to adjust harvesting strategies according to the species or even the cultivation conditions.

The harvesting process of microalgae is a central step in any microalgal production process scheme [7] (Fig. **1**). It is therefore important that harvesting should not interfere with the final product quality, be it whole algae (*e.g.* as animal feed, food or nutritional supplement) or a single extracted product from the microalgal biomass (*e.g.* lipids, pigments). Chemical contamination that increases toxicity risks or that could alter biomass quality or composition should be avoided. This is especially important in a biorefinery concept, where the biomass is refined into different valuable components for various markets and used for human or animal consumption [20, 21]. Secondly, it is important to recycle the water of the spent culture broth after the initial microalgae harvesting process to reduce the overall water footprint [22, 23]. The applied harvesting technologies should therefore not interfere with the recycling of the culture medium, for example by causing contamination of the culture medium that would limit microalgal growth [24, 25].

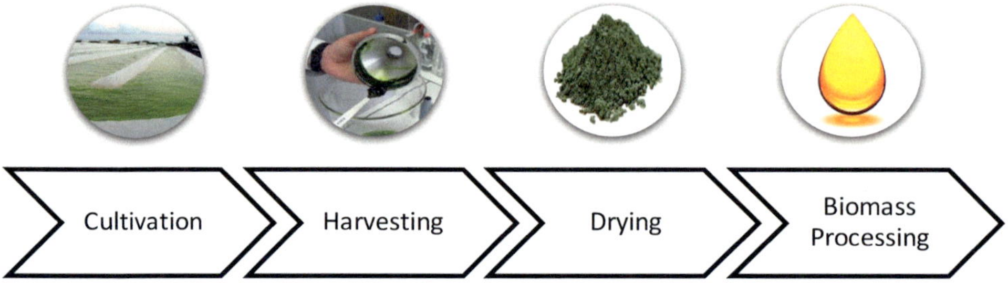

Fig. (1). General microalgae biomass production flow diagram.

MULTI-STAGE APPROACH OF HARVESTING, THICKENING AND DEWATERING

Microalgae harvesting techniques can be implemented individually (single-stage) or in combination (multi-stage). The choice often depends on the microalgal species, and on the desired product concentration and quality [9]. All single-stage solid-liquid separation techniques are based on constraining either the liquid or solid phase (Table **1**). Sedimentation, flotation and centrifugation-based separation methods are liquid-constrained methods. Sedimentation is generally cheap and preceeded by a chemically or biologically induced coagulation step. However, it is often slow and can result in a low CF when not optimized properly. Flotation-based separation is usually faster than sedimentation but requires extra energy and in most conditions it requires chemical coagulants as well.

Centrifugation is most commonly used in current harvesting processes because it is fast and reliable. It results in separation with a high SE and high CF without the use of extra Chemicals. However, centrifugation is an energy intensive separation process and it cannot be used for shear-sentitive species.

Filtration, screening and attachment are examples of particle-constrained separation (Table **1**). Filtration is a well-established separation technology that ensures high SE and can be used to harvest shear-sensitive species. However, it involves the additional cost of pumping and membrane cleaning to minimize fouling issues. Screening, on the other hand, is a simple and proven separation method for species with a large cell size such as *Arthrospira sp.,* but it can cause unwanted selection of smaller cells that are not harvested but recycled in the cultivation process. Novel developments based on cell attachment such as biofilm formation or magnetic separation using nanoparticles can be interesting approaches to decrease separation costs, however, most of these methods have yet to be tested at scale. Generally, different methods will also affect potential output pararmeters including separation efficiency, concentration factor, volumetric throughput, but also the fluid properties (*e.g.* medium composition, ionic strength, salinity, pH, temperature).

Table 1. General microalgae biomass production flow diagram.

Harvesting Method	Advantages	Limitations	References
Liquid-constrained solid-liquid separation			
Sedimentation	Simple, low-cost	Potential need for chemicals, slow, low CF	[6, 26]
Flotation	Limited space requirement, fast	Chemicals, challenging for marine species	[27, 28]
Centrifugation	High SE and CF, fast	Expensive, high shear-forces	[29 - 31]
Solid-constrained solid-liquid separation			
Filtration	High SE, suitable for shear-sentive species	Fouling/cleaning, pumping costs	[19, 32]
Screening	Simple, low-cost	Only for large cell sizes	[8, 9]
Attachment	Low-cost, high CF	To be confirmed at scale	[33 - 35]

The culture broth usually needs to be thickened to a paste with a dry-matter content of about 20% or 200 g L^{-1} (Fig. **2**). This means that a total concentration factor of 400 needs to be achieved, since biomass concentration in culture broths in open raceway ponds rarely exceeds 0.5 g L^{-1} [6, 36]. Multi-stage or sequential harvesting processes can therefore be desirable to increase microalgal concentration by implementing the most economical and sustainable method for each stage which results in an overall reduction of the total energy demand [7, 8, 37 - 39]. The most commonly used set-up is a two- or three-stage harvesting design, in which primary harvesting and thickening concentrates the biomass to 1-4%, followed by a secondary dewatering step (Fig. **2**).

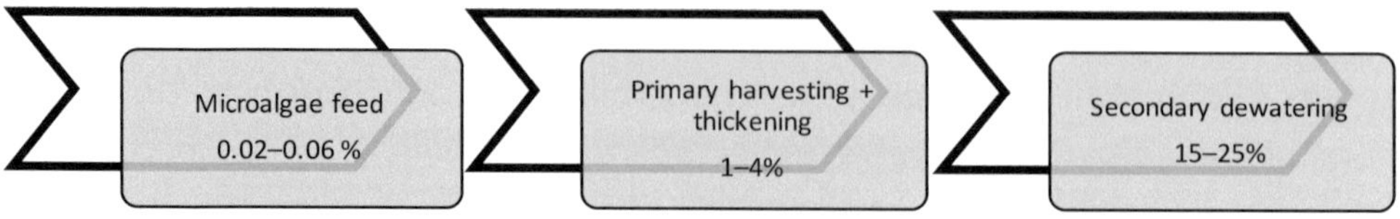

Fig. (2). Two-stage microalgae harvesting process design.

COAGULATION-FLOCCULATION-SEDIMENTATION

Coagulation-flocculation-sedimentation is a separation technology that has been used for decades in different fields of industry, such as in drinking water and wastewater treatment. However, successful transferral of methods across other industries does not guarantee successful implementation in microalgal harvesting. Usually, coagulation and flocculation steps require the addition of chemicals which end up in the waste sludge after separation, which is disposed in most other industrial applications. In microalgal harvesting, most of these chemicals end up in the product instead, contaminating the microalgal biomass. Therefore, the toxicity risks need to be managed more carefully and extensively. The economics are also different across industries. In water treatment processes, the cost of flocculation processes is related to the volume of processed water, rather than to the quantity of removed impurities. In the case of microalgae harvesting, the cost is related to the harvested biomass, and not directly to the volume of processed culture broth.

Coagulation Mechanisms

The spontaneous aggregation of microalgal cells is usually prevented by the negative surface charge of the cells. This negative charge is mainly generated by carboxylic acid surface groups, which are deprotonated at a pH higher than 5 and therefore generate a net negative surface charge [13, 40]. Initially, small flocs can only be formed when the natural electrostatic repulsion is overcome by introducing a coagulant, which then subsequently leads to the formation of bigger flocs, a process better known as flocculation. Coagulation-flocculation of microalgae can generally be attributed to several mechanisms: charge neutralization, electrostatic patching, bridging, sweeping flocculation and bioflocculation (Fig. **3**) [6, 41].

Coagulation of microalgae by charge neutralization is caused by positively charged ions or colloids that can additionally adsorb on the negative surface.

These interactions ultimately result in neutralization of the particle surface charge, and therefore cancel out the electrostatic repulsion. This will then allow coagulation and flocculation (Fig. **3A**). Positively charged polymers can also bind to the surface of several microalgal cells, essentially acting as bridges between the cells and thereby inducing floc formation (Fig. **3B**). Smaller positively charged polymers can bind to the microalgal cell surface, which creates a local positively charged zone. This allows coagulation by an electrostatic patching mechanism (Fig. **3C**). Alternatively, microalgal cells can be trapped by the massive precipitation of mineral salts. This causes flocculation by enmeshment or sweeping (Fig. **3D**). Finally, the interaction of microalgal cells with bacteria, fungi or extracellular polymeric substances can also cause flocculation. This mechanism is generally referred to as bioflocculation (Fig. **3E**).

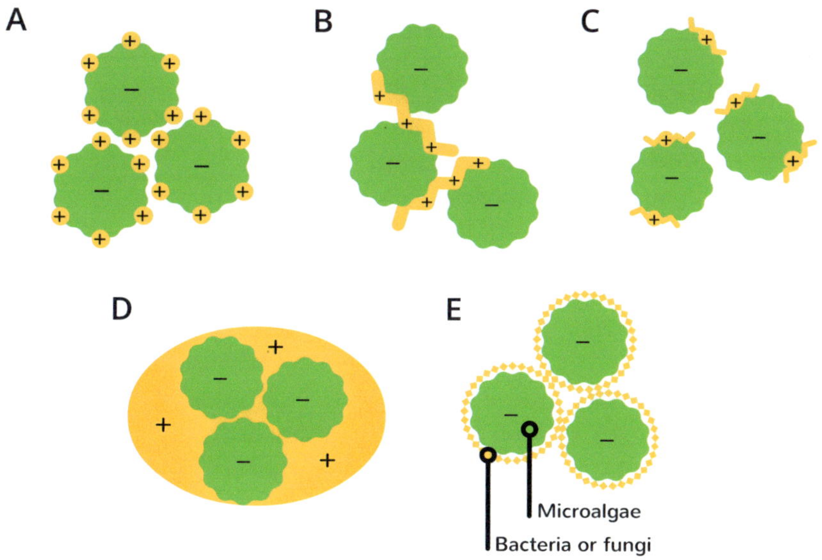

Fig. (3). Microalgae coagulation-flocculation mechanisms: (**A**) charge neutralization, (**B**) bridging, (**C**) electrostatic patching, (**D**) sweeping flocculation and (**E**) bioflocculation.

Chemical Flocculation

Chemical-based flocculation is usually induced using coagulants such as metal salts or synthetic and natural polymers. Metal salts, such as aluminum sulphate or ferric chloride, are widely used coagulants in different industries. Metal salts form metal hydroxides once hydrolyzed in water [42]. These hydroxides can be dissolved and positively charged depending on the total concentration, pH, temperature and medium salinity [43]. Positively charged hydroxide species can cause microalgal flocculation by charge neutralization. Alternatively, when the solubility product is exceeded, precipitates are formed causing flocculation by enmeshment and sweeping [44]. Several studies have reported efficient harvesting of microalgae using metal salts in terms of a high SE. However, in practice, the risk of contamination of the biomass with relatively high concentrations of metals such as aluminum or iron can be a major disadvantage [8, 14, 45, 46]. Secondly, the counter cations such as sulphates remain in the medium and alter the medium composition when recycled. Nevertheless, flocculation using metal salts is useful as a model system to understand the fundamental mechanisms of flocculation of microalgae [47, 48].

As highlighted above, flocculation can also be induced using various types of polymers. Usually, polyectrolytes (polycations or polyanions) are used. In the case of microalgal flocculation, polycations are most popular as they can neutralize the surface charge by electrostatic patching or by forming bridges between the individual microalgal cells [49 - 51]. Synthetic polymers, such as polyacrylamides, have been shown to be very effective at harvesting microalgae, often even at a very low flocculant dosage [52 - 55]. However, these types of polymer flocculants may have toxicity risks [56], which is why current research interest has mainly been focused on the use of biopolymers such as chitosan. Chitosan is produced by deacetylation of chitin [14, 57, 58], but its cost is relatively high. Cheaper biopolymers such as cationic starch or tannin-based biopolymers are interesting alternatives [59 - 61].

An important drawback of polymer flocculants is that they generally perform poorly at harvesting marine microalgae [14, 62 - 65]. High salinity can cause polymer coiling or charge shielding, thereby decreasing the flocculation efficiency. However, more rigid polymers such as tannin-based flocculants can be very effective in marine media [59]. Polymers can also be combined with magneto-responsive iron oxide nanoparticles to separate microalgae in a magnetic field by attachment. In this case, separation often occurs faster than when using traditional gravity sedimentation or flotation methods [66, 67].

Current research efforts focus on the recovery and re-use of polymer flocculants in order to decrease the harvesting costs and potential biomass contamination. When polymers with a reversible pH-responsive surface charge are used, flocculants can be recovered from the harvested microalgae biomass for subsequent re-usage. This has been successfully demonstrated for cellulose nanocrystals [68 - 71].

Auto/Alkaline Flocculation

The term autoflocculation is often used to refer to the spontaneous flocculation of microalgae that occurs upon increase of the culture pH. However, microalgae do not flocculate autonomously at high pH. Rather, flocculation is induced by chemical precipitation of salts such as calcium phosphates, calcium carbonates and magnesium hydroxides [26, 40, 72 - 76]. It is therefore more appropriate to refer to pH-induced or alkaline flocculation to describe this phenomenon [25, 77, 78]. Alkaline flocculation can be induced spontaneously as the result of a photosynthetically induced increase of pH. In other cases, an artificial increase of pH through the addition of base is required [79, 80]. In principle, the calcium and magnesium salts that are formed at high pH are not toxic, but they can increase the total ash content of the harvested microalgal biomass, especially for marine species. However, these salts can be removed quite easily after harvesting by dissolving them through mild acidification [81]. Moreover, alkaline flocculation can be very useful as a primary harvesting method in two-stage cultivation strategies in which the biomass is subsequently resuspended in nutrient-depleted conditions to boost lipid accumulation [82].

Electro-coagulation

Electro-coagulation (EC) has recently gained a lot of interest with regard to reducing microalgal harvesting costs, as it has a low energy demand and avoids the use of flocculant addition [83 - 86]. EC has been widely applied in treatments of wastewater and drinking water for the removal of suspended solids, small colloidal particles or dyes [87 - 89]. In EC, electrolysis is applied by passing an electric current through two electrodes which are submerged in solution. Metal ions will be released into the solution from a sacrificial anode by electrolytic oxidation and, similarly to metal salts, they will act as coagulants. However, unlike for metal salts, EC does not result in the release of counter anions. Furthermore, oxygen and hydrogen bubbles are generated and these optimize separation by flotation rather than sedimentation. In microalgal harvesting, EC is an effective harvesting method for both freshwater and marine species. In marine conditions, the energy requirements are generally much lower, as the electric resistance is lower because of the higher conductivity of the saline water [90 - 94]. Nonetheless, biomass contamination with metals such as aluminum or iron can occur when sacrificial anodes are used. Recent studies have focused on the optimization of EC-based separation by maximizing SE while minimizing energy demands and biomass contamination. This can be done through EC designs that use polarity exchange or inert electrodes [95 - 98].

Bioflocculation

Microalgal flocculation processes that involve various interactions with microorganisms (self-flocculating algae, bacteria, fungi or yeasts), with extracellular polymeric substances (EPS), or with bioflocculants/infochemicals are referred to as bioflocculation processes [99]. Bioflocculation has often been observed in high-rate algal pond systems for wastewater treatment [100, 101].

These ponds are colonized with complex communities of microalgal species, filamentous algae and bacteria [102, 103], but the detailed mechanisms by which multiple interactions between algae, bacteria and EPS induce bioflocculation are not fully understood [104]. However, recent laboratory studies have successfully demonstrated controlled bioflocculation of microalgae by algal-algal interactions [105 - 108]. Bacteria can induce the formation of microalgal-bacteria flocs, either by direct interaction or through the release of EPS that interacts with the microalgae cell surface [109, 110]. Filamentous fungi can also be used to induce microalgal bioflocculation [111, 112].

Recent developments in genetic strain engineering are particularly promising in terms of engineering self-flocculating microalgae. In 2015, the gene responsible for flocculating yeast has been successfully transferred into *Chlamydomonas,* resulting in a significant increase in the bioflocculation efficiency of this microalgal species [113]. Despite the fact that the control of bioflocculation at scale is still very challenging, it is a very promising low-cost flocculation method that deserves further study.

Enhanced Settling

The majority of the studies on flocculation as a microalgae harvesting method have used gravity sedimentation as a separation method because it requires very little energy and only low-cost capital investments such as settling tanks [114]. However, such tanks can have a relative large footprint at scale. After sedimentation, the separated microalgal biomass still contains a lot of water [12]. Therefore, sedimentation is usually combined with another dewatering technology such as centrifugation or filtration [6].

The settling rate is a critical design parameter in any sedimentation process. Microalgae have a very low intrinsic settling velocity of about 1 cm h^{-1} since most microalgae are very small (up to 20 µm) and have a density similar to that of water. This makes conventional gravity settling very slow, which can be problematic because it increases the risk of biomass quality deterioration [114, 115]. Inclined settlers or lamella separators consist of a series of stacked plates, wich increase the effective area available for settling [116, 117]. When combined with a flocculation method prior to settling to increase the floc size, inclined settling can reduce the footprint of the separation process [118].

FLOTATION

Flotation can be seen as inverted sedimentation with the additional advantages of a higher CF, a smaller footprint and lower operation times [27]. Generally, flocs are attached to small air bubbles, allowing them to rise and float to the surface where they concentrate. In order to obtain efficient bubble attachment, coagulants are used to create neutralized flocs with increased hydrophobicity [119]. Air bubbles can be introduced in the system using an air diffuser (dispersed air flotation, DiAF) or *via* pressurization (dissolved air flotation, DAF). Between these two methods, DAF results in better separation but it has a higher energy demand because of the air pressurization requirement [120]. Combining flotation with electro-coagulation (EC) can overcome the need for pressurization as the EC process generates oxygen and hydrogen gas bubbles [92].

The flotation efficiency depends heavily on the interaction between the microalgal cells and the air bubbles. Traditional flocculants can be used to increase the bubble attachment, but this increases the cost [28, 119, 121]. Moreover, it is currently not possible to specify a universally applicable method due to our limited understanding of the specific interactions of different microalgal species. Especially in marine conditions, flotation can be less efficient. Coagulation based on magnesium hydroxide formation/precipitation is effective for flotation of freshwater and marine species. In this approach, the magnesium that is initially present in the cultivation medium is mobilized to form magnesium hydroxides, which in turn form microalgal flocs prior to flotation [122, 123].

CENTRIFUGATION

Centrifugation is based on the generation of a centrifugal force, which accelerates the separation of microalgae based on the density difference between the microalgae cells and the medium. It is the most widely used harvesting technology in microalgae production processes for high-value products. Different types of industrial centrifuges are available. A disc bowl centrifuge is used for separation of microalgae cultures with a low concentration (from 0.01% to 20% algal dry weight), while a decanter centrifuge is usually used for more concentrated suspensions (from 10% to 50% algal dry weight) and can be operated continuously [29, 39]. Centrifugation has several advantages: there is no need for chemicals, a high dry-matter concentration can be achieved in a single-step process, and it is fast, which minimizes the risk for biomass deterioration. However, a major drawback of centrifugation is that it is energy demanding [7, 85].

For low-value microalgal applications, centrifugation should be replaced by more energy-efficient harvesting methods. However, it can still be implemented in the final dewatering of microalgal slurries that are produced after primary concentration by flocculation/sedimentation or filtration [6]. The primary concentration of the biomass results in a reduction in the culture volume by at least an order of magnitude. As a result, the energy demand for subsequent centrifugation will be much lower. Primary concentration by flocculation also increases the particle size, which reduces the required centrifugal force to obtain separation by centrifugation.

FILTRATION

Screening

Simple screening is often used for separation of large microalgal/cyanobacterial species that form filaments (100-200 μm), such as *Arthrospira sp.*. Nylon screens can simply collect the microalgal biomass when the broth is pumped over the screen. This straightforward method requires low capital investment, but unfortunately, it is not applicable for the majority of microalgae that are suitable for biomass production.

Membrane Filtration

Micro- and ultrafiltration membranes are the most studied membrane types for filtration of microalgae. Both are able to retain the microalgae at a SE of nearly 100% [19, 124]. Micro-filtration allows higher fluxes at short term, but ultrafiltration performs better in the long term because of higher fouling resistance [125, 126]. Fouling has a negative impact on the flux through membrane/solutes interaction and cake formation. The critical process parameters that affect fouling are pore size, surface charge, membrane hydrophobicity and the composition of the medium. EPS-like polysaccharides and intracellular products released after cell lysis contribute significantly to fouling. Fouling can be partially controlled (in the case of reversible fouling) by creating turbulent flow near the membrane surface, or by backwashing or chemical cleaning. Irreversible fouling, however, can still occur especially for hydrophobic membranes [18, 127, 128]

Membrane filtration can be designed in different configurations. In conventional filtration or dead-end filtration, the water is perpendicular to the filter and all particles are forced to settle on the filter. However, this approach is considered to be economically non-viable for microalgae because of extensive fouling issues [7]. Cross-flow filtration or tangential flow filtration is less susceptible to fouling, but the shear forces can cause release of additional algal organic matter and even cell lysis, which in turn increases fouling [128]. Submerged filtration is a more recent approach in which membranes are placed directly in the microalgae culture. It shows great potential for the primary concentration of microalgae [129 - 131].

COMPARING HARVESTING METHODS

In Table **2**, the concentration after separation, capital investment costs, operational cost and energy requirements are compared for different harvesting methods.

Flocculation-based separation requires less energy, but the final concentration is generally lower than 5 g L^{-1}, which means subsequent dewatering is still needed. Crossflow filtration is capable of concentrating microalgal biomass in a wide concentration range up to 27 g L^{-1}, but it has a high energy demand. It would be more beneficial to use filtration as a primary harvesting method, combined with centrifugation for secondary dewatering.

Table 2. Comparison between harvesting methods based on final total suspended solids concentration (TSS), capital costs (CAPEX), operational costs (OPEX) and energy demand.

Harvesting Method	TSS (g L^{-1})	CAPEX (€ m^{-3})	OPEX (€ m^{-3})	Energy (kWh m^{-3})
Sedimentation	0.5-3	0.03	0.04-0.3	0.05-0.1
Bioflocculation	1.4-5	0.03	0.05-1.16	0.02-0.2
Metal salts flocculation	1.2-7	0.28	0.41-1.74	0.1-2.85
Polymer flocculation	0.6-15	0.2	0.08-16.5	0.1-14.81
Electro-coagulation sedimentation	3-5	0.04-2.22	0.09	0.2-2
Dissolved air flotation (DAF)	1-8	1.13	0.2-1.39	0.6-20
Electro-coagulation flotation	3-5	0.83	0.5	0.3-2
Dispersed air flotation (DiAF)	1.4-5	0.8	0.5	0.003-0.02
Decanter centrifuge	12-30	0.45-1.35	0.3-0.87	3.6-10.8
Disk stack centrifugation	10-22	0.37-0.43	0.16-1.26	0.7-1.4
Crossflow filtration	2-27	1.44	0.27	0.012-10

CONCLUSION

The development of a low-cost and energy-efficient harvesting strategy remains one of the major challenges to achieve large-scale production of microalgae for bulk biomass production. The most promising strategies are two-stage processes, which use either filtration or flocculation as a primary concentration step. Depending on the microalgal species and the downstream application, the harvesting strategy may need to meet different requirements. The most common ones are to avoid biomass contamination and biomass quality deterioration. Finally, an effective harvesting strategy is also important to reduce the water footprint by allowing the re-use of spent culture medium.□Different microalgal species behave differently during harvesting, and therefore it is crucial to screen species not only based on their productivity but also their potential for energy-efficient harvesting.

CONFLICT OF INTEREST

The author declares no conflict of interest, financial or otherwise.

ACKNOWLEDGEMENTS

D. V. is a postdoctoral Researcher funded by the Research Foundation - Flanders Belgium (FWO - 12D8917N). The manuscript was edited by Highlight Science Writing (www.highlight-science-writing.com).

REFERENCES

[1] J. Ruiz Gonzalez, "Towards industrial products from microalgae", *Energy Environ. Sci.*, vol. 4, no. 4, p. 1364, 2016.

[2] G. Markou, and E. Nerantzis, "Microalgae for high-value compounds and biofuels production: a review with focus on cultivation under stress conditions", *Biotechnol. Adv.*, vol. 31, no. 8, pp. 1532-1542, 2013.
[http://dx.doi.org/10.1016/j.biotechadv.2013.07.011] [PMID: 23928208]

[3] A.C. Guedes, H.M. Amaro, and F.X. Malcata, "Microalgae as sources of high added-value compounds--a brief review of recent work", *Biotechnol. Prog.*, vol. 27, no. 3, pp. 597-613, 2011.
[http://dx.doi.org/10.1002/btpr.575] [PMID: 21452192]

[4] J. Desmet, C. Meunier, and E. Danloy, "Green and sustainable production of high value compounds via a microalgae encapsulation technology that relies on CO 2 as a principle reactant", *J. Mater.*, 2014.

[5] H.C. Greenwell, L.M. Laurens, R.J. Shields, R.W. Lovitt, and K.J. Flynn, "Placing microalgae on the biofuels priority list: a review of the technological challenges", *J. R. Soc. Interface*, vol. 7, no. 46, pp. 703-726, 2010.
[http://dx.doi.org/10.1098/rsif.2009.0322] [PMID: 20031983]

[6] D. Vandamme, I. Foubert, and K. Muylaert, "Flocculation as a low-cost method for harvesting microalgae for bulk biomass production", *Trends Biotechnol.*, vol. 31, no. 4, pp. 233-239, 2013.
[http://dx.doi.org/10.1016/j.tibtech.2012.12.005] [PMID: 23336995]

[7] A.I. Barros, A.L. Gonçalves, M. Simões, and J.C. Pires, "Harvesting techniques applied to microalgae: A review", *Renew. Sustain. Energy Rev.*, vol. 41, pp. 1489-1500, 2015.
[http://dx.doi.org/10.1016/j.rser.2014.09.037]

[8] G. Shelef, A. Sukenik, and M. Green, *Microalgae harvesting and processing: a literature review*, .
[http://dx.doi.org/10.2172/6204677]

[9] S.L. Pahl, A.K. Lee, T. Kalaitzidis, P.J. Ashman, S. Sathe, and D.M. Lewis, "Harvesting, Thickening and Dewatering Microalgae Biomass", In: *Algae for biofuels and Energy, B. Michael A. and M.*, R. Navid, Ed., Springer Netherlands: Dordrecht, 2013, pp. 165-185.
[http://dx.doi.org/10.1007/978-94-007-5479-9_10]

[10] E. Molina Grima, E.H. Belarbi, F.G. Acién Fernández, A. Robles Medina, and Y. Chisti, "Recovery of microalgal biomass and metabolites: process options and economics", *Biotechnol. Adv.*, vol. 20, no. 7-8, pp. 491-515, 2003.
[http://dx.doi.org/10.1016/S0734-9750(02)00050-2] [PMID: 14550018]

[11] A.K. Lee, D.M. Lewis, and P.J. Ashman, "Microbial flocculation, a potentially low-cost harvesting technique for marine microalgae for the production of biodiesel", *J. Appl. Phycol.*, vol. 21, no. 5, pp. 559-567, 2009.
[http://dx.doi.org/10.1007/s10811-008-9391-8]

[12] D. Vandamme, K. Muylaert, I. Fraeye, and I. Foubert, "Floc characteristics of Chlorella vulgaris:

Influence of flocculation mode and presence of organic matter", *Bioresour. Technol.,* vol. 151, pp. 383-387, 2014.
[http://dx.doi.org/10.1016/j.biortech.2013.09.112]

[13] S. Hadjoudja, V. Deluchat, and M. Baudu, "Cell surface characterisation of Microcystis aeruginosa and Chlorella vulgaris", *J. Colloid Interface Sci.,* vol. 342, no. 2, pp. 293-299, 2010.
[http://dx.doi.org/10.1016/j.jcis.2009.10.078] [PMID: 20004408]

[14] S. Lama, K. Muylaert, T.B. Karki, I. Foubert, R.K. Henderson, and D. Vandamme, "Flocculation properties of several microalgae and a cyanobacterium species during ferric chloride, chitosan and alkaline flocculation", *Bioresour. Technol.,* vol. 220, pp. 464-470, 2016.
[http://dx.doi.org/10.1016/j.biortech.2016.08.080] [PMID: 27611030]

[15] A. Ozkan, and H. Berberoglu, "Physico-chemical surface properties of microalgae", *Colloids Surf. B Biointerfaces,* vol. 112, pp. 287-293, 2013.
[http://dx.doi.org/10.1016/j.colsurfb.2013.08.001] [PMID: 24001448]

[16] A.L. Gonçalves, and C. Ferreira, "J. a Loureiro, J. C. Pires, and M. Simões, "Surface physicochemical properties of selected single and mixed cultures of microalgae and cyanobacteria and their relationship with sedimentation kinetics", *Bioresour. Bioprocess.,* vol. 2, no. 1, 2015.
[http://dx.doi.org/10.1186/s40643-015-0051-y]

[17] D. Vandamme, A. Beuckels, E. Vadelius, O. Depraetere, W. Noppe, A. Dutta, I. Foubert, L. Laurens, and K. Muylaert, "Inhibition of alkaline flocculation by algal organic matter for Chlorella vulgaris", *Water Res.,* vol. 88, pp. 301-307, 2016.
[http://dx.doi.org/10.1016/j.watres.2015.10.032] [PMID: 26512808]

[18] V. Discart, M.R. Bilad, D. Vandamme, I. Foubert, K. Muylaert, and I.F. Vankelecom, "Role of transparent exopolymeric particles in membrane fouling: Chlorella vulgaris broth filtration", *Bioresour. Technol.,* vol. 129, pp. 18-25, 2013.
[http://dx.doi.org/10.1016/j.biortech.2012.11.034] [PMID: 23238336]

[19] M.R. Bilad, H.A. Arafat, and I.F. Vankelecom, "Membrane technology in microalgae cultivation and harvesting: a review", *Biotechnol. Adv.,* vol. 32, no. 7, pp. 1283-1300, 2014.
[http://dx.doi.org/10.1016/j.biotechadv.2014.07.008] [PMID: 25109678]

[20] L. Zhu, "Biorefinery as a promising approach to promote microalgae industry: An innovative framework", *Renew. Sustain. Energy Rev.,* vol. 41, pp. 1376-1384, 2015.
[http://dx.doi.org/10.1016/j.rser.2014.09.040]

[21] M.L. Gerardo, S. Van Den Hende, H. Vervaeren, T. Coward, and S.C. Skill, "Harvesting of microalgae within a biorefinery approach: A review of the developments and case studies from pilot-plants", *Algal Res.,* vol. 11, pp. 248-262, 2015.
[http://dx.doi.org/10.1016/j.algal.2015.06.019]

[22] W. Farooq, W.I. Suh, M.S. Park, and J.W. Yang, "Water use and its recycling in microalgae cultivation for biofuel application", *Bioresour. Technol.,* vol. 184, pp. 73-81, 2015.
[http://dx.doi.org/10.1016/j.biortech.2014.10.140] [PMID: 25465788]

[23] C.F. Murphy, and D.T. Allen, "Energy-water nexus for mass cultivation of algae", *Environ. Sci. Technol.,* vol. 45, no. 13, pp. 5861-5868, 2011.
[http://dx.doi.org/10.1021/es200109z] [PMID: 21671675]

[24] W. Farooq, "Effect of harvesting methods on the reusability of water for cultivation of Chlorella vulgaris, its lipid productivity and biodiesel quality", *Algal Res.,* vol. 8, pp. 1-7, 2015.
[http://dx.doi.org/10.1016/j.algal.2014.12.007]

[25] M. Castrillo, L.M. Lucas-Salas, C. Rodríguez-Gil, and D. Martínez, "High pH-induced flocculation-sedimentation and effect of supernatant reuse on growth rate and lipid productivity of Scenedesmus obliquus and Chlorella vulgaris", *Bioresour. Technol.,* vol. 128, pp. 324-329, 2013.
[http://dx.doi.org/10.1016/j.biortech.2012.10.114] [PMID: 23201513]

[26] A. Schlesinger, D. Eisenstadt, A. Bar-Gil, H. Carmely, S. Einbinder, and J. Gressel, "Inexpensive non-toxic flocculation of microalgae contradicts theories; overcoming a major hurdle to bulk algal production", *Biotechnol. Adv.,* vol. 30, no. 5, pp. 1023-1030, 2012. [http://dx.doi.org/10.1016/j.biotechadv.2012.01.011] [PMID: 22306161]

[27] C.A. Laamanen, G.M. Ross, and J.A. Scott, "Flotation harvesting of microalgae", *Renew. Sustain. Energy Rev.,* vol. 58, pp. 75-86, 2016. [http://dx.doi.org/10.1016/j.rser.2015.12.293]

[28] S. Garg, L. Wang, and P.M. Schenk, "Flotation separation of marine microalgae from aqueous medium", *Separ. Purif. Tech.,* vol. 156, pp. 636-641, 2015. [http://dx.doi.org/10.1016/j.seppur.2015.10.059]

[29] A.J. Dassey, and C.S. Theegala, "Harvesting economics and strategies using centrifugation for cost effective separation of microalgae cells for biodiesel applications", *Bioresour. Technol.,* vol. 128, pp. 241-245, 2013. [http://dx.doi.org/10.1016/j.biortech.2012.10.061] [PMID: 23196245]

[30] J.J. Milledge, and S. Heaven, "Disc Stack Centrifugation Separation and Cell Disruption of Microalgae: A Technical Note", *Environ. Nat. Resour. Res.,* vol. 1, no. 1, pp. 17-24, 2011. [http://dx.doi.org/10.5539/enrr.v1n1p17]

[31] M. Brocken, "Case study: Spiral plate technology for totally dewatering algae alive", In: *Microalgal Biotechnology.* Potential and Production, 2013, p. 266.

[32] M. Rickman, J. Pellegrino, and R. Davis, "Fouling phenomena during membrane filtration of microalgae", *J. Membr. Sci.,* vol. 423-424, pp. 33-42, 2012. [http://dx.doi.org/10.1016/j.memsci.2012.07.013]

[33] Y-C. Lee, K. Lee, and Y-K. Oh, "Recent nanoparticle engineering advances in microalgal cultivation and harvesting processes of biodiesel production: a review", *Bioresour. Technol.,* vol. 184, pp. 63-72, 2015. [http://dx.doi.org/10.1016/j.biortech.2014.10.145] [PMID: 25465786]

[34] J.K. Lim, D.C. Chieh, S.A. Jalak, P.Y. Toh, N.H. Yasin, B.W. Ng, and A.L. Ahmad, "Rapid magnetophoretic separation of microalgae", *Small,* vol. 8, no. 11, pp. 1683-1692, 2012. [http://dx.doi.org/10.1002/smll.201102400] [PMID: 22438107]

[35] M. Kesaano, and R.C. Sims, "Algal biofilm based technology for wastewater treatment", *Algal Res.,* vol. 5, pp. 231-240, 2014. [http://dx.doi.org/10.1016/j.algal.2014.02.003]

[36] P. Singh, S. Kumari, A. Guldhe, R. Misra, I. Rawat, and F. Bux, "Trends and novel strategies for enhancing lipid accumulation and quality in microalgae", *Renew. Sustain. Energy Rev.,* vol. 55, pp. 1-16, 2016. [http://dx.doi.org/10.1016/j.rser.2015.11.001]

[37] N. Uduman, Y. Qi, M.K. Danquah, G.M. Forde, and A. Hoadley, "Dewatering of microalgal cultures: A major bottleneck to algae-based fuels", *J. Renew. Sustain. Energy,* vol. 2, no. 1, p. 12701, 2010. [http://dx.doi.org/10.1063/1.3294480]

[38] S. Salim, L. Gilissen, A. Rinzema, M.H. Vermuë, and R.H. Wijffels, "Modeling microalgal flocculation and sedimentation", *Bioresour. Technol.,* vol. 144, pp. 602-607, 2013. [http://dx.doi.org/10.1016/j.biortech.2013.07.026] [PMID: 23899573]

[39] J.J. Milledge, and S. Heaven, "A review of the harvesting of micro-algae for biofuel production", *Rev. Environ. Sci. Biotechnol.,* vol. 12, no. 2, pp. 165-178, 2012. [http://dx.doi.org/10.1007/s11157-012-9301-z]

[40] P.V. Brady, P.I. Pohl, and J.C. Hewson, "A coordination chemistry model of algal autoflocculation", *Algal Res.,* vol. 5, pp. 226-230, 2014. [http://dx.doi.org/10.1016/j.algal.2014.02.004]

[41] K. Muylaert, D. Vandamme, I. Foubert, and P.V. Brady, "Harvesting of Microalgae by Means of Flocculation", [http://dx.doi.org/10.1007/978-3-319-16640-7_12]

[42] J. Duan, and J. Gregory, "Coagulation by hydrolysing metal salts", *Adv. Colloid Interface Sci.,* vol. 100-102, pp. 475-502, 2003. [http://dx.doi.org/10.1016/S0001-8686(02)00067-2]

[43] J. Gregory, "Flocculation fundamentals", In: *Encyclopedia of Colloid and Interface Science.,* T. Tadros, Ed., Springer Berlin Heidelberg: Berlin, Heidelberg, 2013. [http://dx.doi.org/10.1007/978-3-642-20665-8_17]

[44] J. Bratby, *Coagulation and Flocculation in Water and Wastewater Treatment*, 2016.

[45] N.B. Wyatt, L.M. Gloe, P.V. Brady, J.C. Hewson, A.M. Grillet, M.G. Hankins, and P.I. Pohl, "Critical conditions for ferric chloride-induced flocculation of freshwater algae", *Biotechnol. Bioeng.,* vol. 109, no. 2, pp. 493-501, 2012. [http://dx.doi.org/10.1002/bit.23319] [PMID: 21882173]

[46] A.J. Garzon-Sanabria, R.T. Davis, and Z.L. Nikolov, "Harvesting Nannochloris oculata by inorganic electrolyte flocculation: effect of initial cell density, ionic strength, coagulant dosage, and media pH", *Bioresour. Technol.,* vol. 118, pp. 418-424, 2012. [http://dx.doi.org/10.1016/j.biortech.2012.04.057] [PMID: 22705965]

[47] N.B. Wyatt, T.J. O'Hern, B. Shelden, L.G. Hughes, and L.A. Mondy, "Size and structure of Chlorella zofingiensis/FeCl(3) flocs in a shear flow", *Biotechnol. Bioeng.,* vol. 110, no. 12, pp. 3156-3163, 2013. [http://dx.doi.org/10.1002/bit.24996] [PMID: 23842762]

[48] Y. Cui, *Fundamentals in Microalgae Harvesting: From Flocculation to Self-attachment,* 2013.

[49] J. Gregory, *Particles in Water. Properties and Processes., no. 20.* International Water Association, 2006.

[50] J. Gregory, and S. Barany, "Adsorption and flocculation by polymers and polymer mixtures", *Adv. Colloid Interface Sci.,* vol. 169, no. 1, pp. 1-12, 2011. [http://dx.doi.org/10.1016/j.cis.2011.06.004] [PMID: 21762869]

[51] C.M. Fellows, and W.O. Doherty, "Insights into bridging flocculation", *Macromol. Symp.,* vol. 231, no. 1, pp. 1-10, 2006. [http://dx.doi.org/10.1002/masy.200590012]

[52] L. Chen, C. Wang, W. Wang, and J. Wei, "Optimal conditions of different flocculation methods for harvesting Scenedesmus sp. cultivated in an open-pond system", *Bioresour. Technol.,* vol. 133, pp. 9-15, 2013. [http://dx.doi.org/10.1016/j.biortech.2013.01.071] [PMID: 23410531]

[53] J. Mikulec, G. Polakovičová, and J. Cvengroš, "Flocculation Using Polyacrylamide Polymers for Fresh Microalgae", *Chem. Eng. Technol.,* vol. 38, no. 4, pp. 595-601, 2015. [http://dx.doi.org/10.1002/ceat.201400639]

[54] F. Roselet, D. Vandamme, M. Roselet, K. Muylaert, and P. Cesar, "Screening of commercial natural and synthetic cationic polymers for fl occulation of freshwater and marine microalgae and effects of molecular weight and charge density", *Algal Res.,* vol. 10, pp. 183-188, 2015. [http://dx.doi.org/10.1016/j.algal.2015.05.008]

[55] G.P. 't Lam, M.H. Vermuë, G. Olivieri, L.A. van den Broek, M.J. Barbosa, M.H. Eppink, R.H. Wijffels, and D.M. Kleinegris, "Cationic polymers for successful flocculation of marine microalgae", *Bioresour. Technol.,* vol. 169, pp. 804-807, 2014. [http://dx.doi.org/10.1016/j.biortech.2014.07.070] [PMID: 25113884]

[56] R. Costa, J.L. Pereira, J. Gomes, F. Gonçalves, D. Hunkeler, and M.G. Rasteiro, "The effects of acrylamide polyelectrolytes on aquatic organisms: relating toxicity to chain architecture",

Chemosphere, vol. 112, pp. 177-184, 2014.
[http://dx.doi.org/10.1016/j.chemosphere.2014.03.096] [PMID: 25048904]

[57] N. Rashid, S.U. Rehman, and J-I. Han, "Rapid harvesting of freshwater microalgae using chitosan", *Process Biochem.,* vol. 7, pp. 1107-1110, 2013.
[http://dx.doi.org/10.1016/j.procbio.2013.04.018]

[58] M.S. Farid, A. Shariati, A. Badakhshan, and B. Anvaripour, "Using nano-chitosan for harvesting microalga Nannochloropsis sp", *Bioresour. Technol.,* vol. 131, pp. 555-559, 2013.
[http://dx.doi.org/10.1016/j.biortech.2013.01.058] [PMID: 23415940]

[59] F. Roselet, D. Vandamme, M. Roselet, K. Muylaert, and P.C. Abreu, "Effects of pH, Salinity, Biomass Concentration, and Algal Organic Matter on Flocculant Efficiency of Synthetic Versus Natural Polymers for Harvesting Microalgae Biomass", *BioEnergy Res.,* 2016.

[60] D. Vandamme, I. Foubert, B. Meesschaert, and K. Muylaert, "Flocculation of microalgae using cationic starch", *J. Appl. Phycol.,* vol. 22, no. 4, pp. 525-530, 2010.
[http://dx.doi.org/10.1007/s10811-009-9488-8]

[61] P.A. Hansel, R. Guy Riefler, and B.J. Stuart, "Efficient flocculation of microalgae for biomass production using cationic starch", *Algal Res.,* vol. 5, no. 1, pp. 133-139, 2014.
[http://dx.doi.org/10.1016/j.algal.2014.07.002]

[62] G.P. 't Lam, J.B. Giraldo, M.H. Vermuë, G. Olivieri, M.H. Eppink, and R.H. Wijffels, "Understanding the salinity effect on cationic polymers in inducing flocculation of the microalga Neochloris oleoabundans", *J. Biotechnol.,* vol. 225, pp. 10-17, 2016.
[http://dx.doi.org/10.1016/j.jbiotec.2016.03.009] [PMID: 27002231]

[63] L.M. Lubián, "Concentrating cultured marine microalgae with chitosan", *Aquacult. Eng.,* vol. 8, no. 4, pp. 257-265, 1989.
[http://dx.doi.org/10.1016/0144-8609(89)90013-7]

[64] J. Morales, J. De La Noüe, and G. Picard, "Harvesting marine microalgae species by chitosan flocculation", *Aquacult. Eng.,* vol. 4, pp. 257-270, 1985.
[http://dx.doi.org/10.1016/0144-8609(85)90018-4]

[65] S. Şirin, E. Clavero, and J. Salvadó, "Potential pre-concentration methods for Nannochloropsis gaditana and a comparative study of pre-concentrated sample properties", *Bioresour. Technol.,* vol. 132, pp. 293-304, 2013.
[http://dx.doi.org/10.1016/j.biortech.2013.01.037] [PMID: 23416616]

[66] K. Lee, S.Y. Lee, J.G. Na, S.G. Jeon, R. Praveenkumar, D.M. Kim, W.S. Chang, and Y.K. Oh, "Magnetophoretic harvesting of oleaginous Chlorella sp. by using biocompatible chitosan/magnetic nanoparticle composites", *Bioresour. Technol.,* vol. 149, pp. 575-578, 2013.
[http://dx.doi.org/10.1016/j.biortech.2013.09.074] [PMID: 24128604]

[67] Y. Xu, S. Purton, and F. Baganz, "Chitosan flocculation to aid the harvesting of the microalga Chlorella sorokiniana", *Bioresour. Technol.,* vol. 129, pp. 296-301, 2013.
[http://dx.doi.org/10.1016/j.biortech.2012.11.068] [PMID: 23262003]

[68] S. Eyley, D. Vandamme, S. Lama, G. Van den Mooter, K. Muylaert, and W. Thielemans, "CO2 controlled flocculation of microalgae using pH responsive cellulose nanocrystals", *Nanoscale,* vol. 7, no. 34, pp. 14413-14421, 2015.
[http://dx.doi.org/10.1039/C5NR03853G] [PMID: 26248574]

[69] K.L. Morrissey, C. He, M.H. Wong, X. Zhao, R.Z. Chapman, S.L. Bender, W.D. Prevatt, and M.P. Stoykovich, "Charge-tunable polymers as reversible and recyclable flocculants for the dewatering of microalgae", *Biotechnol. Bioeng.,* vol. 112, no. 1, pp. 74-83, 2015.
[http://dx.doi.org/10.1002/bit.25340] [PMID: 25060233]

[70] S. Ge, P. Champagne, H-D. Wang, P.G. Jessop, and M.F. Cunningham, *Microalgae Recovery from Water for Biofuel Production Using CO2-Switchable Crystalline Nanocellulose*, 2016.

[71] P-R. Liu, H-L. Zhang, T. Wang, W-L. Yang, Y. Hong, and Y-L. Hou, "Functional graphene-based magnetic nanocomposites as magnetic flocculant for efficient harvesting of oleaginous microalgae", *Algal Res.,* vol. 19, pp. 86-95, 2016. [http://dx.doi.org/10.1016/j.algal.2016.07.008]

[72] D. Vandamme, P.I. Pohl, A. Beuckels, I. Foubert, P.V. Brady, J.C. Hewson, and K. Muylaert, "Alkaline flocculation of Phaeodactylum tricornutum induced by brucite and calcite", *Bioresour. Technol.,* vol. 196, pp. 656-661, 2015. [http://dx.doi.org/10.1016/j.biortech.2015.08.042] [PMID: 26310384]

[73] C. González-Fernández, and M. Ballesteros, "Microalgae autoflocculation: an alternative to high-energy consuming harvesting methods", *J. Appl. Phycol.,* vol. 25, no. 4, pp. 991-999, 2013. [http://dx.doi.org/10.1007/s10811-012-9957-3]

[74] A. Lavoie, and J. de la Noüe, "Harvesting of Scenedesmus obliquus in wastewaters: auto- or bioflocculation?", *Biotechnol. Bioeng.,* vol. 30, no. 7, pp. 852-859, 1987. [http://dx.doi.org/10.1002/bit.260300707] [PMID: 18581520]

[75] A. Sukenik, and G. Shelef, "Algal autoflocculation--verification and proposed mechanism", *Biotechnol. Bioeng.,* vol. 26, no. 2, pp. 142-147, 1984. [http://dx.doi.org/10.1002/bit.260260206] [PMID: 18551700]

[76] A. Sukenik, W. Schröder, J. Lauer, G. Shelef, and C. Soeder, "Coprecipitation of microalgal biomass with calcium and phosphate ions", *Water Res.,* vol. 19, no. 1, pp. 127-129, 1985. [http://dx.doi.org/10.1016/0043-1354(85)90333-1]

[77] D. Vandamme, I. Foubert, I. Fraeye, B. Meesschaert, and K. Muylaert, "Flocculation of Chlorella vulgaris induced by high pH: role of magnesium and calcium and practical implications", *Bioresour. Technol.,* vol. 105, pp. 114-119, 2012. [http://dx.doi.org/10.1016/j.biortech.2011.11.105] [PMID: 22182473]

[78] S. Şirin, R. Trobajo, C. Ibanez, and J. Salvadó, "Harvesting the microalgae Phaeodactylum tricornutum with polyaluminum chloride, aluminium sulphate, chitosan and alkalinity-induced flocculation", *J. Appl. Phycol.,* vol. 24, no. 5, pp. 1067-1080, 2011. [http://dx.doi.org/10.1007/s10811-011-9736-6]

[79] K. Spilling, J. Seppälä, and T. Tamminen, "Inducing autoflocculation in the diatom Phaeodactylum tricornutum through CO2 regulation", *J. Appl. Phycol.,* vol. 23, no. 6, pp. 959-966, 2011. [http://dx.doi.org/10.1007/s10811-010-9616-5]

[80] T.D. Nguyen, M. Frappart, P. Jaouen, J. Pruvost, and P. Bourseau, "Harvesting Chlorella vulgaris by natural increase in pH: effect of medium composition", *Environ. Technol.,* vol. 35, no. 9-12, pp. 1378-1388, 2014. [http://dx.doi.org/10.1080/09593330.2013.868531] [PMID: 24701936]

[81] D. Vandamme, A. Beuckels, G. Markou, I. Foubert, and K. Muylaert, "Reversible Flocculation of Microalgae using Magnesium Hydroxide", *BioEnergy Res.,* vol. 8, no. 2, pp. 716-725, 2015. [http://dx.doi.org/10.1007/s12155-014-9554-1]

[82] G. Sidney Alémán-Nava, "Two-stage cultivation of Nannochloropsis oculata for lipid production using reversible alkaline flocculation", *Bioresour. Technol.,* 2016. [PMID: 27988475]

[83] C.G. Alfafara, K. Nakano, N. Nomura, T. Igarashi, and M. Matsumura, "Operating and scale-up factors for the electrolytic removal of algae from eutrophied lakewater", *J. Chem. Technol. Biotechnol.,* vol. 77, no. 8, pp. 871-876, 2002. [http://dx.doi.org/10.1002/jctb.649]

[84] C. Feng, N. Sugiura, S. Shimada, and T. Maekawa, "Development of a high performance electrochemical wastewater treatment system", *J. Hazard. Mater.,* vol. 103, no. 1-2, pp. 65-78, 2003. [http://dx.doi.org/10.1016/S0304-3894(03)00222-X] [PMID: 14568697]

[85] L. Gouveia, "Microalgae biomass production using wastewater: Treatment and costs: Scale-up considerations", *Algal Res.,* vol. 16, pp. 167-176, 2016. [http://dx.doi.org/10.1016/j.algal.2016.03.010]

[86] M.Y. Mollah, R. Schennach, J.R. Parga, and D.L. Cocke, "Electrocoagulation (EC)--science and applications", *J. Hazard. Mater.,* vol. 84, no. 1, pp. 29-41, 2001. [http://dx.doi.org/10.1016/S0304-3894(01)00176-5] [PMID: 11376882]

[87] M.Y. Mollah, P. Morkovsky, J.A. Gomes, M. Kesmez, J. Parga, and D.L. Cocke, "Fundamentals, present and future perspectives of electrocoagulation", *J. Hazard. Mater.,* vol. 114, no. 1-3, pp. 199-210, 2004. [http://dx.doi.org/10.1016/j.jhazmat.2004.08.009] [PMID: 15511592]

[88] T. Kim, C. Park, E. Shin, and S. Kim, "Decolorization of disperse and reactive dyes by continuous electrocoagulation process", *Desalination,* vol. 150, no. 2, pp. 165-175, 2002. [http://dx.doi.org/10.1016/S0011-9164(02)00941-4]

[89] J. Koren, and U. Syversen, "State-of-the-art electroflocculation", *Filtr. Sep.,* vol. 32, no. 2, pp. 146-156, 1995. [http://dx.doi.org/10.1016/S0015-1882(97)84039-6]

[90] N. Uduman, V. Bourniquel, M.K. Danquah, and A.F. Hoadley, "A parametric study of electrocoagulation as a recovery process of marine microalgae for biodiesel production", *Chem. Eng. J.,* vol. 174, no. 1, pp. 249-257, 2011. [http://dx.doi.org/10.1016/j.cej.2011.09.012]

[91] D. Vandamme, S.C. Pontes, K. Goiris, I. Foubert, L.J. Pinoy, and K. Muylaert, "Evaluation of electro-coagulation-flocculation for harvesting marine and freshwater microalgae", *Biotechnol. Bioeng.,* vol. 108, no. 10, pp. 2320-2329, 2011. [http://dx.doi.org/10.1002/bit.23199] [PMID: 21557200]

[92] A. Golzary, S. Imanian, M.A. Abdoli, A. Khodadadi, and A. Karbassi, "A cost-effective strategy for marine microalgae separation by electro-coagulation-flotation process aimed at bio-crude oil production: Optimization and evaluation study", *Separ. Purif. Tech.,* 2015. [http://dx.doi.org/10.1016/j.seppur.2015.04.011]

[93] A.K. Lee, D.M. Lewis, and P.J. Ashman, "Harvesting of marine microalgae by electroflocculation: The energetics, plant design, and economics", *Appl. Energy,* vol. 108, pp. 45-53, 2013. [http://dx.doi.org/10.1016/j.apenergy.2013.03.003]

[94] R. Misra, A. Guldhe, P. Singh, I. Rawat, T.A. Stenström, and F. Bux, "Evaluation of operating conditions for sustainable harvesting of microalgal biomass applying electrochemical method using non sacrificial electrodes", *Bioresour. Technol.,* vol. 176, pp. 1-7, 2015. [http://dx.doi.org/10.1016/j.biortech.2014.11.014] [PMID: 25460977]

[95] R. Misra, A. Guldhe, P. Singh, I. Rawat, and F. Bux, "Electrochemical harvesting process for microalgae by using nonsacrificial carbon electrode: A sustainable approach for biodiesel production", *Chem. Eng. J.,* vol. 255, pp. 327-333, 2014. [http://dx.doi.org/10.1016/j.cej.2014.06.010]

[96] A. Guldhe, R. Misra, P. Singh, I. Rawat, and F. Bux, "An innovative electrochemical process to alleviate the challenges for harvesting of small size microalgae by using non-sacrificial carbon electrodes", *Algal Res.,* 2015.

[97] J. Kim, B.G. Ryu, B.K. Kim, J.I. Han, and J.W. Yang, "Continuous microalgae recovery using electrolysis with polarity exchange", *Bioresour. Technol.,* vol. 111, pp. 268-275, 2012. [http://dx.doi.org/10.1016/j.biortech.2012.01.104] [PMID: 22397823]

[98] J. Kim, B-G. Ryu, Y-J. Lee, J-I. Han, W. Kim, and J-W. Yang, "Continuous harvest of marine microalgae using electrolysis: effect of pulse waveform of polarity exchange", *Bioprocess Biosyst. Eng.,* 2013.

[PMID: 24322506]

[99] M.A. Alam, "Bioflocculation as an innovative harvesting strategy for microalgae", *Rev. Environ. Sci. Biotechnol.*, vol. 15, no. 4, pp. 573-583, 2016. [http://dx.doi.org/10.1007/s11157-016-9408-8]

[100] J.B. Park, R.J. Craggs, and A.N. Shilton, "Wastewater treatment high rate algal ponds for biofuel production", *Bioresour. Technol.*, vol. 102, no. 1, pp. 35-42, 2011. [http://dx.doi.org/10.1016/j.biortech.2010.06.158] [PMID: 20674341]

[101] S. Van Den Hende, H. Vervaeren, S. Desmet, and N. Boon, "Bioflocculation of microalgae and bacteria combined with flue gas to improve sewage treatment", *N. Biotechnol.*, vol. 29, no. 1, pp. 23-31, 2011. [http://dx.doi.org/10.1016/j.nbt.2011.04.009] [PMID: 21565287]

[102] P.P. Assemany, "Algae/bacteria consortium in high rate ponds: Influence of solar radiation on the phytoplankton community", *Ecol. Eng.*, vol. 77, pp. 154-162, 2015. [http://dx.doi.org/10.1016/j.ecoleng.2015.01.026]

[103] V.V. Unnithan, A. Unc, and G.B. Smith, "Mini-review: A priori considerations for bacteria-algae interactions in algal biofuel systems receiving municipal wastewaters", *Algal Res.*, 2013.

[104] T. Ndikubwimana, "Microalgae biomass harvesting by bioflocculation-interpretation by classical DLVO theory", *Biochem. Eng. J.*, vol. 101, pp. 160-167, 2015. [http://dx.doi.org/10.1016/j.bej.2015.05.010]

[105] S. Salim, R. Bosma, M.H. Vermuë, and R.H. Wijffels, "Harvesting of microalgae by bio-flocculation", *J. Appl. Phycol.*, vol. 23, no. 5, pp. 849-855, 2011. [http://dx.doi.org/10.1007/s10811-010-9591-x] [PMID: 21957329]

[106] S. Salim, M.H. Vermuë, and R.H. Wijffels, "Ratio between autoflocculating and target microalgae affects the energy-efficient harvesting by bio-flocculation", *Bioresour. Technol.*, vol. 118, pp. 49-55, 2012. [http://dx.doi.org/10.1016/j.biortech.2012.05.007] [PMID: 22695146]

[107] S. Salim, N.R. Kosterink, N.D. Tchetkoua Wacka, M.H. Vermuë, and R.H. Wijffels, "Mechanism behind autoflocculation of unicellular green microalgae Ettlia texensis", *J. Biotechnol.*, vol. 174, no. Jan, pp. 34-38, 2014. [http://dx.doi.org/10.1016/j.jbiotec.2014.01.026] [PMID: 24480568]

[108] C. Yoo, H-J. La, S-C. Kim, and H-M. Oh, "Simple processes for optimized growth and harvest of Ettlia sp. by pH control using CO2 and light irradiation", *Biotechnol. Bioeng.*, vol. 112, no. 2, pp. 288-296, 2015. [http://dx.doi.org/10.1002/bit.25362] [PMID: 25182602]

[109] J. Lee, D-H. Cho, R. Ramanan, B-H. Kim, H-M. Oh, and H-S. Kim, "Microalgae-associated bacteria play a key role in the flocculation of Chlorella vulgaris", *Bioresour. Technol.*, vol. 131, pp. 195-201, 2013. [http://dx.doi.org/10.1016/j.biortech.2012.11.130] [PMID: 23347927]

[110] R. Ramanan, "Phycosphere bacterial diversity in green algae reveals an apparent similarity across habitats", *Algal Res.*, vol. 8, pp. 140-144, 2015. [http://dx.doi.org/10.1016/j.algal.2015.02.003]

[111] S.K. Prajapati, P. Kumar, A. Malik, and P. Choudhary, "Exploring Pellet Forming Filamentous Fungi as Tool for Harvesting Non-flocculating Unicellular Microalgae", *BioEnergy Res.*, 2014. [http://dx.doi.org/10.1007/s12155-014-9481-1]

[112] S. Xie, S. Sun, S.Y. Dai, and J.S. Yuan, "Efficient coagulation of microalgae in cultures with filamentous fungi", *Algal Res.*, vol. 2, no. 1, pp. 28-33, 2013. [http://dx.doi.org/10.1016/j.algal.2012.11.004]

[113] E. Díaz-Santos, "Study of bioflocculation induced by Saccharomyces bayanus var. uvarum and

flocculating protein factors in microalgae", *Algal Res.,* vol. 8, pp. 23-29, 2015. [http://dx.doi.org/10.1016/j.algal.2014.12.013]

[114] R. Gutiérrez, J. García, E. Uggetti, C. Arnabat, H. Salvadó, and I. Ferrer, "Settling velocities distribution of microalgal biomass from urban wastewater high rate algal ponds", *Algal Res.,* vol. 16, pp. 409-417, 2016. [http://dx.doi.org/10.1016/j.algal.2016.03.037]

[115] Y. Nurdogan, and W.J. Oswald, "Tube settling of high-rate pond algae", *Water Sci. Technol.,* vol. 33, no. 7, pp. 229-241, 1996. [http://dx.doi.org/10.1016/0273-1223(96)00358-7]

[116] B.T. Smith, J. Halperin, A. Darzins, and R.H. Davis, "Enhanced sediment flow in inclined settlers via surface modification or applied vibration for harvesting microalgae", *Algal Res.,* vol. 2, no. 4, pp. 369-377, 2013. [http://dx.doi.org/10.1016/j.algal.2013.05.004]

[117] Z. Wang, J. Hou, D. Bowden, and J.M. Belovich, "Evaluation of an inclined gravity settler for microalgae harvesting", *J. Chem. Technol. Biotechnol.,* vol. 89, no. 5, pp. 714-720, 2014. [http://dx.doi.org/10.1002/jctb.4174]

[118] B.T. Smith, and R.H. Davis, "Particle concentration using inclined sedimentation via sludge accumulation and removal for algae harvesting", *Chem. Eng. Sci.,* vol. 91, pp. 79-85, 2013. [http://dx.doi.org/10.1016/j.ces.2013.01.007]

[119] S. Garg, L. Wang, and P.M. Schenk, "Effective harvesting of low surface-hydrophobicity microalgae by froth flotation", *Bioresour. Technol.,* vol. 159, pp. 437-441, 2014. [http://dx.doi.org/10.1016/j.biortech.2014.03.030] [PMID: 24690467]

[120] H.A. Kurniawati, S. Ismadji, and J.C. Liu, "Microalgae harvesting by flotation using natural saponin and chitosan", *Bioresour. Technol.,* vol. 166, pp. 429-434, 2014. [http://dx.doi.org/10.1016/j.biortech.2014.05.079] [PMID: 24935003]

[121] X. Zhang, J.C. Hewson, P. Amendola, M. Reynoso, M. Sommerfeld, Y. Chen, and Q. Hu, "Critical evaluation and modeling of algal harvesting using dissolved air flotation", *Biotechnol. Bioeng.,* vol. 111, no. 12, pp. 2477-2485, 2014. [http://dx.doi.org/10.1002/bit.25300] [PMID: 24889919]

[122] A. Besson, and P. Guiraud, "High-pH-induced flocculation-flotation of the hypersaline microalga Dunaliella salina", *Bioresour. Technol.,* vol. 147, no. August, pp. 464-470, 2013. [http://dx.doi.org/10.1016/j.biortech.2013.08.053] [PMID: 24012843]

[123] X. Zhang, L. Wang, M. Sommerfeld, and Q. Hu, "Harvesting microalgal biomass using magnesium coagulation-dissolved air flotation", *Biomass Bioenergy,* vol. 93, pp. 43-49, 2016. [http://dx.doi.org/10.1016/j.biombioe.2016.06.024]

[124] I.L. Drexler, and D.H. Yeh, "Membrane applications for microalgae cultivation and harvesting: a review", *Rev. Environ. Sci. Biotechnol.,* 2014. [http://dx.doi.org/10.1007/s11157-014-9350-6]

[125] T. De Baerdemaeker, B. Lemmens, C. Dotremont, J. Fret, L. Roef, K. Goiris, and L. Diels, "Benchmark study on algae harvesting with backwashable submerged flat panel membranes", *Bioresour. Technol.,* vol. 129, pp. 582-591, 2013. [http://dx.doi.org/10.1016/j.biortech.2012.10.153] [PMID: 23274222]

[126] M.K. Danquah, L. Ang, N. Uduman, N. Moheimani, and G.M. Forde, "Dewatering of microalgal culture for biodiesel production: exploring polymer flocculation and tangential flow filtration", *J. Chem. Technol. Biotechnol.,* vol. 84, no. 7, pp. 1078-1083, 2009. [http://dx.doi.org/10.1002/jctb.2137]

[127] W. Mo, L. Soh, J.R. Werber, M. Elimelech, and J.B. Zimmerman, "Application of membrane dewatering for algal biofuel", *Algal Res.,* vol. 11, pp. 1-12, 2015.

[http://dx.doi.org/10.1016/j.algal.2015.05.018]

[128] N. Rossignol, L. Vandanjon, P. Jaouen, and F. Quéméneur, "Membrane technology for the continuous separation microalgae/culture medium: compared performances of cross-flow microfiltration and ultrafiltration", *Aquacult. Eng.,* vol. 20, no. 3, pp. 191-208, 1999. [http://dx.doi.org/10.1016/S0144-8609(99)00018-7]

[129] M.R. Bilad, D. Vandamme, I. Foubert, K. Muylaert, and I.F. Vankelecom, "Harvesting microalgal biomass using submerged microfiltration membranes", *Bioresour. Technol.,* vol. 111, pp. 343-352, 2012. [http://dx.doi.org/10.1016/j.biortech.2012.02.009] [PMID: 22391586]

[130] M.R. Bilad, V. Discart, D. Vandamme, I. Foubert, K. Muylaert, and I.F. Vankelecom, "Coupled cultivation and pre-harvesting of microalgae in a membrane photobioreactor (MPBR)", *Bioresour. Technol.,* vol. 155, pp. 410-417, 2014. [http://dx.doi.org/10.1016/j.biortech.2013.05.026] [PMID: 24559585]

[131] Y. Luo, P. Le-Clech, and R.K. Henderson, "Simultaneous microalgae cultivation and wastewater treatment in submerged membrane photobioreactors: A review", *Algal Res.,* 2016.

CHAPTER 9

Oil Extraction Processes in Microalgae

Mengyue Gong, Yulin Hu, Shreyas Yedahalli and **Amarjeet Bassi***

Department of Chemical and Biochemical Engineering, Faculty of Engineering, Western University, London ON N6A 5B9, Canada

Abstract: Microalgae are an attractive source for biofuels due to their high growth rates and lipid content. However, up to the present time, the extraction of lipids from wet microalgae biomass remains an energy-intensive step, and an impediment for the economically viable large scale bio-oil production. During the extraction process, cell disruption and drying are two of the most energy-consuming steps. Several studies on genetic modification of microalgae towards lipid accumulation enhancement, or to reduce the harvesting and lipid extraction cost have been carried out successfully. This chapter aims to summarize the recent developments in microalgal oil extraction processes, including drying methods, cell disruption methods, conventional or supercritical solvent extraction methods, and recent approaches for direct biodiesel production. The current developments in metabolic engineering for lipids production are also reviewed. Hydrothermal liquefaction is also included as an alternative route to convert wet biomass into bio-crude oil.

Keywords: Biofuel, Cell disruption, Extraction, Genetic engineering, Hydrothermal liquefaction, Lipids, Microalgae.

INTRODUCTION

Lipids, currently are an attractive source of biofuels, and a diverse group of compounds with the common property of solubility in organic solvents. Lipids are generally classified into two categories: neutral lipids (fats and oils) and polar or membrane lipids. Microalgae have been shown to be an attractive and potential source of biofuels and oil as they have a higher neutral lipid content and can outcompete terrestrial plants due to a faster growth rate and year-round production [1]. In microalgal cells, the function of neutral lipids is primarily for energy storage, while polar lipids form the bi-layer cell membranes [2]. Lipid extraction from microalgae for oil and fuel production, is therefore a promising approach for environmental sustainability.

* **Corresponding author Amarjeet Bassi:** Department of Chemical and Biochemical Engineering, Faculty of Engineering, Western University, London ON N6A 5B9, Canada; Tel: +1-519-661-2111; Fax: +1-519-661-3498; E-mail: abassi@uwo.ca

The lipid extraction process is usually carried out after the harvesting of microalgae. The extraction is an energy-intensive downstream processing step for biofuel production; therefore, considerable attention has been focused on both cell disruption and solvent extraction methods. These aspects have been extensively reviewed within the last decade [2 - 5]. Microalgae possess rigid cell walls, which are mainly composed of complex carbohydrates and glycoproteins with high mechanical strength and chemical resistance. The cell wall structure, the location and content of the lipids and the size of the algal cell can all pose a significant challenge for inexpensive microalgae oil production [6]. Microalgal lipid content can vary considerably among different species, and may range from 5 to 70 wt.% [7]. A fraction of total oil content may contain unsaponifiable substances, such as hydrocarbons, sterols, ketones, pigments (carotenes and chlorophylls) that are not converted in biodiesel; these substances should be noticed and controlled [2]. Biomass drying, the distillation costs of organic solvent or supercritical fluids, temperature and pressure conditions, and the cell disruption processes represent the major energy and cost constraints [4]. Beside these considerations, the extraction method, whether chemical or physical, should be fast, easily scalable, effective, selective and should not modify the extracted product [8].

In this chapter, recent advances for oil extraction are presented. Firstly, recent studies on genetic manipulation of microalgae for lipid synthesis is summarized. Next, lipid extraction approaches such as, conventional solvent or supercritical fluid extraction technologies, efficiency-promoting cell disruption technologies, with recent advances in extraction methods are discussed. Finally, the direct conversion of microalgae to bio-oil using hydrothermal liquefaction is reviewed.

GENETIC ENGINEERING OF MICROALGAE FOR ENHANCEMENT OF LIPID PRODUCTION

Several microalgae have been identified for potential biofuel production. The biofuel yield is typically 40- 60%, and this may be enhanced by genetic manipulation of microalgae. Due to the complex nature of microalgae however, it is more difficult to carry out the genetic engineering work when compared to bacterial systems such as *E. coli.* The genes in the microalgae are present in three places: nucleus, chloroplast and mitochondria. *Chlamydomonas reinhardtii* is one of the most extensively studied microalgae till date and is applied as a model system for microalgal study. Several genetic engineering techniques have been developed in this strain. The gene sequence is available and the size of nuclear genome is 120 Mbp, chloroplast is 203 kbp and mitochondria is 16 kbp. *Chlamydomonas reinhardtii* is also known as photosynthetic yeast [9].

Lipid Production in Microalgae - Mechanism

Prior to undertaking any metabolic engineering work in microalgae, it is necessary to understand the key enzymes involved in the metabolic pathway, and the rate limiting enzymes. Many advances have been made towards understanding lipid metabolism and regulatory factors in *Arabidopsis thaliana*, soy bean, and rapeseed, but the lipid production in microalgae at molecular level is very poorly understood till now. The fatty acid metabolism in microalgae resembles the plant model organism *Arabidopsis thaliana,* and this metabolism can be used as a template to explore microalgae lipid production at molecular level.

The first step in *de novo* synthesis of triacylglycerol in microalgae starts in the plastid, where pyruvate is produced from glycolysis and Calvin cycle. The pyruvate is converted into acetyl-CoA by Pyruvate dehydrogenase complex (PDC). Acetyl-CoA is converted into malonyl-CoA by acetyl-CoA carboxylase (ACCase). Acetyl-CoA carboxylase is the rate limiting enzyme for lipid biosynthesis [10]. The malonyl-CoA is converted into malonyl-ACP by malonyl-CoA transacylase (MAT) [11]. The malonyl-ACP and Acyl-ACP is converted into 3-ketoacyl-ACP by 3-ketoacyl-ACP reductase (KAS) in fatty acid synthesis cycle. The 3-ketoacyl-ACP is converted into 3-hydroxyacyl-ACP by 3-ketoacyl-ACP reductase (KAR). The 3-hydroxyacyl-ACP is converted into trans-Enoyl-ACP by 3-hydroxyacyl-ACP dehydratase (HD). The trans-Enoyl-ACP is converted into Acyl-ACP by enoyl-ACP reductase (ENR). The Acyl-ACP is converted into free fatty acids (FFAs) by fatty acyl-ACP thioesterase (FAT) [12, 13].

The FFAs is transferred into cytosol and then to endoplasmic reticulum for conversation into TAG in the microalgae. The free fatty acids are converted into Acyl-CoA by long-chain acyl-CoA synthetase. The Acyl-CoA with glycerol 3 phosphate are converted into lysophosphatidic acid by glycerol 3-phosphate acyltransferase (GPAT). The lysophosphatidic acid is converted to phosphatidic acid by lysophosphatidic acid acyltransferase (LPAT). The phosphatidic acid is converted into diacylglycerol by phosphatidic acid phosphatase (PAP). The diacylglycerol is converted into triacylglycerol (TAG) by diacylglycerol acyltransferase (DGAT). The triacylglycerol forms TAG lipid body [12, 14, 15].

Genetically modified strains that can secrete lipids directly into the media have previously been carried out in hosts such as *E. coli*, *S. cerevisiae* and cyanobacteria. Similar studies can potentially be carried out in microalgae to greatly reduce the harvesting and lipid extraction costs.The most recent studies in genetic modification of microalgae are summarized in Table **1**. It can be seen from Table **1** that efforts have been made towards metabolic engineering of microalgae for the enhancement of lipid accumulation, modification of lipid chain length, and successful approaches for knocking out genes in the starch pathway to channelize TAG accumulation. Several gene transfer techniques have been well established in the model microalgae *C. reinhardtii*, the complete genome sequence and EST are available at (http://www.chlamycollection.org/). More studies on better understanding of pathways, protein interactions, limiting enzymes need further investigation in microalgae.

Table 1. Genetic engineering work carried out towards lipid production.

Gene	Origin of Gene	Protein Expressed	Host	Genomic Location	Transformation Technique	Comment	Reference
*Cr*DGAT 2a, b and c	*C. reinhardtii CC3491*	Acylglycerol acyltransferases (DGAT)	*C. reinhardtii* CC3491	Nucleus	Glass bead method	Increase in mRNA level (7-29.1times) but no significant change in fatty acid profiles	[16]
Malic enzyme gene	*P. tricornutum CCMP2561*	Malic enzyme (ME)	*P. tricornutum* CCMP2561	Nucleus	Electroporation	3.5-fold increase in Malic enzyme and 2.5 fold increase in total lipid content	[16]
slm1 starch less mutant	*S. obliquus*	Blocking of starch pathway	*S. obliquus*	-	-	51% TAG increase in mutant strain as compared to wild type under light	[17]
Acc 1 gene	*C. cryptica T13L*	ACCase enzyme - acetyl-CoA to malonyl-CoA FA	*C. cryptica* CYCLOI	Nucleus	Microprojectile bombardment	A 2-3 fold increase in ACC activity	[18, 19]
Acc 1 gene	*C. cryptica T13L*	Generates malonyl-CoA from acetyl-CoA for FA biosynthesis	*N. saprophila* NAVIC1	Nucleus	Microprojectile bombardment	Maybe 2-3 fold increase in ACC activity but no work lipid content is carried out	[18, 19]
Fatty acid acyl carrier protein gene and thioesterase gene	*U. californica, C. hookeriana, and C. reinhardtii*	protein-protein interactions between the fatty acid acyl carrier protein (ACP) and thioesterase (TE)	*C. reinhardtii*	Chloroplast	Microprojectile bombardment	Increase in C18:1 fatty acids	[20]
Thioesterase (myristic acid gene and lauric acid gene)	*C. camphora and U. californica*	Myristyl-ACP TE - Hydrolyzes acyl-ACP to release free FAs and control carbon chain length and properties of the final fatty alkane	*P. tricornutum*	Nucleus	Biolistic transformation	15% increase in C 14:0 fatty acid	[21]
Uc FatB2	*U. californica*	Lauryl-ACP TE - Hydrolyzes Lauryl-ACP to release free FAs and control carbon chain length and properties of the final fatty alkane	*P. tricornutum*	Nucleus	Biolistic transformation	6.2% increase in C 12:0 fatty acid	[21]

(Table 1) contd.....

Gene	Origin of Gene	Protein Expressed	Host	Genomic Location	Transformation Technique	Comment	Reference
CrFAB2	*C. reinhardtii*	Catalyzes the conversion of stearic acid (18:0) to oleic acid (18:1) and 18:2	*C. reinhardtii* cc-125	Nucleus	Glass bead method	28% increase in total fatty acid and 2.4-fold increase in oleic acid (18:1)	[22]
NoD12 gene (Δ12-desaturase)	*N. oceanica*	Catalyzes synthesis of linoleic acid from oleic acid,	*N. oceanica*	Nucleus	Electroporation	50%-75% increase in arachidonic acid (20:4 n − 6) in TAG	[23]
DGAT2	*P. tricornutum*	Tri-acylglyceride (TAG) biosynthesis	*P. tricornutum*	Nucleus	Electroporation	76.2% increase in Eicosa-pentaenoic acid	[24]
DGAT2-1 to 5 and *DGAT1*	*C. reinhardtii*	Catalyzes covalent linkage of diacylglycerol (DAG) to long-chain fatty acyl-CoAs	C. reinhardtii	-	RNAi silencing	20% and 44% increase of total lipid content in the DGAT2-1 and DGAT2-5 overexpression lines, respectively	[25]
DGAT2	*T. pseudonana*	catalyzes the last step of triacylglyceride (TAG) biosynthesis	*T. pseudonana*	Chloroplast	Biolistic transformation	1.52 to 1.95-fold increase in TAG accumulation without detriment to growth	[26]
DGTT4	*C. reinhardtii* CC-408	enhance TAG accumulation	*C. reinhardtii* CC-400	Nucleus	Electroporation	About 1.5 to 2.5-fold increase in TAG content	[27]
G3PDH + GPAT + LPAT + PAP + DGAT	*S. cerevisiae INVSC1 and Y. lipolytica Po1g*	Five-gene cassette encoding Kennedy pathway enzymes	*C. minutissima* UTEX 2219	Nucleus	Electroporation	Two-fold increase in TAG content	[28]
PDAT1	*S. cerevisiae and C. reinhardtii*	Catalyzes conversion of phospholipids into TAG	*C. reinhardtii*	-	-	25% decrease in TAG content of PDAT1	[29]
GPAT	*Lobosphaera incisa*	Catalyzes the first committed step of the Kennedy pathway	*C. reinhardtii*	Nucleus	Glass bead	50% increase in the TAG content without compromising growth	[30]
MLDP (major lipid droplet protein)	*C. reinhardtii*	Implicated in lipid droplet formation or maintenance	*C. reinhardtii*	Nucleus	Glass-bead	Increased lipid droplet size with no changes in TAG content in the MLDP knockdown lines	[31]
PNPLA3	*P. tricornutum*	Membrane bound protein associated with lipid droplets	*P. tricornutum*	Nucleus	Electroporation	70% increase in neutral lipid content	[32]
Nitrogen regulatory protein PII protein	*C. reinhardtii*	Negatively controls TAG accumulation in LBs during acclimation to nitrogen starvation	*C. reinhardtii*	Nucleus	Glass beads	Over-accumulation of LBs and a 1.5-fold increase in TAG level in the PII knockdown lines	[33]

(Table 1) contd.....

Gene	Origin of Gene	Protein Expressed	Host	Genomic Location	Transformation Technique	Comment	Reference
Phosphoenolpyruvate carboxylase 1, CrPEPC1	*Chlamydomonas reinhardtii CC425*	Channelizes carbon flux coming from glycolysis towards protein or FA	*C. reinhardtii*	Nucleus	Glass beads	20% increase in TAG level in the PEPC1 knockdown lines	[34]
Citrate synthase (CIS)	*Chlamydomonas reinhardtii CC425*	Regulates carbon flux to FA synthesis or the TCA cycle	*C. reinhardtii*	Nucleus	Glass beads	A 169.5% increase in TAG content in the CIS knockdown lines	[35]
NAD(H) kinase	*A. thaliana*	Catalyzes synthesis of NADPH	*C. pyrenoidosa* FACHB-9	Nucleus	Electroporation	110.4% increase in cellular lipid content without reducing growth rate	[36]
NRR1	*S. cerevisiae and C. reinhardtii*	Associated with transcriptional regulation of lipid biosynthetic genes under nitrogen-deprivation conditions	*C. reinhardtii*	-	-	50% decrease in TAG content in the insertional mutant of nrr1-1 under nitrogen-deprivation	[29]
DOF-type transcription factor	*Soybean*	Transcriptional regulator for genes involved in FA and glycerolipid biosynthesis	*C. reinhardtii*	Nucleus	*Agrobacterium*	2-fold increase of total lipids	[37]

EXTRACTION OF LIPIDS FROM MICROALGAE

Solvent extraction is commonly utilized for lipid extraction from microalgae. The solvent extraction process can be classified into five steps. During the extraction, the solvent penetrates through the cell membrane into the cytoplasm (step 1) and interacts with the neutral lipids *via* similar van der Waals forces, or it may bind with the polar lipids *via* hydrogen bonds (step 2) to form an organic solvent-lipids complex (step 3). The solvent-lipids complex, driven by a concentration gradient, diffuses across or dissociates away from the cell membrane (step 4) into the static organic solvent film surrounding the cell (step 5) and dissolves into the bulk organic solvent. The solvent interacts with cell wall or cell membrane to form a static solvent film that remains undisturbed by solvent flow or agitation [2]. The Bligh & Dyer method and Soxhlet were most mature protocols in the small scale chemical engineering processing [38, 39]. In addition to the conventional solvent extraction approaches, the development of super-/sub-critical extraction offers a new concept of operation while their cost greatly limits the use.

Cell disruption and drying are two of the most energy-consuming steps. Attempts are focused on elimination of these steps by direct extraction using wet algae. Traditionally, freeze drying is a more preferred drying method over oven drying due to its mild conditions comparing with spray or oven drying, which often lead to degradation of thermal-liable products and loss of volatile lipids coupled with the nonuniform particle size [40]. Wet extraction methods were recently investigated to overcome the bottleneck of high energy consumption [40]. Using wet algae would affect extraction efficiency; one hypothesis is that the presence of residual water will adversely affect the extraction efficiency as water forms a barrier that prohibits the solvent mass transfer from the inside the cell to the outside. The presence of water may cause many problems especially in supercritical extraction, including potential catalytic effects for the hydrolytic-based reactions in the presence of water. The presence of water may result in flow impedance and restrictor plugging and channeling, and formation of highly compacted bed within the vessel [41]. An alternative hypothesis is that residual water in the biomass will improve the extraction, as the presence of water swells the cell and facilitates the lysis of cell wall, allowing better chance of solvent to access the inner cell content and enhances mass transfer. Du *et al.* [42] suggested that the existence of water may swell the center of cell matrix and act as a polar co-solvent to facilitate extraction.

It was confirmed that the drying process consumes a significant amount of energy [43]. The conventional Bligh & Dyer method is very capable to deal with wet biomass due to the presence of alcohols. However, pure alcohols by themselves are not able to extract over 90% of total lipids [42]. Another high efficient wet (84% moisture) lipid extraction procedure *via* acid and base hydrolysis was developed and is capable to capture 79% of transesterifiable lipids [44]. Supercritical CO_2 extraction method is not as captable to treat wet biomass, as discussed above. However, Soh and Zimmerman [45] studied the effect of moisture content (up to 20%) on the extraction efficiency; no obvious change was observed in their study. The study by Halim *et al.* [46] showed reduced efficiency for lipid extraction from wet *Chlorococcum* sp. using non-polar system, while in the polar system, the extraction efficiency was enhanced with increased moisture content. Potential energy benefits are demonstrated *via* a similar Fatty Acid Methyl Ester (FAME) extraction efficiency for wet and dry algae biomass in sub-critical CO_2 [45]. Unlike SFE, most mechanical cell disruption methods and solvent extraction are highly effective with wet extraction. Jiménez Callejón *et al.* [47] harvested tripled amount of lipids than the conventional Bligh & Dyer method. Direct conversion of wet algal biomass to biodiesel has been demonstrated as a more advanced method, but due to limitations in scalability, large scale application remains to be developed [48]. Based on current available technologies, the dry extraction, when combined with waste heat recycling, has higher fossil fuel ratio, but the wet processing has more potential in the future [43].

Conventional Solvent Extraction

Since some of the lipids are membrane-associated *via* hydrogen bonds and would not be extracted by van der Waals forces, the addition of a polar organic solvent to a non-polar solvent system facilitates the lipid extraction. Halim *et al.* [2] extracted three times more total lipid from *Chlorococcum* sp. by employing isopropanol as a co-solvent to pure hexane system. However, the co-extraction of polar lipids and other compounds such as sugars, amino acids, and pigments is undesirable as their presence affects the oil quality. The shifting of pH can disrupt the strong ionic forces, if applicable. Therefore, the solvent should be inexpensive, non-toxic, volatile, and selective to the neutral lipids [8, 49]. Since "like dissolves like", the selectivity of the solvent is dominating for efficient recovery of target compounds from a mixture. Chloroform/methanol, hexane/isopropanol, ethanol and other solvent systems were widely used for efficient extraction of lipids. Although not as efficient as chlorine system, hexane/isopropanol (3:2 v/v) is the suggested solvent for lipid extraction considering the selectivity, toxicity and volatile. In addition, Soxhlet extraction can be used to reduce the sensitivity to solvent polarity during the extraction [50].

The lipid extraction follows the diffusion-driven mechanism, yet, it is not well understood and require further research [51]. The extraction rate is dominantly controlled by the concentration gradient between the cells and the solvent, rather than the bulk convection [51]. Extraction is most rapid in the beginning when concentration gradient is at its maximum. The majority of the lipids (60-70%) can be extracted within the first eight hours. As the gradient becomes smaller, the extraction rate slows down. The extended extraction time did not significantly increase the yield. Following the same reason, a higher microalgal biomass (s/b) value benefits the extraction but results in excessive consumption of organic solvent, while too less solvent lowers the extraction yield and leads to handling difficulties. The appropriate solvent to biomass varies for microalgae species depending on their lipid content and intrinsic solvent-cellular interactions. Increasing the temperature would accelerate the mass transfer, thus increasing the extraction rate. The extraction yield can be enhanced by increasing temperature to near boiling point [2].

Continuous extraction processes can overcome the time and energy limitation but this requires a large amount of organic solvent. Therefore, most laboratory-scale solvent extractions reported were conducted in batch processes. Batch extraction is largely limited by lipid mass transfer equilibrium, hence is quite time-consuming. Soxhlet extraction is a special type of batch extraction, it continuously replenishes cells with fresh solvent through the solvent recycles while minimizes solvent consumption. The Soxhlet apparatus consists of a solvent flask to hold the solvent and extracted contents, the extractor and the condenser to recycle the solvent. However, the high energy requirement for continuous distillation and the prolonged heating time at solvent boiling point limits its application. Superior efficiency of Soxhlet extraction was reported for lipid extraction from microalgae by Halim *et al.* [2]. Over 95% recovery of the total lipids in the algae is possible [8]. It is also noted that Soxhlet system may cause PUFA degradation when contrasted with batch extractions [2].

Bligh and Dyer's is found to be most efficient for lipid extraction (Lam and Lee 44), it yields over 95% of total lipids and can be used for both dry and wet route. The critical ratios of methanol, chloroform and water should be 2:1:1.8 and that of solvent to tissue should be 3:1. For wet route, the water content of microalgae should be considered [8].

The solvent extraction kinetics of microalgal lipids is observed to follow a first-order kinetics equation:

$$m_e = m_{s,o} (1 - e^{-kt}) \quad \textbf{(1)}$$

$$k = f(ag, s/b, T) \quad \textbf{(2)}$$

where m_e (g lipid/g dry biomass) is the amount of lipid extracted in the organic solvent at time t (min), $m_{s,o}$ (g lipid/g dry biomass) is the amount of lipid originally present in the cells, and the parameter k is the lipid mass transfer coefficient from the microalgal cells into the organic solvent (min^{-1}). It is a function of the degree of agitation, ag (revolution per minute), s/b (mL solvent/g dried biomass), and the extraction temperature, T (°C).

The conventional solvent extraction is cheap and easy to scale up, the concern is these processes may take long time, require further treatment like evaporation to concentrate the extract, and cost large amount of toxic organic solvents which brings environmental burden, so other more advanced extraction method have been developed [2].

Super-/Sub-Critical Solvent Extraction

Supercritical fluid extraction (SFE) has been well documented in literature for valuable compounds recovery particularly for decaffeination and essential oil extraction, the application has also extended to microalgal compounds recoveries [52 - 54]. This method is developed and has replaced accelerated solvent extraction, which refers to the extraction using organic solvents at temperatures and pressures above their boiling point [5]. Supercritical fluids have a similar density as fluid, but a similar viscosity as gas. The high pressure forces supercritical liquid into the microalgae cells, as the supercritical liquid have a diffusion rate similar to gas, the mass transfer is greatly enhanced, thus the extraction time can be much shorter than for conventional solvent extraction. Also, as the polarity of solvent is tunable, SFE can be more selectively used to extract the target product than conventional solvent extraction [55]. Yen *et al.* [56] have recently reviewed the advantages and challenges facing SFE extraction of lipids from microalgae biomass.

In SFE, chemical solvent usage is minimized or eliminated. Supercritical carbon dioxide ($SCCO_2$), for its relatively cheap price, safety, non-toxicity, and chemical inertness, and suitable critical temperature and pressure, became the most popular solvent [57, 58]. In this case, the extractant can be separated easily from solvent if no co-solvent is added since CO_2 is gaseous under normal condition [59]. The CO_2 rich flue gas may easily serve as an easy and cheap CO_2 source. High infrastructure and operational cost associated with temperature and pressure control are its main disadvantages. A typical supercritical extraction unit used for the of microalgal lipids recovery is described in Fig. (**1**). A typical extraction unit starts from feed pumps for transportation of liquid CO_2 and if needed, modifier/co-solvent from their source to the extractor through the mixing chamber and heating unit. In the extraction vessel where microalgae powder is located, the compressed CO_2 is heated above its critical temperature and converted to its supercritical state. The supercritical CO_2 travels through the microalgal biomass packed inside the extraction vessel and performs lipid extraction. The biomass are retained in the vessel by frits assembled at both ends of the vessel while the $SCCO_2$ flows to the subsequent fractional collectors and depressurized to precipitate out the extracted lipids [2, 8].

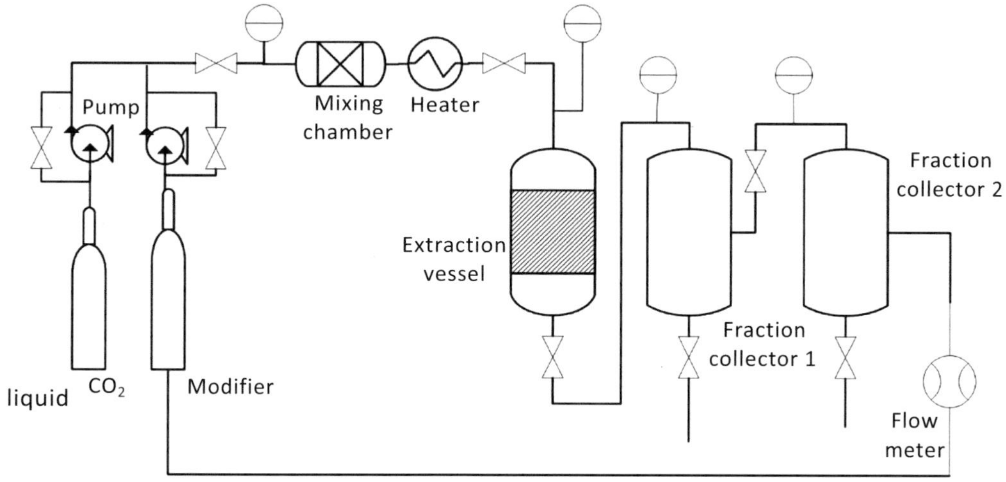

Fig. (1). industrial scale $SCCO_2$ extraction process for lipid extraction from microalgae [2, 8].

Studies have compared various lipid extraction methods including Bligh & Dyer method, solvent extraction, sonication, direct saponification, and $SCCO_2$ extraction [45, 60]. The supercritical extraction was proven to be the most effective, especially for long chain unsaturated fatty acids [60]. The optimized extraction condition was reported to be with pre-treatment, 414 bar; 100 °C. Increased selectivity was revealed in the extracted components consist with decreased pigment, nitrogen and phospholipid fractions [45].

Water is another environmental-friendly solvent for SFE use. The change of liquid parameters would turn water into a less polar solvent having similar dielectric constant as ethanol [52]. Water has a critical temperature of 374 °C and a critical pressure of 22.1 MPa. The condition is much hasher than CO_2, thus requires higher cost. Sub-critical fluid extraction is more economic due to less strict condition control requirement than SFE. Sub-critical refers to a condition at which the temperature ranges from boiling point to critical point, and pressure sufficient to maintain the fluid state. Pressure, temperature, fluid flow rate, and the addition of co-solvents are the critical factors affect the extraction yield [61]. In general, the higher the extraction pressure is, the higher fluid density and hence solvent power. The solvent power is also related to temperature as temperature increase leads to decrease in fluid density but increased lipid mass transfer. However, increasing extraction pressure and temperature also increases operation cost, lowers selectivity, and often results in the co-extraction of unwanted cellular components [2]. High concentrations of hydronium (H^+) and hydroxide (OH^-) ions exist under subcritical conditions. The acid- or base- catalyzed degradation and repolymerization reactions would be accelerated in such environment. Lipid, protein and carbohydrate fractions are usually the products yielded after the extraction [62]. Complete extraction of lipids was reported at the temperature of 205 °C by microwave assisted heating, which resulted 30% higher recovery than conventional heating [48].

Other solvents were not studied as much for lipid extraction studies. Supercritical methanol and ethanol were investigated for the extraction as well. The optimal condition that recovered 90.8% FAMEs was found to be 320 °C, 152 bar, 19:1 molar ratio, 31 min, 7.5 wt% water content [63]. Subcritical co-solvents extraction of algal lipid from hexane and ethanol ratio was also studied to recover an optimum 88% of the total lipids [61]. Two-stage operation of SFE is also suggested for more selective outcome, with first stage focusing on neutral lipids and the second stage focusing on medium-polar lipids or carotenoids [59]. The pilot plant design is available in Soto's review [64].

Ionic Liquid Extraction

Extraction using ionic liquids has recently gained much attention for microalgal lipids. Ionic liquids are liquid phase organic salts, they consist of anions and cations, which usually are nitrogen containing structure with a broad range of functional side groups. Ionic liquids (ILs) are considered green solvents to volatile organic solvents as they are non-volatile and thermal stable. High extraction efficiency is expected due to their chemical nature, since ILs have superior solubility of biomass and acts as cell destabilizers in algae suspensions [40],

Choi *et al.* [65] studied the synergistic effects of the ionic liquid mixtures with different anions to improved the lipid extraction yield of *C. vulgaris*. The anion structure of ILs significantly affects the extraction efficiency of lipids. Generally, hydrophobic ILs such as [Bmim][PF_6] and [Bmim] [Tf_2N] have lower extraction efficiency than hydrophilic and water miscible ILs such as [Bmim] [CF3SO3, Bmim] [$MeSO_4$], and [Emim] [$MeSO_4$]. Exception is observed for [Bmim] [Cl] and [Emim] [Ac], which are attributed to the higher solubility of lipids in these ILs [8]. Imidazolium, ammonium, phosphonium, and pyridinium derivatives, were screened by Orr *et al.* [66] to increase lipids extraction efficiency from wet biomass at ambient temperature. Comparative yield as solvent extraction was achieved by IL extraction with lower chlorophyll content [65 - 67]. Methanol can be used to adjust the viscosity of a ionic liquid. *C. vulgaris* biomass can be dissolved in a mixture of volume ratio 1:1 [Bmim][CF_3SO_3] and methanol, leaving lipids insoluble hence separated [67]. Moreover, ILs would act as biocatalysts for biodiesel production reactions [68]. The simulation of the IL extraction process was conducted for bio-oil recovery [68]. Moderate temperature is suggested for higher extraction yield, like 110°C for [$HNEt_3$][HSO_4] [69]. IL extraction represents a low energy consuming extraction method and capable to deal with wet biomass [66], but the high price of ionic liquid based solvents, high energy requirement, and toxicity for humans currently has prevented their industrial applications.

Some ionic liquids are able to reverse their polarity with CO_2 presence, these solvents are known as CO_2-triggered switchable solvents. Lipids can be extracted by the non-polar state of solvent while recovered by adding CO_2 to tune the solvent to more polar state (Fig. **2**), resulting in the formation of a separate non-polar lipids phase, which is easy to recover [42, 70]. Alternatively, the switchable solvent can be washed away from the extract by water saturated with carbon dioxide and recovered by de-carbonation. A typical example of switchable solvent lipid extraction is the non-polar N,N-dimethylcyclohexylamine (DMCHA). DMCHA would react with CO_2, forming a polar hydrogen carbonate ammonium salt, thus the extracted lipids would float on top of it after polarity switch [71]. Up to 22wt.% crude lipid extraction was reported for this solvent [70]. 1,8-diazabicyclo-[5.4.0]-undec-7-ene (DBU) and alcohol system was even earlier introduced to recover lipids from microalgae. However, the result was not as satisfying than the amines [72]. Secondary amine solvents were investigated for lipid extraction from wet *Desmodesmus* sp., and the yield is comparable with Bligh & Dyer method [73]. A series of tertiary amines have also been studied for their functions as switchable solvents [74]. The structure of amines in relation to their function has been identified, synthesized, and characterized [75, 76]. The mathematical model was also established to assist the optimization of switchable solvent lipid extraction from microalgae [77]. Since it is easy to reuse the solvent and harvest lipids, switchable solvent extraction is a new strategy for minimizing energy and material consumption [74]. This emerging method requires neither distillation nor the use of volatile, flammable or chlorinated organic solvents [70]. Among the various lipid extraction methods from wet microalgae, including organic solvents, supercritical CO_2 and switchable solvents, only switchable solvent showed significant positive energy balance [42]. Although the solvent itself is expensive to synthesize, recently several tertiary amines are becoming more easily prepared and even commercially available [74].

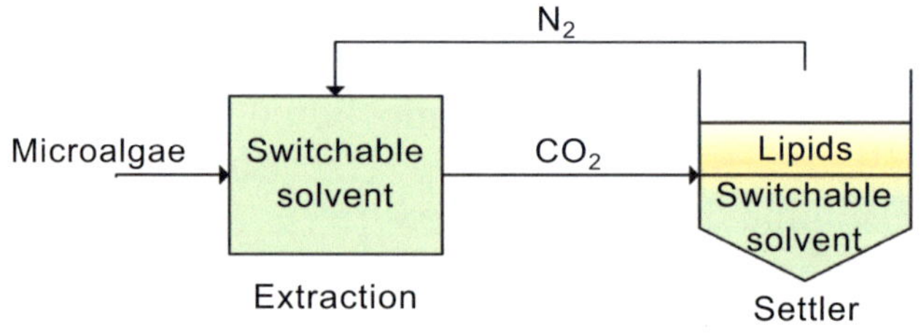

Fig. (2). -Illustration of using switchable solvent to extract lipids from microalgae.

Novel Approaches

The typical microalgal concentration before harvesting/de-watering is about 0.1-1.0% (w/v). The energy consumption for having a reasonable concentrated form of algae is high. Therefore, recently the lipid extraction method that eliminates drying and toxic organic solvent use is of particular interest. These methods could lead to significant energy and cost savings. Simultaneous distillation and extraction is compared with the Soxhlet extraction and Bligh & Dyer method for N. *oculata* lipids extraction, and resulted slightly lower lipid yields to B&D method, thus this may not be of great interest. Bio-solvents like terpenes that widely exist in the citrus fruits or other plants have been studied [38]. The use of H_2O_2 with $FeSO_4$ can lead to a increase of lipid yield from 6.9 to 17.4% (wt) together with reported FAMEs quality improvement [78]. A single-step electromagnetic field process is developed by OriginOil (USA). It is a one step dewatering, cell disruption, and lipid recovery method, and is referred to as OriginOil Single-Step Extraction™. Its Quantum Fracturing™ technology is based on the fluid shear plus pulsed electromagnetic field [79].

Another novel method using ethanol was proposed. This method extracts lipids from wet *Picochlorum* sp. biomass at room temperature and normal pressure, yet achieved high lipid yield comparable to that from Bligh-Dyer method (33.04% wt) in 37 minutes. The FAME composition was similar from the two extraction methods [80]. Therefore, the mild conditions made this a potential method for further scale-up.

Direct transesterification is recently developed for biodiesel production. It omits the oil extraction step, and is a simple and energy efficient process that is capable for wet microalgal biomass [40]. Almost complete biodiesel conversion was obtained through direct transesterification of the disrupted *Chlamydomonas* sp. biomass with 68.7 wt% water content [3]. Direct transesterification reached the similar lipid yield as the conventional Bligh and Dyer or Soxhlet extraction methods for *Phaeodactylum tricornutum, Nannochloropsis gaditana,* and *Chaetoceros calcitrans*; moreover, it saves time, cost and solvent usage and simplify the operation by step elimination [81]. Microwave-assisted transesterification (EHMT) of lipids in wet microalgae is a development of direct transesterification. Different parameters, including temperature, time, solvent and catalyst usage, were investigated to optimize biodiesel yield [3]. The EHMT lipids yield from wet microalgae is comparable to that of direct transesterification from dried microalgae biomass, but the EHMT improved the FAME profile and less polar pigments is extracted, which leads to a 50% reduction in nitrogen content [82].

CELL DISRUPTION

As a pre-treatment method, cell disruption can enhance lipid recovery efficiency by several folds since the cells are not readily broken by suspending in the extracting solvent. Therefore, this step is often considered necessary even though additional processing cost and energy request are introduced to the process. According to the literature, there is no universally best cell disruption method as the efficiency of method varies significantly for different microalgae species [83]. For the most investigated algae strains, the lipids are encapsulated within the cellular structures including the cell membrane and cell wall; the extraction can be inefficient without proper pre-treatment [84]. Operational conditions like temperature, pressure, biomass growth stage and scale significantly affects the efficiency of methods, therefore, accurate comparison of various methods is very difficult [85]. Lee *et al.* [86] estimated the energy for cell disruption. The efficiency and advantages/disadvantages for different cell disruption methods as well as extraction methods are also listed in Fig. (**3**) and compared in Table **2**. It is important to keep in mind that the cell disruption is highly dependent on the type of biomass, state of biomass and operational scale.

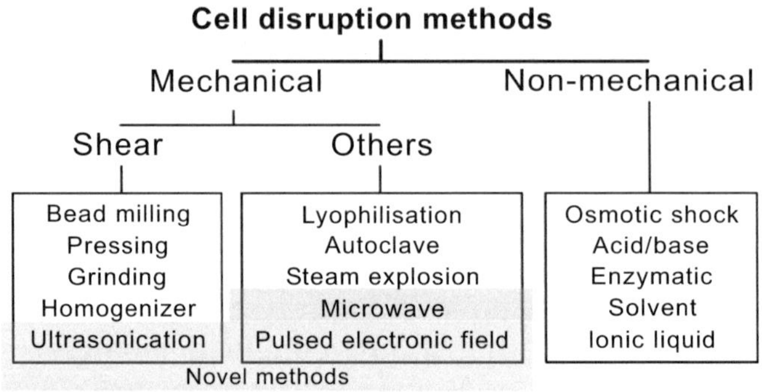

Fig. (3). -An illustration of cell disruption methods.

Table 2. Comparison of different cell disruption methods for lipid extraction from microalgae.

Step	Methodology	Efficiency	Advantages and Disadvantages	References
Cell disruption	Grinding	++	Time-consuming	[87]
	Cryogenic grinding	+++	Expensive	[88, 89]
	Bead milling	+++	Most efficient in some studies; not as efficient in several studies; the inconsistent result may due to the treatment is strain specific; generates heat	[2, 90 - 93]
	High pressure homogenizer	+++	Comparable with bead milling and ultrasound assisted extraction	[46, 85, 88]
	Autoclave	+	Damage of carotenoids occurs	[91]
	Microwave	+++	Comparable efficient with bead milling; Low energy consumption; Simple method; Generates heat	[83, 86, 94]
	Ultrasonication	+++ / +	Most efficient in some studies; Not efficient in other studies	[5, 46, 83, 95, 96]
	Enzymatic hydrolysis	+	Highly selective; mild condition; Expensive, Strict condition maintenance; Long treatment time	[89, 97, 98]

(Table 2) contd.....

Step	Methodology	Efficiency	Advantages and Disadvantages	References
	Pulsed electric field	++	Highly selective; Retain bioactivity of carotenoids Short treatment time; Small solvent requirements	[99 - 102]
	Osmotic shock, Acid/ alkaline treatment,	+ ++	Not efficient; Cause carotenoids to degrade	[2] [2]
	Ionic liquids/ Switchable solvent	++	High price; Toxicity; Cause carotenoids to degrade	[40]
Solvent extraction	Conventional solvent extraction	++	Cheap and easy to scale up; Long extraction time; Multi-step operation; Use large amount of solvents	[52, 59, 92]
	Super-/sub-critical solvent extraction	+++	Polarity of solvent is tunable; Fast; Safe; Easy separation of carotenoids; Expensive	[2, 42, 56, 59]

Mechanical Disruption Methods

Those methods that directly uses physical forces such as solid and liquid shear or thermal/pressure shock to break the cells are categorized as mechanical methods. The mechanical cell disruption techniques utilizing shear forces such as grinding, bead-milling and high pressure homogenizers are the most conventional methods for biomass break down; they can be applied to any biomass and are easier to scale up than other novel methods [4, 92]. However, the high energy requirement for these approaches is a significant limitation during bioprocess scale-up. They generate a large amount of heat, so temperature control may be desired. In addition, they cause non-selective release of inter contents with cell debris, which would complicate the subsequent separation processes [100].

Grinding

Grinding using pestle and mortar is capable of extracting lipids from dry or wet microalgae [87]. However, it is a manual process, it requires longer treatment time, only works at limited scale, and the efficiency may vary upon operator [103]. Cryogenic grinding, most commonly using liquid nitrogen, is another very efficient method that breaks wet algae cell wall instantly. This low temperature treatment is able to recover over 29 wt % lipids from *C. vulgaris*, but the high cost of liquid nitrogen make it not suitable for industrial oil production [88, 89].

Bead Milling

The working principle of bead-milling is similar to grinding. The solid shear by high speed spinning beads offers effective cell disruption for any kind of biomass. Currently, there are two category of bead mills, one is the smaller scale bead beater for laboratory use, which involves shaker vessels (bead beating), and the other can be adapted for larger scale, which involves agitated beads. The high speed spinning beads generates liquid shear and collide with the biomass. Generally high efficiency can be achieved for both types by selecting proper bead size and material. In particular, bead beater has minimized contamination risk from the environment due to the use of separate vials, hence ideal for DNA extraction [5]. The disruption efficiency is determined by the container's shape, shaking speed, bead diameter, bead material, bead quantity and biomass amount. As compared to grinding, bead milling offers a better potential for larger scale operations. The challenge facing this approach is the reduced efficiency for small cells with rigid cell wall, like the microalgae *C. vulgaris* [90]. However, this strain is equally difficult to handle for other cell disruption methods. The small size of beads required by bead milling would request additional separation cost.

High Pressure Homogenizer

The high pressure homogenizer (HPH) is widely used for laboratory-scale microalgal cell disruption, and is also commonly employed for large-scale disruption of bacteria or yeast cells. It generally deals with slurry algal suspension, and can operate in continuous mode, which allows easy scale-up. The microalgal cell suspension is forced to collide against a narrow nozzle outlet by high pressure pumping and subsequently released into a low-pressure chamber. Alternatively, the rotor-stator style homogenizer can also treat biomass in low viscosity liquids. Although yet to be better understood, the mechanism of cell disruption in HPH is believed to be caused by the turbulence, viscosities and high pressure shear, and impingement of the cells on the set surface, cavitation in the liquid, coupled with the sudden pressure drop between the nozzle and the outside environment [84, 104]. The critical factors for HPH system control include the pressure, cell diameter, and nozzle size. High pressure homogenizers showed superior results than traditional pre-treatment methods like osmotic shock or enzymatic hydrolysis [88].

Physical or Chemical Methods

Steam Explosion

Steam explosion first treats biomass with hot steam (160°C to 260°C) under pressure (1 to 3.5 MPa) for 2 to 6000 seconds, then the explosive decompression of the biomass ruptures of the rigid fibers of the cell wall. The sudden pressure drop defibrillates the cellulose bundles. This process differs from hydrothermal liquefaction in its lower operation temperature, generally between 100 to 300°C. Pre-treating microalgae biomass with steam technology should be performed at lower temperatures in order to prevent lipid degradation. The advantages of steam explosion processes include: the high recovery yield, water is used as an green solvent, production of hemicelluloses and the solid residues are broken down from bundles to individual fibers [105]. The optimal lipid yield may be higher than the lipid content of microalgae (10-15% higher) due to the polymerization of proteins and carbohydrates into oily composites, but it changes according to biomass and process specifications [106]. Therefore, steam explosion is an efficient and economical pre-treatment for modifying lignocellulosic cell wall for lipid extraction, yet requires further research [107].

Autoclave

Autoclave (steam autoclave) operates at a temperature of 121°C and pressure of 15 psig. This method is used as sterilizer, which indicates that it can efficiently break the microbial cells. The microbial extracellular cell membrane rupture under high thermal stress, and the intracellular lipids can be extracted. Autoclave showed good results in a study by Lee *et al.* [108] for lipid extraction from *C. vulgaris*. However, autoclave requires a relative long treatment time, and would require high energy consumptions, hence not applicable for large-scale treatment [107].

Enzymatic Hydrolysis

Enzymatic hydrolysis is an effective but expensive process, largely due to the high cost of enzymes and the relative long reaction time. The enzymes facilitate the cleavage of bonds in cell wall forming molecules. The major advantage of enzymatic hydrolysis is by decreasing the activation energy of the cleavage reactions, providing one of the mildest conditions for cell disruption. In addition, increased selectivity is observed, hence less undesired by-products. Corrosion of the machine or lipid degradation are avoided as well. Both lysozyme and cellulase are efficient to break cell walls for lipid production from microalgae, with a higher yield of 16.6% by lysozyme [109]. However, in larger industrial scale, enzymatic hydrolysis typically require long residence times and large reaction tanks, the requirement to maintain stable condition for prolonged time largely limits the industrial applications of this approach for cell disruption [89, 97]. Chen *et al.* [85] suggested the co-cultivation of bacteria with algae to lyse cell walls, which would be an interesting alternative to traditional enzymatic hydrolysis.

Osmotic Shock, Acid/ Alkaline Treatment

Physical or chemical treatment methods such as osmotic shock, acid/alkaline treatment are extremely attractive in the operational aspect, since they only require the addition of certain chemicals and temperature control, if necessary. They are easy to scale-up, and hold low energy input, modest capital cost methods [85, 110]. However, chemical cost, bio-toxicity and lipid degradation are the issues prevented their further use [111]. In addition, selectivity and suitability of the chemical are largely dependent on the microalgae to be targetted, *e.g.* osmotic shock alone is not enough to extract lipids from those microalgae cells with rigid cell wall [2].

Membrane technology (cationic polymer) is also used to save space, energy and time towards lipids production. Efficient functional membrane is found to be simple and efficient to lyse wet cells through direct contact with the tertiary-amine cations on the surface of the membrane [112].

Recent Approaches for Lipid Extraction from Microalgae

Microwave

Microwave assisted extraction (MAE) is a novel method developed within last few decades. It can efficiently rupture cells and is currently applied for vegetable or animal oil extraction [90]. Electromagentic radiation of wavelength ranging from 0.001 m to 1 m are generated and transmitted to the algal slurry. As the medium absorbs the transmitted energy, the algal slurry is homogeneously heated up, the cells are then ruptured by heat shock. Furthermore, the vaporized water content inside the cell would produce an extra pressure shock from inside the cell [52]. The combination of heat and pressure shock can efficiently disrupt the microalgae cells, especially those strains with strong mechanical resistance [113]. The operational factors including power, working volume, temperature, and the polarity, dissipation factor and heat capacity of the solvent, are the significant factors of the extraction process design. It is important to understand the property of target compound before the design of the cell disruption process [89]. In general, solvents having high heat dissipation factors and high dielectric constant would facilitate the extraction [52]. MAE utilizes less energy than other mechanical or thermal methods. Having no thermogradient in MAE, the heat transfer requirements are eliminated, which is an advantage over other treatment methods especially ultrasonication. Yet, proper temperature control may be desired [97]. As a simple and efficient method for lipids extraction, MAE requires less solvents than conventional methods, and have the potential for further scale up. Overall, the large-scale reactor design is yet to be developed and the working mechanism is to be further investigated. It is suggested that the microwave reflection tank may be employed for cell disruption, provided a more efficient microwave absorbent [94]. Chen *et al.* [3] has designed a continuous extraction system based on conventional microwave oven for lipid recovery, about 96.2% oil recovery was achieved in 80 minutes.Various methods were modified by microwave irradiation for extraction of lipids. Hara and Radin *et al.* [114] showed the highest lipid yield for *Tetraselmis* sp. while Folch method is best for *Nannochloropsis* sp.

Ultrasonication

Ultrasound assisted extraction (UAE) has been widely studied for the extraction of lipids, proteins, and sugars. Microbubble cavitation in the suspending medium is considered the underlying mechanism for UAE to break the cell wall, where micro-bubbles form and collapse near the cells, the mechanical energy is released as shock waves, creating micro turbulence, high liquid shear and pressure shock [84]. UAE is reported to have the advantages of higher efficiency, shorter extraction time, simpler handling, and lower cost. the most critical factors affecting UAE yield are the frequency and working power [115]. High frequency is more effective and energy efficient in disruption of microalgal cells, while combination of low frequency was also an option to save energy [116]. Moreover, the extractor size and shape is essential in UAE since the energy dissipates easily through transmittance [89]. UAE is proved to be scalable by the recent developments in installations, which is realized by arranging multiple devices together. It is used by OriginOil (USA) coupled with electromagnetic pulses to disrupt cells for lipids production [5].

Pulsed Electric Field

The pulsed electric field (PEF) assisted cell disruption came out during the 1990s. This non-thermal method has much lower energy cost as less energy is wasted as heat. PEF assisted extraction improves the microalgal cell membrane permeability by electroporation, where high intensity intermittent electric field is applied to the cells in a plused manner (time scale of micro-seconds) [117]. The electroporation process can be either reversible or irreversible, by simply adjusting the electric field intensity. Researchers have shown that PEF is efficient in extraction of lipids and bioactive compounds from microalgae or plant tissues [5, 113]. The advantages in treatment time, energy and solvent requirement, and selectivity of intracellular lipids in microalgae makes PEF an attractive novel method [101, 118]. However, the efficiency of PEF in lipids extraction is yet to be improved, especially when compared with some robust mechanical methods like HPH [102]. Similar but more effectively, high voltage electric discharges provide a solution for electrically resistant strains. As the pulsed streamer is applied, secondary phenomena like with production of localized high electric fields, shockwaves, cavitation, liquid turbulences, ozone and hydroxide radicals are usually accompanied, accelerating the cell damage [100].

The harvested algae suspensions are concentrated and PEF treated. For all pulse parameters applied, the PEF induced cell disintegration resulted in the release of soluble intracellular matter into the suspension. The disintegration efficiency increased with increasing specific treatment energy, whereas the field strength hardly had any influence. For suspensions with a biomass content of 100 g dry weight per kg suspension the electrical energy input necessary for considerable cell rupture was in the range of 1 MJ/kg dried algae. This is equivalent to 4.8% of the upper heating value of the algae. Although the treated algae contained lipids, PEF treatment only led to the spontaneous release of soluble components. The selectiveness of the process might offer the opportunity to use PEF treatment in a biorefinery concept, where soluble algae ingredients are extracted before solvent extraction of lipids is performed [119].

Comparison of Various Methods

It is difficult to identify the best method for lipid extraction as the method to be used is largely dependent on species. Cell disruption may be non-essential if performed in multi-stage extraction, regardless of the microalgal species as all lipids can be extracted within a few stages [120]. However, the additional stages would add cost as well as the disruption step.

Bead beater was most frequently reported to be the most efficient approach to treat the outmost rigid cells like C. *vulgaris, Scenedesmus dimorphus* or *C. protothecoides* when compared among none-treatment, autoclaving, homogenization, microwaves, sonication, osmosis shock, French press, freeze and thaw, and lyophilisation [5, 91, 92, 120]. However, in several studies bead milling was reported to be not as efficient as some other treatments like autoclaving, UAE or MAE [93].

Lipid extraction can be accelerated by MAE when combined with other methods [121]. As the most simple, easy, and effective way, MAE have good potential for lipid extraction [90]. MAE got highest lipid yield from *Scenedesmus* sp. against other treatments of sonication, bead milling, and osmotic shock, which all got similar result [8]. MAE showed similar good result as bead milling when compared to other methods to treat species like *Botryococcus* sp., and in the treatment of N. *oculata* cells, MAE used 13 fold less energy to achieve same yield [83, 86]. Electric methods by alternating current such as Pulsed electric field (PEF) have higher total lipid yield [122], it can achieve high efficiency of cell lyse independent to temperature [118]. In the lipid extraction, PEF resulted 90%

rupture of wet *Nannochloropsis* sp. or *Ankistrodesmus falcatus* cells, but taking both efficiency and cost into consideration, MAE with Sun drying could be a processing that is the most energy efficient over waterbath, ultrasonication, blender and laser [100, 123]. *Via* cell counting and colony diameter measurements of *Chlorococcum* sp., high pressure homogenization has the highest cell disruption rate (73.8%) than sulphuric acid treatment (33.2%), bead beating (33.2%), and UAE (4.5%) [46, 85]. However, for *B. braunii* cells, the high pressure homogenizer was not as efficient [8]. In addition, although being one of the most effective method, HPH requires the highest energy consumption [100, 118]. Steam explosion can facilitate the lipid extraction efficiencies regardless of types of microalgae when compared to autoclaving, UAE or MAE techniques [107]. Yet, steam explosion is not extensively studied and is limited at laboratory scale [107]. Good oil recoveries from *C. vulgaris* and *Nostoc* sp. are sometimes achieved with autoclave assisted extraction, but not even half of that from MAE [93, 107].

None of the various approaches, *i.e.*, Soxhlet extraction, the Bligh and Dyer method, and sonication achieved complete disruption of the cells [51]. Ultrasonication assisted extraction (UAE) was reported to be the most efficient in extracting lipids from *B. braunii*, *Chlorella sp.*, *Crypthecodinium cohnii,* and *Nostoc sp* [5, 88, 95, 118, 124]. In some situations, UAE is also reported to be insufficient to break microalgae cell wall [96]. For instance, McMillan *et al.* [83] revealed that UAE is an inefficient method, not even as efficient as heating in water bath; both high pressure homogenizer and bead milling have way better results than UAE [46]. However, since the comparative experiments are often conducted at the different energy output, the contradictory results from various studies may have an reason (*e.g.* Ultrasonicator 40-130 W; bead beater 850 W; MAE 1000 W) [46]. The different sample volume (*e.g.* 350 mL for UAE versus 10 mL in MAE) would result in different solid to liquid ratio, contributing to another possible reason for the inconsistent results [83]. Therefore, whether UAE for microalgae cell disruption requires further study. It is crucial to consider the energy requirements and economic considerations for microalgae cell disruption, as this is a critical step in many industrial microorganism related operations. The evaluation of different process configurations is also important [83, 125].

CONVERSION OF MICROALGAE TO BIO-OIL USING HYDROTHERMAL LIQUEFACTION

Hydrothermal liquefaction (HTL) is a thermochemical process in water where high temperature (200-350 °C) and high pressure (15-20 MPa) is applied to convert wet biomass into bio-crude oil, gas, aqueous phase, and solid residue. Compared to conventional techniques (*e.g.*, pyrolysis), HTL is regarded as a more suitable conversion route for feedstock with high moisture content (*e.g.*, microalgae) because it requires no dewatering of the biomass [126]. In most cases, researchers have investigated the effects of the operating parameters, such as reaction temperature, retention time, pressure, catalyst, and organic solvent, on the quantity and quality of liquefaction products. Jena *et al.* [127] conducted HTL experiments of *S. platensis* at temperatures of 200-380 °C, retention times of 0-120 min, and solid concentrations of10-15%. The results showed that the maximum bio-crude oil yield obtained at 350 °C, 60 min retention time and 20% solid concentration.

Hot compressed water can either acts as a reactant or a catalyst. This is due to the changes of water properties with temperature, including solubility, density, ionic product and dielectric constant. The dielectric constant of water decreases dramatically with the increasing temperature, thereby making water similar with organic solvent, while, in contrast, the ionic product is proportional to temperature, which could facilitate the acid/base-catalyzed reactions. Therefore, the hot compressed water can be used as an effective medium for thermochemical conversion [128].

In a previous study conducted by Sawayama *et al.* [129], the order for contributions and tendencies to the oil yield from microalgae was: lipid > protein > carbohydrates. Therefore, the high-lipid microalgae are preferably selected for bio-crude oil production. Several studies have investigated the kinetics and particular trend of lipid decomposition under hydrothermal environment. King *et al.* [130] reported that almost all soybean oil could be transformed to free fatty acid at 330-340 °C for 10-15 min. Protein is considered as the most abundant microalgal components compared to lipid and carbohydrates. In some cases, protein content could exceed 50 wt.%, such as *Chlorella vulgaris*. It should be noted that the protein is the major contributor to the high-nitrogen bio-crude oil. The nitrogen-containing compounds generated in the bio-crude include heterocylics (*e.g.*, pyridine, pyrrole, pyrrolidine, piperidine and indole) and non-heterocyclics (*e.g.*, amines and amides). Several studies have proposed a two-step HTL of microalgae combining with a nitrogen-removal pre-treatment. Jazrawi *et al.* [131] developed a two-step HTL approach with a mild (< 200 °C) dilute acid pre-treatment. The pre-treatment was found to remove as high as 50 wt.% of the input nitrogen and, following a severe (250-350 °C) HTL condition (Step II), reductions in the nitrogen content of the oil product was up to 55%, compared to that obtained from direct liquefaction of microalgae. Other major microalgal components, carbohydrates, are present in the form of cellulose, hemicellulose, starch, and other polysaccharides. Cellulose is one of the structural components of the microalgal cell walls. It consists of glucose units, linked by β-1,4-glycosidic bonds. Cellulose is insoluble in water at room temperature due to its high degree of crystallinity. However, at high temperatures cellulose can be hydrolyzed to smaller water-soluble products, like glucose.

HTL products include a bio-crude oil, a solid residue, an aqueous phase, and a gaseous product [132]. Bio-crude oil is similar in nature to conventional crude oil, but bio-crude oil contains higher oxygen and nitrogen contents than petroleum-based fuels [131]. Anastasakis and Ross [133] reported that the nitrogen and oxygen contents of bio-crude oil from *L. Saccharina* were 4.9 wt.% and 5.4 wt.%, respectively. The aqueous phase from microalgae HTL has a large fraction of organics and nutrients (PO43-, NH4+, CH3COO-, K^+, Na^+, and Mg^{2+}), thereby making it a promising growth medium for microalgae. Biller *et al.* [126] reported that microalgae could be cultivated in the process water obtained from liquefaction at 300 °C and 350 °C. However, the heavy dilution of this process water was necessary to reduce the negative effects of harmful compounds, like phenol. Alternatively, the aqueous phase can be used as the feedstock for gasification to produce a hydrogen-rich fuel gas. Gases phase as co-products are also produced during liquefaction of microalgae [132]. CO_2 is one of the major gaseous products, followed by H_2. The produced CO_2 can be recycled for microalgae growth, and the H_2 could be used for subsequent hydrotreatment. The solid products obtained from microalgal HTL contains a high content of ash, and very little amounts of hydrogen, sulfur, and nitrogen, thus, making them a potential soil fertilizer. The high amounts of heteroatoms (*e.g.*, N, S, and O) are commonly identified in the microalgae-based bio-crude oil. In addition, the long-chain fatty acids are usually observed in the bio-crude oil, increasing the viscosity of oil. Therefore, the further upgrading treatment (*e.g.,* denitrogenation, deoxygenation, and catalytic cracking) is needed to make bio-crude oil more similar with transportation fuel. Duan and Savage [134] reported that bio-crude oil from HTL of *Nannochloropsis sp.* could be effectively upgraded in the supercritical water with the addition of Pt/C catalyst. The bio-crude oil obtained after upgrading contained lower acid number, less long-chain fatty acids, and lower concentrations of nitrogen and oxygen. In another study, Zhou and Lawal [135] showed that the bio-crude oil from microalgae can be upgraded to green diesel with the presence of Pt/Al_2O_3, Rh/Al_2O_3, and $NiMo/Al_2O_3$ when subjected to hydrothermal environment.

CONCLUSION

Oil extraction from microalgae can be enhanced using metabolic engineering or improving current downstream processing operations. To date, no efficient cell disruption method has been developed at a large scale for direct lipid or oil extraction from wet microalgae biomass [4, 136]. One of the most significant challenges for the oil extraction process development is the lack of a reasonable way to compare the different methods due to large energy consumption, material cost, extraction time, scale, and post-extraction processes. Thus, the entire economic feasibility of a lipid extraction method should be assessed. The other challenge is the complex lipid quantification technique, thus easier and faster lipid content determination methods such as Nile red staining should be investigated. When it comes to large scale production, increasing standards required for solvent usage should be noticed, which include environmental impact and safety risk precautions evolved with hazardous materials storage and handling. It is suggested to produce biodiesel with other value-added bioactive compounds simultaneously to be economically successful. Thermochemical conversion, such as hydrothermal liquefaction, is an alternative method to overcome the extraction problems and directly produce bio-oil and may be another direction for future development.

CONFLICT OF INTEREST

The authors declare no conflict of interest, financial or otherwise.

ACKNOWLEDGEMENTS

Funding from the NSERC Discovery Grant to Amarjeet Bassi is gratefully acknowledged.

REFERENCES

[1] I. Rawat, R. Ranjith Kumar, T. Mutanda, and F. Bux, "Biodiesel from microalgae: A critical evaluation from laboratory to large scale production", *Appl. Energy,* vol. 103, pp. 444-467, 2013. [http://dx.doi.org/10.1016/j.apenergy.2012.10.004]

[2] R. Halim, M.K. Danquah, and P.A. Webley, "Extraction of oil from microalgae for biodiesel production: A review", *Biotechnol. Adv.,* vol. 30, no. 3, pp. 709-732, 2012. [http://dx.doi.org/10.1016/j.biotechadv.2012.01.001] [PMID: 22266377]

[3] C.L. Chen, C.C. Huang, K.C. Ho, P.X. Hsiao, M.S. Wu, and J.S. Chang, "Biodiesel production from wet microalgae feedstock using sequential wet extraction/transesterification and direct transesterification processes", *Bioresour. Technol.,* vol. 194, pp. 179-186, 2015. [http://dx.doi.org/10.1016/j.biortech.2015.07.021] [PMID: 26196418]

[4] J. Kim, G. Yoo, H. Lee, J. Lim, K. Kim, C.W. Kim, M.S. Park, and J.W. Yang, "Methods of downstream processing for the production of biodiesel from microalgae", *Biotechnol. Adv.,* vol. 31, no. 6, pp. 862-876, 2013. [http://dx.doi.org/10.1016/j.biotechadv.2013.04.006] [PMID: 23632376]

[5] P. Mercer, and R.E. Armenta, "Developments in oil extraction from microalgae", *Eur. J. Lipid Sci. Technol.,* vol. 113, no. 5, pp. 539-547, 2011. [http://dx.doi.org/10.1002/ejlt.201000455]

[6] L. Brennan, and P. Owende, "Biofuels from microalgae—A review of technologies for production, processing, and extractions of biofuels and co-products", *Renew. Sustain. Energy Rev.,* vol. 14, no. 2, pp. 557-577, 2010. [http://dx.doi.org/10.1016/j.rser.2009.10.009]

[7] Y. Chisti, "Biodiesel from microalgae", *Biotechnol. Adv.,* vol. 25, no. 3, pp. 294-306, 2007. [http://dx.doi.org/10.1016/j.biotechadv.2007.02.001] [PMID: 17350212]

[8] N. Pragya, K.K. Pandey, and P.K. Sahoo, "A review on harvesting, oil extraction and biofuels production technologies from microalgae", *Renew. Sustain. Energy Rev.,* vol. 24, pp. 159-171, 2013. [http://dx.doi.org/10.1016/j.rser.2013.03.034]

[9] S.S. Merchant, S.E. Prochnik, O. Vallon, E.H. Harris, S.J. Karpowicz, G.B. Witman, A. Terry, A. Salamov, L.K. Fritz-Laylin, L. Maréchal-Drouard, W.F. Marshall, L.H. Qu, D.R. Nelson, A.A. Sanderfoot, M.H. Spalding, V.V. Kapitonov, Q. Ren, P. Ferris, E. Lindquist, H. Shapiro, S.M. Lucas, J. Grimwood, J. Schmutz, P. Cardol, H. Cerutti, G. Chanfreau, C.L. Chen, V. Cognat, M.T. Croft, R. Dent, S. Dutcher, E. Fernández, H. Fukuzawa, D. González-Ballester, D. González-Halphen, A. Hallmann, M. Hanikenne, M. Hippler, W. Inwood, K. Jabbari, M. Kalanon, R. Kuras, P.A. Lefebvre, S.D. Lemaire, A.V. Lobanov, M. Lohr, A. Manuell, I. Meier, L. Mets, M. Mittag, T. Mittelmeier, J.V. Moroney, J. Moseley, C. Napoli, A.M. Nedelcu, K. Niyogi, S.V. Novoselov, I.T. Paulsen, G. Pazour, S. Purton, J.P. Ral, D.M. Riaño-Pachón, W. Riekhof, L. Rymarquis, M. Schroda, D. Stern, J. Umen, R. Willows, N. Wilson, S.L. Zimmer, J. Allmer, J. Balk, K. Bisova, C.J. Chen, M. Elias, K. Gendler, C. Hauser, M.R. Lamb, H. Ledford, J.C. Long, J. Minagawa, M.D. Page, J. Pan, W. Pootakham, S. Roje, A. Rose, E. Stahlberg, A.M. Terauchi, P. Yang, S. Ball, C. Bowler, C.L. Dieckmann, V.N. Gladyshev, P. Green, R. Jorgensen, S. Mayfield, B. Mueller-Roeber, S. Rajamani, R.T. Sayre, P. Brokstein, I. Dubchak, D. Goodstein, L. Hornick, Y.W. Huang, J. Jhaveri, Y. Luo, D. Martínez, W.C. Ngau, B. Otillar, A. Poliakov, A. Porter, L. Szajkowski, G. Werner, K. Zhou, I.V. Grigoriev, D.S. Rokhsar, and A.R. Grossman, "The Chlamydomonas genome reveals the evolution of key animal and plant functions", *Science,* vol. 318, no. 5848, pp. 245-250, 2007. [http://dx.doi.org/10.1126/science.1143609] [PMID: 17932292]

[10] S. Reverdatto, V. Beilinson, and N.C. Nielsen, "A multisubunit acetyl coenzyme A carboxylase from soybean", *Plant Physiol.,* vol. 119, no. 3, pp. 961-978, 1999. [http://dx.doi.org/10.1104/pp.119.3.961] [PMID: 10069834]

[11] J.W. Chen, W.J. Liu, D.X. Hu, X. Wang, S. Balamurugan, A. Alimujiang, W.D. Yang, J.S. Liu, and H.Y. Li, "Identification of a malonyl CoA-acyl carrier protein transacylase and its regulatory role in fatty acid biosynthesis in oleaginous microalga Nannochloropsis oceanica", *Biotechnol. Appl. Biochem.,* 2016. [PMID: 27572053]

[12] J. Fan, C. Andre, and C. Xu, "A chloroplast pathway for the de novo biosynthesis of triacylglycerol in Chlamydomonas reinhardtii", *FEBS Lett.,* vol. 585, no. 12, pp. 1985-1991, 2011. [http://dx.doi.org/10.1016/j.febslet.2011.05.018] [PMID: 21575636]

[13] L. Wan, J. Han, M. Sang, A. Li, H. Wu, S. Yin, and C. Zhang, "De novo transcriptomic analysis of an oleaginous microalga: pathway description and gene discovery for production of next-generation biofuels", *PLoS One,* vol. 7, no. 4, p. e35142, 2012. [http://dx.doi.org/10.1371/journal.pone.0035142] [PMID: 22536352]

[14] S.K. Lenka, N. Carbonaro, R. Park, S.M. Miller, I. Thorpe, and Y. Li, "Current advances in molecular, biochemical, and computational modeling analysis of microalgal triacylglycerol biosynthesis", *Biotechnol. Adv.,* vol. 34, no. 5, pp. 1046-1063, 2016. [http://dx.doi.org/10.1016/j.biotechadv.2016.06.004] [PMID: 27321475]

[15] L. Wan, J. Han, M. Sang, A. Li, H. Wu, S. Yin, and C. Zhang, "De novo transcriptomic analysis of an oleaginous microalga: pathway description and gene discovery for production of next-generation biofuels", *PLoS One,* vol. 7, no. 4, p. e35142, 2012. [http://dx.doi.org/10.1371/journal.pone.0035142] [PMID: 22536352]

[16] J. Xue, Y.F. Niu, T. Huang, W.D. Yang, J.S. Liu, and H.Y. Li, "Genetic improvement of the microalga Phaeodactylum tricornutum for boosting neutral lipid accumulation", *Metab. Eng.,* vol. 27, pp. 1-9, 2015. [http://dx.doi.org/10.1016/j.ymben.2014.10.002] [PMID: 25447640]

[17] G. Breuer, L. de Jaeger, V.P. Artus, D.E. Martens, J. Springer, R.B. Draaisma, G. Eggink, R.H. Wijffels, and P.P. Lamers, "Superior triacylglycerol (TAG) accumulation in starchless mutants of Scenedesmus obliquus: (II) evaluation of TAG yield and productivity in controlled photobioreactors", *Biotechnol. Biofuels,* vol. 7, no. 1, p. 70, 2014. [http://dx.doi.org/10.1186/1754-6834-7-70] [PMID: 24883102]

[18] T.G. Dunahay, E.E. Jaruis, and P.G. Roessler, *Genetic transformation of the diatoms Cyclotella cryptica and Navicula saprophila*, 1995.

[19] T.G. Dunahay, E.E. Jarvis, S.S. Dais, and P.G. Roessler, *Manipulation of microalgal lipid production using genetic engineering.* vol. Vol. 57-58. Appl. Biochem. Biotechnol. - Part A Enzym. Eng. Biotechnol, 1996. [http://dx.doi.org/10.1007/978-1-4612-0223-3_20]

[20] J.L. Blatti, J. Beld, C.A. Behnke, M. Mendez, S.P. Mayfield, and M.D. Burkart, "Manipulating fatty acid biosynthesis in microalgae for biofuel through protein-protein interactions", *PLoS One,* vol. 7, no. 9, p. e42949, 2012. [http://dx.doi.org/10.1371/journal.pone.0042949] [PMID: 23028438]

[21] R. Radakovits, P.M. Eduafo, and M.C. Posewitz, "Genetic engineering of fatty acid chain length in Phaeodactylum tricornutum", *Metab. Eng.,* vol. 13, no. 1, pp. 89-95, 2011. [http://dx.doi.org/10.1016/j.ymben.2010.10.003] [PMID: 20971205]

[22] K. Hwangbo, J-W. Ahn, J-M. Lim, Y-I. Park, J.R. Liu, and W-J. Jeong, "Overexpression of stearoyl-ACP desaturase enhances accumulations of oleic acid in the green alga Chlamydomonas reinhardtii", *Plant Biotechnol. Rep.,* vol. 8, no. 2, pp. 135-142, 2014. [http://dx.doi.org/10.1007/s11816-013-0302-3]

[23] Y. Kaye, O. Grundman, S. Leu, A. Zarka, B. Zorin, S. Didi-Cohen, I. Khozin-Goldberg, and S. Boussiba, "Metabolic engineering toward enhanced LC-PUFA biosynthesis in Nannochloropsis oceanica: Overexpression of endogenous δ12 desaturase driven by stress-inducible promoter leads to enhanced deposition of polyunsaturated fatty acids in TAG", *Algal Res.,* 2015. [http://dx.doi.org/10.1016/j.algal.2015.05.003]

[24] Y.F. Niu, M.H. Zhang, D.W. Li, W.D. Yang, J.S. Liu, W.B. Bai, and H.Y. Li, "Improvement of neutral lipid and polyunsaturated fatty acid biosynthesis by overexpressing a type 2 diacylglycerol acyltransferase in marine diatom Phaeodactylum tricornutum", *Mar. Drugs,* vol. 11, no. 11, pp. 4558-4569, 2013. [http://dx.doi.org/10.3390/md11114558] [PMID: 24232669]

[25] X.D. Deng, B. Gu, Y.J. Li, X.W. Hu, J.C. Guo, and X.W. Fei, "The roles of acyl-CoA: diacylglycerol acyltransferase 2 genes in the biosynthesis of triacylglycerols by the green algae Chlamydomonas reinhardtii", *Mol. Plant,* vol. 5, no. 4, pp. 945-947, 2012. [http://dx.doi.org/10.1093/mp/sss040] [PMID: 22511604]

[26] K. Manandhar-Shrestha, and M. Hildebrand, "Characterization and manipulation of a DGAT2 from the diatom Thalassiosira pseudonana: Improved TAG accumulation without detriment to growth, and implications for chloroplast TAG accumulation", *Algal Res.,* 2015. [http://dx.doi.org/10.1016/j.algal.2015.09.004]

[27] M. Iwai, K. Ikeda, M. Shimojima, and H. Ohta, "Enhancement of extraplastidic oil synthesis in Chlamydomonas reinhardtii using a type-2 diacylglycerol acyltransferase with a phosphorus starvation-inducible promoter", *Plant Biotechnol. J.,* vol. 12, no. 6, pp. 808-819, 2014. [http://dx.doi.org/10.1111/pbi.12210] [PMID: 24909748]

[28] H.J. Hsieh, C.H. Su, and L.J. Chien, "Accumulation of lipid production in Chlorella minutissima by triacylglycerol biosynthesis-related genes cloned from Saccharomyces cerevisiae and Yarrowia lipolytica", *J. Microbiol.,* vol. 50, no. 3, pp. 526-534, 2012. [http://dx.doi.org/10.1007/s12275-012-2041-5] [PMID: 22752918]

[29] N.R. Boyle, M.D. Page, B. Liu, I.K. Blaby, D. Casero, J. Kropat, S.J. Cokus, A. Hong-Hermesdorf, J.

Shaw, S.J. Karpowicz, S.D. Gallaher, S. Johnson, C. Benning, M. Pellegrini, A. Grossman, and S.S. Merchant, "Three acyltransferases and nitrogen-responsive regulator are implicated in nitrogen starvation-induced triacylglycerol accumulation in Chlamydomonas", *J. Biol. Chem.,* vol. 287, no. 19, pp. 15811-15825, 2012.
[http://dx.doi.org/10.1074/jbc.M111.334052] [PMID: 22403401]

[30] U. Iskandarov, S. Sitnik, N. Shtaida, S. Didi-Cohen, S. Leu, I. Khozin-Goldberg, Z. Cohen, and S. Boussiba, "Cloning and characterization of a GPAT-like gene from the microalga Lobosphaera incisa (Trebouxiophyceae): overexpression in Chlamydomonas reinhardtii enhances TAG production", *J. Appl. Phycol.,* 2016.
[http://dx.doi.org/10.1007/s10811-015-0634-1]

[31] E.R. Moellering, and C. Benning, "RNA interference silencing of a major lipid droplet protein affects lipid droplet size in Chlamydomonas reinhardtii", *Eukaryot. Cell,* vol. 9, no. 1, pp. 97-106, 2010.
[http://dx.doi.org/10.1128/EC.00203-09] [PMID: 19915074]

[32] X. Wang, Y.H. Liu, D.X. Hu, S. Balamurugan, Y. Lu, W.D. Yang, J.S. Liu, and H.Y. Li, "Identification of a putative patatin-like phospholipase domain-containing protein 3 (PNPLA3) ortholog involved in lipid metabolism in microalga Phaeodactylum tricornutum", *Algal Res.,* 2015.
[http://dx.doi.org/10.1016/j.algal.2015.09.005]

[33] Z. Zalutskaya, N. Kharatyan, K. Forchhammer, and E. Ermilova, "Reduction of PII signaling protein enhances lipid body production in Chlamydomonas reinhardtii", *Plant Sci.,* vol. 240, pp. 1-9, 2015.
[http://dx.doi.org/10.1016/j.plantsci.2015.08.019] [PMID: 26475183]

[34] X. Deng, J. Cai, Y. Li, and X. Fei, "Expression and knockdown of the PEPC1 gene affect carbon flux in the biosynthesis of triacylglycerols by the green alga Chlamydomonas reinhardtii", *Biotechnol. Lett.,* vol. 36, no. 11, pp. 2199-2208, 2014.
[http://dx.doi.org/10.1007/s10529-014-1593-3] [PMID: 24966045]

[35] X. Deng, J. Cai, and X. Fei, "Effect of the expression and knockdown of citrate synthase gene on carbon flux during triacylglycerol biosynthesis by green algae Chlamydomonas reinhardtii", *BMC Biochem.,* vol. 14, p. 38, 2013.
[http://dx.doi.org/10.1186/1471-2091-14-38] [PMID: 24373252]

[36] J. Fan, K. Ning, X. Zeng, Y. Luo, D. Wang, J. Hu, J. Li, H. Xu, J. Huang, M. Wan, W. Wang, D. Zhang, G. Shen, C. Run, J. Liao, L. Fang, S. Huang, X. Jing, X. Su, A. Wang, L. Bai, Z. Hu, J. Xu, and Y. Li, "Genomic Foundation of Starch-to-Lipid Switch in Oleaginous Chlorella spp", *Plant Physiol.,* vol. 169, no. 4, pp. 2444-2461, 2015.
[PMID: 26486592]

[37] A. Ibáñez-Salazar, S. Rosales-Mendoza, A. Rocha-Uribe, J.I. Ramírez-Alonso, I. Lara-Hernández, A. Hernández-Torres, L.M. Paz-Maldonado, A.S. Silva-Ramírez, B. Bañuelos-Hernández, J.L. Martínez-Salgado, and R.E. Soria-Guerra, "Over-expression of Dof-type transcription factor increases lipid production in Chlamydomonas reinhardtii", *J. Biotechnol.,* vol. 184, pp. 27-38, 2014.
[http://dx.doi.org/10.1016/j.jbiotec.2014.05.003] [PMID: 24844864]

[38] C. Dejoye Tanzi, M. Abert Vian, and F. Chemat, "New procedure for extraction of algal lipids from wet biomass: a green clean and scalable process", *Bioresour. Technol.,* vol. 134, pp. 271-275, 2013.
[http://dx.doi.org/10.1016/j.biortech.2013.01.168] [PMID: 23500584]

[39] E. Hita Peña, A. Robles Medina, M.J. Jiménez Callejón, M.D. Macías Sánchez, L. Esteban Cerdán, P.A. González Moreno, and E. Molina Grima, "Extraction of free fatty acids from wet Nannochloropsis gaditana biomass for biodiesel production", *Renew. Energy,* vol. 75, pp. 366-373, 2015.
[http://dx.doi.org/10.1016/j.renene.2014.10.016]

[40] J-Y. Park, M.S. Park, Y-C. Lee, and J-W. Yang, "Advances in direct transesterification of algal oils from wet biomass", *Bioresour. Technol.,* vol. 184, pp. 267-275, 2015.
[http://dx.doi.org/10.1016/j.biortech.2014.10.089] [PMID: 25466997]

[41] S. E. G. V Barbosa-cánovas, *High Pressure Fluid Technology for Green Food Processing.*, 2015.

[42] Y. Du, B. Schuur, S.R. Kersten, and D.W. Brilman, "Opportunities for switchable solvents for lipid extraction from wet algal biomass: An energy evaluation", *Algal Res.*, vol. 11, pp. 271-283, 2015. [http://dx.doi.org/10.1016/j.algal.2015.07.004]

[43] L. Xu, D.W. Wim Brilman, J.A. Withag, G. Brem, and S. Kersten, "Assessment of a dry and a wet route for the production of biofuels from microalgae: energy balance analysis", *Bioresour. Technol.*, vol. 102, no. 8, pp. 5113-5122, 2011. [http://dx.doi.org/10.1016/j.biortech.2011.01.066] [PMID: 21334889]

[44] A. Sathish, and R.C. Sims, "Biodiesel from mixed culture algae via a wet lipid extraction procedure", *Bioresour. Technol.*, vol. 118, pp. 643-647, 2012. [http://dx.doi.org/10.1016/j.biortech.2012.05.118] [PMID: 22721684]

[45] L. Soh, and J. Zimmerman, "Biodiesel production: The potential of algal lipids extracted with supercritical carbon dioxide", *Green Chem.*, vol. 13, no. 6, pp. 1422-1429, 2011. [http://dx.doi.org/10.1039/c1gc15068e]

[46] R. Halim, R. Harun, M.K. Danquah, and P.a. Webley, "Microalgal cell disruption for biofuel development", *Appl. Energy,* vol. 91, no. 1, pp. 116-121, 2012. [http://dx.doi.org/10.1016/j.apenergy.2011.08.048]

[47] M.J. Jiménez Callejón, A. Robles Medina, M.D. Macías Sánchez, E. Hita Peña, L. Esteban Cerdán, P.A. González Moreno, and E. Molina Grima, "Extraction of saponifiable lipids from wet microalgal biomass for biodiesel production", *Bioresour. Technol.*, vol. 169, pp. 198-205, 2014. [http://dx.doi.org/10.1016/j.biortech.2014.06.106] [PMID: 25058294]

[48] H.K. Reddy, T. Muppaneni, Y. Sun, Y. Li, S. Ponnusamy, P.D. Patil, P. Cooke, and S. Deng, "Subcritical water extraction of lipids from wet algae for biodiesel production", *Fuel,* vol. 133, pp. 73-81, 2014. [http://dx.doi.org/10.1016/j.fuel.2014.04.081]

[49] T.M. Mata, A.a. Martins, and N.S. Caetano, "Microalgae for biodiesel production and other applications: A review", *Renew. Sustain. Energy Rev.*, vol. 14, no. 1, pp. 217-232, 2010. [http://dx.doi.org/10.1016/j.rser.2009.07.020]

[50] A. Sicaire, M.A. Vian, A. Filly, Y. Li, A. Bily, and F. Chemat, *Alternative Solvents for Natural Products Extraction.*, 2014, pp. 205-219.

[51] A. Ranjan, C. Patil, and V.S. Moholkar, "Mechanistic assessment of microalgal lipid extraction", *Ind. Eng. Chem. Res.*, vol. 49, no. 6, pp. 2979-2985, 2010. [http://dx.doi.org/10.1021/ie9016557]

[52] G.J. Gil-Chávez, J.a. Villa, J. Fernando Ayala-Zavala, J. Basilio Heredia, D. Sepulveda, E.M. Yahia, and G.a. González-Aguilar, "Technologies for Extraction and Production of Bioactive Compounds to be Used as Nutraceuticals and Food Ingredients: An Overview", *Compr. Rev. Food Sci. Food Saf.*, vol. 12, no. 1, pp. 5-23, 2013. [http://dx.doi.org/10.1111/1541-4337.12005]

[53] P.D. Patil, H. Reddy, T. Muppaneni, T. Schaub, F.O. Holguin, P. Cooke, P. Lammers, N. Nirmalakhandan, Y. Li, X. Lu, and S. Deng, "In situ ethyl ester production from wet algal biomass under microwave-mediated supercritical ethanol conditions", *Bioresour. Technol.*, vol. 139, pp. 308-315, 2013. [http://dx.doi.org/10.1016/j.biortech.2013.04.045] [PMID: 23665692]

[54] B-C. Liau, C-T. Shen, F-P. Liang, S-E. Hong, S-L. Hsu, T-T. Jong, and C-M. Chang, "Supercritical fluids extraction and anti-solvent purification of carotenoids from microalgae and associated bioactivity", *J. Supercrit. Fluids,* vol. 55, no. 1, pp. 169-175, 2010. [http://dx.doi.org/10.1016/j.supflu.2010.07.002]

[55] A.C. Guedes, M.S. Gião, A.A. Matias, A.V. Nunes, M.E. Pintado, C.M. Duarte, and F.X. Malcata,

"Supercritical fluid extraction of carotenoids and chlorophylls a, b and c, from a wild strain of Scenedesmus obliquus for use in food processing", *J. Food Eng.,* vol. 116, no. 2, pp. 478-482, 2013. [http://dx.doi.org/10.1016/j.jfoodeng.2012.12.015]

[56] H.W. Yen, S.C. Yang, C.H. Chen, Jesisca, and J.S. Chang, "Supercritical fluid extraction of valuable compounds from microalgal biomass", *Bioresour. Technol.,* vol. 184, pp. 291-296, 2015. [http://dx.doi.org/10.1016/j.biortech.2014.10.030] [PMID: 25455085]

[57] L.S. Daintree, A. Kordikowski, and P. York, "Separation processes for organic molecules using SCF Technologies", *Adv. Drug Deliv. Rev.,* vol. 60, no. 3, pp. 351-372, 2008. [http://dx.doi.org/10.1016/j.addr.2007.03.024] [PMID: 18006179]

[58] B. Wang, Y. Li, N. Wu, and C.Q. Lan, "CO(2) bio-mitigation using microalgae", *Appl. Microbiol. Biotechnol.,* vol. 79, no. 5, pp. 707-718, 2008. [http://dx.doi.org/10.1007/s00253-008-1518-y] [PMID: 18483734]

[59] E. Reverchon, and I. De Marco, "Supercritical fluid extraction and fractionation of natural matter", *J. Supercrit. Fluids,* vol. 38, no. 2, pp. 146-166, 2006. [http://dx.doi.org/10.1016/j.supflu.2006.03.020]

[60] Y. Li, F. Ghasemi Naghdi, S. Garg, T.C. Adarme-Vega, K.J. Thurecht, W.A. Ghafor, S. Tannock, and P.M. Schenk, "A comparative study: the impact of different lipid extraction methods on current microalgal lipid research", *Microb. Cell Fact.,* vol. 13, p. 14, 2014. [http://dx.doi.org/10.1186/1475-2859-13-14] [PMID: 24456581]

[61] M. Chen, T. Liu, X. Chen, L. Chen, W. Zhang, J. Wang, L. Gao, Y. Chen, and X. Peng, "Subcritical co-solvents extraction of lipid from wet microalgae pastes of Nannochloropsis sp", *Eur. J. Lipid Sci. Technol.,* vol. 114, no. 2, pp. 205-212, 2012. [http://dx.doi.org/10.1002/ejlt.201100120] [PMID: 22745570]

[62] S. Thiruvenkadam, S. Izhar, H. Yoshida, M.K. Danquah, and R. Harun, "Process application of Subcritical Water Extraction (SWE) for algal bio-products and biofuels production", *Appl. Energy,* vol. 154, pp. 815-828, 2015. [http://dx.doi.org/10.1016/j.apenergy.2015.05.076]

[63] Y. Nan, J. Liu, R. Lin, and L.L. Tavlarides, "Production of biodiesel from microalgae oil (Chlorella protothecoides) by non-catalytic transesterification in supercritical methanol and ethanol: Process optimization", *J. Supercrit. Fluids,* vol. 97, pp. 174-182, 2015. [http://dx.doi.org/10.1016/j.supflu.2014.08.025]

[64] E. Rosello-Soto, O. Parniakov, Q. Deng, A. Patras, M. Koubaa, and N. Grimi, "Application of Non-conventional Extraction Methods: Toward a Sustainable and Green Production of Valuable Compounds from Mushrooms", *Food Eng. Rev.,* vol. 8, no. 2, pp. 214-234, 2016. [http://dx.doi.org/10.1007/s12393-015-9131-1]

[65] H.S. Cho, Y.K. Oh, S.C. Park, J.W. Lee, and J.Y. Park, "Effects of enzymatic hydrolysis on lipid extraction from Chlorella vulgaris", *Renew. Energy,* vol. 54, pp. 156-160, 2013. [http://dx.doi.org/10.1016/j.renene.2012.08.031]

[66] V.C. Orr, N.V. Plechkova, K.R. Seddon, and L. Rehmann, "Disruption and Wet Extraction of the Microalgae Chlorella vulgaris Using Room-Temperature Ionic Liquids", *ACS Sustain. Chem.& Eng.,* vol. 4, no. 2, pp. 591-600, 2016. [http://dx.doi.org/10.1021/acssuschemeng.5b00967]

[67] Y.H. Kim, Y.K. Choi, J. Park, S. Lee, Y.H. Yang, H.J. Kim, T.J. Park, Y. Hwan Kim, and S.H. Lee, "Ionic liquid-mediated extraction of lipids from algal biomass", *Bioresour. Technol.,* vol. 109, pp. 312-315, 2012. [http://dx.doi.org/10.1016/j.biortech.2011.04.064] [PMID: 21601447]

[68] V. Piemonte, L. Di Paola, G. Iaquaniello, and M. Prisciandaro, "Biodiesel production from microalgae: Ionic liquid process simulation", *J. Clean. Prod.,* vol. 111, pp. 62-68, 2016. [http://dx.doi.org/10.1016/j.jclepro.2015.07.089]

[69] X. Chen, L. Hu, R. Xing, S. Liu, H. Yu, Y. Qin, K. Li, R. Li, and P. Li, "Ionic liquid-assisted subcritical water promotes the extraction of lipids from wet microalgae Scenedesmus sp", *Eur. J. Lipid Sci. Technol.,* vol. 117, no. 8, pp. 1192-1198, 2015. [http://dx.doi.org/10.1002/ejlt.201400189]

[70] A.R. Boyd, P. Champagne, P.J. McGinn, K.M. MacDougall, J.E. Melanson, and P.G. Jessop, "Switchable hydrophilicity solvents for lipid extraction from microalgae for biofuel production", *Bioresour. Technol.,* vol. 118, pp. 628-632, 2012. [http://dx.doi.org/10.1016/j.biortech.2012.05.084] [PMID: 22721685]

[71] C. Samorì, D. López Barreiro, R. Vet, L. Pezzolesi, D.W. Brilman, P. Galletti, and E. Tagliavini, "Effective lipid extraction from algae cultures using switchable solvents", *Green Chem.,* vol. 15, no. 2, pp. 353-356, 2013. [http://dx.doi.org/10.1039/c2gc36730k]

[72] C. Samorì, C. Torri, G. Samorì, D. Fabbri, P. Galletti, F. Guerrini, R. Pistocchi, and E. Tagliavini, "Extraction of hydrocarbons from microalga Botryococcus braunii with switchable solvents", *Bioresour. Technol.,* vol. 101, no. 9, pp. 3274-3279, 2010. [http://dx.doi.org/10.1016/j.biortech.2009.12.068] [PMID: 20071168]

[73] Y. Du, B. Schuur, C. Samorì, E. Tagliavini, and D.W. Brilman, "Secondary amines as switchable solvents for lipid extraction from non-broken microalgae", *Bioresour. Technol.,* vol. 149, pp. 253-260, 2013. [http://dx.doi.org/10.1016/j.biortech.2013.09.039] [PMID: 24121240]

[74] P.G. Jessop, S.M. Mercer, and D.J. Heldebrant, "CO2-triggered switchable solvents, surfactants, and other materials", *Energy Environ. Sci.,* vol. 5, p. 7240, 2012. [http://dx.doi.org/10.1039/c2ee02912j]

[75] A.D. Wilson, and F.F. Stewart, "Structure-function study of tertiary amines as switchable polarity solvents", *RSC Advances,* vol. 4, no. 22, pp. 11039-11049, 2014. [http://dx.doi.org/10.1039/C3RA47724J]

[76] Y. Xie, R. Parnas, B. Liang, Y. Liu, C. Tao, and H. Lu, "Synthesis and characterization of switchable ionic compound based on DBU, CH3OH, and CO2", *Chin. J. Chem. Eng.,* vol. 23, no. 10, pp. 1728-1732, 2015. [http://dx.doi.org/10.1016/j.cjche.2015.08.005]

[77] J. Durelle, J.R. Vanderveen, and P.G. Jessop, "Modelling the behaviour of switchable-hydrophilicity solvents", *Phys. Chem. Chem. Phys.,* vol. 16, no. 11, pp. 5270-5275, 2014. [http://dx.doi.org/10.1039/C3CP55391D] [PMID: 24492636]

[78] A. Steriti, R. Rossi, A. Concas, and G. Cao, "A novel cell disruption technique to enhance lipid extraction from microalgae", *Bioresour. Technol.,* vol. 164, pp. 70-77, 2014. [http://dx.doi.org/10.1016/j.biortech.2014.04.056] [PMID: 24836708]

[79] OriginOil, Algae harvesting, dewatering and extraction a breakthrough technology to transform algae into oil. Amsterdam, The Netherlands: Worldbiofuels Markets; 2010. March, pp. 15-17.

[80] F. Yang, W. Xiang, X. Sun, H. Wu, T. Li, and L. Long, "A novel lipid extraction method from wet microalga Picochlorum sp. at room temperature", *Mar. Drugs,* vol. 12, no. 3, pp. 1258-1270, 2014. [http://dx.doi.org/10.3390/md12031258] [PMID: 24663114]

[81] S.D. Ríos, J. Castañeda, C. Torras, X. Farriol, and J. Salvadó, "Lipid extraction methods from microalgal biomass harvested by two different paths: screening studies toward biodiesel production", *Bioresour. Technol.,* vol. 133, pp. 378-388, 2013. [http://dx.doi.org/10.1016/j.biortech.2013.01.093] [PMID: 23434816]

[82] J. Cheng, R. Huang, T. Li, J. Zhou, and K. Cen, "Biodiesel from wet microalgae: extraction with hexane after the microwave-assisted transesterification of lipids", *Bioresour. Technol.,* vol. 170, pp. 69-75, 2014.

[http://dx.doi.org/10.1016/j.biortech.2014.07.089] [PMID: 25125194]

[83] J.R. McMillan, I.a. Watson, M. Ali, and W. Jaafar, "Evaluation and comparison of algal cell disruption methods: Microwave, waterbath, blender, ultrasonic and laser treatment", *Appl. Energy,* vol. 103, pp. 128-134, 2013.
[http://dx.doi.org/10.1016/j.apenergy.2012.09.020]

[84] R. Halim, T.W. Rupasinghe, D.L. Tull, and P.A. Webley, "Mechanical cell disruption for lipid extraction from microalgal biomass", *Bioresour. Technol.,* vol. 140, pp. 53-63, 2013.
[http://dx.doi.org/10.1016/j.biortech.2013.04.067] [PMID: 23672939]

[85] D-Y. Kim, D. Vijayan, R. Praveenkumar, J-I. Han, K. Lee, J-Y. Park, W.S. Chang, J.S. Lee, and Y.K. Oh, "Cell-wall disruption and lipid/astaxanthin extraction from microalgae: Chlorella and Haematococcus", *Bioresour. Technol.,* vol. 199, pp. 300-310, 2016.
[http://dx.doi.org/10.1016/j.biortech.2015.08.107] [PMID: 26342788]

[86] A.K. Lee, D.M. Lewis, and P.J. Ashman, "Disruption of microalgal cells for the extraction of lipids for biofuels: Processes and specific energy requirements", *Biomass Bioenergy,* vol. 46, pp. 89-101, 2012.
[http://dx.doi.org/10.1016/j.biombioe.2012.06.034]

[87] Y-R. Hu, F. Wang, S-K. Wang, C-Z. Liu, and C. Guo, "Efficient harvesting of marine microalgae Nannochloropsis maritima using magnetic nanoparticles", *Bioresour. Technol.,* vol. 138, pp. 387-390, 2013.
[http://dx.doi.org/10.1016/j.biortech.2013.04.016] [PMID: 23639490]

[88] E. Molina Grima, E.H. Belarbi, F.G. Acién Fernández, A. Robles Medina, and Y. Chisti, "Recovery of microalgal biomass and metabolites: process options and economics", *Biotechnol. Adv.,* vol. 20, no. 7-8, pp. 491-515, 2003.
[http://dx.doi.org/10.1016/S0734-9750(02)00050-2] [PMID: 14550018]

[89] H. Zheng, J. Yin, Z. Gao, H. Huang, X. Ji, and C. Dou, "Disruption of Chlorella vulgaris cells for the release of biodiesel-producing lipids: a comparison of grinding, ultrasonication, bead milling, enzymatic lysis, and microwaves", *Appl. Biochem. Biotechnol.,* vol. 164, no. 7, pp. 1215-1224, 2011.
[http://dx.doi.org/10.1007/s12010-011-9207-1] [PMID: 21347653]

[90] J.Y. Lee, C. Yoo, S.Y. Jun, C.Y. Ahn, and H.M. Oh, "Comparison of several methods for effective lipid extraction from microalgae", *Bioresour. Technol.,* vol. 101, no. 1, suppl. Suppl. 1, pp. S75-S77, 2010.
[http://dx.doi.org/10.1016/j.biortech.2009.03.058] [PMID: 19386486]

[91] M.M Chan, S.S Ho, D.D Lee, C.C Chen, C-C. Huang, and J-S. Chang, "Characterization, extraction and purification of lutein produced by an indigenous microalga Scenedesmus obliquus CNW-N", *Biochem. Eng. J.,* vol. 78, pp. 24-31, 2013.
[http://dx.doi.org/10.1016/j.bej.2012.11.017]

[92] J. Taucher, S. Baer, P. Schwerna, D. Hofmann, M. Hümmer, R. Buchholz, and A. Becker, "Cell Disruption and Pressurized Liquid Extraction of Carotenoids from Microalgae", *Thermodyn. Catal.,* vol. 7, no. 1, pp. 1-7, 2016.

[93] P. Prabakaran, and A.D. Ravindran, "A comparative study on effective cell disruption methods for lipid extraction from microalgae", *Lett. Appl. Microbiol.,* vol. 53, no. 2, pp. 150-154, 2011.
[http://dx.doi.org/10.1111/j.1472-765X.2011.03082.x] [PMID: 21575021]

[94] H. Li, Y. Qu, Y. Yang, S. Chang, and J. Xu, "Microwave irradiation--A green and efficient way to pretreat biomass", *Bioresour. Technol.,* vol. 199, pp. 34-41, 2016.
[http://dx.doi.org/10.1016/j.biortech.2015.08.099] [PMID: 26342787]

[95] G. Cravotto, L. Boffa, S. Mantegna, P. Perego, M. Avogadro, and P. Cintas, "Improved extraction of vegetable oils under high-intensity ultrasound and/or microwaves", *Ultrason. Sonochem.,* vol. 15, no. 5, pp. 898-902, 2008.
[http://dx.doi.org/10.1016/j.ultsonch.2007.10.009] [PMID: 18093864]

[96] V. Pasquet, J-R. Chérouvrier, F. Farhat, V. Thiéry, J-M. Piot, J-B. Bérard, R. Kaas, B. Serive, T. Patrice, J-P. Cadoret, and L. Picot, "Study on the microalgal pigments extraction process: Performance of microwave assisted extraction", *Process Biochem.,* vol. 46, no. 1, pp. 59-67, 2011. [http://dx.doi.org/10.1016/j.procbio.2010.07.009]

[97] S.U. Kadam, B.K. Tiwari, and C.P. O'Donnell, "Application of novel extraction technologies for bioactives from marine algae", *J. Agric. Food Chem.,* vol. 61, no. 20, pp. 4667-4675, 2013. [http://dx.doi.org/10.1021/jf400819p] [PMID: 23634989]

[98] A. Deenu, S. Naruenartwongsakul, and S.M. Kim, "Optimization and economic evaluation of ultrasound extraction of lutein from Chlorella vulgaris", *Biotechnol. Bioprocess Eng.; BBE,* vol. 18, no. 6, pp. 1151-1162, 2013. [http://dx.doi.org/10.1007/s12257-013-0213-8]

[99] C. Sánchez-Moreno, L. Plaza, P. Elez-Martínez, B. De Ancos, O. Martín-Belloso, and M.P. Cano, "Impact of high pressure and pulsed electric fields on bioactive compounds and antioxidant activity of orange juice in comparison with traditional thermal processing", *J. Agric. Food Chem.,* vol. 53, no. 11, pp. 4403-4409, 2005. [http://dx.doi.org/10.1021/jf048839b] [PMID: 15913302]

[100] N. Grimi, A. Dubois, L. Marchal, S. Jubeau, N.I. Lebovka, and E. Vorobiev, "Selective extraction from microalgae Nannochloropsis sp. using different methods of cell disruption", *Bioresour. Technol.,* vol. 153, pp. 254-259, 2014. [http://dx.doi.org/10.1016/j.biortech.2013.12.011] [PMID: 24368274]

[101] X. Yu, T. Gouyo, N. Grimi, O. Bals, and E. Vorobiev, "Pulsed electric field pretreatment of rapeseed green biomass (stems) to enhance pressing and extractives recovery", *Bioresour. Technol.,* vol. 199, pp. 194-201, 2016. [http://dx.doi.org/10.1016/j.biortech.2015.08.073] [PMID: 26341008]

[102] Y.S. Lai, P. Parameswaran, A. Li, M. Baez, and B.E. Rittmann, "Effects of pulsed electric field treatment on enhancing lipid recovery from the microalga, Scenedesmus", *Bioresour. Technol.,* vol. 173, pp. 457-461, 2014. [http://dx.doi.org/10.1016/j.biortech.2014.09.124] [PMID: 25311186]

[103] R. P. Utomo, Y.-R. Chang, D.-J. Lee, and J.-S. Chang, Lutein recovery from *Chlorella* sp. ESP-6 with coagulants, Bioresource technology, vol. 139. pp. 176–80, Jul-2013. [http://dx.doi.org/10.1016/j.biortech.2013.04.025]

[104] C.W. Ho, W.S. Tan, and W.B. Yap, "Comparative Evaluation of Different Cell Disruption Methods for the Release of Recombinant Hepatitis B Core Antigen from Escherichia coli", *Biotechnol. Bioprocess Eng.; BBE,* no. 13, pp. 577-583, 2008. [http://dx.doi.org/10.1007/s12257-008-0020-9]

[105] W. Stelte, *Steam explosion for biomass pre-treatment.* Danish Technol. Inst., 2013, pp. 1-15.

[106] A. M. Aguirre, *Study of high pressure steaming on lipid recovery from microalgae,* 2014.

[107] "M. A. hattab Abdel Ghaly, "Microalgae Oil Extraction Pre-treatment Methods: Critical Review and Comparative Analysis", *J. Fundam. Renew. Energy Appl,* vol. 5, no. 4, 2015.

[108] S-H. Lee, Z-J. Qian, and S-K. Kim, "A novel angiotensin I converting enzyme inhibitory peptide from tuna frame protein hydrolysate and its antihypertensive effect in spontaneously hypertensive rats", *Food Chem.,* vol. 118, no. 1, pp. 96-102, 2010. [http://dx.doi.org/10.1016/j.foodchem.2009.04.086]

[109] H. Taher, S. Al-Zuhair, A.H. Al-Marzouqi, Y. Haik, and M. Farid, "Effective extraction of microalgae lipids from wet biomass for biodiesel production", *Biomass Bioenergy,* vol. 66, pp. 159-167, 2014. [http://dx.doi.org/10.1016/j.biombioe.2014.02.034]

[110] G. Yoo, W-K. Park, C.W. Kim, Y-E. Choi, and J-W. Yang, "Direct lipid extraction from wet Chlamydomonas reinhardtii biomass using osmotic shock", *Bioresour. Technol.,* vol. 123, pp. 717-

722, 2012.
[http://dx.doi.org/10.1016/j.biortech.2012.07.102] [PMID: 22939599]

[111] E. Günerken, E. D'Hondt, M.H. Eppink, L. Garcia-Gonzalez, K. Elst, and R.H. Wijffels, "Cell disruption for microalgae biorefineries", *Biotechnol. Adv.,* vol. 33, no. 2, pp. 243-260, 2015.
[http://dx.doi.org/10.1016/j.biotechadv.2015.01.008] [PMID: 25656098]

[112] G. Yoo, Y. Yoo, J-H. Kwon, C. Darpito, S.K. Mishra, K. Pak, M.S. Park, S.G. Im, and J-W. Yang, "An effective, cost-efficient extraction method of biomass from wet microalgae with a functional polymeric membrane", *Green Chem.,* vol. 16, no. 1, p. 312, 2014.
[http://dx.doi.org/10.1039/C3GC41695J]

[113] F.J. Barba, N. Grimi, and E. Vorobiev, "New Approaches for the Use of Non-conventional Cell Disruption Technologies to Extract Potential Food Additives and Nutraceuticals from Microalgae", *Food Eng. Rev.,* vol. 7, no. 1, pp. 45-62, 2015.
[http://dx.doi.org/10.1007/s12393-014-9095-6]

[114] C.L. Teo, and A. Idris, "Enhancing the various solvent extraction method via microwave irradiation for extraction of lipids from marine microalgae in biodiesel production", *Bioresour. Technol.,* vol. 171, pp. 477-481, 2014.
[http://dx.doi.org/10.1016/j.biortech.2014.08.024] [PMID: 25201293]

[115] M. Wang, W. Yuan, X. Jiang, Y. Jing, and Z. Wang, "Disruption of microalgal cells using high-frequency focused ultrasound", *Bioresour. Technol.,* vol. 153, pp. 315-321, 2014.
[http://dx.doi.org/10.1016/j.biortech.2013.11.054] [PMID: 24374364]

[116] M. Wang, W. Yuan, X. Jiang, Y. Jing, and Z. Wang, "Disruption of microalgal cells using high-frequency focused ultrasound", *Bioresour. Technol.,* vol. 153, pp. 315-321, 2014.
[http://dx.doi.org/10.1016/j.biortech.2013.11.054] [PMID: 24374364]

[117] E. Luengo, S. Condón-Abanto, I. Álvarez, and J. Raso, "Effect of pulsed electric field treatments on permeabilization and extraction of pigments from Chlorella vulgaris", *J. Membr. Biol.,* vol. 247, no. 12, pp. 1269-1277, 2014.
[http://dx.doi.org/10.1007/s00232-014-9688-2] [PMID: 24880235]

[118] M.D. Zbinden, B.S. Sturm, R.D. Nord, W.J. Carey, D. Moore, H. Shinogle, and S.M. Stagg-Williams, "Pulsed electric field (PEF) as an intensification pretreatment for greener solvent lipid extraction from microalgae", *Biotechnol. Bioeng.,* vol. 110, no. 6, pp. 1605-1615, 2013.
[http://dx.doi.org/10.1002/bit.24829] [PMID: 23297018]

[119] M. Goettel, C. Eing, C. Gusbeth, R. Straessner, and W. Frey, "Pulsed electric field assisted extraction of intracellular valuables from microalgae", *Algal Res.,* vol. 2, no. 4, pp. 401-408, 2013.
[http://dx.doi.org/10.1016/j.algal.2013.07.004]

[120] E. Ryckebosch, K. Muylaert, and I. Foubert, ""Optimization of an analytical procedure for extraction of lipids from microalgae," JAOCS", *J. Am. Oil Chem. Soc.,* vol. 89, no. 2, pp. 189-198, 2012.
[http://dx.doi.org/10.1007/s11746-011-1903-z]

[121] M. Soštarič, D. Klinar, M. Bricelj, J. Golob, M. Berovič, and B. Likozar, "Growth, lipid extraction and thermal degradation of the microalga Chlorella vulgaris", *N. Biotechnol.,* vol. 29, no. 3, pp. 325-331, 2012.
[http://dx.doi.org/10.1016/j.nbt.2011.12.002] [PMID: 22178401]

[122] A.P. Florentino de Souza Silva, M.C. Costa, A. Colzi Lopes, and E. Fares Abdala Neto, "R. Carrh?? Leit??o, C. R. Mota, and A. Bezerra dos Santos, "Comparison of pretreatment methods for total lipids extraction from mixed microalgae", *Renew. Energy,* vol. 63, pp. 762-766, 2014.
[http://dx.doi.org/10.1016/j.renene.2013.10.038]

[123] A. Guldhe, B. Singh, I. Rawat, K. Ramluckan, and F. Bux, "Efficacy of drying and cell disruption techniques on lipid recovery from microalgae for biodiesel production", *Fuel,* vol. 128, pp. 46-52, 2014.
[http://dx.doi.org/10.1016/j.fuel.2014.02.059]

[124] M. Plaza, S. Santoyo, L. Jaime, B. Avalo, A. Cifuentes, G. Reglero, G. García-Blairsy Reina, F.J. Señoráns, and E. Ibáñez, "Comprehensive characterization of the functional activities of pressurized liquid and ultrasound-assisted extracts from Chlorella vulgaris", *Lebensm. Wiss. Technol.,* vol. 46, no. 1, pp. 245-253, 2012.
[http://dx.doi.org/10.1016/j.lwt.2011.09.024]

[125] E.M. Spiden, B.H. Yap, D.R. Hill, S.E. Kentish, P.J. Scales, and G.J. Martin, "Quantitative evaluation of the ease of rupture of industrially promising microalgae by high pressure homogenization", *Bioresour. Technol.,* vol. 140, pp. 165-171, 2013.
[http://dx.doi.org/10.1016/j.biortech.2013.04.074] [PMID: 23688668]

[126] P. Biller, A.B. Ross, S.C. Skill, A. Lea-Langton, B. Balasundaram, C. Hall, R. Riley, and C.A. Llewellyn, "Nutrient recycling of aqueous phase for microalgae cultivation from the hydrothermal liquefaction process", *Algal Res.,* vol. 1, no. 1, pp. 70-76, 2012.
[http://dx.doi.org/10.1016/j.algal.2012.02.002]

[127] U. Jena, K.C. Das, and J.R. Kastner, "Effect of operating conditions of thermochemical liquefaction on biocrude production from Spirulina platensis", *Bioresour. Technol.,* vol. 102, no. 10, pp. 6221-6229, 2011.
[http://dx.doi.org/10.1016/j.biortech.2011.02.057] [PMID: 21444202]

[128] S.S. Toor, L. Rosendahl, and A. Rudolf, "Hydrothermal liquefaction of biomass: A review of subcritical water technologies", *Energy,* vol. 36, no. 5, pp. 2328-2342, 2011.
[http://dx.doi.org/10.1016/j.energy.2011.03.013]

[129] S. Sawayama, T. Minowa, and S.Y. Yokoyama, "Possibility of renewable energy production and CO2 mitigation by thermochemical liquefaction of microalgae", *Biomass Bioenergy,* vol. 17, no. 1, pp. 33-39, 1999.
[http://dx.doi.org/10.1016/S0961-9534(99)00019-7]

[130] J.W. King, R.L. Holliday, and G.R. List, "in a Subcritical Water Flow Reactor," no. December, pp. 261–264, 1999.

[131] C. Jazrawi, P. Biller, Y. He, A. Montoya, A.B. Ross, T. Maschmeyer, and B.S. Haynes, "Two-stage hydrothermal liquefaction of a high-protein microalga", *Algal Res.,* vol. 8, pp. 15-22, 2015.
[http://dx.doi.org/10.1016/j.algal.2014.12.010]

[132] Y. Guo, T. Yeh, W. Song, D. Xu, and S. Wang, "A review of bio-oil production from hydrothermal liquefaction of algae", *Renew. Sustain. Energy Rev.,* vol. 48, pp. 776-790, 2015.
[http://dx.doi.org/10.1016/j.rser.2015.04.049]

[133] K. Anastasakis, and A.B. Ross, "Hydrothermal liquefaction of the brown macro-alga Laminaria saccharina: effect of reaction conditions on product distribution and composition", *Bioresour. Technol.,* vol. 102, no. 7, pp. 4876-4883, 2011.
[http://dx.doi.org/10.1016/j.biortech.2011.01.031] [PMID: 21316946]

[134] P. Duan, and P.E. Savage, "Catalytic hydrotreatment of crude algal bio-oil in supercritical water", *Appl. Catal. B,* vol. 104, no. 1-2, pp. 136-143, 2011.
[http://dx.doi.org/10.1016/j.apcatb.2011.02.020]

[135] L. Zhou, and A. Lawal, "Hydrodeoxygenation of microalgae oil to green diesel over Pt, Rh and presulfided NiMo catalysts", *Catal. Sci. Technol.,* vol. 6, pp. 1442-1454, 2016.
[http://dx.doi.org/10.1039/C5CY01307K]

[136] I. Michalak, and K. Chojnacka, "Algae as production systems of bioactive compounds", *Eng. Life Sci.,* vol. 15, no. 2, pp. 160-176, 2015.
[http://dx.doi.org/10.1002/elsc.201400191]

Part II: Production and Economics in Microalgal Applications

CHAPTER 10

Research and Deployment of Renewable Bioenergy Production from Microalgae

Lieve M.L. Laurens* and **Melodie Chen-Glasser**

National Bioenergy Center, National Renewable Energy Laboratory, Golden, CO 80401, USA

Abstract: Recent progress towards the implementation of renewable bioenergy production has included microalgae, which have potential to significantly contribute to a viable future bioeconomy. In a current challenging energy landscape, where an increased demand for renewable fuels is projected and accompanied by plummeting fossil fuels' prices, economical production of algae-based fuels becomes more challenging. However, in the context of mitigating carbon emissions with the potential of algae to assimilate large quantities of CO_2, there is a route to drive carbon sequestration and utilization to support a sustainable and secure global energy future. This chapter places international energy policy in the context of the current and projected energy landscape. The contribution that algae can make, is summarized as both a conceptual contribution as well as an overview of the commercial infrastructure installed globally. Some of the major recent developments and crucial technology innovations are the results of global government support for the development of algae-based bioenergy, biofuels and bioproduct applications, which have been awarded as public private partnerships and are summarized in this chapter.

Keywords: Algae, Biomass, Renewable Fuels, Carbon Capture and Utilization, Future Energy Outlook, Commercial and Research Operations.

INTRODUCTION

In light of an international drive of nations towards energy independence and energy security, researchers and policy makers are increasingly searching for ways to reduce dependence on fossil fuels. The production of both renewable fuels from bio-based feedstocks as well as high-value products displacing petrochemicals will support these goals. Of all the bio-based feedstocks, algae are particularly attractive because of their highly efficient photosynthetic conversion of sunlight and carbon dioxide (CO_2) to organic material or biomass along with an

* **Corresponding author Lieve M.L. Laurens:** National Renewable Energy Laboratory, 15013 Denver West Parkway, Golden, CO 80401, USA; Tel: +1 (303) 384-6196; E-mail: Lieve.Laurens@nrel.gov

José Carlos Magalhães Pires (Ed.)

innate metabolic flexibility to adapt to potentially challenging physiological environments [1]. The biomass can serve as the direct source for liquid transportation fuels, primarily from the lipid fraction of the biomass. Similarly, the inherent richness of algal biomass in the natural products opens up avenues towards a biorefinery approach supporting the commercial deployment of algae-based products beyond fuels.

There are significant barriers currently impeding commercialization and economic production of algae for relatively low value energy and fuel markets. The barriers to commercial deployment of algae bioenergy processes range from incomplete knowledge of algae biology to challenges associated with economical integration of technologies at the demonstration scale. Even though many algae-based technologies have been demonstrated at the laboratory scale, most often this work has been done in isolation and thus a challenge remains in the full integration of algae-based processes with extended multi-season operation. Nevertheless, tremendous progress has been made in improving and demonstrating algae-based production, cultivation and upgrading technologies, even though the prospects for commercial algae-based bioenergy or biofuels production are more challenging today than they were 10 years ago. This is primarily the result of a substantial decline in petroleum prices. As a consequence, companies that were leading commercial development of biofuels have been increasingly redirecting their commercial focus towards the production of higher value food, feed and specialty products. To aid with the future commercial deployment of algae industries, progress is needed in reducing the use of energy, water and land footprints in integrated algae-based operations for larger scale demonstrations. It is in light of this that in recent years, there has been significant government financial support for the transition of laboratory to field or farm deployment of algae technologies.

CURRENT AND PROJECTED FUTURE ENERGY LANDSCAPE

Although worldwide petroleum reserves continue to be depleted and the costs of fossil fuel use to the global environment continue to mount, many countries remain highly dependent upon non-renewable and often imported sources of oil (and other fossil fuels), especially for transportation fuels. The Energy Information Agency (EIA) estimates that global petroleum and other liquid fuels inventory builds averaged 1.8 million b/d in 2015 [2]. Similarly, the IEA projects under its New Policies Scenario that world oil demand will increase from approximately 90 million barrels per day (mb/d) in 2014 to 104 mb/d by 2040 [2]. The EIA forecasts somewhat higher global demand growth over this same period, with total worldwide production rising to almost 120 mb/d by 2040 [2]. However, the price of petroleum has fallen by roughly 50% in recent years. This has greatly increased the techno-economic barrier to achieve cost competitive biofuel production. It has also created great uncertainty about the future petroleum market owing to a variety of on-going and unresolved geopolitical, technical and environmental factors. Thus, while vehicle fuel efficiency technologies continue to reduce fuel use per unit distance travelled, total worldwide demand for liquid fuels is projected to continue to grow over the next several decades.

The United States currently imports approximately one third of its petroleum, mostly from only a few countries around the world [2]. The projected increased demand for energy and liquid fuels worldwide is driven by rising living standards and higher energy diets and lifestyles in more rapidly developing countries, particularly China and India. Nonetheless, the continued extraction and combustion of fossil fuels has created environmental concerns about increased global warming and ocean acidification due to greenhouse gases (GHG) accumulating in the atmosphere. Biofuels and bioenergy remain one of only a few potential options to reduce the world's dependence on using fossil fuels for transportation and chemicals production. In this context, it has to be noted that a concern associated with substantially increasing biofuel production is the limited availability of or competition for suitable land. For example, the GHG mitigation benefits of biofuels can be negated if land with existing high carbon intensity is cleared for the production of biofuel feedstocks. Biofuels that could be produced without large increases in arable land or reductions in tropical rainforests that obviate or greatly diminish land availability and land use change (LUC) concerns, such as those derived from algae, remain highly attractive to develop.

Carbon emissions are continuing to increase to unprecedented levels, and have been linked to climate change. According to the National Research Council, "*Emissions of CO_2 from the burning of fossil fuels have ushered in a new epoch where human activities will largely determine the evolution of Earth's climate. Because CO_2 in the atmosphere is long lived, it can effectively lock Earth and future generations into a range of impacts, some of which could become very severe. Therefore, emission reduction choices made today matter in determining impacts experienced not just over the next few decades, but in the coming centuries and millennia*" [3].

Global CO_2 emissions are dominated by the world's 2 largest emitting countries, China and the US, at 10 GT/yr and 6.4 GT/yr respectively. Of the 6.4 GT generated annually in the US, about 3.3 GT is produced from point sources, such as coal and natural gas power plants, ethanol fermentation, cement, fertilizer and chemical plants [4]. In efforts to limit rising atmospheric CO_2 concentrations, while still meeting the increasing global energy demand, technologies for carbon capture, sequestration, utilization and storage (CCS and CCU) need to be implemented [5]. In addition to deploying and incentivizing clean energy solutions, such as algae cultivation for biofuels and bioproducts, legislative pressure to reducing power plant emissions can help to drive the development of algae-based technologies. However, like other CCU technologies, algae deployment is not without its own challenges.

According to a published roadmap for biofuels use in the transportation sector [6], up to 27% of worldwide transportation fuel could be supplied by biofuels by 2050, in particular replacing diesel, kerosene and jet fuel. This projected use would avoid 2.1 GT of CO_2 emissions per year. To achieve this target it is necessary to create a stable long-term policy framework to increase investor and end-user confidence and allow rapid expansion of this nascent industry. There also needs to be a sustained funding and support infrastructure provided by the governing agencies globally. As part of this financial and policy support for the development of technology, it is critical to link the performance of groups developing and implementing technology to GHG emissions reduction and other sustainability targets. The use of incentives may help spur commercial implementation and help reach the set targets. Finally, because of the magnitude of the global biofuel and bioenergy challenge, there should be international support and collaboration to aid with building out capacity as well as increased technology transfer to promote adoption of sustainable biofuel production pathways globally.

PROMISE AND OPPORTUNITY FOR ALGAE TECHNOLOGY

Because of their rapid rate of adaptation to differing environments and high photosynthetic efficiency, algae are considered excellent feedstocks for future bioenergy production [1, 7 - 10]. When following the stepwise losses of energy during photosynthesis, the assimilation of CO_2 to metabolic energy, a theoretical yield of ~12% can be estimated, at least 2-fold higher compared to terrestrial plants, even though experimentally observed yields are 3-4 times lower [1]. Overall biomass production yields vary with species and with geographical locations. An average yield of between 10 and 15 g m^{-2} day^{-1} is used as a benchmark, with reports of up to 20 g m^{-2} day^{-1} in the literature [11 - 13]. When taking 15 g m^{-2} day^{-1} as the baseline, we can calculate areal yields of 54 T ha^{-1} yr^{-1}.

One of the most challenging aspects for sustainable cultivation of algae for commercial production to supply commodity-scale markets is to ensure the continuous delivery of the enormous amounts of water and nutrients required to grow and process algal feedstocks. Effective water recycling is essential to minimize freshwater and chemical nutrients consumption [14, 15]. Water usage requirements for biomass and biofuel production will vary depending on growth conditions and ultimately the lipid or biofuel yield from the biomass. For example, for a production system growing algae at approximately 1 g L^{-1}, with about 20% oil content of the biomass for biofuel applications, a total of ~5000 L of algae culture would need to be processed to generate 1 kg of biofuel (green or bio-diesel). Algal biomass typically contains 45-50% C, 7.6% N and 1.4% P, though, the elemental composition can vary dramatically based on growth conditions and species of algae used, but on average, the above approximation can be made and is consistent with the Redfield ratio (106:16:1 C:N:P) on a weight basis (40:7:1 C:N:P). Thus, the nutrient requirements to support the same 1 kg of biofuel would be in the range of 0.38 kg N and 0.07 kg P equivalent, corresponding to 0.214 kg phosphate delivered (Fig. **1**). This is consistent with earlier reported estimates in the literature, where 3000 liters of water per kg of microalgae-based biodiesel have been estimated [16], and associated nutrients requirements are reported as 0.18-0.33 kg nitrogen if freshwater without any recycling is used for open pond cultivation [17, 18]. Similarly, based on the carbon content of the biomass, the CO_2 assimilation or sequestration potential can be derived, which amounts to 113 T ha^{-1} yr^{-1} (assuming 2.1 kg CO_2 needed to produce 1 kg algal biomass, at 47.6% C).

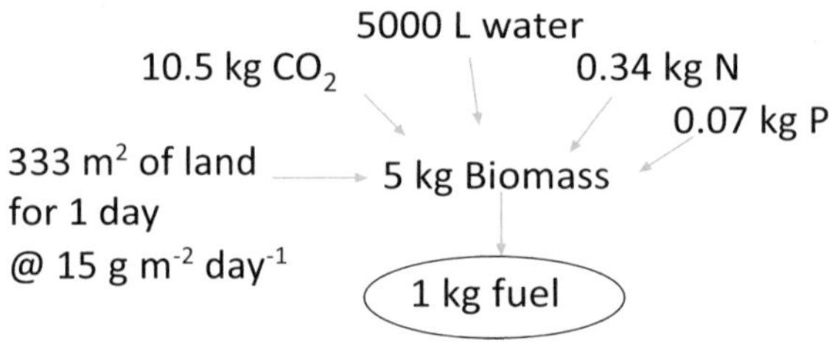

Fig. (1). Illustration of nutrient inputs to produce 1 kg of fuel derived from algal biomass.

The point sources of CO_2, such as ethanol, coal and natural gas plants (or Electric Utility Generating Units, EGU) are distributed throughout the country and thus the geographical dependence of productivity, along with day/night cycles of assimilation, will impact the actual potential of algae to assimilate CO_2. Furthermore, the apparent incompatibility in co-locating an urban situated EGU with the more rural location of the projected algae farm may bring practical challenges. For example, to mitigate the emissions from one 800 MWe power plant, approximately 5 MT yr^{-1}, roughly an area of 442 km^2 of algae ponds would be needed. A recent, thorough analysis of carbon sequestration potential from point sources of CO_2 distributed throughout the US, overlaid with the availability of nutrients, land and water, estimated that of the 3.3 GT CO_2 emitted per year, about 180 MT CO_2 can be mitigated, generating approximately 86 MT of algal biomass. This would be produced on farms spanning a 7,600 km^2 land area [19].

INTERNATIONAL BIOFUELS POLICY

On a global scale, each country or region has a distinct approach to the implementation of bioenergy into its national or regional fuel and energy infrastructure. It is generally accepted that on an international level, renewable energy must play a fundamental role in the transition towards a more competitive, secure and sustainable energy system. This transition will not be possible without a much larger contribution of renewable energy to our current infrastructure. The production and use of biofuels has been mainly driven by government policies in order to reduce oil dependency and in turn mitigate carbon dioxide emissions through deployment of renewable energy technologies. The following section lists some of the current and past international commitments to bioenergy and describes how algae fit into this approach. The main mechanisms for governments supporting biofuel policies are blending mandates and tax exemptions, however, other policies also can be used to support the development of nascent industries; *e.g.* grants to support the installation of production facilities, farmer premiums for the production of energy crops, and supporting R&D funding.

The recent **European Union (EU)** energy strategy has called for a substantial transformation of Europe's energy system based on a more secure, sustainable and low-carbon economy, with the commitment to achieve, by 2030, at least 27% share of renewables and 40% greenhouse gas (GHG) emissions reduction relative to emissions in 1990 [20]. In this context, the EU has set a cap of 7% on the final consumption of biofuels produced from agricultural crops in favor of advanced biofuels produced from non-food materials, including algae [21]. Under the Renewable Energy Directive (RED, 2009/28/EC) [22], the European Commission (EC) promotes the use of biofuels and bioenergy to accomplish various climate and energy targets to be met in the European Union (EU) by 2020 (also known as the 20-20-20 targets). These targets include: i) a reduction in GHG emissions of at least 20% compared to 1990 levels; ii) a final energy consumption of 20% derived from renewable sources, including biofuels and bioenergy, among others; iii) a reduction in primary energy use of 20% compared with the projected levels, to be achieved by improving energy efficiency. To guarantee the sustainable use of biofuels and bioenergy, the RED established mandatory sustainability criteria [23]. Among them, a minimal threshold of GHG saving from the use of biofuels of 35% has to be achieved. From 2017, the GHG emissions saving from the use of biofuels must be at least 50% and, from 2018, it must be at least 60% from the use of biofuels produced in new installations [22]. Algae are among a number of different technologies pursued by EC to contribute to the RED.

In the **United States of America (USA)**, the Energy Independence and Security Act of 2007 (EISA) expanded the national Renewable Fuel Standard (RFS) [24] by increasing and diversifying the qualifying biofuel alternatives as well as increasing the contribution of renewable fuels to the total liquid transportation fuel mix. In particular, EISA aimed to increase the supply of alternative fuels and set a target for the use of 136 billion liters (36 billion gallons) of renewable fuels, including advanced and cellulosic biofuels and biomass-based diesel, by 2022. EISA defines four categories of renewable fuel, with specified minimum GHG reduction thresholds that must be met to qualify under each category. EISA requires $\geq$ 20% GHG reduction for any renewable fuel production facility constructed after 2007, 50% reduction for advanced biofuels, 50% reduction for biomass-based diesel, and 60% reduction for cellulosic biofuels. All of these are measured against the 2005 average petroleum baseline. Having achieved significant success through 2014, primarily through expansion of conventional corn grain (starch-based) ethanol production, the majority of biofuels production growth remaining in the program is to be fulfilled by advanced biofuels, which include biomass-based diesel and cellulosic biofuels. Implementation of the RFS requires the US Environmental Protection Agency (EPA) to estimate the life cycle GHG emissions for renewable fuels pathways to determine their eligibility for the available RFS fuel categories. EISA defines life cycle GHG emissions as "*the aggregate quantity of GHG emissions (including direct emissions and significant indirect emissions such as those from [feedstock production and any associated] land use changes), related to the full fuel life cycle, including all stages of fuel and feedstock production and distribution, from feedstock generation or extraction through the distribution and delivery and use of the finished fuel to the ultimate consumer, where the mass values for all GHGs are adjusted to account for their relative global warming potential*" [25].

In the US, the Clean Power Plan (CPP) is a recently (2015) enacted policy agreement, which is hailed as the first policy designed to set a national limit on carbon emissions from powerplant [26]. The final vision of the plan is to reduce CO_2 emissions by 32 percent from 2005 levels. This legislative rule was established in late 2015, ahead of the **United Nations Climate Change Conference of the Parties (COP 21)**. The agreement will be ratified by 55 countries, representing at least 55 percent of global greenhouse emissions.

China's commitment to renewable energy is stated in its most recent five-year plan (FYP) [27]. The Chinese government strengthened the 12th FYP's efforts to address China's severe environmental degradation and build its clean energy, green manufacturing, and environmental services sectors. Ten out of the 25 priority targets in the 13th FYP are related to the environment, and all ten are included as part of the 13th FYP's 13 binding targets that must be achieved by 2020. These targets establish caps for energy use and ambitious goals for city air quality, carbon dioxide intensity, and reduction of soil and water contamination.

Much of the research done on algae to bioenergy is contributed by partnerships between universities, commercial entities, and research institutes. The National Basic Research Development Program has funded collaborations to investigate energy production from microalgae, and the National Science and Technology and Support program has funded research on the cultivation and feedstock development for algae [28]. Microalgal oil production has been the focus of institutions like the Ocean University of China and Tsinghua University, and industry partners. The Dalian Institute of Chemical Physics has also done work on microalgal hydrogen production [29]. Like South Korea, China investigates using algae for fixation of CO_2. Their CO_2-Microalgae-Fuels Project is funded by the National High-Technology Development Program [28].

South Korea currently sponsors four major research and development projects, which will be in effect until 2019. The Marine Bioenergy Development Project and Green Growth *Via* Marine Algal Biomass project focuses on clean energy production from marine algal biomass. Algae technology is also incorporated in the Global Frontier project, which investigates the mass cultivation of biomass, and the Carbon Capture and Sequestration 2020 project; which investigates technologies for CO_2 capture and storage [30].

In **Japan**, a biofuels program was established as part of the **Kyoto Protocol** [31]. The country has high standards for biofuels, especially bioethanol which must demonstrate a 50% reduction in CO_2 emissions compared to gasoline. Due to high food prices, however, this program is more focused on second and third generation biofuels, with feedstocks such as rice straw, woody biomass, and algae [31, 32]. *Pseudochoricystis* is the main algae used for technology development, and the Ministry of Agriculture, Forestry, and Fisheries is funding a joint research project with firms and universities to produce biofuel from algae. One of the main goals is to produce jet fuel from algae and demonstrate commercialization by 2020 [32].

In 2009, **Taiwan** passed the Renewable Energy Development Act, which allocated government funds to support the generation of 650-1000 MW of renewable energy by 2030. The Bureau of Energy, monitors policies and regulates the energy market accordingly. Funding is likewise distributed to universities, research laboratories, and industry partners. In dealing with policy, Taiwan has three main research areas of interest: selection of microalgal strains, cultivation strategies, and lipid extraction. Entities like National Taiwan Ocean University research feedstock; the university currently has a seed stock for microalgae, macroalgae, prokaryotes and eukaryotes. In terms of industry, Far Easter Bio-Tec Co. partners with Taiwan's China Steel Co. to investigate emissions reduction through fixation of flue gas by algae. The Industrial Technology Research

Institute's Green Energy and Environment Research Labs also studies topics relevant to production of algae biofuel, such as cell disruption and nutrient starvation [33].

In **India**, a National Policy on Biofuels supports research in the first, second, and third generation biofuels [34]. This support can be seen in the creation of the Vivekananda Institute of Algal Technology, which investigates topics such as algal treatment of wastewater and biofuel production from phytoplanktonic algae (diatoms). Universities like the University of Madras, Chennai, research biogas production and algae cultivation [35]. Throughout 2008-2009, a National Algal Biofuels Network was launched to support algal biofuels research, however; research progressed slowly and the program has since diminished in scope [36].

Brazil is known for its production of bioethanol from sugarcane, however it also supports the production of biofuels from algae. The National Research Council and National Fund for Research Projects both finance bioenergy research, and in conjunction with other federal, state, and private initiatives, finance bioenergy research in Brazil. Much of the research on algae has been done at universities, like the University of Campinas. The government funds research through FAPESP 2008/57873-8 on algae growth for its potential to mitigate CO_2 emissions from ethanol fermentations [37]. The Centro de Pesquisas Leopoldo Américo Miguez de Mello, known as CENPES, is the unit of the Research Center of Petrobras responsible for research and development (R & D) and basic engineering of the company and is partnered with the Federal University of Rio Grande do Norte operating a microalgae pilot plant [38].

Mexico's Secretariat of Energy and its National Council for Science and Technology supply the majority of funding for algae biofuels research. Many universities in the country are involved with research on biocatalysts and feedstocks for the second and third generation biodiesel production, which includes microalgae. Funding is also available for the production of hydrogen from algae biomass, and algae cultivation using wastewater [39]. Most of these projects are still at their early stages [40].

GLOBAL SUPPORT FOR RESEARCH AND COMMERCIAL DEPLOYMENT

There has been global government support for the development of algal biomass and algae-derived biofuels and bioproducts, mainly through public-private partnerships. This support is mainly motivated by a need to implement policies to i) reduce each country's dependence on importing fossil fuels (energy security and independence) and ii) mitigate GHG emissions through reduced fossil fuel usage (environmental health and security).

In the US, in the context of the RFS and EISA, there has been considerable support to study the feasibility of algal biofuel production in the US. A large part of this work was financially supported by the US Department of Energy (DOE), where a strategic multiyear program plan (MYPP) [41] has been established to provide a roadmap for an economically viable algae biofuel or algal biorefinery industry. The information presented here is a summary of projects funded and installed, with most of the data provided by media, websites and personal communication.

The American Recovery and Reinvestment Act (ARRA) of 2009 infused a significant amount of funding for supporting the Algae Research Community. As a leading example, in 2009 the **National Alliance for Algal Biofuels and Bioproducts (NAABB)** was funded for $44M. Specific outcomes from this large multi-year project range from basic advances in algal biology such as genomic sequencing of production strains, development of a new open pond cultivation system, demonstrated use of low-energy harvesting technology, to the development of a new hydrothermal liquefaction conversion pathway for algal biomass upgrading. Alongside NAABB, three integrated biorefinery (IBR) demonstration plants were funded, at a combined total of almost $97M, respectively for the algal technology developers Solazyme, Algenol and Sapphire. **Solazyme Inc.** (now **TerraVia**) worked for the development of heterotrophic algae that can convert cellulosic sugars into diesel fuel. This demonstration plant has a capacity to process daily 13 metric tons of dry lignocellulosic feedstocks, including switchgrass, corn stover, wheat straw, and municipal green waste, transforming it through an industrial deconstruction and fermentation process to produce algal oil, which can then be converted into FAME biodiesel or renewable (hydrocarbon) diesel. The biofuels produced by this project target reducing lifecycle GHG emissions by 90%, with the facility having the capacity to produce 300,000 gal yr^{-1} of purified algal oil. **Algenol Biotech LLC**. developed an integrated pilot project involving photosynthesis driven algal conversion of solar energy and CO_2 to ethanol and the delivery of a photobioreactor system that can be economically scaled to enable commercial production. This project utilizes a hybrid cyanobacteria species to directly secrete ethanol within a closed photobioreactor. It is worth noting that following the collapse in oil prices, Algenol has shifted their approach towards algae for food and feed, and increased emphasis on using their approach to achieve carbon capture and fresh water production until global oil prices recover. **Sapphire Energy Inc**. constructed a demonstration-scale project to construct and operate a 120 ha (300 acre) algae cultivation farm and conversion facility in Columbus, New Mexico, to produce renewable bio-crude (for subsequent upgrading to jet and diesel fuels). The target capacity of this plant is 1 million gallons per year of finished product, or 100 barrels of green crude algal oil per day. The biofuels produced are intended to achieve a 60-70% GHG reduction over the traditional fossil fuels being displaced.

Three additional consortia were funded in the US shortly thereafter, totalling another ~$25M: The **Consortium for Algal Biofuels Commercialization (CAB-Comm),** led by the University of California, San Diego, had a main goals of improved algal feedstock protection, nutrient utilization and recycling; and genetic tools. The outcomes of this project included increasing algal biomass productivity, creating new advanced biotechnology tools, and commercializing co-products with industrial partners. The **Sustainable Algal Biofuels Consortium (SABC)**, led by Arizona State University's Arizona Center for Algae Technology and Innovation (AzCATI) in Mesa, AZ worked towards the main goal of developing multiple biochemical routes to hydrolyze and convert untreated or pretreated whole algal biomass, oil extracts, and algal residues; and determine the acceptability of algal biofuels as replacements for petroleum-based fuels. A key outcome from this project was the development of a novel approach for fractionating algal biomass and converting each fraction to higher-value fuel products. The **Cornell Marine Algal Biofuels Consortium,** led by Cornell University and Cellana, Inc., focused on large-scale production of marine microalgae for biofuels and products. This consortium utilized the large-scale algae production facility operated by Cellana in Kona, Hawaii, to develop integrated design cases for the production of higher-value products alongside advanced biofuel production. Highlighted technical accomplishments included developing two novel algal strains well suited for large-scale production and demonstrating an improved annual operating reliability of 350 days per year.

Since 2012, two additional large consortia were funded in the US, with a focus on developing user facilities for long-term cultivation trials across the country in order to support a more rapid transition from lab to outdoor production systems and thereby reduce the risk to budding commercial operations. The **Algae Testbed Public Private Partnership (ATP^3)** is a 5 year (starting in 2012) $15M project, lead by AzCATI in Arizona. The objectives of this project were to establish collaborative open testbeds that increase stakeholder access to scale up facilities as well as collect and publish high-impact data on long-term outdoor cultivation trials. These testbed sites are located at universities and companies across the southern US, specifically at ASU (Arizona), CalPoly (California), Georgia Institute of Technology (Georgia), Florida Algae (Florida) and Cellana (Hawaii). Similarly, a **Regional Algal Feedstock Testbed (RAFT)**, a $5M project, was established in 2013 with the goal of creating long-term cultivation data to understand, de-risk and thereby promote increased algal biomass production. The RAFT's four testbeds are located in Texas, New Mexico, Washington and Arizona.

Similar to the US, the **Canadian government** has actively supported domestic algal biofuel industry activities through funding various projects. In 2008 - 2011, the Canadian government invested approximately 5M Canadian dollars (CAD) in research led by Canada's National Research Council (NRC) to produce algal fuels on a large scale in Canada under a program called the **National Bioproducts Program Algal Biofuels Initiative**. Some of the activities conducted through this program included an international collaboration between NRC Canada, NREL, Sandia National Laboratories, and PNNL on microalgae strain collection, site modelling for the co-deployment of algal biorefineries with Canadian wastewater treatment plants, and early hydrothermal liquefaction (HTL) experiments with sugar kelp biomass collected from the east and west coasts of North America. This bilateral collaboration was conducted from 2010 - 2012 under the framework of Canada-US Clean Energy Dialogue.

The Canadian federal government also invested over 377,000 CAD in November 2010 in R&D conducted by **Solarvest (PEI), Inc**., a subsidiary of Solarvest Bioenergy Inc., to generate hydrogen through algae production. In 2011, a large inter-institutional project, the **Algal Carbon Conversion (ACC) Flagship Program**, was established by Canada's National Research Council to mitigate carbon emissions through the commercial scale cultivation microalgae linked to large final CO_2 emitters. The ACC program addresses a number of factors which influence the commercial potential of ACC technologies, including identifying the most appropriate algae strains for industrial deployment, increasing the productivity and reducing energy costs of photobioreactors, identifying ways to reduce energy costs for processing algal biomass and assisting in the development of high-value, sustainable products from algal biomass. In the spring of 2016, NRC and its industry partners began operation of a pilot scale ACC facility linked to a cement manufacturing plant in Southern Ontario.

European support for the development of algae as bioenergy feedstocks was mainly driven by the European Commission's (EC) FP7 program, and this support has been continued through the Horizon 2020 program. Horizon 2020 was implemented as a means to drive economic growth and create jobs, stating that research and development is an investment in the future, with the science it supports focused on tackling societal challenges such as secure, clean and efficient energy.

The **AQUAFUELS project** (2009) was the very first European Union funded action in the field of algae biomass production for the creation of energy, in this specific case for the production of biofuels. AQUAFUELS was coordinated by the European Biodiesel Board and included partners form 12 European countries. Over the course of 18 months, the partners put together their efforts and their scientific, legal, industrial and technical knowledge in order to understand the real potential of algae and other aquatic biomass as feedstocks for biofuels.

Aligned with the EU's renewable energy targets, the EC is participating in the funding of three large-scale industry-led projects aimed at demonstrating the production of algal biofuels along the whole value chain at a 10 ha scale, spanning algae strain selection to cultivation and production, algal oil extraction, biofuel production and biofuel testing in transportation applications, with a minimum productivity target of 90 dry solid tonnes per hectare per year. The total cost for the three projects is about €31 million and the corresponding EC contribution amounts to about €20 million. The three large currently funded, collaborative R&D projects under the **AlgaeCluster** umbrella (algaecluster.eu) are **InteSusAl**, coordinated by CPI, UK; **All-Gas**, coordinated by Aqualia, Spain; **BioFAT**, coordinated by A4F, Portugal. **InteSusAl** strives to demonstrate an integrated approach to generate biofuels from algae in a sustainable manner on an industrial scale. This project aims to demonstrate an optimized approach to produce biofuels from algae in a sustainable manner on an industrial scale. It will integrate high quality research that has previously been undertaken at national and international levels, both publicly supported and privately financed. InteSusAl has so far built a 1 hectare pilot facility in Olhão, Portugal, based on a mixture of fermenters and photobioreactors. The system concept involves the recycling of waste glycerol from the transesterification stage of biodiesel production, and the fermenters contribute CO_2 to the photobioreactors. **All-Gas** targets 10 ha of microalgae cultivation and use for biofuels production lead by Aqualia, the third-largest private water and wastewater company in the world, consisting of seven partners and supported by a scientific advisory board to provide the knowledge and experience for this challenging endeavor. All-Gas is based around the concept of using a mixture of algae and bacteria to clean wastewater and produce fuel. All-Gas has built a 1 hectare pilot facility in Chiclana de la Frontera, Spain, and will soon expand to 3.5 hectare system. **BioFAT** aims for 10 ha demonstration of cultivation and conversion to biofuels, a microalgae-to-biofuel demonstration project targeted at both biodiesel and ethanol production, and will integrate the whole algae process value chain from algae optimized for growth and starch and/or oil accumulation, to downstream biofuel production processes, including a biorefinery approach. CO_2 from industrial fermentation will be used as a renewable carbon source. BioFAT has so far constructed two ½ hectare pilot facilities, one in Pataias (Portugal) and one in Camporosso (Italy). A 10 ha facility is currently being designed for construction in 2017.

In addition, the EU more recently initiated support for 5 more projects that seek to demonstrate the feasibility of an algae-based bioenergy: **AlgaeBioGas** aims to demonstrate algal treatment of biogas digestate; **DEMA** aims to demonstrate a technology for direct production of ethanol from microalgae; **D-Factory** targets demonstration of large scale (100s ha) cultivation of *Dunaliella salina;* **EnAlgae** operates 9 pilot facilities for micro- and macroalgae cultivation; **Fuel4ME** targets the pilot scale production of biofuels from algal lipids.

Along with numerous government-supported projects, a large number of commercial entities are supporting algae (both micro- and macroalgae) production and research. The strategies employed range widely, from open pond cultivation to photobioreactors in the area of phototrophic cultivation to large-scale aerobic fermentors for heterotrophic production of algae. Commercial facilities are either for algal biomass production, *i.e.*, both phototrophic and heterotrophic microalgae cultivation installations, or for macroalgae production.

Table 1. Summary of commercial and research operations working towards commodity algae-based (both micro- and macroalgae) products globally, separated by region and by commercial installation. Fermentation includes predominantly heterotrophic cultivation companies, Suppliers include cultivation systems, measurement and general equipment manufacturers, Research includes large government supported academic and public private partnerships projects and consortia. N/A = No information available.

	Total	Europe	North America	Asia	Oceania	Middle East
Commercial	306	166	105	26	2	7
PBR	50	32	14	1	N/A	3
Raceway	50	32	11	4	1	2
Combined PBR and Raceway	12	5	7	N/A	N/A	N/A
Fermentation	13	4	5	4	N/A	N/A
Unknown cultivation method	160	85	55	17	1	2
Suppliers	21	8	13	N/A	N/A	N/A
Research	94	50	27	9	5	3
Total	400	216	132	35	7	10
Shut operations	50	28	22	N/A	N/A	N/A

At intermediate scales, there are research projects underway to support the development of a bioenergy industry based on the production of algae. A summary of currently funded research projects and commercial operations worldwide is shown in Table **1** and visually represented in (Fig. **2**). They describe the state of the algae industry globally. The data is divided between research and corporate entities, and is based on the year of last known website update. Incorporated in company data are businesses involved in producing any algae-derived higher volume commercial products, including biofuels, skin care products, nutrients, and animal feedstock. The research subsection incorporates government funded projects, as well as universities and national laboratories. The compilation of this list was achieved through browser searches, and by reading company websites. While this list is undoubtedly incomplete, it gives a good sense of general trends worldwide and within specific geographic regions. This actual number of commercial entities involved with the algae industry is estimated to be > 2000 with more than 600 companies in China (Chinese Algae Association [42]), mostly farming *Spirulina*, *Haematococcus* and *Nannochloropsis* and over 200 small farms in France, covering mostly *Spirulina* production. The data compiled here provides an overview of data that is available publicly and is used to separate out the commercial operations by year of operation and by region. Similarly, the research operations are listed by year and region (Table **1**).

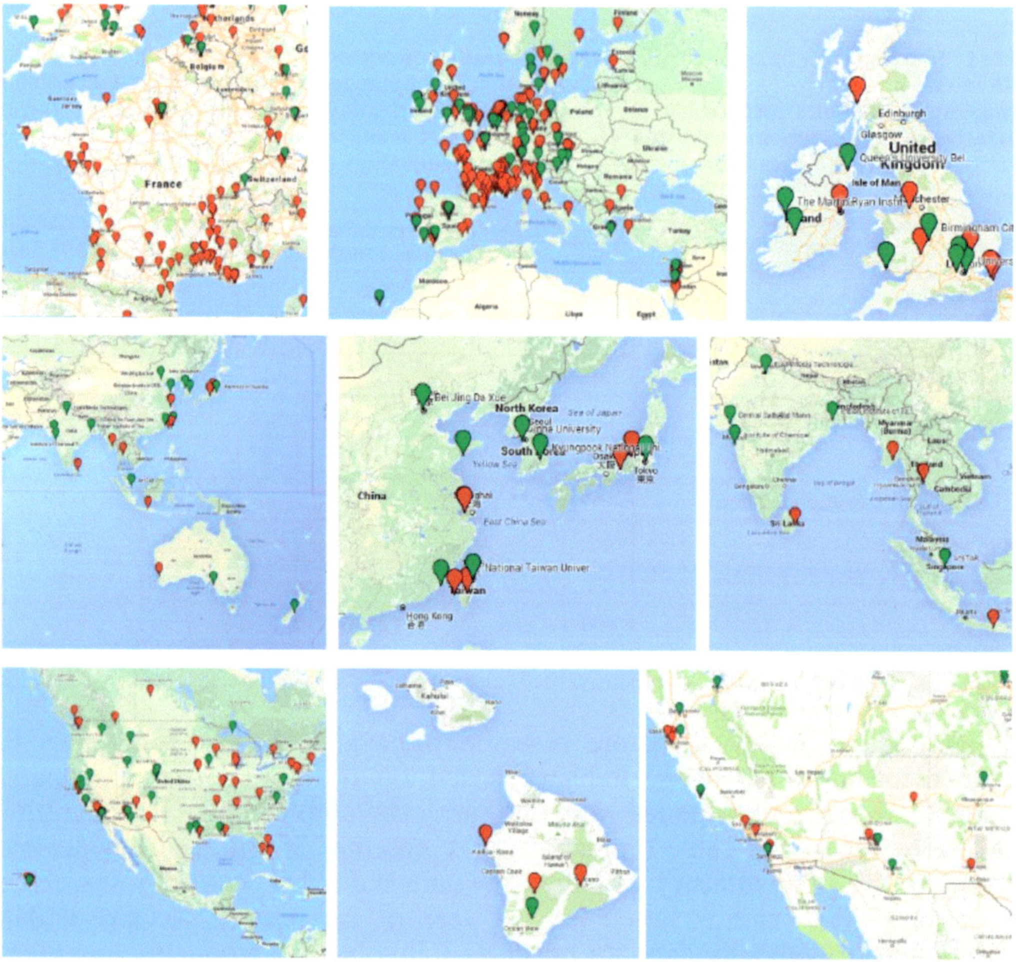

Fig. (2). Overview of global commercial and research operations; red balloons represent commercial operations, green balloons show research and demonstration projects.

For example, in Europe, there are significantly more distinct research projects than in North America, particularly in the United States [43]. In terms of end products, companies based in Europe are more likely to be working with microalgae and producing food, feed, nutraceuticals and biofuels/bioenergy. Though there has been a shift recently where a large number of companies, also in the US are changing their focus towards higher value products such as food and feed, moving away from fuels. In contrast, many businesses in Asian countries focus on the production of seaweed (macroalgae): the majority of algae related corporations in this region produce seaweed as a food crop. In the Asian coastal region, the majority of the algae industry produces seaweed as a food crop. Seaweed has a historically established place in the region's economy and aquaculture industry, however a large percentage of the world's micro algae is grown in the region as well [44 - 46]. There are companies and research groups in Asia working on converting algae into biofuels using both seaweed and microalgae, though the publications and patent applications around algae and biofuels are dominated by European and American groups [47].

CONCLUSION

The challenges encountered by the bioenergy field in the current global energy climate are significant and primarily driven by low petroleum prices accompanied by an increase in the demand. There is potential for algae to contribute to the future bioeconomy both as feedstock for biofuel production, but also as food, feed and other bioproducts. This apparent competition for market applications, along with economical pressure, is reflected in the changing shift in focus of a large number of commercial algae production facilities towards higher value and smaller market size products. A large number of projects and innovations are directly the result of public private partnerships between government funding agencies and commercial global entities and are primarily driven by the quest for energy security and greenhouse gas (in particular CO_2) emissions mitigation. Recent years have seen an increase in the number of commercial production facilities as well as public-private support for research projects. If this trend remaining, significant progress can be made towards bringing algae to the forefront of the future bioeconomy while contributing to the carbon mitigation global challenge.

CONFLICT OF INTEREST

The authors declare no conflict of interest, financial or otherwise.

ACKNOWLEDGEMENTS

We are grateful for fruitful discussions with Dr. James McMillan, NREL, Dr.

Vítor Verdelho, European Algae Biomass Association, Dr. Matt Carr, Algae Biomass Organization and Dr. John Benemann on the potential for large-scale deployment of algae for bioenergy applications and global commercial installed algae infrastructure. This work was supported by the U.S. Department of Energy under contract no. DE-AC36-08- GO28308 with the National Renewable Energy.

REFERENCES

[1] P.J. Williams, and L.M. Laurens, "Microalgae as biodiesel & biomass feedstocks: Review & analysis of the biochemistry, energetics & economics", *Energy Environ. Sci.,* vol. 3, pp. 554-590, 2010. [http://dx.doi.org/10.1039/b924978h]

[2] U.S. Energy Information Administration, International Energy Outlook 2016, 2016. http://www.eia.gov/outlooks/aeo/pdf/0383(2016).pdf

[3] "EPA, National Research Council, Climate Stabilization Targets", p. 3., (n.d.).

[4] US DOE, 2016 Billion-Ton Report: Advancing Domestic Resources for a Thriving Bioeconomy,, Volume 1: Economic Availability of Feedstocks. M. H. Langholtz, B. J. Stokes, and L. M. Eaton, 2016. https://energy.gov/sites/prod/files/2016/12/f34/2016_billion_ton_report_12.2.16_0.pdf

[5] S. Pacala, and R. Socolow, "Stabilization wedges: solving the climate problem for the next 50 years with current technologies", *Science,* vol. 305, no. 5686, pp. 968-972, 2004. [http://dx.doi.org/10.1126/science.1100103] [PMID: 15310891]

[6] International Energy Agency, Technology Roadmap: Biofuels for Transport, Paris, France, 2011.

[7] L.G. Elliott, C. Feehan, L.M. Laurens, P.T. Pienkos, A. Darzins, and M.C. Posewitz, "Establishment of a bioenergy-focused microalgal culture collection", *Algal Res.,* vol. 1, pp. 102-113, 2012. [http://dx.doi.org/10.1016/j.algal.2012.05.002]

[8] H.C. Greenwell, L.M. Laurens, R.J. Shields, R.W. Lovitt, and K.J. Flynn, "Placing microalgae on the biofuels priority list: a review of the technological challenges", *J. R. Soc. Interface,* vol. 7, no. 46, pp. 703-726, 2010.
[http://dx.doi.org/10.1098/rsif.2009.0322] [PMID: 20031983]

[9] R.H. Wijffels, and M.J. Barbosa, "Barbosa, An Outlook on Microalgal Biofuels,", *Science,* vol. 329, no. 5993, pp. 796-799, 2010.

[10] P.T. Pienkos, and A. Darzins, "The promise and challenges of microalgal-derived biofuels", *Biofuels Bioprod. Biorefin.,* vol. 3, pp. 431-440, 2009. [http://dx.doi.org/10.1002/bbb.159]

[11] R. Davis, D. Fishman, E. Frank, and M. Wigmosta, "Wigmosta, Renewable Diesel from Algal Lipids: An Integrated Baseline for Cost,", Emissions, and Resource Potential from a Harmonized Model, Golden, CO, 2012.

[12] M.E. Huntley, Z.I. Johnson, S.L. Brown, D.L. Sills, L. Gerber, and I. Archibald, "Demonstrated large-scale production of marine microalgae for fuels and feed", *Algal Res.,* vol. 10, pp. 249-265, 2015. [http://dx.doi.org/10.1016/j.algal.2015.04.016]

[13] R.L. White, and R.a. Ryan, "Long-Term Cultivation of Algae in Open-Raceway Ponds: Lessons from the Field", *Ind. Biotechnol. (New Rochelle N.Y.),* vol. 11, pp. 213-220, 2015. [http://dx.doi.org/10.1089/ind.2015.0006]

[14] M.K. Lam, and K.T. Lee, "Microalgae biofuels: A critical review of issues, problems and the way forward", *Biotechnol. Adv.,* vol. 30, no. 3, pp. 673-690, 2012. [http://dx.doi.org/10.1016/j.biotechadv.2011.11.008] [PMID: 22166620]

[15] Y. Chisti, "Constraints to commercialization of algal fuels", *J. Biotechnol.,* vol. 167, no. 3, pp. 201-214, 2013.

[http://dx.doi.org/10.1016/j.jbiotec.2013.07.020] [PMID: 23886651]

[16] W. Farooq, and M. Moon, "Effect of harvesting methods on the reusability of water for cultivation of chlorella vulgaris, its lipid productivity and biodiesel quality", *Algal Res.,* vol. 8, pp. 1-7, 2015. [http://dx.doi.org/10.1016/j.algal.2014.12.007]

[17] J. Yang, M. Xu, X. Zhang, Q. Hu, M. Sommerfeld, and Y. Chen, "Life-cycle analysis on biodiesel production from microalgae: water footprint and nutrients balance", *Bioresour. Technol.,* vol. 102, no. 1, pp. 159-165, 2011. [http://dx.doi.org/10.1016/j.biortech.2010.07.017] [PMID: 20675125]

[18] R. Pate, G. Klise, and B. Wu, "Resource demand implications for US algae biofuels production scale-up", *Appl. Energy,* vol. 88, pp. 3377-3388, 2011. [http://dx.doi.org/10.1016/j.apenergy.2011.04.023]

[19] U.S. Department of Energy, 2016 Billion-Ton Report, 2016. https://energy.gov/sites/prod/files/2016/12/f34/2016_billion_ton_report_12.2.16_0.pdf

[20] European Commission, Communication from the Commission to the European Parliament, the Council, the European Economic and Social Committee and the Committee of the Regions, A policy framework for climate and energy in the period from 2020 to 2030, Brussels, Belgium, 2014.

[21] European Commission, DIRECTIVE (EU) 2015/1513 OF THE EUROPEAN PARLIAMENT AND OF THE COUNCIL of 9 September 2015 amending Directive 98/70/EC relating to the quality of petrol and diesel fuels and amending Directive 2009/28/EC on the promotion of the use of energy from renewable,, 2015.

[22] European Commission, Renewable Energy Directive 2009-28-EC Annex 1,, 2009.

[23] A. Agostini, J. Giuntoli, A. Boulamanti, and L. Marelli, *Carbon accounting of forest bioenergy,* 2014.

[24] http://www.epa.gov/otaq/fuels/renewablefuels/index.htm, (n.d.).

[25] EPA, EPA Lifecycle Analysis of Greenhouse Gas Emissions from Renewable Fuels,, EPA-420-F-10-006. (2010) 3.

[26] USEPA, *Clean Power Plan for Existing Power Plants.* (n.d.).

[27] US-China Economic and Security Review Commission, 2016 Report to Congress, Washington, DC, 2016.

[28] F.W. Bai, C.G. Liu, H. Huang, and G.T. Tsao, "Biotechnology in China III: Biofuels and bioenergy", *Adv. Biochem. Eng. Biotechnol.,* p. 128, 2012.

[29] S. Huo, and R. Dong, "Available resources for algal biofuel development in China", *Energies,* vol. 4, pp. 1321-1335, 2011. [http://dx.doi.org/10.3390/en4091321]

[30] J.W. Hong, S.W. Jo, and H.S. Yoon, "Research and development for algae-based technologies in Korea: a review of algae biofuel production", *Photosynth. Res.,* vol. 123, no. 3, pp. 297-303, 2015. [http://dx.doi.org/10.1007/s11120-014-9974-y] [PMID: 24496987]

[31] T. Koizumi, *Biofuel Programs in East Asia: Developments, Perspectives, and Sustainability, Environ.* Impact Biofuels, 2011.

[32] M. Iijima, Biofuels Annual: Japan Focuses on Next Generation Biofuels, USDA Gain Rep. (2013). https://gain.fas.usda.gov/Recent%20GAIN%20Publications/Biofuels%20Annual_Tokyo_Japan_7-2-2013.pdf

[33] R.Y. Yuan, and C.S Hsiung, "Biofuels Production and Microalgal Production in Taiwan", *12th Int. Conf. Clean Energy,* 2012

[34] Government of India, *National Policy on Biofuels.*India, 2008.

[35] M. Kumar, and M.P. Sharma, "Status of biofuel production from microalgae in India", *J. Integr. Sci.*

Technol., vol. 2, pp. 72-75, 2014.

[36] S. Hemaiswarya, R. Raja, I.S. Carvalho, R. Ravikumar, V. Zambare, and D. Barh, "An Indian scenario on renewable and sustainable energy sources with emphasis on algae", *Appl. Microbiol. Biotechnol.,* vol. 96, no. 5, pp. 1125-1135, 2012. [http://dx.doi.org/10.1007/s00253-012-4487-0] [PMID: 23070650]

[37] M. Henrique, L. Silveira, M. Siika-aho, and K. Kruus, *Biofuels in Brazil,* 2014.

[38] L.B. Bacellar Mendes, and A.B. Vermelho, "Allelopathy as a potential strategy to improve microalgae cultivation", *Biotechnol. Biofuels,* vol. 6, no. 1, p. 152, 2013. [http://dx.doi.org/10.1186/1754-6834-6-152] [PMID: 24499580]

[39] G. Montero, M. Stoytcheva, M. Coronado, C. García, J. Cerezo, and L. Toscano, "Biofuels - Status and Perspective", In: *An Overv.* Biodiesel Prod. Mex, 2015, pp. 413-424.

[40] D. Berman, and L. Chavez, *Uncertainty on the Future of Mexican Biofuels,* 2012.

[41] U.S. Department of Energy, Bioenergy Technologies Office Multi-Year Program Plan,, Washington D.C, 2016. https://energy.gov/sites/prod/files/2016/07/f33/mypp_march2016.pdf

[42] http://zaoye.org.cn/a/zaoyezuzhi/20150826/273.html, Chinese Algae Association, (2016).

[43] E.D. Elliott, "Why the United States Does Not Have a Renewable Energy Policy", *Environ. Law Report.,* vol. 43, pp. 10095-10101, 2013.

[44] W.L. Zemke, and M. Ohno, "World seaweed utilisation: An end-of-century summary", *J. Appl. Phycol.,* vol. 11, pp. 369-376, 1999. [http://dx.doi.org/10.1023/A:1008197610793]

[45] X. Li, J. Li, Y. Wang, L. Fu, Y. Fu, and B. Li, "Aquaculture Industry in China: Current State, Challenges, and Outlook", *Rev. Fish. Sci.,* vol. 19, pp. 187-200, 2011. [http://dx.doi.org/10.1080/10641262.2011.573597]

[46] Y.K. Lee, "Commercial production of microalgae in the Asia-Pacific rim", *J. Appl. Phycol.,* vol. 9, pp. 403-411, 1997. [http://dx.doi.org/10.1023/A:1007900423275]

[47] A.A. Adenle, G.E. Haslam, and L. Lee, "Global assessment of research and development for algae biofuel production and its potential role for sustainable development in developing countries", *Energy Policy,* vol. 61, pp. 182-195, 2013. [http://dx.doi.org/10.1016/j.enpol.2013.05.088]

CHAPTER 11

Current Production of Microalgae at Industrial Scale

Luis G. Ramírez-Mérida[1], **Leila Queiroz Zepka**[2] and **Eduardo Jacob-Lopes**[2,*]

[1] *Applied Biotechnology Center, Department of Biology, University of Carabobo, Universidad Avenue, 2002, Valencia, Edo. Carabobo, Venezuela*

[2] *Food Science and Technology Department, Federal University of Santa Maria, UFSM, Roraima Avenue 1000, 97105-900, Santa Maria, RS, Brazil*

Abstract: The microalgae culture has a potential for obtaining bioproducts that can be used in various commercial segments. The microalgae industry generates interest in the investors who recognize good performance and suitability in the production of various microalgal products. In recent years, research projects and microalgal industries worldwide have been increased with a focus on the agriculture, aqualculture, human and animal nutrition, bulk and fine chemicals, cosmetics and biofuels. The microalgae biomass production has a limited scale, due to the problems presented in the scaling of the production process. The industrial production is dominated by open systems, allowing growth only of some microalgal species. The consolidation of microalgae-based products has been focused on improvements in the culture systems and production costs, making this industrial sector more competitive.

Keywords: Microalgal biotechnology, Photobioreactors, Heterotrophic microalgal bioreactors, Bioproducts.

INTRODUCTION

The microalgae cultivation currently has many challenges due to which the microalgal biotechnology is advancing steadily. Microalgae represent a potential source of active compounds, which are used in different types of industries [1]. The production of microalgae in natural lakes has been used by Asian and African populations as a food source for centuries [2]. The industrial production seeks to reproduce, in a controlled manner, the microalgal biomass in open and closed systems. The industrial production at large scales started from 1960 in Asia. These facilities were developed for the purpose of human and animal food. In subsequent years, a great number of microalgal companies were established in the

* **Corresponding author Eduardo Jacob-Lopes:** Food Science and Technology Department, Federal University of Santa Maria, UFSM, Roraima Avenue 1000, 97105-900, Santa Maria, RS, Brazil; Tel/Fax: +55 55 3220 8822; E-mail: jacoblopes@pq.cnpq.br

José Carlos Magalhães Pires (Ed.)

world market to produce solutions for agriculture, aqualculture, human and animal nutrition, bulk and fine chemicals, cosmetics and biofuels [3, 4].

Currently, there are a variety of companies that use different technologies in several scales in operation in the world, including facilities in Asia, North America and Europe [5].

Conversely, microalgae-based products is currently limited to a few thousand tons per year, mainly produced in open ponds. Closed-photobioreactors available in the market do not represent an optimal solution to business needs, because although the volumetric and superficial productivity are higher than in open photobioreactors, performance does not approach the theoretical levels obtained in laboratory scale [6]. Success in the commercial production of large-scale microalgae depends on several factors, mainly used microalgal culture system. In this sense, several research and development (R&D) efforts have been carried out in recent years aiming to circumvent these barriers. So, the aim of this chapter is to present a current view of the commercial production of microalgae cultivation at large scale worldwide.

ELEMENTS OF MICROALGAL CULTURE

The cultivation of photosynthetic microorganisms represents a useful tool in industrial biotechnology, so that adequate supply of light energy is the most important factor to consider. This feature significantly affects the configuration of the bioreactors for the microalgae cultivation. In addition, other parameters such as temperature, gas transfer, pH, nutrient availability and agitation must be provided appropriately in the microalgal culture [7]. Fig. (**1**) summarizes the main aspects related to microalgal cultures.

The photosynthetic microalgae culture depends essentially on the supply of light energy to cells. Photoinhibition phenomena are frequent in culture systems with improper illumination, causing significant losses in kinetic performance in bioreactors. In addition to the quantitative aspects, the qualitative nature of the incident light in culture systems should also be considered. Natural or artificial lighting can be used depending on the characteristics required in culture systems. Aspects such as location, climatic variation, photoperiod are the main aspects to consider in cultivation systems with natural illumination [8]. In culture systems with artificial lighting, the energy source and cost are the main elements to consider for implementation of the lighting system. Alternatively, the heterotrophic cultures fully eliminate all aspects related to energy lighting [9].

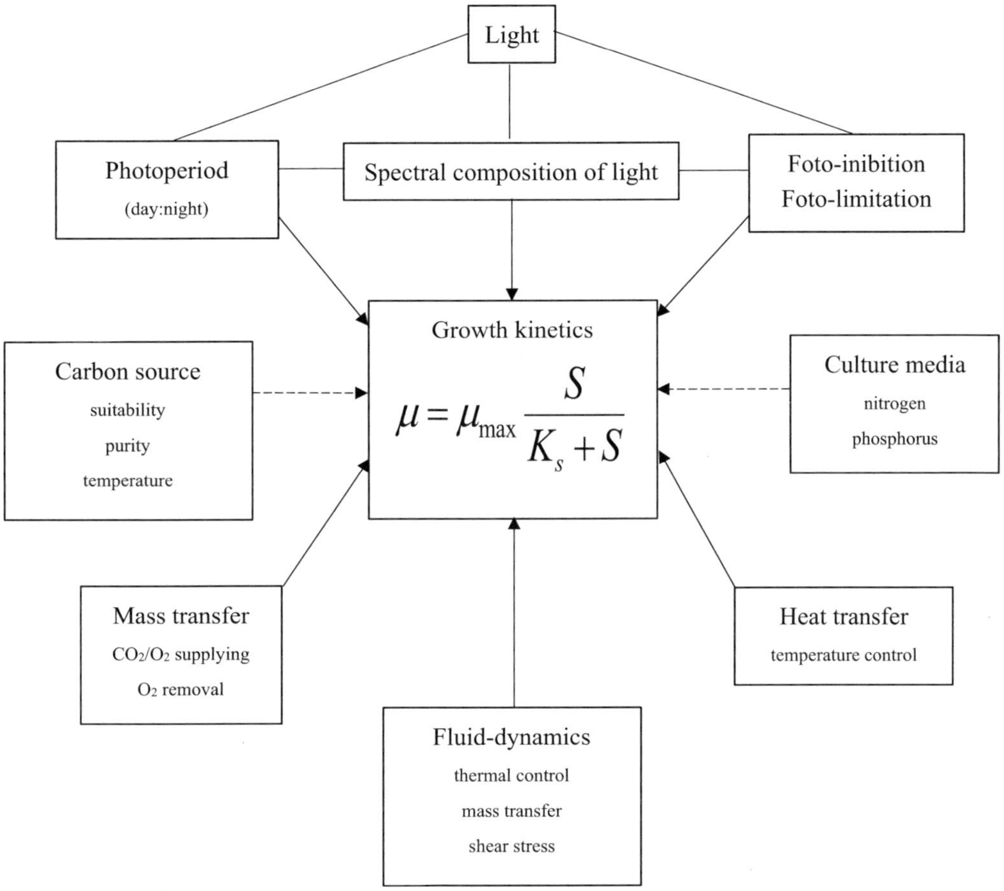

Fig. (1). Elements of microalgal culture.

Temperature is another important factor associated with microalgae culture. Generally, the optimal temperature of microalgal cultivation occurs in mesophilic conditions (25-35°C), however some thermophilic strains are temperature resistant (60 °C). The release of energy during cellular metabolism in the great majority of microalgae culture systems produces variation in the microenvironment. However, heat exchangers can be installed to control and maintain constant temperature in the bioreactors [10].

The chemical composition of microalgae biomass is basically carbon, nitrogen and phosphorus in proportions close to 50, 8 and 1%, respectively. Consequently, the availability of these nutrients is essential for cell growth in bioreactors. CO_2 is the main carbon source used in the photosynthetic culture. The concentration of

CO_2 in the atmosphere (0.04%) is not sufficient to keep intensive cultures. Thus, this gas is supplied through stationary sources (compressed CO_2, primary pattern mixtures and industrial flue gases) that allow obtaining inlet concentrations in photobioreactor between 3 and 15%. There is a wide expectation on the availability and use of CO_2 from industrial flue gases to keep photosynthetic microalgae culture. This approach has faced several difficulties in its implementation because these gases are characterized by high levels of impurities (CO, CH_4, NOx, SOx, H_2 and particulate matter). In addition, the gas temperature can reach over 1000 °C, so these two factors make the biological processes unfeasible [11]. Besides providing the carbon, feeding a carbon dioxide enriched air stream helps in pH control and mixing [12].

In relation to the pH, the microalgal culture is normally favored in neutral/alkaline range (pH 7-9), although some species have optimum growth at pH 4 (*Chlorococcum littorale*) and pH 10 (*Spirulina platensis*). There is a complex relationship between the concentration and type of carbon source (CO_2 or exogenous sources of organic carbon) with the pH in photosynthetic and heterotrophic cultures, respectively. This relationship is extended to the source of nitrogen, because NH_3 and NH_4^+ are strongly affected by the pH of the culture medium. Thus, pH control devices must be integrated into the bioreactors to maintain the optimum conditions of this parameter in the process [13].

Furthermore, the heterotrophic cultures require exogenous carbon sources. Glucose is the substrate traditionally used, although other carbohydrates or organic carbon sources, such as glycerol and agroindustrial wastes, have been efficiently used for this purpose [14]. The carbon source being efficiently assimilable by the microalgae, this selection is based on a purely economic criterion.

After carbon, nitrogen is the second limiting element in microalgal cultures. Nitrate is the usual source of nitrogen although urea, ammonium and organic sources are assimilated by several species. Nitrogen sources from wastewaters have been used efficiently to keep microalgal cultures. Phosphorus is the third element of greater demand in these systems. The reactive phosphorus is the most easily assimilable species, although the acid-hydrolysable phosphorus and organic phosphorus are efficiently used. Like the other elements, the use of wastewaters rich in phosphorus are considered a potential sources. Other trace elements may be required for microalgal growth, depending on the individual requirements of each specie [15].

Finally, agitation is a necessary operation in the microalgae culture, since the spatial uniformity of the reaction vessels is ensured favoring the exposure of cells to light, heat transfer and thermal stratification, and improve the gas exchange. A suitable mixture reduces the formation of cell aggregates, which increase the overall inefficiency of the bioreactor. Although the mixture is essential for proper development of the process, under certain circumstances it may lead to shear stress phenomena which damages and inhibits the microalgal growth. Microalgae bioreactors are normally equipped with pneumatic aeration systems and mechanical agitation or a combination of these systems [16].

MICROALGAL BIOREACTORS

The construction and installation of microalgal facilities should consider the microalgal strain, the cost and availability of land, the labor, energy, water, nutrients, climate conditions and characteristics of the final product. Based on this, microalgae culture systems are classified as follow [17]:

1. Open Systems
   i. Circular ponds
   ii. Raceways ponds
2. Closed Systems
   i. Tubular photobioreactors
   ii. Bubble column
   iii. Air-lift
   iv. Big-bag
   v. Flat-plate
   vi. Fermenters (heterotrophic microalgal bioreactor)

The open systems dominate the commercial operation of large-scale microalgae because they are relatively easier to scale-up and more economic. These culture systems have two main configurations: circular pond and raceway pond. These ponds are 15 to 30 cm deep with circular or oval geometry and mechanical agitation systems allowing constant mixing of the culture medium. Although they are simple to construct, these systems are inefficient due to limitations of light and CO_2 transfer, usually leading to low productivities. Additionally, large losses of water by evaporation lead to significant variations in the culture medium composition, in addition to being susceptible to contamination. All these features condition the use of these systems to microalgae strains that grow in extreme conditions and selective environments [18].

Advances in bioreactor engineering seek to solve the main limitations of open systems through the development of closed photobioreactors. Closed systems allow greater control of biological reaction conditions and can be used for photosynthetic, heterotrophic and mixotrophic cultures [19]. The tubular photobioreactor is the main configuration which dominates the market of the closed systems, although the bubble column, air-lift, big-bag, flat-plate and fermenters (heterotrophic microalgal bioreactors) have been considered with industrial potential.

The tubular photobioreactors consist in a series of transparent tubes to 0.1 m in diameter constructed in borosilicate glass or transparent polymer, arranged in a horizontal or vertical sequence. The culture medium is continuously recirculated along the tubes, allowing the cells to be exposed to light energy. Geometrically, this arrangement aims to achieve increased surface/volume ratio (S/V) to obtain proportions that allows high conversion rates of CO_2 in bioproducts. The length of tubes is determined by the tolerance to accumulation of oxygen in the culture medium by microalgae strains. However, this configuration is limited by factors such as the concentration gradients along the tubes, biofilm formation on the walls of the tubes and requiring high land area [20].

The flat-plate photobioreactors have received much attention for microalgal cultures due to the high S/V ratio. Various arrangements have been constructed to maximize light transfer to the cell, by decreasing the light path. The separation between plates varies between 2 to 4 cm. Although the flat-plate photobioreactors have an adequate kinetic performance, they are expensive and have many compartment and support devices that hinder to scale-up and operation [21].

The bubble column and air-lift photobioreactors are reaction vessels geometrically cylindrical, with height/diameter ratio (H/D) of up to 1:10, allowing to obtain high surface areas for uptake of light energy by the cells. Both systems have a gas diffuser at the base of the column. The air-lift differs in bubble column photobioreactors because it presents a concentric tube to the main tube that allows the liquid flows downward through the outer tube, which substantially improves the mixing and mass transfer rates. From the point of view of kinetics, these systems are efficient, although limited for scale-up due to the low working volume, so the arrangement of several units in parallel can improve the process [22].

As an alternative to reduce costs, the big-bag photobioreactors were developed. These photobioreactors consist in sterile bags plastic of up 0.5 m, with a gas diffusion system and suspended in appropriate structures. Although inexpensive and easy construction, the big-bag photobioreactos have problems in mixing and temperature control [23].

Finally, the fermenters represent an alternative for the heterotrophic microalgae cultivation because certain strains can use organic substrates as carbon source in the total absence of light. This feature prevents virtually all above limitations, as it allows the use of conventional perfect mixing bioreactors, usually bubble columns, with H/D ratio 1.0 to 1.5 [24]. They have a high kinetic performance and are easily scalable. The main disadvantage is that it prevents photosynthesis and require high levels of asepsis, in order to avoid competition with other heterotrophic microorganisms.

OPERATIONAL MODES TO MICROALGAE CULTURE

Whichever type of bioreactor is used, the microalgal bioprocesses are conducted in conventional operation modes (batch, fed-batch, semi-continuous and continuous) of the industrial biotechnology.

The microalgal cultures in batch mode have a dynamic in which cells pass through the stages of adaptation, logarithmic and stationary, over a period ranging from one to four weeks depending on the microalgal strain. The separation of the biomass from the culture medium occurs when the population reaches the maximum cell density. After this period the asepsis and sterilization process of bioreactors is performed to start the next growing cycle. In view of simplicity, operational flexibility, security in terms of external contamination and ease of control, this mode is best used for industrial purposes. However, it is inefficient from the viewpoint of productivity [25].

The fed-batch cultures consist of adding a limiting nutrient for microalgae culture, and where the products remain in the bioreactor until the end of the growing cycle. Thus, it seeks to provide sufficient nutrients to cells to prolong the growth, reducing the limitation for substrate with increased biomass production and bioproducts. These operational strategies have been widely used for feeding heterotrophic microalgal cultures with exogenous organic carbon sources [26].

Additionally, the semicontinuous culture is characterized by periodic removal of a volume of culture medium in the bioreactor, being fed with an equal volume of culture medium. This operation is repeated until the system collapses. This operation mode is easy to operate but unstable and labor intensive, with low potential for large-scale application [27].

Finally, the continuous cultures are characterized by constant addition of culture medium in parallel with the simultaneous removal of an equivalent volume inside the bioreactor. This operational mode allows to maintain the culture in steady-state. Consequently, during the residence time, the cells are in optimal growth, providing high productivities in the process. However, continuous culture need bioreactors and complex accessories, resulting in increased capital expenditure for industry. In addition, the control of the external contamination is the main limitation of this operational mode [28].

INDUSTRIAL PRODUCTION OF MICROALGAL-BASED PRODUCTS

The production of microalgae at industrial scale emerged as a commercial opportunity aiming to gain market share in the bioactive molecules segment. Japan was the first country to cultivate microalgae on a large scale in the 1960s through *Spirulina* and *Chlorella* cultivations [3]. At this stage, the single-cell protein was the main product that targeted the industry, with applications directed to food and prophylactic use. Later, the pigments production emerged through the cultivation of the *Dunaliella salina* and *Haematococcus pluvialis* with focus on β-carotene and astaxanthin applied to human food, animal feed and chemical-pharmaceutical industry. Currently, a huge efforts have been made in research and development to produce biofuels from single-cell oil. The annual production of microalgal biomass dry is about 19,000 ton, generating an estimated USD 5.7 billion [29].

In this sense, the main companies producing microalgae-based products, and technological routes are summarized as following:

Far East Microalgae Industries was founded in 1976 in southern tip of Taiwan in the Pacific Ocean. The industrial facility have a total area of 140,000 m^2 with an annual production capacity of 1,000 ton for *Chlorella*, and 200 ton for *Spirulina*. The microalgae are cultured in open circular ponds using pure mineral water. The bioproducts include the single-cell protein (powder and tablets), microalgae extracts used as ingredients in health food formula as dietary supplements, besides cosmetics products, micro-encapsulated aquaculture feed and special fluorescent diagnostics [30].

Earthrise is the largest *Spirulina* production industry in the world. It is located in Sonoran desert of southeastern California, USA. Culture facilities have 43.7 hectares with 37 raceway ponds of 5,000 m^2 in size. Clean, fresh water and nutrients are daily added to feed the algae, mixed by 50-foot paddlewheels in each pond. Carbon dioxide similar to the one used in carbonated drinking water is pumped directly into the ponds. In addition, nitrogen, potassium, iron and trace elements with high purity are added into culture medium. During the growing season, ponds are harvested every day and a 15-minute journey from ponds to powder is estimated. The post-processing consist in a system composed of two screen filters (nutrient rich water is recycled back to the ponds) followed by spray-drying. The annual production is estimated to be 450 ton [31].

Hainan-DIC microalgae is a company founded in 1977 in Hainan, China. The industrial facility is 100,000 m^2 in size. The production process is based on traditional cultivating of the *Spirulina* into open raceways ponds to produce *Spirulina* powder for food, tablets, coated tablets, food colorant (phycocyanin extract) and *Spirulina* for animal feed. The installed capacity correspond to 350 ton per year [32].

Taiwan *Chlorella* Manufacturing Company was established in 1964. The company has two industrial facilities with 50,000 m^2 of total area in the cities of Chung Li and Lung Chun, Taiwan. The production system is based on conventional open circular ponds. The company has a complete line of *Chlorella*-based products, including powder, tablets, extract fluid, extract powder and noodles. Annual production is estimated to be 320 ton [33].

Cyanotech is a microalgae producing company located in Kona Coast of the island of Hawaii, USA. The company produces astaxanthin and *Spirulina Pacifica*. The microalgae are cultured in shallow, open ponds (approximately 20 cm deep) adjacent to the Pacific Ocean. A combination of fresh water and supplemental deep ocean water (pumped up from depth of 2000 feet) is used to fill the ponds. The other major components required for growing *Spirulina* are food-grade sodium bicarbonate and carbon dioxide. Food grade fertilizers are used and the deep sea water is added as a rich source of minerals and trace elements. Paddle wheels agitate the water, ensuring even exposure of the algae to the sun. The post-processing consist in a system composed by separation screens, vacuum washing system and ocean chill drying [34].

Solazyme is a microalgae-based products industry located in Moema, São Paulo, Brazil. The company grows native microalgae strains in dark fermentation tanks. The platform is feedstock flexible and can utilize a wide variety of renewable plant-based sugars, such as sugarcane-based sucrose, corn-based dextrose, and sugar from other sustainable biomass sources including cellulosics that are converted in various types of single-cell oil. The bioproducts include oleochemicals, functional fluids, fuels (biodiesel and jet fuel), food ingredients (single-cell protein and single-cell oils), feed for aqualculture (DHA rich biomass) and specialty personal care (cosmetic microalgae powders and oils). The annual production is estimated to be 120 ton [35].

AlgaEnergy started its industrial operations in 2009. The production plant is located in Arcos de la Frontera, Cadiz, Spain, and occupied a total area of 1.0 hectare with an installed capacity of 1,000,000 liters. The production system is based on vertical tubular photobioreactors using several microalgae strains that include *Nannochloropsis*, *Isochrysis*, *Spirulina*, *Chlorella*, *Dunaliella* and *Haematococcus*. Industrial flue gases are used as carbon source. The product portfolio included solutions for aquaculture, agriculture, human and animal nutrition, cosmetics and biofuels. Annual production is estimated to be 100 ton of dried biomass [36].

Alltech Algae in Winchester, Kentucky, USA opened in early 2011. The company chose the heterotrophic route for production of biomass of the *Schizochytrium* applied to aquaculture, pet and livestock. The production system is based on yeast components from which the algae gain a number of critical nutrients to grow in fermenters, producing high levels of fat content with a focus on DHA omega-3 fatty acid. Annual production is estimated to be 20,000 ton [37].

Algatechnologies is a microalgae company established in Kibbutz Ketura, Israel. Their cultivation technology include closed and semi-closed bioreactors with *Haematococcus pluvialis*, distributed in 6.0 hectares of arid desert land, hosted in a modular array of tubular glass photobioreactors with more than 500 km long. The entire system is automated and controlled remotely. The bioproducts include the astaxanthin derivates such as oleoresins, beadlets, powders, emulsions and capsules, applied to dietary supplements, food and beverage formulations, cosmetics, animal health and aquaculture. The annual production is estimated to be 3.0 ton of astaxanthin [38].

Cellana is an algae-based bioproducts company located in Kona, Hawaii, USA. The company produces a line of omega-3, EPA and DHA oils, animal feed, and biofuel feedstocks from several microalgae strains. The technological route consist in a hybrid system, using a combination of open seawater ponds and closed-photobioreactors (horizontal tubular photobioreactors) in a two-stage process, requiring only carbon dioxide, sunlight, seawater, and nutrients. Once harvested, conversion technologies are used to obtain the single-cell oils. The residual defatted biomass and carbohydrates is destinated to make a fishmeal or animal feed supplement for the aquaculture and animal feed markets. The production scale is pilot, in a demonstration facility of 2.5 hectares with an annual production close to 2 ton [39].

Algomed began their activities in 1999 in Klötze, Germany. The company has an industrial facility of 1.2 hectare. The production system is based on 500 km long vertical tubular photobioreactors using *Chlorella vulgaris* culture in a greenhouse. The installed capacity is around 600,000 litres, distributed over 20 separate units. The products marketed consist in tablets based on *Chlorella vulgaris* powder free of additives or enriched with iron and vitamin C [40].

Nature Beta Technologies is a company located in Eilat, Israel. The industrial facility is 100,000 m^2 in size. The production process is based on cultivating the unique line of *Dunaliella bardawil* into open raceways ponds to produce carotenoid-based products.

Finally, other relevant commercial operations include the natural harvest of the *Aphanizomenon flos-aquae* in the Lake Klamath, Oregon, USA, with an annual production of 500 ton [41], and the natural harvest of the *Spirulina* in Twyn Taung Crater Lake, Myanmar, with an annual production of 200 ton [42].

In addition to these major producers, dozens of other companies are in operation worldwide, representing industrial small scale operations or pilot/demonstrative operations. The fact is that the algal industrial production is under way.

COMPARISON BETWEEN TECHNOLOGICAL ROUTES (OPEN *VS.* CLOSED SYSTEMS)

Independent of the target product, algae-based products are produced exclusively in open and/or closed systems. Both technological strategies have advantages and disadvantages.

The highest percentage of industrial microalgae processes are carried out in open systems. Although limited only to certain microalgal strains, the open systems have advantages from the economic point of view and are easily scalable. However, many of these facilities are in regions with climatic conditions that do not allow production all year. The real productivity is less than the theoretical, being difficult to control the environment to which the microalgae cultures are exposed. There are projects of raceway pond in greenhouses under artificial light providing greater control of variables. Controlled environment for microalgae culture is in continuing development aiming to reduce the contaminants and improve the culture conditions, leading to better yields [43]. The main types of open systems include circular and raceway ponds. The choice of one of them depends on the land area, climatic conditions, access to water and end use of the product. The susceptibility to contamination with other microorganisms in open systems is reduced by culturing of extremophile strains, that support high salinity (*Dunaliella*), high alkalinity (*Spirulina*) or high nitrogen concentrations (*Chlorella*) [44]. Finally, the pond depth is essential to allow good lighting, because the higher the water column, the less light is available to cells. The ideal depth for open ponds is up to 30 cm, resulting in a low working volume by area unit, limiting the commercial application to several products and processes [45]. These bioreactors are not suitable for CO_2 biofixation or production of commodities products, being cost effective only for high-value products [46].

Conversely, closed photobioreactors have been industrially used, since that many strains of interest need more control to growth. These systems provide certain advantages during the production process. To increase the diversity of microalgae products, there is a great interest in developing photobioreactors for the production of chemical compounds with high added value which require a high degree of assepsy [47].

The main configuration of closed systems include the tubular photobioreactors (vertical or horizontal), that has many advantages because the fundamental principle in its design is to increase the amount of light available to the cell, reducing the light path. However, the low working volume by area unit makes the implementation of these photobioreactors in the industrial difficult. The cost effectiveness is only applicable for high-value products, since the capital expenditure is very high [48]. Aiming to circumvent these barriers, big-bag photobioreactors have been implemented in some demonstrative operations. Although with a low capital expenditure, the dramatic increase in temperature of culture medium significantly reduces the kinetic performance of culture [3].

Another industrial possibility of the closed photobioreactors is the fermenters (heterotrophic microalgal bioreactor). Your operation follows the well-known criteria of the classical biotechnology processes, with basis in an exogenous organic carbon source [9]. The costs of the carbon source, the competition with others heterotrophs and the impossibility of light induced synthesis are the main drawbacks of this technological route. Table **1** show a comparison between open and closed systems for microalgae culture.

Table 1. Comparison of different parameters in microalgae culture systems.

Parameters	Open Systems	Closed Systems
Land space for installation	High	Moderate
Building costs	Low	High
Microalgae strains	Limited to a few species, basically extremophiles	Wide range of use, including genetically modified strains
Contamination risk	High	Low
Efficient use of light	Low Ratio surface-volume (<10m^{-1})	High Ratio surface-volume (>50m^{-1})
Mixing efficiency	Poor	Homogeneous
Dependence on artificial light	Low	Moderate-High
Gas transfer	Poor	Moderate-High
CO2 loss	High	Low-Moderate
Buildup of oxygen	Low	High
Temperature problems	Low	High
Evaporative water loss	High	Low
Residence time	High	Moderate
Energy consumption	Low	High

FRONTIERS IN INDUSTRIAL PHOTOBIOREACTORS

Photobioreactors are a potential and promising technology to conduct industrial importance processes and production of valuable products. The main problem to be solved in the design and configuration of efficient photobioreactors is that to obtain sufficient light that is essential to maintain high S/V ratio [24].

The configurations of photobioreactors (currently in operation) are extrapolations of the conventional chemical reactors. These systems (as well as their variations) have been developed and adapted for meeting the peculiarities of bioprocesses involving microalgae. In laboratory and pilot scale, these configurations have met the basic requirements of the photosynthetic processes. In the meantime, with the increase in scale, they present operational problems in the balance of lighting demand with the working volume. In addition to these limitations, CO_2 loss during injection, inadequate mixing, poor mass transfer and the accumulation of photosynthetic oxygen are the main faults of the conventional photobioreactors [48].

For this reason, the design criteria H/D ratio is a fundamental parameter to build industrial photobioreactors. Despite the development of some existing photobioreactors with good productivity and biomass yields, the technological future of the photobioreactors will be associated with hybrid reactors [49].

The hybrid photobioreactors are a combination of at least two types of different photobioreactors. These configurations aim to compensate the drawbacks caused by the limitation of S/V ratio and scale-up of open and closed conventional photobioreactors. These systems are based on a proper H/D ratio, generating configurations of reactors with heavy workloads in contrast to very long tubes or shallow ponds [49, 50].

For illustration, the comparison of a vertical column photobioreactor with an H/D ratio=10 (ratio used for a bubble column) and other with an H/D ratio=1.5 (ratio ideal) is considered. If we propose the construction of a photobioreactor with a length of 1.5 m, we would have working volumes of 26.5 L (H/D=10) in contrast to 1,178 L (H/D=1.5) [24].

This would lead to some advantages such as less use of land area with more culture volume, lower operating costs, greater stability of the reactor against mild winds or quakes. This type of configuration would generate higher volumetric production, reducing the capital and operational expenditure. So, the keeping of the low H/D ratio (below of 2.0) is a fundamental aspect for future development in the industrial photobioreactors, mainly when the system is applied to CO_2 sequestration, to wastewater treatment and the final product is a commodity.

This type of configuration is only feasible due to the capacity of some microalgae of storing sufficient energy to sustain cell growth for periods in the dark, without affecting the rate of photosynthetic metabolism. The light energy storage capacity can reach up to 2 h [8].

THE BIOECONOMY OF MICROALGAE-BASED PROCESSES AT INDUSTRIAL SCALE

The production costs of the microalgae-based process vary depending on variables such as culture system, microalgae strain, downstream processing operations, type and quality of the end product, among others.

The core of the process is the photobioreactor, and its performance will represent significant portion of the capital and operational expenditure. Table **2** summarizes some companies specializing in the supply of commercial photobioreactors.

The techno-economic analysis of the microalgae-based processes have been shown to be economically infeasible scenarios for commodities products and for processes with environmental purposes (CO_2 mitigation and wastewater treatment) [51 - 53]. This infeasibility is related mainly to the reduced scalability of the photosynthetic and the high operational costs of the heterotrophic processes. According to Wijffels *et al.* [54], the technological routes are immature and need to be fully developed, implying the need for a large effort in R&D. These authors reported that microalgal biotechnology will be competitive and commercially attractive by 2020.

Table 2. Worldwide manufacturers of photobioreactors.

Company	Location	Configuration	Capacity (m^3)	Cost (k USD)	Cost (USD/L)
MicroBio Engineering	USA	Raceway ponds	7 – 30	40 – 60	2 – 6
Schott AG	Germany	Tubular	1 – 10	50 – 400	40 – 50
Industrial Plankton	Canada	Tubular	1 – 2.5	35 – 60	24 – 35
Photon Systems Instruments	Czech Republic	Flat–panel	0.12	18	150
Biodiesel Business Academy	India	Tubular	0.08	8.4	100

In the analysis and cost estimate for designing a new process, almost all the decisions are impacted by the economic factors and, therefore, it is critical to study process economics. The major criteria to judge feasibility are preliminary design and economic potential estimation to be attained, and knowledge of the price of the final product is necessary for covering the costs involved. The feasibility of a process should be based on a benefit-cost ratio. Feasibility indicators such as economic equilibrium, profitability, rentability, and period of return on investment are the main parameters in use [55].

The production costs of the microalgal biomasses have been estimated at USD 5.71/kg for the tubular photobioreactor and US$ 8.18/kg for the flat panel photobioreactor [52]. Additionally, the production cost of the heterotrophic fermenters is close to USD 12.00/kg [51]. According to Wijffels *et al.* [54], the production cost of microalgae biomass may not be higher than USD 0.55 cents/kg (ideal theoretical price) for manufacturing of commodities products.

These issues show that there is still a long way to go until the microalgae-based processes are fully viable.

CONCLUSION

Current production of microalgae at industrial scale is a billionaire business. The microalgae business segment is currently very dynamic with several companies starting every year as well as several going out of business. The products from microalgae have currently three major possible forms, including pastes, dried powders and extracts. The microalgae production platforms are currently dominated by open systems, although this situation will be reversed in near future by massive inclusion of the closed systems at industrial scale. Finally, it should be considered that a new era has started with the interest of multinational companies.

CONFLICT OF INTEREST

The authors declare no conflict of interest, financial or otherwise.

ACKNOWLEDGEMENTS

The authors are grateful to National Council of Technological and Scientific Development of the Brazil (CNPq) by the financial support.

REFERENCES

[1] D.B. Rodrigues, E.M. Flores, J.S. Barin, A.Z. Mercadante, E. Jacob-Lopes, and L.Q. Zepka, "From waste to natural pigments: productions of microalgal carotenoids in agroindustrial wastewater", *Food Res. Int.*, vol. 65, pp. 144-148, 2014. [http://dx.doi.org/10.1016/j.foodres.2014.06.037]

[2] I. Priyadarshani, and B. Rathj, "Commercial and industrial applications of micro algae – A review", *J. Algal Biomass Utln.*, vol. 3, no. 4, pp. 89-100, 2012.

[3] M.A. Borowitzka, "Commercial production of microalgae: ponds, tanks, and fermenters", *J. Biotechnol.*, vol. 70, pp. 313-321, 1999. [http://dx.doi.org/10.1016/S0168-1656(99)00083-8]

[4] E. Stephens, I.L. Ross, and B. Hankamer, "Expanding the microalgal industry--continuing controversy or compelling case?", *Curr. Opin. Chem. Biol.*, vol. 17, no. 3, pp. 444-452, 2013. [http://dx.doi.org/10.1016/j.cbpa.2013.03.030] [PMID: 23587864]

[5] M.A. Borowitzka, "High-value products from micro — algae — their development and commercialization", *J. Appl. Phycol.*, vol. 25, pp. 743-756, 2013. [http://dx.doi.org/10.1007/s10811-013-9983-9]

[6] J. Pruvost, F. Le Borgne, A. Artu, J.F. Cornet, and J. Legrand, "Chapter Five - Industrial Photobioreactors and Scale-Up Concepts", In: *Advances in Chemical Engineering. Photobioreaction Engineering.* vol. Vol. 48. 1st ed. Elsevier: Amsterdam, 2016, pp. 257-310. [http://dx.doi.org/10.1016/bs.ache.2015.11.002]

[7] E. Jacob-Lopes, A.M. Santos, D.B. Rodrigues, M.C. Lui, C. Souza, D. Prudente, and L.Q. Zepka, *Bioprocesso de conversão de efluentes híbridos, biorreator heterotrófico, bioprodutos e seus usos, .*

[8] E. Jacob-Lopes, C.H. Gimenes Scoparo, L.C. Ferreira Lacerda, and T. Teixeira Franco, "Effect of light cycles (night/day) on CO2 fixation and biomass production by microalgae in photobioreactors", *Chem. Eng. Process.*, vol. 48, pp. 306-310, 2009. [http://dx.doi.org/10.1016/j.cep.2008.04.007]

[9] O. Perez-Garcia, F.M. Escalante, L.E. de-Bashan, and Y. Bashan, "Heterotrophic cultures of microalgae: metabolism and potential products", *Water Res.*, vol. 45, no. 1, pp. 11-36, 2011. [http://dx.doi.org/10.1016/j.watres.2010.08.037] [PMID: 20970155]

[10] P.H. Chiu, K. Soong, and C.N. Nathan Chen, *Cultivation of two thermotolerant microalgae under tropical conditions: Influences of carbon sources and light duration on biomass and lutein productivity in four seasons,* 2016. [http://dx.doi.org/10.1016/j.biortech.2016.04.045]

[11] E. Jacob-Lopes, and T. T. Franco, *From oil refinery to microalgal biorefinery,* 2013. [http://dx.doi.org/10.1016/j.jcou.2013.06.001]

[12] L. Ramírez-Mérida, Q.L. Zepka, and E. Jacob-Lopes, "Photobioreactor: tool for mass cultivation of cyanobacteria", *Ciencia y Tecnología,* vol. 6, no. 2, pp. 9-19, 2013.

[13] G. Markou, O. Depraetere, and K. Muylaert, "Effect of ammonia on the photosynthetic activity of Arthrospira and Chlorella: A study on chlorophyll fluorescence and electron transport", *Algal Res.*, vol. 16, pp. 449-457, 2016. [http://dx.doi.org/10.1016/j.algal.2016.03.039]

[14] E.C. Francisco, T.T. Franco, R. Wagner, and E. Jacob-Lopes, "Assessment of different carbohydrates as exogenous carbon source in cultivation of cyanobacteria", *Bioprocess Biosyst. Eng.*, vol. 37, no. 8, pp. 1497-1505, 2014. [http://dx.doi.org/10.1007/s00449-013-1121-1] [PMID: 24445336]

[15] A. Solovchenko, A.M. Verschoor, N.D. Jablonowski, and L. Nedbal, "Phosphorus from wastewater to crops: An alternative path involving microalgae", *Biotechnol. Adv.*, vol. 34, no. 5, pp. 550-564, 2016. [http://dx.doi.org/10.1016/j.biotechadv.2016.01.002] [PMID: 26795876]

[16] C.G. Khooa, M.K. Lam, and K.T. Lee, "Pilot-scale semi-continuous cultivation of microalgae Chlorella vulgaris in bubble column photobioreactor (BC-PBR): Hydrodynamics and gas–liquid mass transfer study", *Algal Res.*, vol. 15, pp. 65-76, 2016. [http://dx.doi.org/10.1016/j.algal.2016.02.001]

[17] R.N. Singh, and S. Sharma, "Development of suitable photobioreactor for algae production–A review", *Renew. Sustain. Energy Rev.*, vol. 16, pp. 2347-2353, 2012. [http://dx.doi.org/10.1016/j.rser.2012.01.026]

[18] P.M. Slegers, M.B. Lösing, R.H. Wijffels, G. van Straten, and A.J. van Boxtel, "Scenario evaluation of open pond microalgae production", *Algal Res.*, vol. 2, pp. 358-368, 2013. [http://dx.doi.org/10.1016/j.algal.2013.05.001]

[19] J.S. Deschênes, A. Boudreau, and R. Tremblay, "Mixotrophic production of microalgae in pilot-scale photobioreactors: Practicability and process considerations", *Algal Res.,* vol. 10, pp. 80-86, 2015. [http://dx.doi.org/10.1016/j.algal.2015.04.015]

[20] P.M. Slegers, R.H. Wijffels, G. van Straten, and A.J. van Boxtel, "Scenario analysis of large scale algae production in tubular photobioreactors", *Appl. Energy,* vol. 105, pp. 395-406, 2013. [http://dx.doi.org/10.1016/j.apenergy.2012.12.068]

[21] P.M. Slegers, R.H. Wijffels, G. van Straten, and A.J. van Boxtel, "Design scenarios for flat panel photobioreactors", *Appl. Energy,* vol. 88, pp. 3342-3353, 2011. [http://dx.doi.org/10.1016/j.apenergy.2010.12.037]

[22] A.P. Carvalho, L.A. Meireles, and F.X. Malcata, "Microalgal reactors: a review of enclosed system designs and performances", *Biotechnol. Prog.,* vol. 22, no. 6, pp. 1490-1506, 2006. [http://dx.doi.org/10.1002/bp060065r] [PMID: 17137294]

[23] M. Morweiser, O. Kruse, B. Hankamer, and C. Posten, "Developments and perspectives of photobioreactors for biofuel production", *Appl. Microbiol. Biotechnol.,* vol. 87, no. 4, pp. 1291-1301, 2010. [http://dx.doi.org/10.1007/s00253-010-2697-x] [PMID: 20535467]

[24] L. Ramírez-Mérida, L.Q. Zepka, and E. Jacob-Lopes, "Current Status, Future Developments and Recent Patents on Photobioreactor Technology", *Recent Patents on Eng.,* vol. 9, no. 2, pp. 80-90, 2015. [http://dx.doi.org/10.2174/1872212109666150206235316]

[25] V. Mortezaeikia, R. Yegani, and O. Tavakoli, *Membrane-sparger vs. membrane contactor as a photobioreactors for carbon dioxide biofixation of Synechococcus elongatus in batch and semi-continuous mode,* . [http://dx.doi.org/10.1016/j.jcou.2016.05.009]

[26] D. Morales-Sánchez, R. Tinoco-Valencia, M.A. Caro-Bermúdez, and A. Martinez, "Culturing Neochloris oleoabundans microalga in a nitrogen-limited, heterotrophic fed-batch system to enhance lipid and carbohydrate accumulation", *Algal Res.,* vol. 5, pp. 61-69, 2014. [http://dx.doi.org/10.1016/j.algal.2014.05.006]

[27] A.A. Henrard, M.G. de Morais, and J.A. Costa, "Vertical tubular photobioreactor for semicontinuous culture of Cyanobium sp", *Bioresour. Technol.,* vol. 102, no. 7, pp. 4897-4900, 2011. [http://dx.doi.org/10.1016/j.biortech.2010.12.011] [PMID: 21295968]

[28] E. Jacob-Lopes, S. Revah, S. Hernández, K. Shirai, and T.T. Franco, "X., “Development of operational strategies to remove carbon dioxide in photobioreactors", *Chem. Eng. J.,* vol. 153, pp. 120-126, 2009. [http://dx.doi.org/10.1016/j.cej.2009.06.025]

[29] M.H. Langholtz, A.M. Coleman, L.M. Eaton, M.S. Wigmosta, C.M. Hellwinckel, and C.C. Brandt, "Potential land competition between open-pond microalgae production and terrestrial dedicated feedstock supply systems in the U.S", *Renew. Energy,* vol. 93, pp. 201-214, 2016. [http://dx.doi.org/10.1016/j.renene.2016.02.052]

[30] FEMICO Far East Microalgae Industries, Co., Ltd., Nan Gang Dist., Taipei, Taiwan. [Online]. Available from: http://www.femico.com.tw/, [Accessed: 12th Jun. 2016].

[31] Earthrise Nutritionals, Irvine, California,, USA [Online]. Available from: http://earthrise.com/, [Accessed: 12th Jun. 2016].

[32] Hainan-DIC Microalgae Co., Ltd., Hainan Province,, People's Republic of China [Online]. Available from: http://www.dlt-spl.co.jp/, [Accessed: 12th Jun. 2016].

[33] Taiwan Chlorella Manufacturing Company, Ltd., Taipei, Taiwan [Online]. Available from: http://www.taiwanchlorella.com/, [Accessed: 12th Jun. 2016].

[34] Cyanotech Corporation kailua kona, Hawaii,, People's Republic of China [Online]. Available from:

http://www.cyanotech.com/, [Accessed: 12th Jun. 2016].

[35] Solazyme, South San Francisco, California, , USA [Online]. Available from: http://solazyme.com/, [Accessed: 12th Jun. 2016].

[36] AlgaEnergy, Alcobendas, Madrid, , Spain [Online]. Available from: http://www.algaenergy.es/, [Accessed: 12th Jun. 2016].

[37] Alltech, Winchester, Kentucky, , USA [Online]. Available from: http://www.alltech.com/, [Accessed: 13th Jun. 2016].

[38] Algatechnologies, Ltd., Kibbutz Ketura,, Israel [Online]. Available from: http://www.algatech.com/, [Accessed: 13th Jun. 2016].

[39] Cellana Inc., kailua kona, Hawaii,, USA [Online]. Available from: http://cellana.com/, [Accessed: 13th Jun. 2016].

[40] Algomed, Klötze,, Germany [Online]. Available from: http://www.algomed.de/, [Accessed: 13th Jun. 2016].

[41] Earths First Foods, [Online]. Available from: http://www.earthsfirstfoods.com/, [Accessed: 12th Jun. 2016].

[42] P. Clemens, and C. Walter, Microalgal Biotechnology: Integration and Economy, Ed. De Gruyter, pp 51-70, 2012.

[43] S. Li, D.H. Willits, C.L. Browdy, M.B. Timmons, and T.M. Losordo, "Thermal modeling of greenhouse aquaculture raceway systems", *Aquacult. Eng.,* vol. 41, no. 1, pp. 1-13, 2009. [http://dx.doi.org/10.1016/j.aquaeng.2009.04.002]

[44] E.M. Radmann, C.O. Reinehr, and J.A. Costa, "Optimization of the repeated batch cultivation of microalga Spirulina platensis in open raceway ponds", *Aquaculture,* vol. 265, no. 1–4, pp. 118-126, 2007.
[http://dx.doi.org/10.1016/j.aquaculture.2007.02.001]

[45] C. Gonzalez-Fernandez, B. Molinuevo-Salces, and M.C. García-González, "Open and enclosed photobioreactors comparison in terms of organic matter utilization, biomass chemical profile and photosynthetic efficiency", *Ecol. Eng.,* vol. 36, pp. 1497-1501, 2010.
[http://dx.doi.org/10.1016/j.ecoleng.2010.07.007]

[46] A. Saeid, and K. Chojnacka, "Evaluation of Growth Yield of Spirulina maxima in Photobioreactors", *Chem. Biochem. Eng. Q.,* vol. 30, no. 1, pp. 127-136, 2016.
[http://dx.doi.org/10.15255/CABEQ.2015.2269]

[47] A. San Pedro, C.V. González-López, F.G. Acién, and E. Molina-Grima, "Outdoor pilot production of Nannochloropsis gaditana: Influence of culture parameters and lipid production rates in flat-panel photobioreactors", *Algal Res.,* vol. 18, pp. 156-165, 2016.
[http://dx.doi.org/10.1016/j.algal.2016.06.011]

[48] L. Ramírez-Mérida, L.Q. Zepka, and E. Jacob-Lopes, "Why does the Photobioreactors Fail?", *J. Bioprocess. Biotech.,* vol. 5, p. 7, 2015.

[49] E. Jacob-Lopes, L.Q. Zepka, L. Ramírez-Mérida, M.M. Maroneze, and C. Neves, *Bioprocess for the conversion of carbon dioxide from industrial emissions, bioproducts use thereof and hybrid photobioreactor,* 2016.

[50] S. Schuessler, E.J. Roebroeck, S.H. Hazewinkel, and O.J. Hazewinkel, *Method and bioreactor for the cultivation of microorganisms,* 2014.

[51] Y.K. Lee, "Microalgal mass culture systems and methods: their limitation and potential", *J. Appl. Phycol.,* vol. 13, pp. 307-315, 2001.
[http://dx.doi.org/10.1023/A:1017560006941]

[52] N.H. Norsker, M.J. Barbosa, M.H. Vermuë, and R.H. Wijffels, "Microalgal production--a close look at

the economics", *Biotechnol. Adv.,* vol. 29, no. 1, pp. 24-27, 2011. [http://dx.doi.org/10.1016/j.biotechadv.2010.08.005] [PMID: 20728528]

[53] F.G. Acién, J.M. Fernández, J.J. Magán, and E. Molina, "Production cost of a real microalgae production plant and strategies to reduce it", *Biotechnol. Adv.,* vol. 30, no. 6, pp. 1344-1353, 2012. [http://dx.doi.org/10.1016/j.biotechadv.2012.02.005] [PMID: 22361647]

[54] R.H. Wijffels, M.J. Barbos, and M.H. Eppink, "Microalgae for the production of bulk chemicals and biofuels", *Biofuels Bioprod. Biorefin.,* vol. 4, pp. 287-295, 2010. [http://dx.doi.org/10.1002/bbb.215]

[55] E. Salehi, S.S. Madaeni, A.A. Shamsabadi, and S. Laki, "Applicability of ceramic membrane filters in pretreatment of coke-contaminated petrochemical wastewater: Economic feasibility study", *Ceram. Int.,* vol. 40, pp. 4805-4810, 2014. [http://dx.doi.org/10.1016/j.ceramint.2013.09.029]

CHAPTER 12

Environmental Applications of Microalgae: CO_2 Capture and Nutrient Recycling

Meng Wang*

Civil and Environmental Engineering, University of South Florida, Tampa, USA

Abstract: Nitrogen (N) and Phosphorus (P) in the wastewater are economical nutrient sources for algae growth. N and P can be assimilated into biomass. The mechanisms of N and P utilization by microalgae was discussed in this chapter. Diverse reactor configurations used for wastewater treatment and algae cultivation were developed to improve the biomass productivity and their settling ability. Enhanced algal-prokaryotic wastewater treatment systems are being designed to improve the wastewater treatment efficiency. A case study of a concentrated animal feeding operation (CAFO) integrating algae cultivation and anaerobic digestion indicated most N ad P are recycled in the form of biomass. The fixation of CO_2 by microalgae can be the alternative methods for the CO_2 mitigation. The CO_2 mitigation rate is affected by the mass transfer and biomass assimilation. Flue gas can be used to reduce the cost of source gases and mitigate greenhouse gases emission. Membrane sparging systems provide fine bubbles and way higher mass transfer efficiency than airlift reactors. Design that improves light penetration should be considered for the scale-up of the complex membrane sparging systems.

Keywords: CO_2 capture, Nutrient removal and recycling, Reactor configuration, Use of biomass, Wastewater treatment, Greenhouse gases emission.

INTRODUCTION

The energy cost of the water and wastewater treatment accounts for about 3-4% of energy consumption in the US, in which aeration is the most energy intense process [1]. Microalgae contain pigments that can use solar energy for their metabolism. When algal biomass undergoes autotrophic metabolism, inorganic carbon is used as the carbon source, and sunlight is the energy source. Sufficient light, either naturally or artificially applied to algae cultivation, is critical for massive algae production.

* **Corresponding author Meng Wang:** Civil and Environmental Engineering, University of South Florida, Tampa, USA; Tel: 813-974-2275; Fax: +1(813)-974-2957; E-mail: mwang3@usf.edu

José Carlos Magalhães Pires (Ed.)

The ability of CO_2, nitrogen (N) and phosphorus (P) uptake by algal photosynthesis draws the attentions of algae applications in environmental engineering. The N and P in the wastewater can be a good source for algae growth. The addition of CO_2 from waste gases can reduce the footprint of greenhouse gases (GHG) emissions.

Algal wastewater treatment system was first utilized as the water stabilization pond [2]. Recently, more engineered systems have been developed to improve the nutrient removal efficiency and biomass production. The typical reactor configurations used in the wastewater treatment include open pond systems closed photobioreactors and algal biofilm reactors. Large open ponds, many closed photobioreactors (PBR) and algal biofilm reactors have been built to cultivate algae by natural or artificial light to treat wastewater and to produce food supplements, lipid, enzymes, and biofuels [3 - 6]. However, several factors limit the algal growth rate. A large reactor volume will reduce light penetration efficiency inside the PBR. Biofilm formed on the surface of a closed PBR will also inhibit light penetration. Proper reactor design is required to improve the biomass production.

N AND P REMOVAL BY ALGAE

With the presence of light and CO_2, algae can utilize CO_2 as a carbon source and light as the source of energy for photosynthesis. During algae growth, N and P can be fixed into biomass. Harvested algae can potentially be used as biofuels, proteins, pharmaceuticals and food supplements [7 - 9]. When ammonium (NH_4^+) is the N source, the photosynthesis of algae in the presence of sunlight is expressed as [10]:

$$106CO_2 + 236H_2O + 16NH_4^+ + HPO_4^{2-} \rightarrow C_{106}H_{181}O_{45}N_{16}P + 118O_2 + 171H_2O + 14H^+ \quad (1)$$

The main mechanisms of N removal from wastewater include algal uptake and stripping of ammonia (NH_3) due to the increase in pH from photosynthesis [11]. The chemical equilibria of inorganic carbon are:

$$H_2O + CO_2 \leftrightarrow H_2CO_3 \leftrightarrow H^+ + HCO_3^- \leftrightarrow 2H^+ + CO_3^{2-} \quad (2)$$

When algae perform photosynthesis, CO_2 is consumed and equilibration shifts towards the left lead to the increase in pH in the surrounding medium. NH_3 may be evaporated at high pH. Algae can also use other types of inorganic N such as nitrate (NO_3^-). Nutrient removal by microalgae, if successfully implemented, may have several advantages over the conventional biological nutrient removal processes used in wastewater treatment plants (WWTPs): low capital and operational costs, no need for additional organic carbon for nutrient removal, simultaneous mitigation of CO_2 output, and production of oxygen during treatment process [11, 12].

Mechanism of N and P Utilization by Algae

N accounts for 1-10% dry weight of algae cell and the metabolism of N is very important for algae growth [13]. Microalgae can assimilate various N sources. In general, NH_4^+ is a preferred N source for algae because less energy is required for its metabolism compared to NO_3^- or urea. The main pathway for NH_4^+ incorporation into algae is catalyzed by glutamine synthase (GS; EC6.3.1.2) and glutamate synthase (GOGAT; EC 1.4.1.14). Another way to incorporate NH_4^+ into algae is catalyzed by glutamate dehydrogenase (GDH, EC1.4.1.2), which is activated under stress conditions [14].

NO_3^- is another major N source for algae growth. NO_3^- is metabolized through the reduction of NO_3^- to NH_4^+ through NO_3^- reductase and NO_2^- reductase. Different metabolite pathways are expressed under different conditions. The NO_3^- reductase is expressed when NO_3^- is the only N source while the activity is repressed when there is sufficient NH_4^+ in the environment [15].

Although NH_4^+ is readily assimilated by algae, free NH_3 at high concentration is toxic to microalgae. *C. vulgaris* is capable of growth at total ammonia nitrogen (TAN) concentrations from 10-1000 mg L^{-1}. However, growth was inhibited under less than 10 mg N L^{-1} or greater than 750 mg N L^{-1} [16].

The preference of N species is also affected by algal strains. NO_3^- instead of NH_4^+ was required for the maximum growth of *Dunaliella tertiolecta* [17]. Similar results were found by Li *et al.* [18] for the growth of *Neochloris oleoabundans,* an oil-rich algae. *Neochloris oleoabundans* grown in NO_3^- had the highest biomass productivity (0.133 g L^{-1} day^{-1}) and the highest lipid content of 38% compared with that grown in urea or ammonium bicarbonate [18]. The medium contained 3 mM $NaNO_3$ resulted in the highest cell lipid content while the NO_3^- concentration of 10 mM had the highest algal productivity [18].

The mechanisms of phosphorus removal are metabolism uptake by algae and adsorption and precipitation of inorganic phosphorus at pH over 8 [19, 20]. Li *et al.* [20] suggested that over 76% of phosphorus was removed by inorganic P sedimentation and less than 23% of phosphorus was assimilated by algae. A case study on N and P mass balance of the integration of algae cultivation and anaerobic digestion will be presented later.

Impact of N on Intercellular and Extracellular Compounds Production

Production of biodiesel requires the accumulation of lipids. With the presence of a catalyst, biodiesel is produced by transesterification of triacylglycerols (TAG) with methanol (or other alcohol) [17], and glycerol as a byproduct. The intercellular compounds of algae are dependent on the nutrient levels. N deficiency is regarded as the primary factor to stimulate the accumulation of lipid or carbohydrates in algal cells [12, 21]. However, phosphorus deprivation had little effect on the lipid accumulation due to the intercellular storage of phosphorus [17]. Limitation of N decreases the content of thylakoid membrane whose main content is glycolipid, activate acyl hydrolase and stimulates the hydrolysis of phospholipid, which lead to the accumulation of fatty acid acyl-CoA. At the same time, N limitation will activates the diacylglycerol acyltransferase, which converts fatty acid acyl-CoA into triglyceride [22]. Therefore, limitation of N not only increases the intercellular lipid content but also the triglyceride of algae [12, 22].

Meanwhile, the intercellular content of algae also affects the anaerobic digestibility of algae because of different methane yields from lipids, proteins, and polysaccharides. Anaerobic digestion of lipids yields the most methane and proteins are the second macromolecule type that have high methane production [23]. Therefore, the control of algae growth conditions is necessary to deplete the N in the medium and to get the desired end product from algae cultivation.

In addition to the intercellular compounds production, the extracellular polymeric substances (EPS) was also affected by the N concentrations in the growth medium. Wang *et al.* investigated the production of EPS by *Chlorella* and *Micratinium* grown in primary effluent and the combination of primary effluent and anaerobically digested centrate. Algae grown under high NH_4^+ concentration produced more extracellular protein than that grown in primary effluent [24, 25].

REACTOR CONFIGURATIONS

Algae used for wastewater treatment can be traced back to 1950s when stabilization ponds (WSPs) and high rate ponds were applied for wastewater treatment [2, 26]. Recently, more and more algal-based wastewater treatment systems are designed to improve the nutrient removal efficiency and biomass productivity. The reactors used for wastewater treatment can be divided into open pond systems, closed photobioreactors and immobilized systems (algal biofilm reactors). Selection of reactor configurations are critical because it is highly related to the biomass productivity, capital costs, effluent quality and biomass end product [27].

Open Pond Systems

The common types of open ponds include WSPs and high rate algal ponds (HRAPs). HRAPs were stemmed WSPs with the addition of paddle wheels to provide turbulence and improve biomass productivity [28]. Compared with the WSPs, the wastewater treatment efficiency was higher in HRAPs especially with N and P removal [29]. In addition to the carbon and nutrient removal, open ponds systems can also remove pathogens [30].

HRAPs are typical very shallow ponds (0.2-1m), lined with PVC or clay to prevent penetration into surrounding environment [5, 31]. The recommended horizontal velocity is between 0.09-0.3 m s^{-1} [5, 29, 32]. Another important parameter of HRAPs is the hydraulic residence time (HRT). Typical HRT for HRAPs is between 3-15 days [33]. Thus, large reactor volume is required for wastewater treatment. A large area is required for HRAPs, given the shallow depth required for light penetration. In full-scale applications, the area ranged from 1000 to 50000 m^2 [28, 34]. HRAPs are appropriate to sanitation in small rural areas due to the simple operation [35]. Besides, HRAPs are also applied in urban, industrial and agricultural wastewater treatment. For example, an HRAP for the treatment of high nitrate wastewater from aquacultural recirculation aquacultural systems improved the fish survival rates [36].

In general, open pond systems have low capital cost but large area requirement. The system controls are also relatively poor [27]. HRAPs usually have high evaporation rates (3-10 L m^{-2} d^{-1}), which could be partially controlled by the turbulence [33]. Besides, the total suspended solids (TSS) concentrations of the effluent are usually high because of the poor settleability of the biomass [33]. Operational strategies such as recycling settled biomass have been applied to HRAPs to improve the settling efficiency [5]. Algal harvesting technologies include chemical flocculation and coagulation, gravity sedimentation, dissolved air floatation (DAF), filtration and centrifugation [31, 37]. Among these technologies, DAF and coagulation are effective methods in algal harvesting [38 - 40].

Closed Photobioreactors

Flat plate and tubular photobioreactors are common modules of closed photobioreactors. The arrangement of modules can be horizontal, vertical, inclined or vertical [41]. Flat plates have larger illumination surface thus higher algal productivity [42]. However, scale-up of flat plate photobioreactors are more complicated than tubular reactors [42, 43]. Scale up of tubular photobioreactors can be achieved by increasing the length of tubular modules or by connecting individual modules in various arrangement [43].

Compared with open ponds, closed photobioreactors have better system control, higher biomass productivity and lower evaporation rates [44]. However, the construction is more expensive and transparent materials (*e.g.* glass or acrylic) are required [44]. The closed photobioreactors may accumulate high dissolved oxygen (DO) during algae growth. DO as high as 400% of saturation is detrimental to algae [45]. The high temperature developed in the closed photobioreactors under peak solar irradiance requires cooling systems to keep the algal culture below 40 °C [41].

Immobilized Algal Cultivation Systems (Algal Biofilm Reactors)

Immobilized algal cultivation systems (*i.e.* algal biofilm reactors) are developed due to the challenges of algal harvesting in suspended systems. Supporting materials are used for the attachment of algae in the biofilm reactors. The attached growth of algae allowed a much longer solids residence time (SRT or mean cell residence time [MCRT]) than HRT. The shorter HRT allowed higher loading rates. Biomass can be scraped from supporting materials [46].

Supporting materials can be polyethylene, polystyrene, polyurethane, loofah, nylon sponges, cardboard and cotton [31, 47, 48]. Cotton canvas or cord are the good supporting medium for algae growth but their poor durability requires the materials to be changed every 2-3 months [47, 49].

The supporting materials can be stationary or mobile. The layout of the biofilm modules could be horizontal or vertical. Algal turf scrubbers (ATS) and twin layer vertical glass fiber biofilm reactors are stationary algal biofilm reactors [50 - 52]. The vertical twin layer biofilm reactor had a smaller footprint than the horizontal ATS. The liquid was distributed to the two layer system by an agricultural dripping system.

Mobile biofilm reactors include rotating algal biofilm reactors (RABRs) and revolving algal biofilm (RAB). In an RABR, the cotton cord is coiled around the cylinder, and 40% of the cylinder was submerged in a raceway pond. A motor was used to drive the rotation of the cylinder. The TN and TP removal rates of RABRs were as high as 14.1 and 2.1 g m^{-2} d^{-1}, respectively. Pilot scale RABRs had higher aerial algal productivity (31 g m^{-2} d^{-1}) than that of bench scale reactor (5.5 g m^{-2} d^{-1}) [49]. While the RAB consisted of a vertical revolving cotton canvas as biofilm carrier.

Combined membrane bioreactors with algal biofilm reactors can further reduce the TSS of the effluent to less than 0.5 mg L^{-1} [53]. Purging algal biofilm reactors with CO_2 can increase the algal productivity and improve the wastewater treatment efficiency. A biomass productivity of 60 g m^{-2} d^{-1} can be achieved by purging 0.5% CO_2 in an outdoor algal biofilm reactor [54].

Enhanced Algal - Prokaryotic Wastewater Treatment Systems (EAPS) for N Removal

It is impractical to sterile a significant amount of wastewater when used as the algal growth medium. Thus, pure algal cultures are hard to maintain when growing in wastewater. Prokaryotes (including bacteria and archaea) always co-exist with algae in algal based wastewater treatment systems. Enhanced algal-prokaryotic wastewater treatment systems (EAPS) have been developed to enhance N removal efficiency by utilizing high N and organic carbon removal rates of prokaryotes and DO production by algae. EAPS provides engineered conditions required to optimize the functions of different groups of microorganisms. In addition, EPS produced by both algae and prokaryotes promote the formation of flocs with good settling ability (van den Hende *et al.* 2016, Wang *et al.* 2014, Wang *et al.* 2015a). Most of the EAPS are still in the bench-scale research stage. Recently, several EAPS have been developed, such as photo-activated sludge process, two-stage anoxic-aerobic photobioreactor, photo-sequencing batch reactor (PSBR) and ALGAMMOX (algal anaerobic ammonium oxidation) process.

Photo-activated sludge process was seeded with algae and activated sludge for the treatment of municipal wastewater [55]. The oxygen production of 0.46 kg m^{-3} d^{-1} in this system was higher than that of HRAPs (0.3-0.38 kg m^{-3} d^{-1}). The NH_4^+ removal were 81-85%, mainly through denitrification.

The two stage anoxic-aerobic photobioreactor with the SRT of 20-30 days removed 95% of organic carbon and 90% of N simultaneously [56]. The longer SRT retained the slower growing nitrifiers in the reactor. The biomass produced in the system also have a high settling rate of 0.28-0.42 m h^{-1}.

Wang *et al.* developed a shortcut N removal process using an algal-prokaryotic process for the treatment of high strength wastewater [57]. A photo sequencing batch reactor (PSBR) operated at an HRT of 4 days and an SRT of 8 days removed over 90% of total N, of which over 80% was removed by nitritation and denitritation. Additional sodium acetate was added during the dark phase to promote the denitritation due to the low concentrations of bioavailable organic carbon in the digested centrate.

ALGAMMOX process was developed to reduce the external carbon sources required for the shortcut N removal process [58]. The combined ammonia oxidation microorganisms (AOM) and anammox granular in a PSBR converted

NH_4^+ to NO_2^- at the rate of 7 mg NH_4^+-N L^{-1} h^{-1}. NO_2^- was removed by anammox during dark phase without the addition of carbon source.

A CASE STUDY ON NUTRIENT RECYCLE IN A CONCENTRATED ANIMAL FEEDING OPERATIONS (CAFOS) ADAPTING ALGAL-BASED WASTEWATER TREATMENT

Due to the high N concentrations in wastewater, the lipid content of algal biomass produced from wastewater ranged from 0.6-10.9% w/w, which is too low for economic biodiesel production [59, 60]. Anaerobic digestion of algal biomass or co-digestion with existing biomass have been proposed as an alternative method to recover energy [61 - 64]. The nutrient can be recycled in the integrated system.

A case study of large swine concentrated animal feeding operation (CAFO) that integrated algal wastewater treatment and anaerobic digestion was conducted by Wang *et al.* to analyze the energy production and nutrient balance [63]. The overview of the integrated system is shown in Fig. (**1**). The system assumes the CAFO consists of 7000 pigs with the average weight of 140 lbs. Anaerobic digestion was applied to stabilize the swine waste and produce energy. Centrate produced from anaerobic digestion was treated in a photo-sequencing batch reactor (PSBR) with the SRT of 24 days. The HRT of PSBR was 12 days. Algae harvested by gravity sedimentation and centrifugation were co-digested with swine waste for energy production. The N and P produced from the CAFO were approximately 93.4 and 79.4 kg d^{-1}. The data of nutrient removal by PSBR and energy production from the anaerobic co-digestion process were obtained from the experimental study conducted at University of South Florida [57, 63]. AOM were naturally accumulated after about 20 days operation due to the high NH_4^+ concentrations (891 mg NH_4^+-N/L) and low DO (<0.2 mg/L) in the system. Thus major N entered the PSBR was removed through the nitritation/denitritation process (81.7%) while about 18.3% of N was assimilated into algal biomass. Mass balance of the integrated PSBR and anaerobic digestion system indicated that the loss of N during the anaerobic digestion was 6.6 kg N/d due to the NH_3 evaporation. Approximately 33 kg N (35%) were removed from the PSBR. The majority of N (58%) and P (98%) were recovered in biosolids after anaerobic digestion. The main N and P recovered by algae were retained in the form of biomass.

CO_2 CAPTURE BY ALGAE

CO_2 emission to the atmosphere is regarded as one of the primary sources of global warming. Algae have high growth rates. Thus, they are more efficient than terrestrial crops on fixation of CO_2 [65]. Biofixation of CO_2 by algae could be the alternative methods for CO_2 mitigation and have several benefits over other physical and chemical methods [66], because sunlight is used as the energy source and harvested algal biomass can produce biofuels or high-value products, such as proteins, pigments, and lipids.

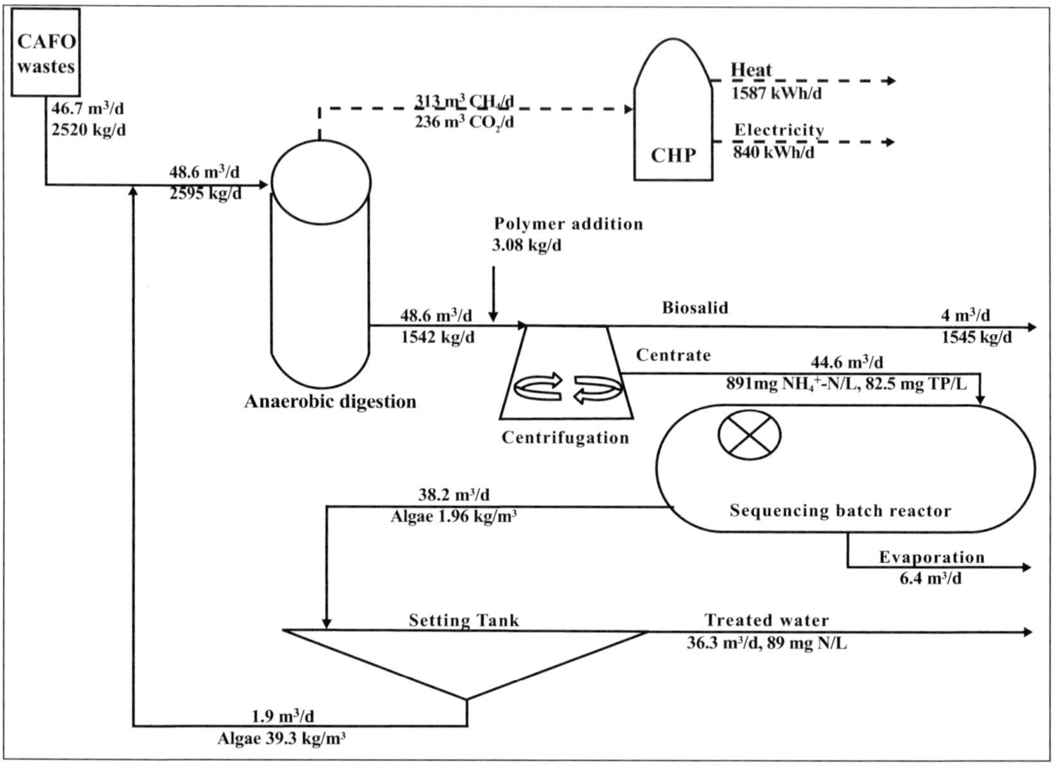

Fig. (1). Overview of the system coupling anaerobic co-digestion with algae cultivation. Values are indicated for a cultivated area of 535 m^3 in sequencing batch bioreactor [63].

Chlorella is favorable due to the abundance in natural environmental, the high biomass productivity and the tolerance to high CO_2 concentrations and pH [67]. *Chlorella* can maintain high growth rate using air containing up to 10-30% CO_2 [67, 81]. Mayo and Noike [68] found no significant effect on the survival of *Chlorella vulgaris* for pH up 10.5. However, the optimal pH for the highest specific growth rate *Chlorella vulgaris* was 6.3-6.8 [69]. The high tolerance of pH enables that algae can survive under acid or base conditions.

Impact of pH and CO_2 on Algae Growth

CO_2 was always the limiting factor for algae growth in natural systems due to the low solubility of CO_2 in water [70]. The utilization of CO_2 by algae leads to the increase of pH, because CO_2 utilization rate is higher than the replenish rates in natural systems.

Although bicarbonate and free CO_2 can be utilized by algae [65], free CO_2 is always recommended to maintain the high algae growth rate [71]. Lam *et al.* found that the growth rate of *Chlorella vulgaris* was 5.2 times higher when bubbling with CO_2 than using sodium bicarbonate as the carbon source [65].

pH direct affects the distribution of carbon dioxide species and available inorganic carbon to algae. The relation of speciation of inorganic carbon and pH at 21 ºC is shown as follow [72]:

$$pH = \log\frac{[HCO_3^-]}{[CO_2]} + 6.38 \quad (3)$$

The methods to increase CO_2 fixation rate include the increase of the light intensity, improving mass transfer efficiency and adjusting the physical and chemical conditions such as pH and temperature [66, 73]. The CO_2 fixation rate is largely dependent on the algae growth rate. High concentration CO_2 gas sources including flue gases draw a lot of attention to eliminate the cost of gas sources for algae cultivation [74]. The flue gas from combustion of natural gas contained about 6-8% of CO_2 and can be used for algae cultivation [75]. Increase the CO_2 concentration from 0.03% to 5% CO_2 increase the algal productivity from 0.03 g L^{-1} d^{-1} to 0.07 g L^{-1} d^{-1}. However, the lipid content was not affected by the concentration of CO_2 [65]. Similar results were found by Chiu *et al.* [76]. Increasing CO_2 concentration to 5% increased the algae productivity, however increasing CO_2 to 15% showed a decrease of biomass productivity [77].

Factors Affecting the Gas Transfer Efficiency

The CO_2 fixation depends on the gas transfer from the gas phase to liquid phase and the assimilation by biomass, which are related to algal species, the operation conditions, such as pH, temperature, light intensity, CO_2 loading, and the reactor configurations [66]. Proper reactor design, sufficient area to volume ratio and countercurrent flow of gas and liquid at the optimal flow rate would be desired to improve the biomass productivity and CO_2 fixation rate [78]. Grima *et al.* found continuous bubbling of 10% CO_2 enriched air at 40 ml min^{-1} enhanced the biomass productivity and CO_2 fixation efficiency [79].

Most CO_2 is consumed by the growth of biomass cells rather than EPS production [66]. Thus, the ratio of CO_2 to algae carbon can be estimated by the stoichiometric relations obtained from Eq. 1. Yun *et al.* indicated that 1.8 kg of CO_2 could be fixed by the production of 1 kg of *Chlorella vulgaris* cultivated in wastewater from a steel-making facility, whose main N content was NH_4^+ [80]. This value is quite similar to the theoretical value predicted by Eq. 1. The biomass carbon to

CO_2 ratio can also be evaluated by experimental work [66]. Judd *et al.* [66] reported the biomass carbon ratio to CO_2 ratio of 1:1.9, which is similar to that reported by Yun *et al.* [80].

CO_2 fixation ratio in the biomass under room temperature (21 °C) can be calculated by Eq. 4 [66]:

$$F\%=110\ PV/(CQ)=100\ P\tau/C \quad (4)$$

Where P is the biomass productivity in g L^{-1} d^{-1}, V is the reactor volume and C is the CO_2 percentage in the feed gas, Q is the gas flow rate in L d^{-1}, and τ is the gas residence time in days.

The equation can be used to estimate the required CO_2 gas flow to achieve desired CO_2 removal efficiency. For example, if 100% CO_2 removal and a biomass productivity of 0.5 g L^{-1} d^{-1} is desired, when the CO_2 concentration of feed gas is 5%, then $\tau = 5/(110\times5) = 0.09$ d or 2.18 h. The gas residence time of 2.18 hours is desired.

Gas Supply Systems

Gas usually is continuously or intermittently injected into photobioreactors. The countercurrent flow has higher gas transfer efficiency than directly bubbling [81]. Various reactors have been developed to assist the gas transfer efficiency, such as airlift reactors, revolving systems and membrane sparge systems [82, 83]. Gas can also be injected into the culture by bubbling column. The mass transfer coefficient (K_La) was determined by the reactor's hydraulics. CO_2 uptake rates had been reported to increase with the increase in mass transfer efficiency [66]. The CO_2 concentrations of feed gas also affect the gas transfer efficiency [75].

Airlift Photobioreactor

Extended mixing and high energy consumption are the major drawbacks of photobioreactors [84]. Pirouzi *et al.* developed a vertical triangular external airlift loop reactor, and the feed gas was supplied by two spargers; one at the bottom of the downcomer and the other located at the bottom of the vertical side. The configuration provided efficient gas-liquid countercurrent flow for the contact between gas and liquid phase. The design also provided lower power consumptions compared with other photobioreactors. The gas transfer coefficient (K_La) ranged from 18-24 h^{-1}.

Membrane Sparged Systems

A combined membrane-ring sparger reactor (BCMR) was developed to provide both fine bubbles and good mixing, thus higher gas transfer efficiency. However, the complex configuration hindered the scale up of BCMR as the module inside reduces the available light regimes inside the reactor [82].

Sparged helical tubular photobioreactor (MSTR) was designed to improve the mass transfer and light utilization efficiencies. The high surface to volume ratio enables high illumination efficiency and biomass productivity. The maximum CO_2 fixation rate was 148 mg L^{-1} d^{-1} when the light intensity was 3600 Lux, and the gas flow rate was 3.6 L min^{-1} with CO_2 concentration of 0.93% [82]. The gas transfer coefficient ranged from 249-464 h^{-1}, which is way higher than that of airlift reactor. Membrane sparging system increased the CO_2 fixation rate from 80 to 260 mg L^{-1} h^{-1} [85].

CONCLUDING REMARKS

Microalgae has the potential to lower the energy cost of wastewater treatment. The N and P in the wastewater can be recycled in the form of biomass. Algal harvesting and wastewater treatment efficiencies are being improved by the advances in treatment process and reactor design. Biofixion of CO_2 by microalgae is a promising technology to mitigate greenhouse gases emission. Deriving valuable products from harvested microalgae will make the algae applications in environmental engineering more economically feasible.

CONFLICT OF INTEREST

The authors declare no conflict of interest, financial or otherwise.

ACKNOWLEDGEMENTS

Declared none.

REFERENCES

[1] EPRI, *Water & Sustainability (Volume 4): U.S. Electricity Consumption for the Water Supply & Treatment-The Next Half Century.*

[2] W.J. Oswald, and H.B. Gotaas, "Photosynthesis in sewage treatment", *Trans. Am. Soc. Civ. Eng.,* vol. 122, pp. 73-97, 1957.

[3] Y. Chisti, "Biodiesel from microalgae beats bioethanol", *Trends Biotechnol.,* vol. 26, no. 3, pp. 126-131, 2008.
[http://dx.doi.org/10.1016/j.tibtech.2007.12.002] [PMID: 18221809]

[4] Y-K. Lee, "Microalgal mass culture systems and methods: Their limitation and potential", *J. Appl. Phycol.,* vol. 13, pp. 307-315, 2001.
[http://dx.doi.org/10.1023/A:1017560006941]

[5] J.B. Park, R.J. Craggs, and A.N. Shilton, "Recycling algae to improve species control and harvest efficiency from a high rate algal pond", *Water Res.,* vol. 45, no. 20, pp. 6637-6649, 2011.
[http://dx.doi.org/10.1016/j.watres.2011.09.042] [PMID: 22048019]

[6] X. Yuan, A. Kumar, A.K. Sahu, and S.J. Ergas, "Impact of ammonia concentration on Spirulina platensis growth in an airlift photobioreactor", *Bioresour. Technol.*, vol. 102, no. 3, pp. 3234-3239, 2011. [http://dx.doi.org/10.1016/j.biortech.2010.11.019] [PMID: 21112776]

[7] A. Carlsson, J. van Beilen, R. Möller, and D. Clayton, *Micro- and macroalgae utility for industrial applications bioproducts,* 2007.

[8] Y. Chisti, "Biodiesel from microalgae", *Biotechnol. Adv.*, vol. 25, no. 3, pp. 294-306, 2007. [http://dx.doi.org/10.1016/j.biotechadv.2007.02.001] [PMID: 17350212]

[9] A. Kumar, S. Ergas, X. Yuan, A. Sahu, Q. Zhang, J. Dewulf, F.X. Malcata, and H. van Langenhove, "Enhanced CO(2) fixation and biofuel production via microalgae: recent developments and future directions", *Trends Biotechnol.*, vol. 28, no. 7, pp. 371-380, 2010. [http://dx.doi.org/10.1016/j.tibtech.2010.04.004] [PMID: 20541270]

[10] J.U. Grobbelaar, "Algal Nutrition – Mineral Nutrition", in Handbook of Microalgal Culture, ed: Blackwell Publishing Ltd, 2007, pp. 95-115.

[11] S. Aslan, and I.K. Kapdan, "Batch kinetics of nitrogen and phosphorus removal from synthetic wastewater by algae", *Ecol. Eng.*, vol. 28, pp. 64-70, 2006. [http://dx.doi.org/10.1016/j.ecoleng.2006.04.003]

[12] L. Xin, H.Y. Hu, G. Ke, and Y.X. Sun, "Effects of different nitrogen and phosphorus concentrations on the growth, nutrient uptake, and lipid accumulation of a freshwater microalga Scenedesmus sp", *Bioresour. Technol.*, vol. 101, no. 14, pp. 5494-5500, 2010. [http://dx.doi.org/10.1016/j.biortech.2010.02.016] [PMID: 20202827]

[13] O. Perez-Garcia, F.M. Escalante, L.E. de-Bashan, and Y. Bashan, "Heterotrophic cultures of microalgae: metabolism and potential products", *Water Res.*, vol. 45, no. 1, pp. 11-36, 2011. [http://dx.doi.org/10.1016/j.watres.2010.08.037] [PMID: 20970155]

[14] B. Lu, Y. Yuan, C. Zhang, J. Ou, W. Zhou, and Q. Lin, *Modulation of key enzymes involved in ammonium assimilation and carbon metabolism by low temperature in rice (Oryza sativa L.) roots,* . [http://dx.doi.org/10.1016/j.plantsci.2004.09.031]

[15] A.C. Cannons, and J. Cannon, "The stability of the Chlorella nitrate reductase mRNA is determined by the secondary structure of the 5′-UTR: implications for posttranscriptional regulation of nitrate reductase", *Planta,* vol. 214, no. 3, pp. 488-491, 2002. [http://dx.doi.org/10.1007/s00425-001-0679-z] [PMID: 11855653]

[16] N.F. Tam, and Y.S. Wong, *Effect of ammonia concentrations on growth of Chlorella vulgaris and nitrogen removal from media*, 1996.

[17] M. Chen, H. Tang, H. Ma, T.C. Holland, K.Y. Ng, and S.O. Salley, *Effect of nutrients on growth and lipid accumulation in the green algae Dunaliella tertiolecta*, 2011.

[18] Y. Li, M. Horsman, B. Wang, N. Wu, and C.Q. Lan, *Effects of nitrogen sources on cell growth and lipid accumulation of green alga Neochloris oleoabundans.* vol. Vol. 81. Applied Microbiology and Biotechnology, 2008, pp. 629-636.

[19] Y. Song, H.H. Hahn, and E. Hoffmann, "Effects of solution conditions on the precipitation of phosphate for recovery. A thermodynamic evaluation", *Chemosphere,* vol. 48, no. 10, pp. 1029-1034, 2002. [http://dx.doi.org/10.1016/S0045-6535(02)00183-2] [PMID: 12227507]

[20] Y. Li, Y-F. Chen, P. Chen, M. Min, W. Zhou, B. Martinez, J. Zhu, and R. Ruan, "Characterization of a microalga Chlorella sp. well adapted to highly concentrated municipal wastewater for nutrient removal and biodiesel production", *Bioresour. Technol.*, vol. 102, no. 8, pp. 5138-5144, 2011. [http://dx.doi.org/10.1016/j.biortech.2011.01.091] [PMID: 21353532]

[21] G.Y. Rhee, *Effects of N:P Atomic Ratios and Nitrate Limitation on Algal Growth, Cell Composition, and Nitrate Uptake*, 1978.

[22] M. Takagi, K. Watanabe, K. Yamaberi, and T. Yoshida, "Limited feeding of potassium nitrate for intracellular lipid and triglyceride accumulation of Nannochloris sp. UTEX LB1999", *Appl. Microbiol. Biotechnol.*, vol. 54, no. 1, pp. 112-117, 2000. [http://dx.doi.org/10.1007/s002530000333] [PMID: 10952013]

[23] D.G. Cirne, X. Paloumet, L. Björnsson, M.M. Alves, and B. Mattiasson, "Anaerobic digestion of lipid-rich waste—Effects of lipid concentration", *Renew. Energy,* vol. 32, pp. 965-975, 2007. [http://dx.doi.org/10.1016/j.renene.2006.04.003]

[24] G.P. Robertson, and J.M. Tiedje, *"Nitrous oxide sources in aerobic soils: Nitrification, denitrification and other biological processes.* vol. 19. Soil Biology and Biochemistry, 1987, pp. 187-193.

[25] M. Wang, W. C. Kuo-Dahab, S. Dolan, and C. Park, *Kinetics of nutrient removal and expression of extracellular polymeric substances of the microalgae, Chlorella sp. and Micractinium sp., in wastewater treatment, .* [http://dx.doi.org/10.1016/j.biortech.2013.12.047]

[26] W. Oswald, "The high-rate pond in waste disposal", *Dev. Ind. Microbiol.*, vol. 4, pp. 112-125, 1963.

[27] R. Muñoz, and B. Guieysse, "Algal-bacterial processes for the treatment of hazardous contaminants: a review", *Water Res.*, vol. 40, no. 15, pp. 2799-2815, 2006. [http://dx.doi.org/10.1016/j.watres.2006.06.011] [PMID: 16889814]

[28] R. Craggs, D. Sutherland, and H. Campbell, "Hectare-scale demonstration of high rate algal ponds for enhanced wastewater treatment and biofuel production", *J. Appl. Phycol.*, vol. 24, pp. 329-337, 2012. [http://dx.doi.org/10.1007/s10811-012-9810-8]

[29] J. Garcia, R. Mujeriego, and M. Hernandez-Marine, "High rate algal pond operating strategies for urban wastewater nitrogen removal", *J. Appl. Phycol.*, vol. 12, pp. 331-339, 2000. [http://dx.doi.org/10.1023/A:1008146421368]

[30] M.E. Verbyla, S.M. Oakley, and J.R. Mihelcic, *Wastewater Infrastructure for Small Cities in an Urbanizing World: Integrating Protection of Human Health and the Environment with Resource Recovery and Food Security*, 2013.

[31] J.P. Hoffmann, "Wastewater treatment with suspended and nonsuspended algae", *J. Phycol.*, vol. 34, pp. 757-763, 1998. [http://dx.doi.org/10.1046/j.1529-8817.1998.340757.x]

[32] J.B. Park, and R.J. Craggs, "Nutrient removal in wastewater treatment high rate algal ponds with carbon dioxide addition", *Water Sci. Technol.*, vol. 63, no. 8, pp. 1758-1764, 2011. [http://dx.doi.org/10.2166/wst.2011.114] [PMID: 21866778]

[33] E. Posadas, A. Muñoz, M-C. García-González, R. Muñoz, and P.A. García-Encina, "A case study of a pilot high rate algal pond for the treatment of fish farm and domestic wastewaters", *J. Chem. Technol. Biotechnol.*, vol. 90, pp. 1094-1101, 2015. [http://dx.doi.org/10.1002/jctb.4417]

[34] A. Abeliovich, *Algae in wastewater oxidation ponds*, 1986.

[35] J. García, B.F. Green, T. Lundquist, R. Mujeriego, M. Hernández-Mariné, and W.J. Oswald, *Long term diurnal variations in contaminant removal in high rate ponds treating urban wastewater*, 2006.

[36] G. Deviller, C. Aliaume, M. A. F. Nava, C. Casellas, and J. P. Blancheton, *High-rate algal pond treatment for water reuse in an integrated marine fish recirculating system: effect on water quality and sea bass growth,* 2004. [http://dx.doi.org/10.1016/j.aquaculture.2004.01.023]

[37] I. Udom, B. H. Zaribaf, T. Halfhide, B. Gillie, O. Dalrymple, and Q. Zhang, *Harvesting microalgae grown on wastewater,* 2013. [http://dx.doi.org/10.1016/j.biortech.2013.04.002]

[38] F.B. Green, T.J. Lundquist, and W.J. Oswald, *Energetics of advanced integrated wastewater pond*

systems. vol. Vol. 31. Water Science and Technology, 1995, pp. 9-20.

[39] N. Uduman, Y. Qi, M.K. Danquah, G.M. Forde, and A. Hoadley, "Dewatering of microalgal cultures: a major bottleneck to algae-based fuels", *J. Renew. Sustain. Energy,* vol. 2, p. 012701, 2010. [http://dx.doi.org/10.1063/1.3294480]

[40] P.E. Wiley, K.J. Brenneman, and A.E. Jacobson, "Improved algal harvesting using suspended air flotation", *Water Environ. Res.,* vol. 81, no. 7, pp. 702-708, 2009. [http://dx.doi.org/10.2175/106143009X407474] [PMID: 19691251]

[41] G. Torzillo, B. Pushparaj, F. Bocci, W. Balloni, R. Materassi, and G. Florenzano, *Production of Spirulina biomass in closed photobioreactors*, 1986.

[42] C. U. Ugwu, H. Aoyagi, and H. Uchiyama, *Photobioreactors for mass cultivation of algae,* 2008. [http://dx.doi.org/10.1016/j.biortech.2007.01.046]

[43] M. A. Borowitzka, *Commercial production of microalgae: ponds, tanks, tubes and fermenters,* . [http://dx.doi.org/10.1016/S0168-1656(99)00083-8]

[44] M.R. Tredici, and G.C. Zittelli, "Efficiency of sunlight utilization: tubular versus flat photobioreactors", *Biotechnol. Bioeng.,* vol. 57, no. 2, pp. 187-197, 1998. [http://dx.doi.org/10.1002/(SICI)1097-0290(19980120)57:2<187::AID-BIT7>3.0.CO;2-J] [PMID: 10099193]

[45] E. Molina, J. Fernández, F. G. Acién, and Y. Chisti, *Tubular photobioreactor design for algal cultures,* 2001. [http://dx.doi.org/10.1016/S0168-1656(01)00353-4]

[46] M. Kesaano, and R. Sims, *Algal biofilm based technology for wastewater treatment.* vol. Vol. 5. Algal Res, 2014.

[47] M. Gross, D. Jarboe, and Z. Wen, "Biofilm-based algal cultivation systems", *Appl. Microbiol. Biotechnol.,* vol. 99, no. 14, pp. 5781-5789, 2015. [http://dx.doi.org/10.1007/s00253-015-6736-5] [PMID: 26078112]

[48] A. C. Wilkie, and W. W. Mulbry, "Recovery of dairy manure nutrients by benthic freshwater algae," Bioresource Technology, vol. 84, pp. 81-91, 8// 2002. [http://dx.doi.org/10.1016/S0960-8524(02)00003-2]

[49] L.B. Christenson, and R.C. Sims, "Rotating algal biofilm reactor and spool harvester for wastewater treatment with biofuels by-products", *Biotechnol. Bioeng.,* vol. 109, no. 7, pp. 1674-1684, 2012. [http://dx.doi.org/10.1002/bit.24451] [PMID: 22328283]

[50] J. Shi, B. Podola, and M. Melkonian, *Application of a prototype-scale Twin-Layer photobioreactor for effective N and P removal from different process stages of municipal wastewater by immobilized microalgae,* 2014. [http://dx.doi.org/10.1016/j.biortech.2013.11.100]

[51] T. Naumann, Z. Çebi, B. Podola, and M. Melkonian, "Growing microalgae as aquaculture feeds on twin-layers: a novel solid-state photobioreactor", *J. Appl. Phycol.,* vol. 25, pp. 1413-1420, 2012. [http://dx.doi.org/10.1007/s10811-012-9962-6]

[52] R. Muñoz, C. Köllner, and B. Guieysse, "Biofilm photobioreactors for the treatment of industrial wastewaters," Journal of Hazardous Materials, vol. 161, pp. 29-34, 1/15/ 2009.

[53] F. Gao, Z.-H. Yang, C. Li, G.-M. Zeng, D.-H. Ma, and L. Zhou, "A novel algal biofilm membrane photobioreactor for attached microalgae growth and nutrients removal from secondary effluent," Bioresource Technology, vol. 179, pp. 8-12, 3// 2015. [http://dx.doi.org/10.1016/j.biortech.2014.11.108]

[54] L. Zhang, L. Chen, J. Wang, Y. Chen, X. Gao, and Z. Zhang, *Attached cultivation for improving the biomass productivity of Spirulina platensis,* 2015. [http://dx.doi.org/10.1016/j.biortech.2015.01.025]

[55] N.G. Karya, N.P. van der Steen, and P.N. Lens, "Photo-oxygenation to support nitrification in an algal-bacterial consortium treating artificial wastewater", *Bioresour. Technol.,* vol. 134, pp. 244-250, 2013.
[http://dx.doi.org/10.1016/j.biortech.2013.02.005] [PMID: 23500581]

[56] I. de Godos, V. A. Vargas, H. O. Guzmán, R. Soto, B. García, and P. A. García, "Assessing carbon and nitrogen removal in a novel anoxic–aerobic cyanobacterial–bacterial photobioreactor configuration with enhanced biomass sedimentation," Water Research, vol. 61, pp. 77-85, 9/15/ 2014.
[http://dx.doi.org/10.1016/j.watres.2014.04.050]

[57] M. Wang, H. Yang, S. J. Ergas, and P. van der Steen, *A novel shortcut nitrogen removal process using an algal-bacterial consortium in a photo-sequencing batch reactor (PSBR),* 2015.
[http://dx.doi.org/10.1016/j.watres.2015.09.016]

[58] N.D. Manser, M. Wang, S.J. Ergas, J.R. Mihelcic, A. Mulder, and J. van de Vossenberg, *Biological Nitrogen Removal in a Photosequencing Batch Reactor with an Algal-Nitrifying Bacterial Consortium and Anammox Granules*, 2016.

[59] W. Mulbry, S. Kondrad, and J. Buyer, "Treatment of dairy and swine manure effluents using freshwater algae: fatty acid content and composition of algal biomass at different manure loading rates", *J. Appl. Phycol.,* vol. 20, pp. 1079-1085, 2008.
[http://dx.doi.org/10.1007/s10811-008-9314-8]

[60] B. Hu, M. Min, W. Zhou, Y. Li, M. Mohr, and Y. Cheng, *Influence of Exogenous CO2 on Biomass and Lipid Accumulation of Microalgae Auxenochlorella protothecoides Cultivated in Concentrated Municipal Wastewater*, 2012.

[61] X. Yuan, M. Wang, C. Park, A.K. Sahu, and S.J. Ergas, "Microalgae growth using high-strength wastewater followed by anaerobic co-digestion", *Water Environ. Res.,* vol. 84, no. 5, pp. 396-404, 2012.
[http://dx.doi.org/10.2175/106143011X13233670703242] [PMID: 22852424]

[62] M. Wang, A.K. Sahu, B. Rusten, and C. Park, "Anaerobic co-digestion of microalgae Chlorella sp. and waste activated sludge", *Bioresour. Technol.,* vol. 142, pp. 585-590, 2013.
[http://dx.doi.org/10.1016/j.biortech.2013.05.096] [PMID: 23770998]

[63] M. Wang, E. Lee, Q. Zhang, and S.J. Ergas, *Anaerobic Co-digestion of Swine Manure and Microalgae Chlorella sp.: Experimental Studies and Energy Analysis.* BioEnergy Research, 2016, pp. 1-12.

[64] W. Zhong, Z. Zhang, Y. Luo, W. Qiao, M. Xiao, and M. Zhang, "Biogas productivity by co-digesting Taihu blue algae with corn straw as an external carbon source", *Bioresour. Technol.,* vol. 114, pp. 281-286, 2012.
[http://dx.doi.org/10.1016/j.biortech.2012.02.111] [PMID: 22459954]

[65] M. K. Lam, and K. T. Lee, "Effect of carbon source towards the growth of Chlorella vulgaris for CO2 bio-mitigation and biodiesel production," International Journal of Greenhouse Gas Control, vol. 14, pp. 169-176, 5// 2013.
[http://dx.doi.org/10.1016/j.ijggc.2013.01.016]

[66] S. Judd, L. J. P. van den Broeke, M. Shurair, Y. Kuti, and H. Znad, *Algal remediation of CO2 and nutrient discharges: A review,* 2015.
[http://dx.doi.org/10.1016/j.watres.2015.08.021]

[67] K-D. Sung, J-S. Lee, C-S. Shin, S-C. Park, and M-J. Choi, "CO2 fixation by Chlorella sp. KR-1 and its cultural characteristics", *Bioresour. Technol.,* vol. 68, pp. 269-273, 1999.
[http://dx.doi.org/10.1016/S0960-8524(98)00152-7]

[68] A.W. Mayo, and T. Noike, "Effect of glucose loading on the growth behavior of Chlorella vulgaris and heterotrophic bacteria in mixed culture", *Water Res.,* vol. 28, pp. 1001-1008, 1994.
[http://dx.doi.org/10.1016/0043-1354(94)90184-8]

[69] A.W. Mayo, "Effects of temperature and pH on the kinetic growth of unialga Chlorella vulgaris

cultures containing bacteria", *Water Environ. Res.*, vol. 69, pp. 64-72, 1997. [http://dx.doi.org/10.2175/106143097X125191]

[70] J.C. Goldman, W.J. Oswald, and D. Jenkins, "The kinetics of inorganic carbon limited algal growth", *J. Water Pollut. Control Fed.*, pp. 554-574, 1974.

[71] Y. Azov, "Effect of pH on inorganic carbon uptake in algal cultures", *Appl. Environ. Microbiol.*, vol. 43, no. 6, pp. 1300-1306, 1982. [PMID: 16346029]

[72] P. Brezonik, and W. Arnold, *Water chemistry: an introduction to the chemistry of natural and engineered aquatic systems.* Oxford University Press, 2011.

[73] A. L. Gonçalves, M. Simões, and J. C. M. Pires, *The effect of light supply on microalgal growth, CO2 uptake and nutrient removal from wastewater,* 2014. [http://dx.doi.org/10.1016/j.enconman.2014.05.085]

[74] A. Concas, G. A. Lutzu, M. Pisu, and G. Cao, "Experimental analysis and novel modeling of semi-batch photobioreactors operated with Chlorella vulgaris and fed with 100% (v/v) CO2," Chemical Engineering Journal, vol. 213, pp. 203-213, 12/1/ 2012. [http://dx.doi.org/10.1016/j.cej.2012.09.119]

[75] J. Doucha, F. Straka, and K. Lívanský, "Utilization of flue gas for cultivation of microalgae (Chlorella sp.) in an outdoor open thin-layer photobioreactor", *J. Appl. Phycol.,* vol. 17, pp. 403-412, 2005. [http://dx.doi.org/10.1007/s10811-005-8701-7]

[76] S-Y. Chiu, C-Y. Kao, C-H. Chen, T-C. Kuan, S-C. Ong, and C-S. Lin, "Reduction of CO2 by a high-density culture of Chlorella sp. in a semicontinuous photobioreactor", *Bioresour. Technol.,* vol. 99, no. 9, pp. 3389-3396, 2008. [http://dx.doi.org/10.1016/j.biortech.2007.08.013] [PMID: 17904359]

[77] Q.-H. Shen, J.-W. Jiang, L.-P. Chen, L.-H. Cheng, X.-H. Xu, and H.-L. Chen, *Effect of carbon source on biomass growth and nutrients removal of Scenedesmus obliquus for wastewater advanced treatment and lipid production,* 2015. [http://dx.doi.org/10.1016/j.biortech.2015.04.053]

[78] W. Kargupta, A. Ganesh, and S. Mukherji, *Estimation of carbon dioxide sequestration potential of microalgae grown in a batch photobioreactor,* 2015. [http://dx.doi.org/10.1016/j.biortech.2015.01.017]

[79] E. Molina Grima, F. G. A. Fernández, F. García Camacho, and Y. Chisti, *Photobioreactors: light regime, mass transfer, and scaleup,* 1999. [http://dx.doi.org/10.1016/S0168-1656(99)00078-4]

[80] Y-S. Yun, S.B. Lee, J.M. Park, C-I. Lee, and J-W. Yang, "Carbon Dioxide Fixation by Algal Cultivation Using Wastewater Nutrients", *J. Chem. Technol. Biotechnol.,* vol. 69, pp. 451-455, 1997. [http://dx.doi.org/10.1002/(SICI)1097-4660(199708)69:4<451::AID-JCTB733>3.0.CO;2-M]

[81] R. Putt, M. Singh, S. Chinnasamy, and K. C. Das, "An efficient system for carbonation of high-rate algae pond water to enhance CO2 mass transfer," Bioresource Technology, vol. 102, pp. 3240-3245, 2// 2011. [http://dx.doi.org/10.1016/j.biortech.2010.11.029]

[82] L.-H. Fan, Y.-T. Zhang, L. Zhang, and H.-L. Chen, "Evaluation of a membrane-sparged helical tubular photobioreactor for carbon dioxide biofixation by Chlorella vulgaris," Journal of Membrane Science, vol. 325, pp. 336-345, 11/15/ 2008. [http://dx.doi.org/10.1016/j.memsci.2008.07.044]

[83] M. Gross, and Z. Wen, *Yearlong evaluation of performance and durability of a pilot-scale revolving algal biofilm (RAB) cultivation system.* vol. Vol. 171. Bioresour Technol, 2014.

[84] A. Pirouzi, M. Nosrati, S. A. Shojaosadati, and S. Shakhesi, "Improvement of mixing time, mass transfer, and power consumption in an external loop airlift photobioreactor for microalgae cultures,"

Biochemical Engineering Journal, vol. 87, pp. 25-32, 6/15/ 2014.
[http://dx.doi.org/10.1016/j.bej.2014.03.012]

[85] L. Cheng, L. Zhang, H. Chen, and C. Gao, *Carbon dioxide removal from air by microalgae cultured in a membrane-photobioreactor,* 2006.
[http://dx.doi.org/10.1016/j.seppur.2005.12.006]

CHAPTER 13

Maximising Value: The Bio-Refinery Concept

Darren L. Oatley-Radcliffe[1,2,*], **Thea Ekins-Coward**[1] and **Robert W. Lovitt**[1,2]

[1] *Centre for Water Advanced Technologies and Environmental Research (CWATER), College of Engineering, Swansea University, Fabian Way, Swansea SA1 8EN, UK*

[2] *Membranology Ltd., Charter Court, Phoenix Way, Enterprise Park, Swansea, SA79FS, UK*

Abstract: The lipids accumulated inside microalgal biomass can be trans-esterified into biodiesel and have brought this type of feedstock to the forefront of the second generation renewable fuels research. Nevertheless, even though energy prices are generally increasing, fuel and energy products are of high volume with low value commodities and the production of microalgae-derived biofuels is not currently economically feasible when compared to fossil fuel energy. Thus, to achieve true sustainability in transferring microalgae into fuels, there is a need for integrated solutions which utilise low-value inputs to the process and exploit the full product portfolio achievable from microalgae. For example, microalgae have been shown to contain quantities of a vast array of biological materials such as exopolysaccharides, phycobiliproteins, potent antioxidants, polyunsaturated fatty acids, and mycosporine-like amino acids that have a range of high-value commercial applications in the cosmetic, pharmaceutical and nutraceutical industries. As a result, several research groups have introduced the microalgae bio-refinery approach to encapsulate this concept and maximise the full potential of an integrated microalgal production facility that harvests the high value components available and produces fuels as a secondary revenue stream.

Keywords: Algae, Biorefinery, Astaxanthin, Biofuels, Membrane technology, Sustainability, Mass efficiency, Fractionation, Optimization, Artificial lighting.

INTRODUCTION

The world's population is growing; the current estimates suggest that the population in mid-2015 being around 7.3 billion and is expected to increases to 8.5 billion by 2030 and 9.7 billion by 2050 [1]. To sustain the population, there are three major resources required: food, water and energy. Each of these resources is intrinsically linked and as the population is growing and advancing, demand for each is also growing. The total global fuel consumption has increased

* **Corresponding author Darren L. Oatley-Radcliffe:** Institute Centre for Water Advanced Technologies and Environmental Research (CWATER), College of Engineering, Swansea University, Fabian Way, Swansea SA1 8EN, UK; Tel/Fax: +44 (0) 1792 606668; E-mail: D.L.Oatley@Swansea.ac.uk

José Carlos Magalhães Pires (Ed.)

from around 4661 Mtoe (million tonnes of oil equivalent) in 1971 to 9425 Mtoe in 2014 [2]. This increase of 102% is significant and the increasing trend of energy consumption shows no sign of reduction. The energy mix for consumption is currently 40% oil, 18% electricity, 15% natural gas, 12% biofuels and waste, 11.5% coal and 3% others. By very nature, oil, natural gas, coal and the majority of electricity supplies are unsustainable in the long term and great attention is now being placed on developing sustainable supplies of energy. First generation supplies of renewable biofuel focused on land based crops, such as oil seed, but these are limited in their ability to achieve the targets required for biofuel production, climate change mitigation and economic growth [3]. Thus, alternative sources of biofuel feedstocks are being actively pursued and microalgae in particular are gaining significant interest.

Microalgae are a diverse group of prokaryotic and eukaryotic photosynthetic microorganisms that grow rapidly due to their simplistic structure [4]. They have been investigated for the production of a number of different biofuels including biodiesel, bio-oil, bio-syngas, and bio-hydrogen. Microalgae have a varied composition and some microalgae contain relatively large quantities of lipids that can be trans-esterified into biodiesel. For example, the dry weight lipid composition of *Chaetoceros calcitrans* is 40%, of *Pavlova lutheri* it is 36% and of *Nitzschia palea* it is 47% [5]. Microalgae are able to convert carbon dioxide and nutrients into life products and lipids *via* photosynthesis with oxygen as a waste product. Microalgae are able to capture up to 1.8 kg CO_2 kg^{-1} of algal biomass [6] and given that arable land is not required for cultivation, the environmental benefits of this form of biomass surpass any other feedstock used in renewable energy production (see Table **1**). Microalgal biodiesel is the only renewable biodiesel that has the potential to completely displace liquid transport fuels derived from petroleum [7]. The scope for fast growth while converting carbon dioxide into fuels is a major factor of interest in such biomass. The normal productivity of microalgae exceeds any other feedstock at 50-120 tonnes ha^{-1} yr^{-1}. However, the biochemical composition of algae is highly different and species-dependent. In addition, the proportion of lipids that can be converted to fuels is greatly reliant on the cultivation strategy employed; such as nutrient supply, composition of the growth media, technology used and the quality and intensity of the light provided [8]. Thus, once the microalgae has been cultivated, the lipid content contained within the cells must be harvested and processed into a feedstock for the biodiesel production process. This multi-step production process is expensive and leads to an end fuel cost significantly higher than that obtained from fossil fuels. A study on the economics of algal biodiesel in the USA [9] suggests that the production rate of biodiesel from this source would not be sufficient to meet the supply needs of the transport sector without significant technical change, of the order of 30% per annum, and would require heavy

Table 1. Comparison of the oil content and productivity metrics for microalgae and other crops used in the production of biodiesel. (adapted from [3]).

Plant Source	Seed Oil Content	Oil Yield	Land Use	Biodiesel Productivity
	(% oil by wt in biomass)	(L oil/ha year)	(m^2 year/kg biodiesel)	(kg biodiesel/ha year)
Corn/Maize	44	172	66	152
Hemp	33	363	31	321
Soybean	18	636	18	562
Jatropha	28	741	15	656
Camelina	42	915	12	809
Canola/Rapeseed	41	974	12	862
Sunflower	40	1,070	11	946
Castor	48	1,307	9	1,156
Palm oil	36	5,366	2	4,747
Microalgae (low oil content)	30	58,700	0.2	51,927
Microalgae (medium oil content)	50	97,800	0.1	86,515
Microalgae (high oil content)	70	136,900	0.1	121,104

government subsidies to be competitive. Life-cycle assessment studies have also shown that biodiesel production from microalgae under current operating technology is not sustainable [10 - 12]. Moreover, improvements are needed in algae production technologies, harvesting and downstream processing as well as the realisation of cheaper nutrients and colocation of facilities to avoid the transport of intermediate or final products from each of the biodiesel production stages in order to reduce costs to similar levels as that of diesel obtained from petroleum sources.

THE QUEST FOR SUSTAINABLE ALGAL MANUFACTURING

One of the major drivers used when improving the sustainability of processes and products is the concept of mass efficiency [13]. In basic terms, mass efficiency can be thought of as the total quantity of products generated from a given feedstock. The higher the mass efficiency, the lower is the quantity of waste for a given process. In the ideal world, a sustainable product or process would have a mass efficiency approaching 100%; however, many industrial processes are far from this ideal value. For example, the mass efficiency of a typical pharmaceutical product is around 1-6% indicating that there is a great deal of

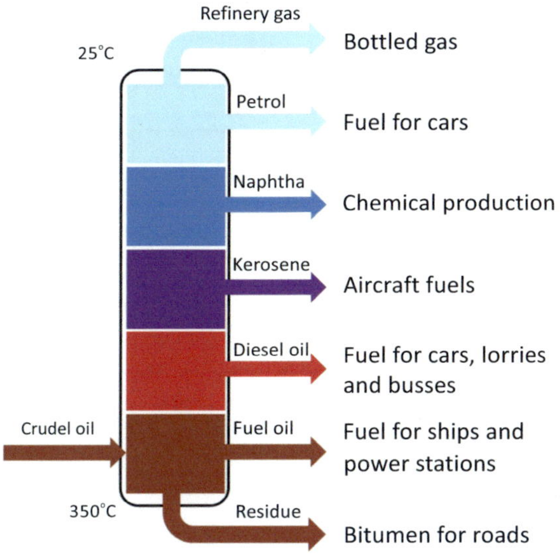

Fig. (1). Diagram of a crude oil refinery fractional distillation column.

waste involved in the production of drugs. Whereas the mass efficiency achieved in crude oil processing is close to 100%, indicating that the barrel of crude oil is almost entirely converted into usable products. The high mass efficiency of the crude oil process is the result of a technology known as fractional distillation that splits the crude oil into constituent components of the original oil in a production facility known as a refinery (see Fig. **1**) [14]. The refinery approach renders the crude oil process highly cost effective even though the energy demands for the process are significant. As a result of the large scale production of crude oil around the world, fuel from petroleum is a high volume low cost commodity. In order for biodiesel obtained from microalgae to compete with petroleum fuel, a similar production methodology must be adopted. The merits of microalgae as a sustainable source of valuable chemicals with commercial applications in energy, health, and cosmetics are widely acknowledged [3, 15, 16]. Microalgae generally consist of carbohydrates, proteins and lipids. However, there are many other materials integrated into the microalgal cell in lower quantities, such as exopolysaccharides [17], phycobiliproteins [18], potent antioxidants [19], polyunsaturated fatty acids [20], and mycosporine-like amino acids [21] that have a range of high-value commercial applications. Cuellar-Bermudez *et al.* [22] reviewed some of the organisms, products and application areas that are currently known for high value compound extraction from microalgae (see Table **2**). Significant differences can be found between the biochemical characteristics of microalgal species and this determines the specific application of algae species in a commercial setting. Thus, biomass with lipid content approaching 50% may be

Table 2. High value compounds available for extraction from various microalgae and potential commercial sector (taken from [22]).

Species	Product	Application Areas
Chlorella vulgaris	Biomass, pigments	Health food, food supplement
Chlorella spp.		
Chlorella ellipsodea		
Coccomyxa acidophila	Lutein, β-carotene	Pharmaceuticals, nutrition
Coelastrella striolata var. multistriata	Canthaxanthin, astaxanthin, ß-carotene	Pharmaceuticals, nutrition, cosmetics
Crypthecodinium conhi	Docosahexaenoic acid	Pharmaceuticals, nutrition
Diacronema vlkianum	Fatty acids	Pharmaceuticals, nutrition
Dunaliella salina	Carotenoids, ß-carotene	Health food, food supplement, feed
Galdiera suphuraria	Phycocyanin	Pharmaceuticals, nutrition
Haematococcus pluvialis	Carotenoids, astaxanthin, canthaxanthin, lutein	Health food, food supplement, feed additives
Isochrysis galbana	Fatty acids, carotenoids, fucoxanthin	Pharmaceuticals, nutrition, cosmetics, animal nutrition
Lyngbya majuscule	Immune modulators	Pharmaceuticals, nutrition
Muriellopsis sp.	Lutein	Pharmaceuticals, nutrition
Nannochloropsis gaditana	Eicosapentaenoic acid	Pharmaceuticals, nutrition
Nannochloropsis sp.		
Odontella aurita	Fatty acids	Pharmaceuticals, cosmetics, baby food
Parietochloris incise	Arachidonic acid	Nutritional supplement
Phaedactylum tricornutum	Lipids, eicosapentaenoic acid, fatty acids	Nutrition, fuel production
Porphyridium cruentum	Arachidonic acid, polysaccharides	Pharmaceuticals, nutrition, cosmetics
Scenedesmus almeriensis	Lutein, ß-carotene	Pharmaceuticals, nutrition, cosmetics
Schizochytrium sp.	Docosahexaenoic acid	Pharmaceuticals, nutrition
Spirulina platensis	Phycocyanin, Linolenic acid, biomass protein	Health food, cosmetics
Ulkenia spp.	Docosahexaenoic acid	Pharmaceuticals, nutrition

more suited for the production of biodiesel, while pharmaceutical and biomedical applications would be reliant on specific carbohydrate, pigment or protein content. This leads to the concept of a value chain, *i.e.* different materials have different value and the possibility to obtain multiple products from one organism becomes feasible. For example, in Europe the typical cost of diesel is around 1 Euro per kg (depending on local taxes *etc.*); however, high value oils such as docosahexaenoic acid (DHA or Omega-3) can be worth as much as 100 Euro per kg and are far more attractive as a product. By the same token, β-carotene and astaxanthin (pigments) can worth 2000 to 5000 Euro per kg, respectively. Other pigments such as phycoerythrin and phycocyanin worth as much as 36000 to 72000 Euro per kg. Fig. (**2**) illustrates this value chain concept where base commodities such as fuels are potentially plentiful in microalgae but have intrinsically low value, whereas speciality products have enormous price tags but are in much lower quantities. Thus, in order to drive up the sustainability of microalgal derived fuels, the natural way forward is to increase the mass efficiency of the production process and develop a bio-refinery approach that separates the raw products into several fractions, maximising the product portfolio available and optimises the potential revenue derived from the algae biomass. Unlike the relative simplicity of the crude oil fractionation process where all of the components are in the liquid phase and have a wide range of volatility, refining algae is no simple task. The constituents of the algal cell are multiphase (typically solids and liquids), many are thermally labile (heat sensitive) and there are varying degrees of hydrophilic and hydrophobic materials. These physical and chemical properties are difficult to work with and lead to a highly complex series of unit operations in the downstream process that can be summarised as cell harvesting, cell disruption, bulk separation and high resolution separation. Clearly low value high volume materials like fuels only require bulk separation, but high value low volume materials like pigments require high resolution separation and the selling price of the final product is directly related to the quality and purity achieved. So while the value of these materials is very high, so is the cost of refinement in order to obtain them and, depending on the ultimate application, these materials may be needed in high purity form – even more cost. A potential outline for an algal bio-refinery is illustrated in Fig. (**3**). Starting at the top right of the diagram is the algal production unit, in this case a photobioreactor is selected but this could be changed to an open pond or raceway facility. The production of algae requires nutrients in the form of nitrates, phosphates, light and a carbon source (along with some trace elements and other minerals). These inputs are shown in the orange boxes along with some options for refinement or post processing. For example, low value inputs like farm or municipal waste may be separated to liquid and solid fractions. The solids may undergo pyrolysis and the combustion products carbon dioxide and nitrates can be fed to a gas absorber which will deliver a relatively clean gas stream to the reactor. The spent ash will contain trace elements that are also required and can be mixed with the liquid effluent that also contains nitrates and phosphates and is fed to the reactor. The other essential inputs are an algal inoculum that can be taken from cell line and

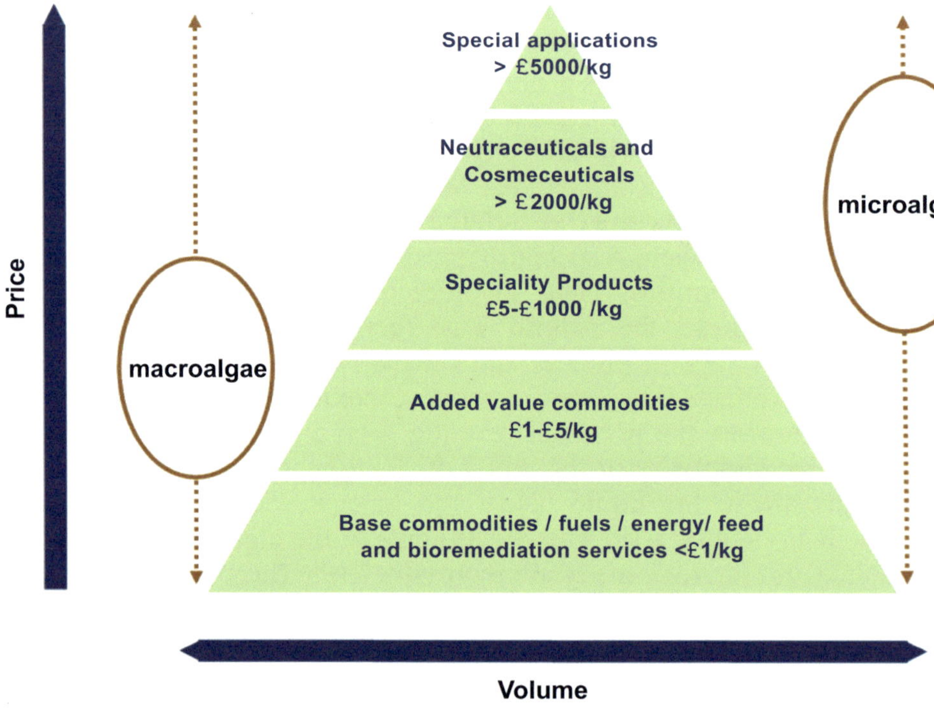

Fig. (2). Capacity and pricing of products from algal sources (taken from [47]).

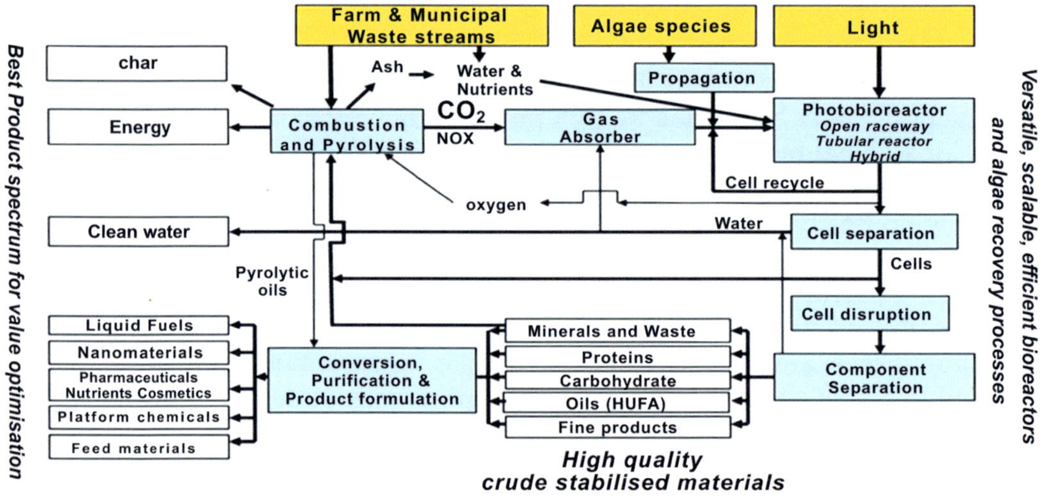

Fig. (3). Potential options for a microalgae bio-refinery.

grown up to provide a seed input to the reactor and lastly the reactor is illuminated with light, which may be sunlight or artificial or a combination of both. Once the reactor has grown a significant biomass loading there are now two main operational considerations. The reactor maybe completely harvested and the cells sent into the downstream process – batch processing, or a portion of the cells may be removed from the reactor and the contents topped up with fresh nutrients and water – semi-continuous processing. Semi-continuous processing allows for a daily 'batch' or series of batches of algae to be removed from the reactor. This has several advantages; firstly the timing of the removal can be optimised such that the algal species is maintained in a constant state of exponential growth. This will maximise the production rate of the algae and ensure high productivity. Secondly, a constant daily removal rate facilitates a more simplistic and smaller footprint for the downstream process delivering consistent product outputs. This lines up with modern processing strategies and just-in-time delivery models. There are some downsides to this form of production that should be mentioned. Over time, if good aseptic conditions are not maintained, then foreign species may well enter the production system and propagate. If this is the case, then a periodic clear-out will be required to ensure product quality. Similarly, the algae will form biofilms on the surface of the reactor and this will eventually reach a critical level where light will no longer penetrate which sufficiently impacts on productivity, again a clear-out will be required and cleaning and maintenance will be needed (see [23] for more details on algae cultivation strategies). Once harvested, algae biomass is typically in the concentration range of around 1 to 5 g L^{-1} or 0.1 to 0.5 wt.%. This is very dilute and any processing on these low level concentrations would be economically non-viable; so the first task following harvesting is always cell separation and recovery. The algae may be dewatered using traditional separation technologies such as sedimentation or flocculation when the species exhibits high settling rates or flocculent aids are viable. More commonly, the algae will be dewatered using either membrane technology or centrifugation. In principle, membranes operate best at low algae concentrations (typically 0.01 to 10wt.%, but 15 wt.% can be achieved) and centrifugation operates best at higher concentrations (typically 5 to 15 wt.%, but 20wt% can be achieved) [24, 25]. For small scale operations, the preferred option would be the use of membrane technologies as the capital and operating costs are lower than that for centrifugation and the efficiency of recovery is very high. However, for larger operations where increased capital investment is justified, then a combination of membranes and a centrifuge would be advisable. In this case, the membranes will concentrate the very dilute algae feed stream and the centrifuge can then take this pre-concentrated algae stream and produce higher slurry concentrations than membranes are capable of while working in the high efficiency range of the centrifuge. Following cell recovery, the water effluent may be discharged to a

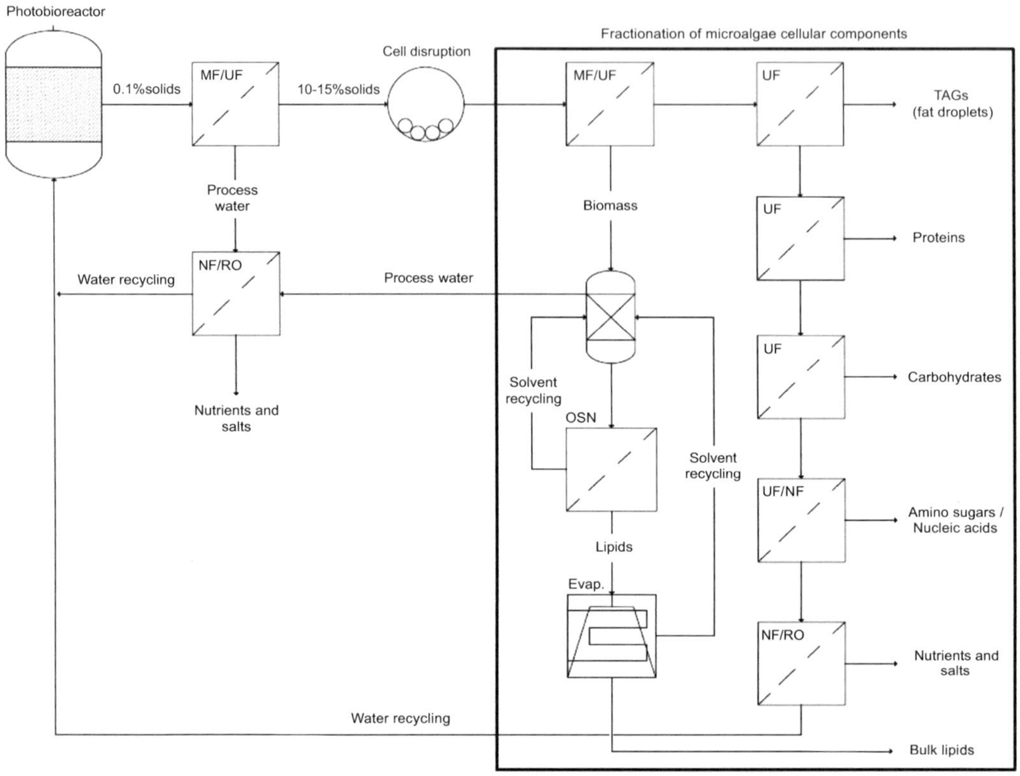

Fig. (4). A potential fractionation methodology for the bulk components contained in the algal cell.

local water course or recycled back to the upstream process as appropriate. The cells themselves will be fed to the cell disruption or cell breakage stage. The purpose of this operation is to open up the cells and release the valuable materials inside. Depending on the physical properties of algal species, mechanical strength and elasticity, this can be a simple operation such as passing the material through a high shear pump or for more difficult cases a bead mill or high pressure homogeniser may be required [26, 27]. The cell disruption stage combined with the dewatering stage are key operations in the downstream process as these two activities are highly energy intensive and may represent as much as 50% of the downstream costs. Thus, selection of the most appropriate technology and optimisation is essential. Following cell disruption, the resultant cell fractions must be harvested to deliver the bulk product streams from the algae process. These are typically proteins, carbohydrates, lipids and high value (fine) products. Several fractions may be obtained in aqueous format from the soluble materials using membrane technology in sequence, such as microfiltration followed by high molecular weight cut off (MWCO) ultrafiltration, followed by low MWCO ultrafiltration, followed by nanofiltration and reverse osmosis. Further fractions may be obtained from the biomass by solvent extraction processes. The production sequence outlined was explored by [28] and is illustrated in Fig. (**4**) with some of the resultant products highlighted in Fig. (**5**). The final stage of the algal biorefinery is further processing of the resultant bulk products to usable end products. This may include transesterification of the lipids to biodiesel or other liquid fuels, conversion of the proteins and carbohydrates to nutritional or cosmetic products, further refining to produce high quality pharmaceuticals or platform chemicals and potentially some nanomaterials.

Fig. (5). Products obtained from bulk separation of the algal biomass. Left to right; spray-dried algae, proteins, carbohydrates, lipid rich biomass and lipids obtained *via* solvent extraction cell.

ARE HIGH VALUE PRODUCTS FROM ALGAL SOURCES REALISABLE?

Currently, there are very few microalgae derived products available on the open market. Although there has been a significant development towards the use of microalgae in the food, feed, chemical and pharmaceutical sectors. The future for microalgae appears to need integrated solutions which exploit low-value inputs such as free energy (*e.g.* sunlight), waste effluents rich in nitrogen (*e.g.* wastewater), and flue gases rich in carbon dioxide (*e.g.* power station gasses) with a range of high-value commercial applications [28]. The exploitation of low-value inputs such as sunlight and waste effluents introduces some significant issues when aspiring to deliver high-value products, particularly for the pharmaceutical sector. The 2008 Waste Framework Directive (Directive 2008/98/EC) describes waste as 'any substance or object which the holder discards'. Thus, waste effluents by very nature are clearly defined as waste and this has serious processing implications for the resulting algal products, which may also be classified as waste and will therefore be unsuitable for pharmaceutical or nutritional use. Recovery of wastes for recycling and reuse purposes is technically possible; Gerardo *et al.* [29] have shown that the recovery of nitrates and phosphates from manure digestate is perfectly feasible and the resultant materials are suitable as feedstock for algae production. However, the legal framework for reclassification of such wastes to become viable feedstock is convoluted to say the least and, as a result, there needs to be clear regulatory guidance in this area to ensure product quality and integrity. These issues are obviously easily bypassed by using high quality nutrient feedstock but this comes at a price, it will reduce the end product margins and will have some impact on the process sustainability. There is significant and increasing pressure within the cosmetic, pharmaceutical, biopharmaceutical and nutraceutical sectors to deliver so called 'green pharmaceuticals', *i.e.* those from a production process that is more environmentally friendly and sustainable [30]. This demand is the result of an increased consumer awareness that drug manufacturing processes have a significant environmental impact [31]. In order to realise the value of these potential algal products, the industry must adopt control systems which guarantee quality, purity, potency, efficacy, efficiency, and safety. The quality standards required for the production of active pharmaceutical ingredients are commonly referred to as Good Manufacturing Practice (GMP). These guidelines are outlined for the European Union, Japan and the USA by the ICH guidelines [32] and include: establishing and maintaining strong quality management systems, training and qualification of personnel, record keeping, appropriate premises and equipment, appropriate quality raw materials, establishing robust operating procedures, detecting and investigating product quality deviations, and reliable testing laboratories [33]. Adhering to such standards is a significant investment in time and money and is effectively an operating philosophy that is quite foreign to current algal producers. One of the greatest challenges for the algal industry is perhaps the reproducibility of the operating conditions which leads to consistency and reliability of the product. The reproducibility of bioprocesses is mainly dependent on the continuous reproduction of the environmental conditions for the cells and hence is dependent on the robustness of the equipment used. The major limitation associated with open bioreactors (currently used in large scale algal cultivation) is the large open surface required for exposure to sunlight rendering the equipment totally susceptible to contamination. Similarly, other parameters that require control (such as mixing, aeration rates and temperature) are difficult to accurately maintain during normal operation. However, a suitably designed closed photobioreactor that contains artificial lighting (as well as other control systems for essential parameters such as pH, dissolved oxygen, and temperature) is not only capable of GMP manufacture, but a level of quality control that is far superior to open systems which are simply not capable of such high quality production. The sensible question to ask at this point is what is the cost of this increase in manufacturing control? Clearly, both closed photobioreactors and artificial lighting are more complex than the basic alternative which increases the initial capital costs when compared to a simple open pond. However, sensible design can minimise these additional costs. Furthermore, in the event that reactor designs of this type prove successful, then there is no reason why mass production should not bring the capital expenditure down to a very low level as the major components can be simple plastics and LEDs, both of which are essentially inexpensive. Moreover, there is also an opportunity to design these reactors as disposable units and completely avoid cross contamination and cleaning issues. Then, the real question becomes what is the market value and demand for the end product and does this justify the increased capital expense? When this is the case, closed photobioreactors with artificial light will succeed and will eliminate the variability associated with autotrophic biomass production under natural light and open environments.

ASTAXANTHIN AS A MODEL PRODUCT

Astaxanthin is a xanthophylls carotenoid and has been dubbed the 'king of carotenoids' due to significant antioxidant activity and diverse application for the improvement of human health [34]. As a result, the annual growth and demand for natural pigments is increasing by 10-15%, with the main molecules of interest being astaxanthin and canthaxanthin [35]. In 2013 the global market size for natural astaxanthin for the human market was estimated to be $200 million, with a substantial increase expected and the market to reach $700 million by 2017 [36]. The increase in global demand is mainly due to the extensive research demonstrating an array of health benefits from the consumption of natural astaxanthin. These include applications for anti-inflammatory treatments, anti-diabetic activity, cardiovascular disease prevention, anti-cancer activity, and improvement in eye health. The potential market for natural astaxanthin could easily reach billions of dollars if one or more of these applications were to be employed as a main stream medical treatment. In order for this potential to be realised, bio-astaxanthin needs robust and reliable production methods capable of meeting the strict pharmaceutical regulations. The annual production of *Haematococcus* from commercial facilities is around 300 tons dry weight [37] containing 1.5–3.0% astaxanthin [38]. Therefore, assuming a total astaxanthin yield of 2.25%, the annual production of natural astaxanthin is only in the region of 6.75 tons. Clearly without rapid expansion these production facilities are unlikely to meet the growing demand required for this algal product.

The bottleneck in the current production systems is the technology used. The majority of large-scale cultivations of *Haematococcus* are performed outdoors in a two-stage process utilising natural light. In the first stage the green motile cells are cultivated under near optimal growth conditions, to ensure a maximum biomass yield. During the second stage astaxanthin accumulation is induced, on a commercial scale this is commonly achieved by subjecting the cells to nitrate and phosphate starvation, increasing temperature, increasing light, or increasing the salinity of the culture medium. These cultivation methods produce haematocysts two to three days after the cell has been subjected to the stress environment. *Haematococcus* commonly contains 1.5-3% astaxanthin by weight and can be harvested three to five days after the formation of the haematocysts. Commercially, the first stage is always in closed photobioreactors which can then be either transferred into a low nutrient open pond (Cyanotech USA) or can be transferred into a second closed photobioreactor (Algatechnologies, Isreal) [39, 40]. A number of studies have demonstrated that tailored light wavelengths can also induce significant accumulation of astaxanthin [41, 42], but these methodologies have not yet been implemented at commercial scale. The reliance on natural sunlight has proved to be a significant disadvantage for Cyanotech, the producer of the market-leading astaxanthin brand for human consumption. For the fiscal year 2014, astaxanthin production was well below expected and historical levels, with a reduction in productivity of 16.5% when compared to the fiscal year 2013. Cyanotech cited high wind, rain, and cloud coverage as the reason for low production levels. This reduction in productivity resulted in a $4.1 million shortfall in bulk business when compared to the previous year. The Cyanotech Corporation has also highlighted the need to reduce the required footprint necessary for production and a need to increase the frequency and volume of output from the early stages of production [43]. The quality and quantity of light are critical variables in the control of growth and product formation in microalgae making a significant issue for all commercial production systems reliant on solar illumination. The diurnal fluctuations of light intensity associated with solar illumination results in a decrease in the total biomass concentration by as much as 35% [44]. For high-value products such as natural astaxanthin, a loss of 35% biomass equates to a significant financial loss that may indeed mitigate the additional capital and operating costs of artificial light. Light-emitting diodes (LEDs) offer an important modern lighting technique that is relatively cheap, efficient, and is still rapidly advancing. There are major benefits in using artificial light for the cultivation of microalgae, these are mostly related with the supply of a controlled light source [42]. Obviously, power is needed to drive a lighting system, but the majority of this power could be provided from renewable sources. A comprehensive and thorough economic assessment of astaxanthin production by large scale cultivation of *Haematococcus pluvialis* is given by Li *et al.* [45]. In this assessment, the authors made several assumptions related to the current industrial cultivation and downstream process and concluded that the cost of biomass production was $18 per kg and the cost of astaxanthin product was $718 per kg. If one were to use the most recent estimations of the capital and operating costs for a closed photobioreactor using artificial lighting based on LED technology, then for a similar production capacity the cost of biomass could be $30-39 per kg and the cost of astaxanthin could be $508-647 per kg. These cost estimates are within the range cited by Blanken *et al.* [46]. The cost of biomass is higher, but the increased control delivered by the technology delivers a higher astaxanthin yield and hence a slightly reduced specific production cost for the pigment. Delivered within a biorefinery, astaxanthin of this type could form part of a sustainable product portfolio that includes biofuels from algae.

CONCLUSION

The single aim of producing biofuels from algae using current technologies is not economically feasible when compared to the petroleum industries. By further development of current algal technologies to utilise low value inputs, to co-locate production facilities with allied industries and by significantly improving the mass efficiency of the process through an integrated biorefinery approach, potentially algal biofuels may come to fruition. The biorefinery concept is critical to make this technology work by realising a different product portfolio that harvests both the high value compounds and bulk commodities synergistically. Delivering high value products to industries such as the pharmaceutical sector will require a step change in thinking and control strategies, however, this is not an impossible task and high quality products are achievable from algae. A sustainable future for biofuels is possible but development time and effort are needed.

CONFLICT OF INTEREST

The authors declares no conflict of interest, financial or otherwise.

ACKNOWLEDGEMENTS

Declared none

REFERENCES

[1] Department of Economic and Social Affairs, *Population Division.*. Population Division. World Population Prospects: The 2015 Revision, Key Findings and Advance Tables. Working Paper No. ESA/P/WP.241.

[2] Key World Energy Statistics,, *International Energy Agency*. 2016. www.iea.org.

[3] T.M. Mata, A.A. Martins, and N.S. Caetano, "Microalgae for biodiesel production and other applications: a review", *Renew. Sustain. Energy Rev.,* vol. 14, no. 1, pp. 217-232, 2010. [http://dx.doi.org/10.1016/j.rser.2009.07.020]

[4] Y. Li, M. Horsman, N. Wu, C.Q. Lan, and N. Dubois-Calero, "Biofuels from microalgae", *Biotechnol. Prog.,* vol. 24, no. 4, pp. 815-820, 2008. [PMID: 18335954]

[5] M.J. Griffiths, and T.L. Susan, "Harrison. "Lipid productivity as a key characteristic for choosing algal species for biodiesel production", *J. Appl. Phycol.,* vol. 21, no. 5, pp. 493-507, 2009. [http://dx.doi.org/10.1007/s10811-008-9392-7]

[6] J.R. Benemann, "CO2 mitigation with microalgae systems", *Energy Convers. Manage.,* vol. 38, pp. S475-S479, 1997. [http://dx.doi.org/10.1016/S0196-8904(96)00313-5]

[7] Y. Chisti, "Biodiesel from microalgae beats bioethanol", *Trends Biotechnol.,* vol. 26, no. 3, pp. 126-131, 2008.
[http://dx.doi.org/10.1016/j.tibtech.2007.12.002] [PMID: 18221809]

[8] J.J. Mayers, K.J. Flynn, and R.J. Shields, "Rapid determination of bulk microalgal biochemical composition by Fourier-Transform Infrared spectroscopy", *Bioresour. Technol.,* vol. 148, pp. 215-220, 2013.
[http://dx.doi.org/10.1016/j.biortech.2013.08.133] [PMID: 24050924]

[9] N. Ghimire, Economics of biodiesel production from microalgae., 2008.

[10] P. Collet, L. Lardon, A. Hélias, S. Bricout, I. Lombaert-Valot, B. Perrier, O. Lépine, J-P. Steyer, and O. Bernard, "Biodiesel from microalgae–Life cycle assessment and recommendations for potential improvements", *Renew. Energy,* vol. 71, pp. 525-533, 2014.
[http://dx.doi.org/10.1016/j.renene.2014.06.009]

[11] H. Passell, H. Dhaliwal, M. Reno, B. Wu, A. Ben Amotz, E. Ivry, M. Gay, T. Czartoski, L. Laurin, and N. Ayer, "Algae biodiesel life cycle assessment using current commercial data", *J. Environ. Manage.,* vol. 129, pp. 103-111, 2013.
[http://dx.doi.org/10.1016/j.jenvman.2013.06.055] [PMID: 23900083]

[12] J.C. Quinn, and R. Davis, "The potentials and challenges of algae based biofuels: a review of the techno-economic, life cycle, and resource assessment modeling", *Bioresour. Technol.,* vol. 184, pp. 444-452, 2015.
[http://dx.doi.org/10.1016/j.biortech.2014.10.075] [PMID: 25453439]

[13] A.D. Curzons, and J.C. David, "Constable, David N. Mortimer, and Virginia L. Cunningham. "So you think your process is green, how do you know?—Using principles of sustainability to determine what is green–a corporate perspective", *Green Chem.,* vol. 3, no. 1, pp. 1-6, 2001.
[http://dx.doi.org/10.1039/b007871i]

[14] J-P. Wauquier, *Petroleum Refining: Crude oil, petroleum products, process flowsheets.* vol. Vol. 1. Éditions Technip, 1995.

[15] M. Vanthoor-Koopmans, R.H. Wijffels, M.J. Barbosa, and M.H. Eppink, "Biorefinery of microalgae for food and fuel", *Bioresour. Technol.,* vol. 135, pp. 142-149, 2013.
[http://dx.doi.org/10.1016/j.biortech.2012.10.135] [PMID: 23186688]

[16] H-W. Yen, "I-Chen Hu, Chun-Yen Chen, Shih-Hsin Ho, Duu-Jong Lee, and Jo-Shu Chang. "Microalgae-based biorefinery–from biofuels to natural products", *Bioresour. Technol.,* vol. 135, pp. 166-174, 2013.
[http://dx.doi.org/10.1016/j.biortech.2012.10.099] [PMID: 23206809]

[17] A. Mishra, and B. Jha, "Isolation and characterization of extracellular polymeric substances from micro-algae Dunaliellasalina under salt stress", *Bioresour. Technol.,* vol. 100, no. 13, pp. 3382-3386, 2009.
[http://dx.doi.org/10.1016/j.biortech.2009.02.006] [PMID: 19272770]

[18] A.N. Glazer, "Phycobiliproteins—a family of valuable, widely used fluorophores", *J. Appl. Phycol.,* vol. 6, no. 2, pp. 105-112, 1994.
[http://dx.doi.org/10.1007/BF02186064]

[19] E. Roselló-Soto, C.M. Galanakis, M. Brnčić, V. Orlien, F.J. Trujillo, R. Mawson, K. Knoerzer, B.K. Tiwari, and F.J. Barba, "Clean recovery of antioxidant compounds from plant foods, by-products and algae assisted by ultrasounds processing. Modeling approaches to optimize processing conditions", *Trends Food Sci. Technol.,* vol. 42, no. 2, pp. 134-149, 2015.
[http://dx.doi.org/10.1016/j.tifs.2015.01.002]

[20] S. Bellou, M.N. Baeshen, A.M. Elazzazy, D. Aggeli, F. Sayegh, and G. Aggelis, "Microalgal lipids biochemistry and biotechnological perspectives", *Biotechnol. Adv.,* vol. 32, no. 8, pp. 1476-1493, 2014.

[http://dx.doi.org/10.1016/j.biotechadv.2014.10.003] [PMID: 25449285]

[21] S.C. Fernandes, A. Alonso-Varona, T. Palomares, V. Zubillaga, J. Labidi, and V. Bulone, "Fernandes, Susana CM, Ana Alonso-Varona, Teodoro Palomares, Verónica Zubillaga, Jalel Labidi, and Vincent Bulone. "Exploiting mycosporines as natural molecular sunscreens for the fabrication of UV-absorbing green materials", *ACS Appl. Mater. Interfaces,* vol. 7, no. 30, pp. 16558-16564, 2015. [http://dx.doi.org/10.1021/acsami.5b04064] [PMID: 26168193]

[22] S.P. Cuellar-Bermudez, I. Aguilar-Hernandez, D.L. Cardenas-Chavez, N. Ornelas-Soto, M.A. Romero-Ogawa, and R. Parra-Saldivar, "Extraction and purification of high-value metabolites from microalgae: essential lipids, astaxanthin and phycobiliproteins", *Microb. Biotechnol.,* vol. 8, no. 2, pp. 190-209, 2015.
[http://dx.doi.org/10.1111/1751-7915.12167] [PMID: 25223877]

[23] D. White, A. Silkina, S. Skill, D. Oatley-Radcliffe, S. Van Den Hende, A. Ernst, C. De Viser, W. Van Dijk, M. Davey, and J. Day, *Best Practices for the Pilot-Scale Cultivation of Microalgae.,* no. 4, p. 34, 2015.

[24] M.L. Gerardo, D.L. Oatley-Radcliffe, and R.W. Lovitt, "Minimizing the energy requirement of dewatering scenedesmus sp. by microfiltration: performance, costs, and feasibility", *Environ. Sci. Technol.,* vol. 48, no. 1, pp. 845-853, 2014.
[http://dx.doi.org/10.1021/es4051567] [PMID: 24341825]

[25] T-S. Sim, A. Goh, and E.W. Becker, "Comparison of centrifugation, dissolved air flotation and drum filtration techniques for harvesting sewage-grown algae", *Biomass,* vol. 16, no. 1, pp. 51-62, 1988. [http://dx.doi.org/10.1016/0144-4565(88)90015-7]

[26] J. Kim, G. Yoo, H. Lee, J. Lim, K. Kim, C.W. Kim, M.S. Park, and J-W. Yang, "Methods of downstream processing for the production of biodiesel from microalgae", *Biotechnol. Adv.,* vol. 31, no. 6, pp. 862-876, 2013.
[http://dx.doi.org/10.1016/j.biotechadv.2013.04.006] [PMID: 23632376]

[27] R. Halim, R. Harun, M.K. Danquah, and P.A. Webley, "Microalgal cell disruption for biofuel development", *Appl. Energy,* vol. 91, no. 1, pp. 116-121, 2012.
[http://dx.doi.org/10.1016/j.apenergy.2011.08.048]

[28] M.L. Gerardo, D.L. Oatley-Radcliffe, and R.W. Lovitt, "Integration of membrane technology in microalgae biorefineries", *J. Membr. Sci.,* vol. 464, pp. 86-99, 2014.
[http://dx.doi.org/10.1016/j.memsci.2014.04.010]

[29] M.L. Gerardo, N.H. Aljohani, D.L. Oatley-Radcliffe, and R.W. Lovitt, "Moving towards sustainable resources: Recovery and fractionation of nutrients from dairy manure digestate using membranes", *Water Res.,* vol. 80, pp. 80-89, 2015.
[http://dx.doi.org/10.1016/j.watres.2015.05.016] [PMID: 25996755]

[30] J. Ottman, *The new rules of green marketing: Strategies, tools, and inspiration for sustainable branding.* Berrett-Koehler Publishers, 2011.

[31] K.V. Gernaey, A.E. Cervera-Padrell, and J.M. Woodley, "A perspective on PSE in pharmaceutical process development and innovation", *Comput. Chem. Eng.,* vol. 42, pp. 15-29, 2012.
[http://dx.doi.org/10.1016/j.compchemeng.2012.02.022]

[32] ICH, *Harmonised Tripartite Guideline.* Q7: Good Manufacturing Practice Guide for Active Pharmaceutical Ingredients, Current Step 4 (2005).

[33] World Health Organization, WHO Expert Committee on Specifications for Pharmaceutical Preparations, "Thirty seventh report, Geneva", *Annex,* vol. 2, pp. 77-135, 2003.

[34] B. Capelli, D. Bagchi, and G.R. Cysewski, "Synthetic astaxanthin is significantly inferior to algal-based astaxanthin as an antioxidant and may not be suitable as a human nutraceutical supplement", *Nutrafoods,* vol. 12, no. 4, pp. 145-152, 2013.
[http://dx.doi.org/10.1007/s13749-013-0051-5]

[35] D.M. Pereira, P. Valentão, and P.B. Andrade, "Marine natural pigments: chemistry, distribution and analysis", *Dyes Pigments,* vol. 111, pp. 124-134, 2014. [http://dx.doi.org/10.1016/j.dyepig.2014.06.011]

[36] Y. Kuroda, *AstaReal set to increase production of natural astaxanthin to meet global demand.* Fuji Health Science, Inc.: Burlington, 2013.

[37] D. Bagchi, F.C. Lau, D.K. Ghosh, Ed., *Biotechnology in functional foods and nutraceuticals.* CRC Press, 2010. [http://dx.doi.org/10.1201/9781420087123]

[38] R.T. Lorenz, and G.R. Cysewski, "Commercial potential for Haematococcus microalgae as a natural source of astaxanthin", *Trends Biotechnol.,* vol. 18, no. 4, pp. 160-167, 2000. [http://dx.doi.org/10.1016/S0167-7799(00)01433-5] [PMID: 10740262]

[39] O. Pulz, and W. Gross, "Valuable products from biotechnology of microalgae", *Appl. Microbiol. Biotechnol.,* vol. 65, no. 6, pp. 635-648, 2004. [http://dx.doi.org/10.1007/s00253-004-1647-x] [PMID: 15300417]

[40] M.A. Borowitzka, "High-value products from microalgae—their development and commercialisation", *J. Appl. Phycol.,* vol. 25, no. 3, pp. 743-756, 2013. [http://dx.doi.org/10.1007/s10811-013-9983-9]

[41] T. Katsuda, H. Shiraishi, N. Ishizu, R. Ranjbar, and S. Katoh, "Effect of light intensity and frequency of flashing light from blue light emitting diodes on astaxanthin production by Haematococcus pluvialis", *J. Biosci. Bioeng.,* vol. 105, no. 3, pp. 216-220, 2008. [http://dx.doi.org/10.1263/jbb.105.216] [PMID: 18397771]

[42] P.S. Schulze, L.A. Barreira, H.G. Pereira, J.A. Perales, and J.C. Varela, "Light emitting diodes (LEDs) applied to microalgal production", *Trends Biotechnol.,* vol. 32, no. 8, pp. 422-430, 2014. [http://dx.doi.org/10.1016/j.tibtech.2014.06.001] [PMID: 25012573]

[43] AIM, (2014). AlgaeIndustryMagazine.com, http://www.algaeindustrymagazine.com/cyanotech-improving-astaxanthin-production/, (accessed September 2016).

[44] C-Y. Chen, K-L. Yeh, R. Aisyah, D-J. Lee, and J-S. Chang, "Cultivation, photobioreactor design and harvesting of microalgae for biodiesel production: a critical review", *Bioresour. Technol.,* vol. 102, no. 1, pp. 71-81, 2011. [http://dx.doi.org/10.1016/j.biortech.2010.06.159] [PMID: 20674344]

[45] J. Li, D. Zhu, J. Niu, S. Shen, and G. Wang, "An economic assessment of astaxanthin production by large scale cultivation of Haematococcus pluvialis", *Biotechnol. Adv.,* vol. 29, no. 6, pp. 568-574, 2011. [http://dx.doi.org/10.1016/j.biotechadv.2011.04.001] [PMID: 21497650]

[46] W. Blanken, M. Cuaresma, R.H. Wijffels, and M. Janssen, "Cultivation of microalgae on artificial light comes at a cost", *Algal Res.,* vol. 2, no. 4, pp. 333-340, 2013. [http://dx.doi.org/10.1016/j.algal.2013.09.004]

[47] B. Schlarb-Ridley, and B. Parker, *A UK roadmap for algal technologies.* NERC-TSB Algal Bioenergy Special Interest Group, 2013.

CHAPTER 14

Energy and Economic Challenges in the Growth and Harvesting of Microalgae. The *Chlorella vulgaris* Case

Daissy Lorena Restrepo Serna, Juan Carlos Higuita Vásquez and **Carlos Ariel Cardona Alzate***

Instituto de Biotecnología y Agroindustria, Departamento de Ingeniería Química, Universidad Nacional de Colombia, Manizales, Colombia

Abstract: A description of the importance of microalgae at industrial level and especially the case of *Chlorella vulgaris* are presented in this chapter. The cost and some existing technologies for microalgae oil production were analyzed. Furthermore, some of these technologies were selected for the design of microalgae oil production process using the simulation software Aspen Plus.

Keywords: Microalgae, *Chlorella vulgaris*, Microalgae culture, Microalgae harvest.

CULTURE OF MICROALGAE TODAY

Microalgae and their Importance

Microalgae are very important not only for their role in nature but also for the fact that they contain different bioactive components [1]. These components can be used in different industries such as; food industry, pharmaceutical [2] and especially biofuels [3]. With this wide variety of usage, microalgae culture at industrial levels is projected as a viable option but the challenges remain in terms of obtaining higher yields and productivities for the extraction of their bioactive components.

Wastewater treatment and carbon dioxide (CO_2) mitigation are two examples of the most relevant applications that involve microalgae [4]. This implicates that microalgae grow with components such as, phosphates, nitrates and some metals,

* **Corresponding author Carlos Ariel Cardona Alzate:** Instituto de Biotecnología y Agroindustria, Departamento de Ingeniería Química, Universidad Nacional de Colombia, Manizales campus, Manizales, Colombia; Tel: (+57) (6) 8879300 ext 55354; E-mail: ccardonaal@unal.edu.co

José Carlos Magalhães Pires (Ed.)

which can be found in wastewater for the microalgae to assimilate them [4], reducing their concentration in the water. Furthermore, they can help reducing greenhouse gases (mainly CO_2) emissions, which, in recent years, has been increasing in the atmosphere due to continuous emissions by different industries like the transport sector.

Microalgae are a potential source for the production of many products, which make them the center of biotechnology research. As shown in Table **1**, many products with a wide variety of applications can be obtained from microalgae [5].

Among the most marketed microalgae, products derived from the following genera *Isochrysis, Chaetoceros, Chlorella, Arthorospira (Spirulina)* and *Dunaliella* are mainly used as nutritional supplements [6].

Table 1. Biotechnological products and applications from microalgae [5, 6].

Product	Applications	Microalgae Producers
Eicosapentaenoic acid (EPA)	Nutritional supplements, aquaculture feed	*Pavlova, Nannochloropsis, Monodus & Phaeodactytum*
Docohecaenoic acid (DHA)	Infant formula, nutritional supplements, aquaculture feed	*Crypthecodiuimu & Schizochytrium*
γ - linoleic acid (GLA)	Nutritional supplements	*Spirulina*
Arachidonic acid (AA)	Nutritional supplements	*Porphyridium*
Phycocyanin	Natural dye health food and cosmetics (lipsticks and eyeliners) antioxidants	*Spirulina platensis*
Phycoerythrin	Fluorescent agent, tool for biomedical research, diagnostic tool	Red algae (*e.g. Porphyridium cruentum*)
Carotenoids	Health food, food supplement	*Muriellopsis sp., Dunaliella salina*
β – carotene	Food colorant; antioxidant; cancer-preventive properties	*Dunaliella salina*
Astaxanthin	Pigmenter for salmon, antioxidant	*Haematococcus pluvialis*
Mycosporine-like amino acids (MAA)	UV-screening agent; sunscreen	*Aphannizomenon flos-aquae*
Polysaccharides	Viscosifiers, lubricants and flocculants for industrial applications; antiviral agent	*Porphyridium cruentum*
Phycotoxins – okadaic acid, gonyautoxins, & yessotosins	Experimental tools for investigations on neurodegenerative diseases	Dinoflagellates (*e.g. Amphidnium, Prorocentrum & Dinophysis*)
Lipids – triglycerides and hydrocarbons	Biofuels	*Chlorella protothecoides Botrycoccus braunii*

Microalgae *Chlorella vulgaris*

Among the different species of microalgae that exist, one of the most used in research is *Chlorella vulgaris* due to its high capacity for CO_2 fixation, production of oils (especially those used for the production of biodiesel) and other products [7, 8].

The characteristics which make *Chlorella vulgaris* a viable alternative for biomass production, are as follows [9]:

- High growth rate
- It can grow in different culture media and is easily adapted to different conditions
- Resistant to adverse conditions
- It has the possibility for a high lipid and starch contents, if the required conditions of nitrogen and phosphate concentrations are presented.

Applications

Biofuels

Given the growing demand for fuels nowadays, microalgae-derived biofuels are considered one of the most viable alternatives. Microalgae have a rapid growth, they do not compete with food crops and present higher yields of oil production used for obtaining biofuels [9] (Table **2**).

Table 2. Comparison of CO_2 fixing capacities and biodiesel productivity of microalgae and higher plants [10, 11].

Plants	CO_2 Capture Capacity (%)	Productivity ($g/m^2/d$)	Lipid Content of Biomass (%)	Biodiesel Productivity ($g/m^2/d$)
Chlorella	22 – 26	17 – 20	40 – 60	5.4 – 6
Miscanthus giganteus	10 – 22	8 – 17	N/A	2.4 – 5.4
Oilseed rape	1 – 2	0.8 – 1.6 (of seed)	40 – 44 (of seed)	0.24 – 5.4
Soya	N/A	0.65 (of seed)	20 (of seed)	0.2
Jatropha	N/A	2.2 (of seed)	30 (of seed)	0.66

CO_2 Fixation

One of the most important characteristics possessed by microalgae is the ability to fix CO_2, either directly from the atmosphere or from combustion gases which can contain between 5- 30% CO_2 that can be used by the microalgae as a carbon source for the production of biomass [10]. This capability makes them a viable alternative for decreasing CO_2 in the atmosphere, considering their higher CO_2 fixation rates compared to other plants (Table **2**). Table **3** presents some microalgae and their capacities for CO_2 fixation.

Table 3. Capacities of CO_2 fixation for some microalgae [12, 13].

Microalgae	CO_2 Fraction in Inlet Gas [% vol]	CO_2 Fixation [g/L/day]
Anabaena sp.	Air	1.45
Aphanothece microscopica Nageli	15	5.435
Chlorococcum littorale	40	1
Chlorella kessleri	6	0.164
Chlorella kessleri	18	0.163
Chlorella vulgaris	Air	0.075
Chlorella vulgaris	0.09	3.45
Chlorella vulgaris	0.8 - 1.0	6.24
Chlorella vulgaris	4	0.71
Chlorella vulgaris	8	0.55
Chlorella vulgaris	15	0.0625
Chlorella sp.	10	1.767
Chlorella sp.	40	1
Spirulina sp.	12	0.413
Dunaliella	3	0.313

Human Nutrition

The microalgae biomass contains three main components: proteins, carbohydrates and lipids. They are also a good source of vitamins (A, C, B1, B2, and B6) and minerals (iodine, potassium, iron, magnesium and calcium). By possessing these compositions they have been used as a food source in Asian countries for hundreds of years. Given its nutritional value, they are today widely consumed around the world [1].

Due to the high amounts of proteins, lipids, polysaccharides, pigments and vitamins present in *C. vulgaris*, this microalgae is one of the few microalgae that has the possibility of being used as a food supplement or additive, colorant and food emulsion [9]. Among the major economically important products found in

Chlorella, several by-products that are used as fruit and vegetable preservatives are considered.

Use in Pharmaceuticals

Microalgae are considered a rich source of primary and secondary metabolites, because they have the potential to produce bioactive compounds in culture, which are difficult to produce by chemical synthesis. These metabolites may be potential bioactive compounds of interest in the pharmaceutical industry [1].

CHALLENGES IN THE PROCESS OF OIL EXTRACTION FROM MICROALGAE

Fig. (**1**) shows some products that can be obtained from the cultivation of *Chlorella vulgaris*. The extraction process is paramount in the determination of the product quality. The selected microalgae culture conditions can also influence the diversity of products that can be obtained thus affecting the purification and extraction processes.

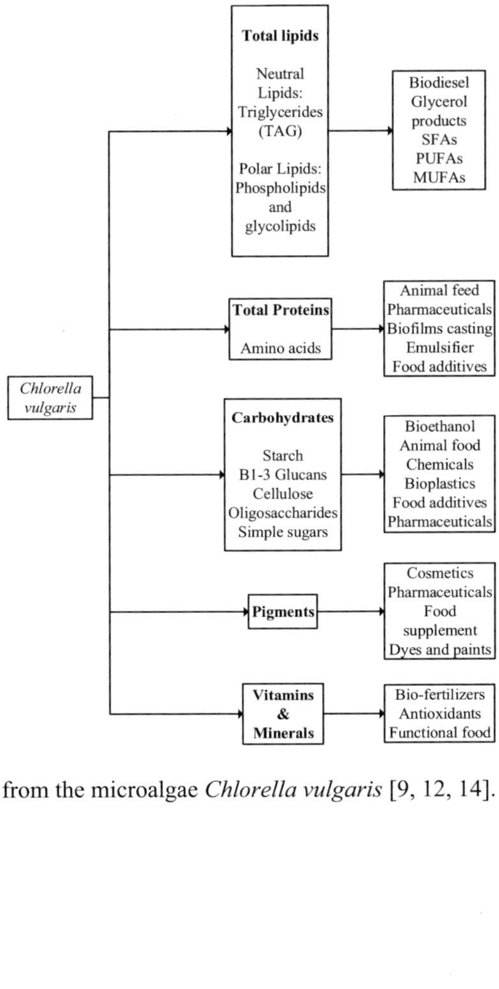

Fig. (1). Products obtained from the microalgae *Chlorella vulgaris* [9, 12, 14].

Culture Stage

In order to provide the sufficient conditions for microalgae growth, the following nutrients must be added to the culture media: carbon, nitrogen, phosphate, sulfur, inorganic salts, trace elements and some vitamins (Table **4**).

Table 4. Nutrient functions during microalgae cultivation [10].

Nutrition	Function
Carbon source	Provide C to the hole cell
Nitrogen source	Provide N to the hole cell
Phosphorus	Provide P to every reaction in cells
Sulfur	Provide S to proteins and reactions
Inorganic salts	Maintain cell structure and activity
Trace elements	Be coenzyme factors
Vitamin	Aid to cell division

Microalgae consume in different proportions all the nutrients found in the culture medium being the carbon source the highest which presents different yields depending whether it is of an organic or inorganic nature. On the other hand, different consumption rates of the nitrogen, phosphorus and sulfur sources are also presented (Table **5**).

Table 5. Consumption of nutrients [2].

Component	Yield %	$Y_{s/b}$
CO_2	74.5	2.30
Sulfate	69.2	0.19
Nitrate	88.2	0.33
Potassium	17.1	0.01
Calcium	18.8	0.03
Magnesium	86.3	0.10
Phosphorus	65.8	0.03

Depending on the used carbon source, the cultivation of microalgae can be carried out under different conditions and depending on these conditions, the culturing system (*e.g.*, open systems or photobioreactors) can be selected (Table **6**).

Although oil production by microalgae depends on the strain used, factors such as the medium appear to influence easily. Thus, when this process is conducted in a heterotrophic medium, higher oil productivities compared to other conditions can be obtained. Furthermore, the phototrophic medium is the most frequently used because it is presented as an alternative for the mitigation of CO_2, but compared to the heterotrophic medium, it has less oil productivity due to a lower biomass productivity [15].

Table 6. Comparison of the characteristics of different cultivation conditions [16 - 18].

Cultivation Condition	Energy Source	Carbon Source	Cell Density	Reactor Scale-up	Cost	Issues Associated with Scaled with Scale-up
Phototrophic	Light	Inorganic	Low	Open pond or photobioreactor	Low	Low cell density High condensation cost
Heterotrophic	Organic	Organic	High	Conventional fermenter	Medium	Contamination High substrate cost
Mixotrophic	Light and organic	Inorganic and organic	Medium	Closed photobioreactor	High	Contamination High equipment cost High substrate cost
Photoheterotrophic	Light	Organic	Medium	Closed photobioreactor	High	Contamination High equipment cost High substrate cost

Harvesting Stage

Depending on the conditions under which microalgae are cultivated, they will reach their maximum growth at different time rates [19]. Then, it is necessary to ensure that the concentration of microalgae in the medium is the highest possible thereby the calculation of the best growing conditions and the selection of the optimal carbon source must be carefully selected.

To determine the technique for harvesting microalgae, it is necessary to know the characteristics of the microalgae, likewise the products that are desired to be obtained to prevent contamination of the components during the process. Generally, the harvesting process is conducted in two stages (*i.e.*, bulk harvesting and thickening) as shown in Table **7** [20].

Based on the culture volume, costs and other factors, it is possible to select several harvesting techniques making necessary the definition of these conditions so the best option for harvesting biomass is achieved (Table **8**).

Table 7. Stage of harvesting process [20, 21].

Stage	Description	Techniques
Bulk harvesting	The concentration factors for this operation are generally 100-800 times to reach 2-7% total solid matter.	Flocculation Flotation
	Depend of the initial biomass concentration	Gravity sedimentation
Thickening	The proposed is to concentrate the slurry	Centrifugation
	Is generally a more energy intensive step than bulk harvesting	Filtration

Table 8. Description of harvesting techniques [12, 20, 22].

Technique	Description	Observation	
Flocculation	Increase the effective particle size	The efficacy of the method was very sensitive to pH	Ferric chloride, aluminum sulfate and ferric sulfate
	Is a preparatory step prior to other harvesting methods	Filtration, flotation or gravity sedimentation	
Flotation	Are based on the trapping of algae cells using dispersed micro-air bubbles Does not require any addition of chemicals	There is very limited evidence of its technical or economic viability	
Gravity sedimentation	Are based on Stoke´s law	The is only suitable for large (ca. >70 µm) microalgae	Spirulina
	Is the most common harvesting technique for algae biomass in wastewater treatment	Allows the treatment of large volumes of water at low prices	
Centrifugation	Is preferred for harvesting of high-value metabolites	High energy costs	
	The process is rapid and energy intensive		
	The biomass recovery depends on the settling characteristics of the cells, slurry residence time in the centrifuge, and settling depth	Potentially higher maintenance requirements due to freely moving parts	
Filtration	It is suitable for relatively large microalgae (>70 µm)	For small microalgae is necessary to use alternative to conventional filtration techniques	Microfiltration or ultrafiltration

(Table 8) contd.....

Technique	Description	Observation	
Filtration	Processing small volumes can be more cost effective compared with centrifugation.	For large-scale processes is more cost effective use of centrifugation due to pumping costs and replacement of the membrane	

Among the presented harvesting techniques, those that display the best alternatives are centrifugation and flocculation. The former being the most viable option from the economic point of view [12]. This is explained by the fact that this technique uses the minimum amount of chemicals and energy, and in some cases, it helps with cell disruption which improves the liberation of inner cellular components [15].

Extraction Stage

Once biomass is harvested, decomposition occurs so it is necessary to carry out a rapid extraction of the bioactive components. Among the different extraction processes dehydration or drying can be found (Table **9**).

Table 9. Extraction techniques based on dehydrated or dried biomass [23 - 25].

Technique	Observations	Advantage - Disadvantage
Sun drying	The cheapest dehydration method	Long drying times
		Large drying surfaces
		The risk of material loss
Spray drying	Commonly used for extraction of high value products	Relatively expensive
		Significant deterioration of some algal pigments
Freeze drying	Equally expensive that spray drying	This eases extraction of oils

Other ways to carry out the extraction of bioactive components are [26]:

- High-pressure homogenizers
- Autoclaving
- Addition of hydrochloric acid, sodium hydroxide or alkaline lysis

For the extraction of metabolites such as astaxanthin, β-carotene and fatty acids from algal biomass, the use of solvents is widely utilized, which cause alterations in the cell membrane leading to the liberation of these components [27]. Table **10** shows some popular methods used for the extraction of oil from microalgae. Some of the most common limitations encountered in these methods are high power consumption of the solvent used and the expensive/difficult to scaling up the process.

PROCESS DESIGN FOR THE PRODUCTION OF OIL FROM MICROALGAE

In order to obtain the desired final product (microalgae oil), it is necessary to define the steps - stages that will enable this purpose [31]. In this sense, the definition of variables such as, the conditions of the culture medium, operating conditions, the harvesting and the extraction methods is necessary.

Table 10. Advantages and disadvantages of popular extraction methods for recovering oil from microalgae [28 - 30].

Extraction Method	Advantages	Limitations
Pressing	Easy to use, no solvent involved	Large amount of sample required, slow process
Solvent extraction	Solvents used are relatively inexpensive; results are reproducible	Most organic solvents are highly flammable and/or toxic; solvent recovery is expensive and energy intensive; large volume of solvent is required
Supercritical fluid extraction	Non-toxic (no organic solvent residue in extracts); 'green solvent'; non-flammable and simple operation	High power consumption; expensive/difficult to scale up at this time
Ultrasonic-assisted	Reduced extraction time; reduced solvent consumption; greater penetration of solvent into cellular materials; improved release of cell contents into bulk medium	High power consumption; difficult to scale up

Conditions for the Culture Medium

For the preparation of the culture medium, there are nutrients and concentrations used for the preparation of a Basal medium [17], which provides high concentrations of nitrogen and other nutrients that in turn are the influential factors for microalgae growth (Table **11**). The operating conditions are presented in Table **12**.

Table 11. Compositions of Basal medium [17].

Component	Composition [g/L]
$CaCl_2 * 2H_2O$	0.1106
KNO_3	0.25

(Table 11) contd.....

Component	Composition [g/L]
$MgSO_4 * 7H_2O$	1.00
KH_2PO_4	1.25
$FeSO_4 * 7H_2O$	0.0498
$EDTA * 2Na$	0.50
H_3BO_3	0.1142
$MnCl_2 * 4H_2O$	0.0144
$ZnSO_4 * 7H_2O$	0.0882
$Na_2MoO_4 * 2H_2O$	0.0119
$CuSO_4 * 5H_2O$	0.0157
$CO(NO_3)_2 * 6H_2O$	0.0049

Table 12. Conditions of operation [17].

Temperature	25 °C
pH	6.2
Agitation	150 rpm

Definition of the Harvesting Process

In order to improve processes such as harvesting and extraction, it is necessary to perform a combination of different extraction methods to make the process more efficient, thus enabling greater recovery of bioactive components contained in microalgae. For this reason, flocculation and then centrifugation are chosen as the extraction techniques.

Oil Extraction Method

The main purpose of the cultivation process using *Chlorella vulgaris* is oil production. The selected extraction method is solvent extraction using butanol since it showed the best yields [32].

ECONOMIC ANALYSIS OF *CHLORELLA VULGARIS* PRODUCTION

Table **13** presents different production costs for microalgae oil. The cost of microalgae oil production is variable mainly due to the technology used for the cultivation, harvesting and extraction. Based on this, it is possible to obtain a large variety of production costs, depending on growing conditions and production volumes [35].

Table 13. Production costs of microalgae oil.

Price	Reference
0.56 USD/Kg	[33]
8.52 USD/gal	[32]
18.10 USD/gal	[32]
3.80 USD/Kg	[34]
2.95 USD/Kg	[34]

Since the results for the production costs of microalgae oil can be uncertain and have a great variability, it is necessary to perform a simulation process that would enable the evaluation of the process viability by means of matter and energy balances.

METHODOLOGY

In order to obtain different matter and energy balances for process design and economic analysis, the Aspen plus and Aspen economic analyzer software were used, respectively. Since the purpose of the suggested process is the production of microalgae oil using a specific carbon source, an analysis that enables the comparison of carbon usage, biomass and oil production is performed.

All the scenarios and analyzes presented here are based on results obtained from the simulations carried out using Aspen plus. Each of the aspects considered for the process simulation are presented below.

Definition of Variables for the Simulation Procedure

The model proposed by Yoo *et al.* [36] was used for the determination of glucose consumption when microalgae was grown under heterotrophic conditions. This model allows the calculation of biomass production, glucose and nutrients (such as nitrogen) consumption (Table **14**). The parameters used are those reported in the article.

Table 14. Kinetic growth of microalgae [36].

Biomass production	$\frac{dx}{dt}$	$\mu * x - xD$
Substrate consumption	$\frac{dS_1}{dt}$	$-\rho x + S_1^i * \frac{f_1^i}{V} - S_1 D$

(Table 14) contd.....

Nitrogen consumption	$\frac{dS_2}{dt}$	$-\frac{1}{Y_{xs}}\mu x - \frac{1}{Y_{is}\pi x} + S_2^i * \frac{f_2^i}{V} - S_2 D$
Intracellular nitrogen concentration	$\frac{dQ}{dt}$	$\rho x - \frac{1}{Yxq}\mu x - QD$
Intracellular lipid concentration.	$\frac{dL}{dt}$	$\pi x - LD$
Volume	$\frac{dV}{dt}$	$VD - f_o$

With the selected techniques shown in the previous section, a process to obtain oil from microalgae was designed. This process involves a photobioreactor or a conventional fermenter (depending on the carbon source used), a feeding stream of nutrients (Table **11**) that takes into account the evaporation of either the CO_2 that is not consumed or the gases expected to be generated during the process. Then a flocculation stage is considered in which aluminum sulfate is added as coagulant. Subsequently the resulting current is transported to a centrifuge to recover a large portion of the biomass. The resulting liquid stream (which still contains nutrients) is then recycled to the bioreactor. Biomass is then transferred to an extraction step in which the solvent (butanol) is added. For solvent recovery, the stream containing the remaining butanol is carried to an evaporator and the recovered butanol is subsequently recycled to the extraction process (Fig. **2**).

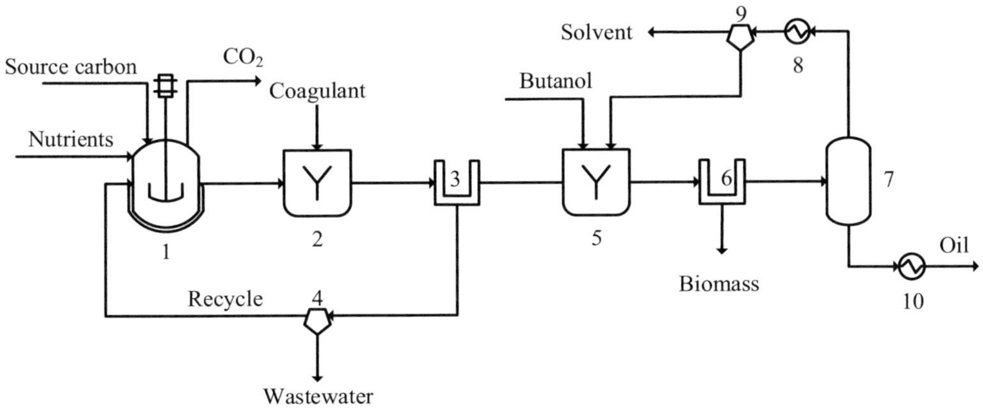

Fig. (2). Process scheme for the production of microalgae oil.
1. Photobioreactor. 2. Flocculator. 3. Centrifuge 1. 4. Divider of water. 5. Extractor 6. Centrifuge 2. 7. Evaporator. 8. Heat exchanger for the solvent. 9. Divider of solvent. 10. Heat exchanger for the oil

Equipment Description

-Bioreactor: In order to avoid variations in the productivities when using heterotrophic conditions and to assure a better control over the operating conditions

[37], a bioreactor was selected as the best cultivating option. The main function of this unit is to provide the necessary conditions for the cultivation of *Chlorella vulgaris* (*e.g.*, temperature and agitation) using glucose as the carbon source.

-Flocculator: In order to increase the effective size of the "particles", and to improve the centrifugation process, a flocculator unit is added to the process. The coagulant used is aluminum sulfate, which has shown good yields when used with *Chlorella* [22].

-Centrifuge 1: This equipment is necessary since an almost total recovery of the microalgae biomass generated in the bioreactor is needed. Once the stream containing the microalgae cells is centrifuged, a second stream containing the remaining nutrients is produced which can be recycled to the bioreactor.

-Extractor: In this equipment is then carried out the recovery of the oil present in the microalgae biomass, which is accomplished through the addition of a solvent (butanol).

-Centrifuge 2: In this part of the process, the separation of the oil obtained from the remaining biomass is attained.

-Evaporator: In order to recover the solvent an evaporator is used. The solvent is vaporized and recovered by the top stream and recirculated to the extraction process.

-Heat exchanger: The function of this equipment for both cases (cooling solvent and microalgae oil), is to accommodate the temperature either for the solvent (butanol) or the oil obtained for the reusability of the former and final disposition of the latter.

-Splitter: This equipment has the function of dividing the liquid stream obtained in the first centrifuge, from which one portion is recycled to the bioreactor and the other one is discarded in the process.

ANALYSIS OF THE YIELDS COMPARED WITH LITERATURE

Mass and energy balances for both processes are shown in Table **15** and **16**, respectively. Where the stream "biomass" represents the biomass obtained after the separation processes and oil extraction.

According to the results shown in Tables **15** and **16**, the microalgae oil obtained using CO_2 as carbon source, contains a higher energy value compared to the oil obtained using glucose. This fact could be explained by the energetic differences present in the substrates.

Table 15. Results from the mass balances.

Carbon Source	Carbon Dioxide		Glucose	
Stream	Microalgae oil	Biomass	Microalgae oil	Biomass
Mass Flow [Kg/h]	7.8	25.1	9.8	31.2
Composition (%)				
Biomass	0.00	0.98	0.00	0.98
Water	0.03	0.00	0.05	0.00
Palmitic Acid	0.28	0.01	0.23	0.00
Palmitoleic Acid	0.00		0.03	0.00
Stearic Acid	0.13	0.00	0.20	0.00
Oleic Acid	0.18	0.00	0.05	0.00
Linoleic Acid	0.39	0.01	0.45	0.01

Table 16. Results from the energy balances.

Stream	Enthalpy [KJ/h]
CO_2	-357,660
Glucose	-283,250
Oil microalgae (autotrophic)	-32,234
Oil microalgae (heterotrophic)	-23,724
Biomass (autotrophic)	-297,280
Biomass (heterotrophic)	-152,530

Economic Analysis

For the economic analysis, several variables namely Raw Materials Costs, Utilities Cost, Operating Labor Cost, Maintenance Cost, Operating Charges, Plant Overhead, General and Administrative Cost and the Depreciation Expense (DEP) are considered. The results of the economic analysis are presented in Figs. (**3** and **4**), where the differences between using organic or inorganic compounds as carbon sources can be appreciated.

- The cost of raw materials: the CO_2 cost is not considered here since it can be obtained from an anaerobic digestion of biomass after the extraction process.

- In general all costs involved in the process had similar shares, considering that the cost of raw materials is the item that shows the biggest impact on the final price of the product.

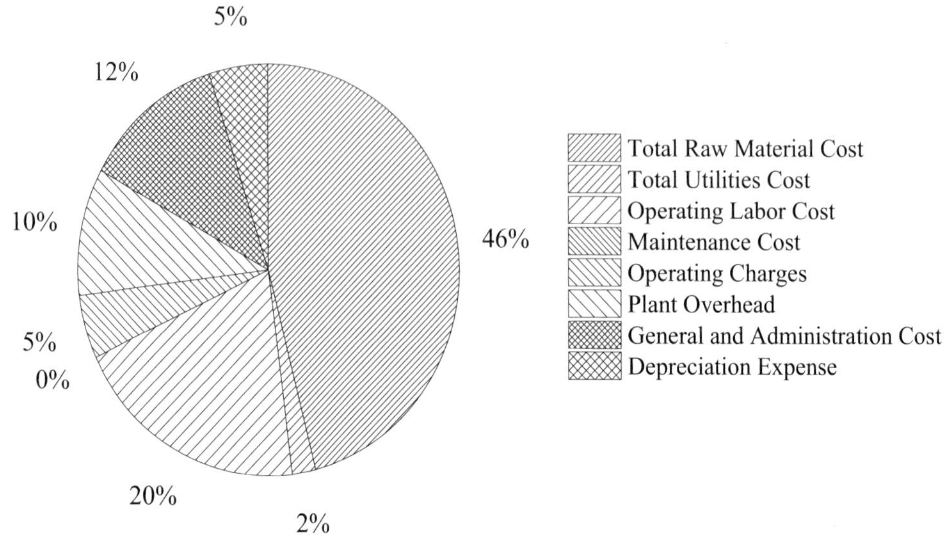

Fig. (3). Economic analysis of oil production using glucose as substrate.

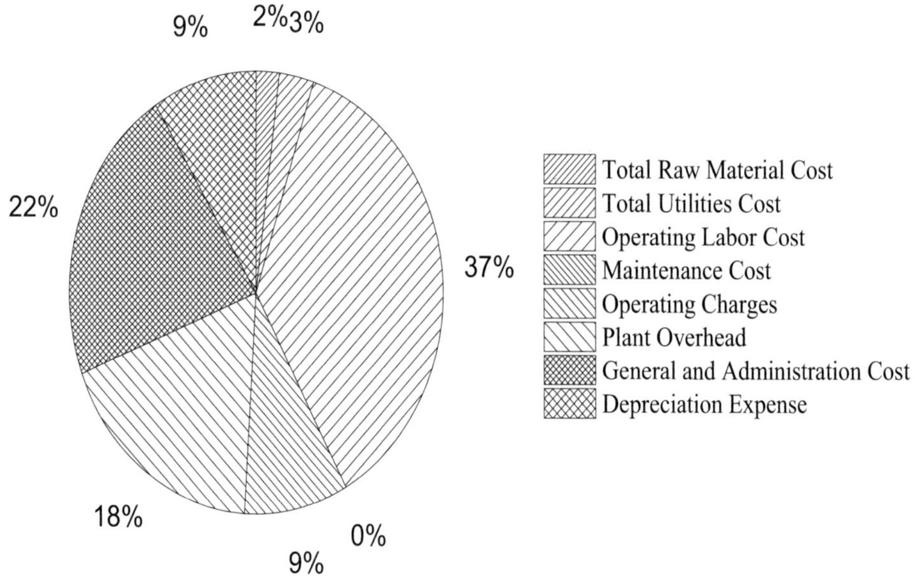

Fig. (4). Economic analysis of oil production using CO_2 as substrate.

Table **17** shows the costs of producing microalgae oil using either CO_2 or glucose as carbon sources. Glucose presents a higher cost for the production of oil. In general, the microalgae oil production showed high costs which highlights the necessity of implementing an optimization strategy. Several products can be obtained from microalgae (not only oil) that in turn will help to decrease the overall cost of the process. Implementing a biorefinery concept to the process will take into consideration other value-added compounds that can be obtained from microalgae and that will decrease the oil production costs. Furthermore, those valued-added compounds will also expand the range of products obtained in the process thus increasing their market value.

Table 17. Production cost oil microalgae.

Substrate	Total Production Cost [USD/Kg]
CO_2	19.39
Glucose	27.55

Table **18** shows that despite the high costs involved in utilizing commercial glucose as substrate, the yields of both oil and biomass per kilogram of substrate are higher than when CO_2 was used as the carbon source.

Table 18. Yields.

Source of Carbon	CO_2	Glucose
Yps [Kg oil/Kg Substrate]	0.19	0.25
Yxs [Kg biomass/Kg Substrate]	0.83	1.03

CONCLUSION

The differences observed in microalgae oil prices through each of the analyzed processes are very significant (even at small scales). For that reason, the difference is much more noticeable at higher scales. Therefore, it is necessary to design a process that uses as raw material a substance that can be easily obtained at lower prices. The use of different kinds of waste is a promising alternative since it will not only decrease the share of raw materials to the process, but also decrease the pollution in the environment.

In order to improve and make the process more efficient, it is necessary to analyze other substrates for the cultivation of microalgae; this search can also focus on reducing gases in the atmosphere such as CO_2, which can be assimilated by *Chlorella vulgaris*. In this way it would help the reduction of this particular gas in the atmosphere, thus helping to decrease the greenhouse effect and the process of growing this microalga will also improve. CO_2 can be obtained by anaerobic digestion where it also serves in the energy production using a turbine.

CONFLICT OF INTEREST

The authors declares no conflict of interest, financial or otherwise.

ACKNOWLEDGEMENTS

The authors express their acknowledgments to the Universidad Nacional de Colombia for the financial support at projects HERMES 23132, 23069, 30928 and 21798.

REFERENCES

[1] I. Priyadarshani, and B. Rath, "Commercial and industrial applications of microalgae - A review", *J. Algal Biomaa Utln.,* vol. 3, no. 4, pp. 89-100, 2012.

[2] F.G. Acién, J.M. Fernández, J.J. Magán, and E. Molina, "Production cost of a real microalgae production plant and strategies to reduce it", *Biotechnol. Adv.,* vol. 30, no. 6, pp. 1344-1353, 2012. [http://dx.doi.org/10.1016/j.biotechadv.2012.02.005] [PMID: 22361647]

[3] A.P. Abreu, B. Fernandes, A.A. Vicente, J. Teixeira, and G. Dragone, "Mixotrophic cultivation of Chlorella vulgaris using industrial dairy waste as organic carbon source", *Bioresour. Technol.,* vol. 118, pp. 61-66, 2012. [http://dx.doi.org/10.1016/j.biortech.2012.05.055] [PMID: 22705507]

[4] A.A. Darmaki, L. Govindrajan, S. Talebi, S. Al-Raijhi, T. Al-Barwani, and Z. Al-Bulashi, "Cultivation and Characterization of Microalgae for Wastewater Treatment", *Proceedings of the World Congress on Engineering,* vol. I, 2012 pp. 1-4

[5] W-L. Chu, "“Biotechnology applications of microalgae,” International e-Journal of Science", *Med. Educ.,* vol. 6, no. 1, pp. 24-37, 2012.

[6] A.S. Carlsson, J.B. Beilen, R. Möler, and D. Clayton, *Micro - and Macro - Algae: Utility for Industrial Applications, CPL Press.* Science Publishers, 2007.

[7] T.M. Mata, A.A. Martins, and N.S. Caetano, "Microalgae for biodiesel production and other applications: A review", *Renew. Sustain. Energy Rev.,* vol. 14, pp. 217-232, 2010. [http://dx.doi.org/10.1016/j.rser.2009.07.020]

[8] P. Rattanapoltee, "Cultivation of microalga, Chlorella vulgaris under different auto-hetero-mixo trophic growths as a raw material during biodiesel production and cost evaluation", *Energy,* vol. 78, pp. 4-8, 2014. [http://dx.doi.org/10.1016/j.energy.2014.06.049]

[9] C. Safi, B. Zebid, O. Merah, P-Y. Pontalier, and C. Vaca-Garcia, "Morphology, composition, production, processing and applications of Chlorella vulgaris: A review", *Renew. Sustain. Energy Rev.,* vol. 35, pp. 265-278, 2014. [http://dx.doi.org/10.1016/j.rser.2014.04.007]

[10] X. Zeng, M.K. Danquah, X.D. Chen, and Y. Lu, "Microalgae bioengineering: From CO2 fixation to biofuel production", *Renew. Sustain. Energy Rev.,* vol. 15, pp. 3252-3260, 2011. [http://dx.doi.org/10.1016/j.rser.2011.04.014]

[11] J.A. Quintero, E.R. Feliz, L.E. Rincón, M. Crisspín, J. Fernandez Baca, Y. Khwaja, and C.A. Cardona, "Social and techno-economic analysis of biodiesel production in Peru", *Energy Policy,* vol. 43, pp. 427-435, 2012. [http://dx.doi.org/10.1016/j.enpol.2012.01.029]

[12] S-H. Ho, C-Y. Chen, D-J. Lee, and J-S. Chang, "Perspectives on microalgal CO2-emission mitigation systems--a review", *Biotechnol. Adv.,* vol. 29, no. 2, pp. 189-198, 2011.

[http://dx.doi.org/10.1016/j.biotechadv.2010.11.001] [PMID: 21094248]

[13] M. Adamczyk, J. Lasek, and A. Skawińska, "CO2 Biofixation and Growth Kinetics of Chlorella vulgaris and Nannochloropsis gaditana", *Appl. Biochem. Biotechnol.,* vol. 179, no. 7, pp. 1248-1261, 2016.
[http://dx.doi.org/10.1007/s12010-016-2062-3] [PMID: 27052208]

[14] O. Pulz, and W. Gross, "Valuable products from biotechnology of microalgae", *Appl. Microbiol. Biotechnol.,* vol. 65, no. 6, pp. 635-648, 2004.
[http://dx.doi.org/10.1007/s00253-004-1647-x] [PMID: 15300417]

[15] C-Y. Chen, K-L. Yeh, R. Aisyah, D-J. Lee, and J-S. Chang, "Cultivation, photobioreactor design and harvesting of microalgae for biodiesel production: a critical review", *Bioresour. Technol.,* vol. 102, no. 1, pp. 71-81, 2011.
[http://dx.doi.org/10.1016/j.biortech.2010.06.159] [PMID: 20674344]

[16] M.A. Borowitzka, "Commercial production of microalgae: ponds, tanks, tubes and fermenters", *J. Biotechnol.,* vol. 70, no. 1-3, pp. 313-321, 1999.
[http://dx.doi.org/10.1016/S0168-1656(99)00083-8]

[17] K-L. Yeh, and J-S. Chang, "Effects of cultivation conditions and media composition on cell growth and lipid productivity of indigenous microalga Chlorella vulgaris ESP-31", *Bioresour. Technol.,* vol. 105, pp. 120-127, 2012.
[http://dx.doi.org/10.1016/j.biortech.2011.11.103] [PMID: 22189073]

[18] Y. Liang, N. Sarkany, and Y. Cui, "Biomass and lipid productivities of Chlorella vulgaris under autotrophic, heterotrophic and mixotrophic growth conditions", *Biotechnol. Lett.,* vol. 31, no. 7, pp. 1043-1049, 2009.
[http://dx.doi.org/10.1007/s10529-009-9975-7] [PMID: 19322523]

[19] L. Xin, H.Y. Hu, G. Ke, and Y.X. Sun, "Effects of different nitrogen and phosphorus concentrations on the growth, nutrient uptake, and lipid accumulation of a freshwater microalga Scenedesmus sp", *Bioresour. Technol.,* vol. 101, no. 14, pp. 5494-5500, 2010.
[http://dx.doi.org/10.1016/j.biortech.2010.02.016] [PMID: 20202827]

[20] L. Brennan, and P. Owende, "Biofuels from microalgae - A review of technologies for production, processing, and extractions of biofuels and co-products", *Renew. Sustain. Energy Rev.,* vol. 14, pp. 557-577, 2010.
[http://dx.doi.org/10.1016/j.rser.2009.10.009]

[21] M. Olaizola, "Commercial development of microalgal biotechnology: from the test tube to the marketplace", *Biomol. Eng.,* vol. 20, no. 4-6, pp. 459-466, 2003.
[http://dx.doi.org/10.1016/S1389-0344(03)00076-5] [PMID: 12919832]

[22] E. Molina Grima, E-H. Belarbi, F.G. Acién Fernández, A. Robles Medina, and Y. Chisti, "Recovery of microalgal biomass and metabolites: process options and economics", *Biotechnol. Adv.,* vol. 20, no. 7-8, pp. 491-515, 2003.
[http://dx.doi.org/10.1016/S0734-9750(02)00050-2] [PMID: 14550018]

[23] E.H. Belarbi, E. Molina, and Y. Chisti, "A process for high yield and scaleable recovery of high purity eicosapentaenoic acid esters from microalgae and fish oil", *Enzyme Microb. Technol.,* vol. 26, no. 7, pp. 516-529, 2000.
[http://dx.doi.org/10.1016/S0141-0229(99)00191-X] [PMID: 10771055]

[24] A.B. Culaba, A. Ubando, R. Tan, J.M. Biona, N.S. Lopez, J. Tanchuco, S.S. Garibay, N.A. Toledo, C.N. Jimenez, I.G. Pahila, and L. Ami, "A Mathematical Model for the Drying Characteristics of Microalgae (Tetraselmis sp.)", *Philipp. Sci. Lett.,* vol. 6, no. 2, pp. 206-219, 2013.

[25] H. Desmorieux, and N. Decaen, "Convective drying of spirulina in thin layer", *J. Food Eng.,* vol. 66, pp. 497-503, 2005.
[http://dx.doi.org/10.1016/j.jfoodeng.2004.04.021]

[26] E. Ibañez, M. Herrero, J.A. Mendiola, and M. Castro-Puyana, "Extraction and Characterization of Bioactive Compounds with Health benefits from Marine Resources: Macro and Micro Algae, Cyanobacteria, and Invertebrates", In: *Marine bioactive compounds sources, Characterization and Applications*. Springer, 2012, pp. 55-98. [http://dx.doi.org/10.1007/978-1-4614-1247-2_2]

[27] M. Mendes-Pinto, M. Raposo, J. Bowen, A. Young, and R. Morais, "Evaluation of different cell disruption processes on encysted cells of Haematococcus pluvialis: effects on astaxanthin recovery and implications for bio-availability", *J. Appl. Phycol.,* vol. 13, pp. 19-24, 2001. [http://dx.doi.org/10.1023/A:1008183429747]

[28] P. Mercer, and R.E. Armenta, "Developments in oil extraction from microalgae", *Eur. J. Lipid Sci. Technol.,* vol. 113, pp. 539-547, 2011. [http://dx.doi.org/10.1002/ejlt.201000455]

[29] J.A. López, V.M. Trejos, and C.A. Cardona, "Parameters estimation and VLE calculation in asymmetric binary mixtures containing carbon dioxide +n-alkanols", *Fluid Phase Equilib.,* vol. 275, pp. 1-7, 2009. [http://dx.doi.org/10.1016/j.fluid.2008.09.013]

[30] M. Herrero, E. Ibáñez, A. Cifuentes, and J. Señoráns, "Pressurized liquid extracts from Spirulina platensis microalga. Determination of their antioxidant activity and preliminary analysis by micellar electrokinetic chromatography", *J. Chromatogr. A,* vol. 1047, no. 2, pp. 195-203, 2004. [PMID: 15460249]

[31] Ó.J. Sánchez, and C.A. Cardona, "Conceptual design of cost-effective and environmentally-friendly configurations for fuel ethanol production from sugarcane by knowledge-based process synthesis", *Bioresour. Technol.,* vol. 104, pp. 305-314, 2012. [http://dx.doi.org/10.1016/j.biortech.2011.08.125] [PMID: 22137752]

[32] R. Davis, A. Aden, and P.T. Pienkos, "Techno-economic analysis of autotrophic microalgae for fuel production", *Appl. Energy,* vol. 88, pp. 3524-3531, 2011. [http://dx.doi.org/10.1016/j.apenergy.2011.04.018]

[33] J.J. Jaramillo, J.M. Naranjo, and C.A. Cardona, "Growth and Oil Extraction from Chlorella vulgaris: A Techno-economic and Environmental Assessment", *Ind. Eng. Chem. Res.,* vol. 51, pp. 10503-10508, 2012. [http://dx.doi.org/10.1021/ie300207x]

[34] L. Christenson, and R. Sims, "Production and harvesting of microalgae for wastewater treatment, biofuels, and bioproducts", *Biotechnol. Adv.,* vol. 29, no. 6, pp. 686-702, 2011. [http://dx.doi.org/10.1016/j.biotechadv.2011.05.015] [PMID: 21664266]

[35] J. Moncada, L.G. Matallana, and C.A. Cardona, "Selection of Process Pathways for Biorefineries Design Using Optimization Tools. A Colombian Case for Conversion of Sugarcane Bagasse to Ethanol, PHB and Energy", *Ind. Eng. Chem. Res.,* vol. 52, no. 11, pp. 4132-4145, 2013. [http://dx.doi.org/10.1021/ie3019214]

[36] S.J. Yoo, J.H. Kim, and J.M. Lee, "Dynamic modelling of mixotrophic microalgal photobioreactor systems with time-varying yield coefficient for the lipid consumption", *Bioresour. Technol.,* vol. 162, pp. 228-235, 2014. [http://dx.doi.org/10.1016/j.biortech.2014.03.128] [PMID: 24755320]

[37] I.C. Paz Astudillo, and C.A. Cardona Alzate, "Importance of stability study of continuous systems for ethanol production", *J. Biotechnol.,* vol. 151, no. 1, pp. 43-55, 2011. [http://dx.doi.org/10.1016/j.jbiotec.2010.10.073] [PMID: 21034786]

CHAPTER 15

Techno-economics of a Microalgal Route to Green Diesel

Lin Zhou, James L. Manganaro* and **Adeniyi Lawal**

Department of Chemical Engineering and Materials Science, Stevens Institute of Technology, Hoboken, NJ 07030, USA

Abstract: The environmental impacts associated to the use of fossil fuels have greatly accelerated the research and development of renewable fuels. This chapter reviews the technical and economic challenges for biofuel production from different feedstocks. Second generation lipid-based biomass feedstocks are considered to be cost-effective and eco-friendly for biofuel production, with microalgae being the most promising. The upgrading techniques of lipid-based feedstocks are also reviewed in terms of biodiesel and green diesel production, focusing on production chemistries and approaches, product application, and associated key technical challenges and opportunities. Due to the high polyunsaturated fatty acids (PUFAs) content, microalgae oil can be used to produce biodiesel through the insertion as a middle distillate feed stream into a hydrotreating unit of a petroleum refinery. The techno-economic assessment was performed for the production of hydrotreated algal oil (HTAO), animal feed and nutraceuticals. Base case calculations were for 10 000 barrels per day of HTAO. It was considered that nutraceuticals represented only 0.05% (w/w) of the raw algae oil. Algae doubling time has a strong influence on the sales price, which might be reduced if culture agitation is increased. Process economics can be improved taking into account the algal oil content, CAPEX, and moisture content of post-extracted algal residue. Considering the current operating parameter values and co-product credits, plant gate price was estimated by ~$10/gal. However, several improvements were also identified to enhance the process economics. The trade-off between oil content and area weight productivity favors oil content at constant oil area productivity ($gal \cdot acre^{-1} yr^{-1}$). For the limit of 100% oil content, (*i.e.,* no solid co-product), the sale price was estimated to be $7.90/gal.

Keywords: Biofuel, Green diesel, Hydrotreating, Lipid-based biomass feedstocks, Microalgae oil, Renewable fuel, Techno-economic analysis.

* **Corresponding author James L. Manganaro:** Department of Chemical Engineering and Materials Science, Stevens Institute of Technology, Hoboken, NJ 07030, USA; Tel/Fax: 201-417-9048 ; E-mail: jljm11@verizon.net

RENEWABLE FUELS

Economic growth (evaluated by the Gross Domestic Product - GDP) is the key indicator for the prediction of energy demand. Considering 3.3% per year rise of the world's average GDP (in terms of purchasing power parity) from 2012 to 2040, as well as the improvements in energy efficiency due to the development of technologies, total world's social and economic energy demand is estimated to increase by 48%, from 549 quadrillion Btu in 2012 to 815 quadrillion Btu, with fossil fuels still accounting for 78% of the energy consumption in 2040 [1]. However, although no consistent estimation data for the fossil fuel depletion timeline has been reported, it is well accepted that the reserves of fossil fuel are finite. Moreover, environmental concerns caused by the fossil fuels combustion (air pollution that negatively affects human health; greenhouse gas - GHG - emission that is associated to climate change; NO_x and SO_x emissions that acidify the rain), have received extensive attention. Rigid government regulations and tax exemption policies have been established to encourage the utilization of renewable fuels. Thus, the research and development of sustainable and alternative renewable energy resources are highly encouraged to reduce the dependence on petroleum-based fuels and to meet the ever-growing energy demand due to the growth of population and economy [2].

According to International Energy Agency (IEA), world demand on liquid fuels will expand from 90 million bbl/day in 2012 to 121 million bbl/day in 2040 due to the transportation and industrial sectors, with the transportation sector accounting for 62% of this increase [1]. Biofuels are a renewable source of carbon [3], and are proposed to be the only effective alternative to petroleum-based fuels due to their potential for larger volume production [4].

BIOFUEL FEEDSTOCKS

Among all currently well explored biomass feedstocks for liquid fuel production, oleaginous compounds are reported to be the easiest feedstocks to be converted to liquid fuel compared with amorphous sugars and lignocellulosics [5]. However, the selection of feedstocks to produce biofuel commercially is a complex problem, as the chemical and physical characteristics, as well as the supply, cost, storage and engine performance need to be considered [6].

Lipid-based Biomass Feedstocks

Lipids refer to substances that are soluble in alcohol, ether and chloroform, but insoluble in water, including fatty acids, neutral fats, waxes and steroids, which can be converted is to liquid fuels through various upgrading processes. Based on the source, lipid-based biomass feedstocks can be categorized as:

First Generation Lipid-Based Biomass Feedstocks

The first generation lipid-based biomass feedstocks refer to edible oils, or cooking oils. They are mainly composed of acylglycerides (~90-98 wt %), with triacylglycerides (TAG) being predominant. As shown in Fig. (**1**), a triglyceride is composed of a glycerol backbone and three fatty acids. R_n stands for the hydrocarbon portion of the fatty acids, and the three R_n can all be the same or different in each triglyceride molecule. If one or two of the $-COR_n$ groups are replaced by hydrogen atoms, then the compound is called diacylglyceride (DAG) and monoacylglyceride (MAG) respectively. The naturally occurring acylglycerides contain only even-numbered carbon fatty acids, especially 14, 16, 18, and 20 carbon atoms. For example, as listed in Table **1**, 95 wt % of soybean oil was identified as fatty acids.

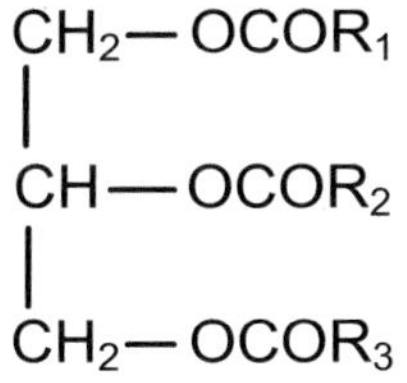

Fig. (1). Generic formula of triglycerides.

Table 1. Fatty Acid Profile of Soybean Oil [5].

Fatty Acid Profile (wt %)				
$C_{16:0}$	$C_{18:0}$	$C_{18:1}$	$C_{18:2}$	$C_{18:3}$
10	4	23	51	7

Edible oils are commonly extracted from oilseeds, tree fruits and kernels. Among the general edible oil plants, oil palm and coconut tree have the highest oil productivity, but require specific growing conditions in temperature, rainfall, and sunshine. That means their usage as biofuel feedstocks is geographically limited to certain areas, such as Indonesia, Malaysia, Thailand, Columbia and Nigeria, which were also ranked as the top 5 countries in palm oil production in 2011 [7]. Olive trees, rapeseeds and sunflower seeds are mainly produced in the European Union; and Asia accounts for more than 80% of the peanut oil production. Despite the low lipid content, soybean oil is the most common biodiesel feedstock in the United States, due to its highest total worldwide production. However, all these edible oils are not economically competitive as feedstocks for biofuels production. It should be noted that, at the current stage, due to insufficient supply beyond the household and industrial demands, the utilization of edible oil as biofuel feedstock will increase the price of edible oils, and is not economically feasible.

Second Generation Lipid-Based Biomass Feedstocks,

The second generation lipid-based biomass feedstocks refer to oils that cannot be used as human food or animal feed, due to their high FFA (Free Fatty Acid) content, different fatty acids composition or the presence of toxic compounds [8]. They include:

1. Non-edible oils, which are oils derived from plant sources. One of the emerging candidates as a non-food biofuel feedstock is Jatropha oil, and its composition is presented in Table **2**. It is extracted from the seed of Jatropha curcas tree, which has high oil content (27-40%) [10], and can be grown in degraded or semi-arid lands that cannot be used for cultivating food crops. Non-edible oil plants have been shown to produce less oil than edible-oil plants, due to the imperfect development of screening and domestication, as well as the inadequate nutrition and water supply since they are typically grown in degraded lands [7]. Therefore, further research on the selective breeding and cultivation technologies is required to increase their oil productivity.

Table 2. Fatty Acid Profile of Jatropha Oil [9].

Fatty Acid Profile (wt %)				
$C_{16:0}$	$C_{16:1}$	$C_{18:0}$	$C_{18:1}$	$C_{18:2}$
12.5	0.9	7.5	42.1	37

2. Waste edible oils, namely spent cooking oils. Their composition reflects that of the parent oils, but varies slightly due to the thermolytic (pyrolysis), hydrolytic (hydrolysis) and oxidative reactions that the virgin oils undergo during cooking [11]. The main contaminants found in the waste cooking oils that may exert adverse effect on the biodiesel production *via* homogeneous catalytic process are soluble salts, FFA (2-50 wt %) and water (0.5-20 wt %) [12]. Therefore, a pretreatment process or advanced and compatible upgrading technologies are required. Besides, the productivity of waste cooking oils varies geographically, and its unpredictability is another obstacle in its usage as a biofuel feedstock.

3. Animal fats, which are the non-food portions obtained from animal slaughter, such as beef tallow, chicken fat and duck tallow [7]. Their fatty acid profiles are found to be slightly different from both edible and non-edible plant oils [7], as indicated in Table **3**. Similar to waste cooking oils, the high contents of water and FFA are the key technical barriers that need to be overcome.

Table 3. Fatty Acid Profile of Beef Tallow [13].

Fatty Acid Profile (wt %)					
$C_{14:0}$	$C_{16:0}$	$C_{18:0}$	$C_{18:1}$	$C_{18:2}$	$C_{18:3}$
2.69-4.00	22.30-28.00	19.20-41.28	28.82-48.90	0.58-3.00	0.50-2.00

4. Industrial wastes and by-products, which are obtained from the industrial manufacturing processes, for examples, waste yellow grease and brown grease produced from water treatment [14], and tall oil from the paper pulping process [15]. According to the reported information on the composition of currently available industrial wastes and by-products, they contain both the major fatty acid components of plant oils, like palmitic (C16:0), oleic (C18:1) and linoleic (C18:2) acids, and the fatty acids that are rarely present in plant oils, such as palmitoleic (C16:1) and eicosatrienoic (C20:3) acids. Depending on the source of the feedstocks, contaminants associated with their production processes need to be properly and specifically treated prior to the upgrading process.

5. Lipids derived from insects. Some insects can convert vegetal and food waste into lipids to achieve the green waste to energy conversion. Due to the simplicity of their life-cycle and high lipids content, insect farming can be conducted in both small scale and industrial scale [7]. Lipid extraction from insects involves the drying, milling and extracting units, and is energy-intensive and prohibitively expensive. Therefore, an integrated biodiesel production process, which combines lipids extraction and upgrading, has been proposed to make the whole upgrading process more cost-effective [16].

6. Lipids derived from microorganisms. It has been reported that microbially produced biofuels are projected to account for major proportion of the future fuels market [17]. For example, many oleaginous yeasts and fungi species are found to be able to convert both lignocellulosic and sugar-based biomass into lipids in the absence of light. Therefore, many agriculture, livestock and food and household wastes can be used as the carbon source for lipids production [7]. Bacteria have also been considered as one of the potential candidates for lipids production due to the high lipids content in dry cell, and their fatty acid compositions depend much on the carbon source [7]. Besides, the current most promising biofuel feedstocks are microalgae oil, which will be further reviewed in the following section.

Microalgae Oil

Microalgae are photosynthetic microorganisms that can be found in oceans and seas (covering 71% of the earth's surface), freshwater bodies, soil, rocks, ice, snow, plants, and animals [18]. Their growth does not compete with food or oil production. Their growth rate is 10 – 200 times faster than that of terrestrial oil crops, and some species can double their biomass within 24 h [19, 20]. Phototropic microalgae use carbon dioxide as carbon source and light as energy source. Oleaginous species present an average lipids content varying between 1% and 70%, and under specific culture conditions the lipid percentage of 90% on dry weight basis can be achieved [19, 21, 22]. The accumulated oil in almost all microalgae is mainly triglycerides (>80%) with C16 and C18 being the dominant in its fatty acid profile [19], which is a substantial energy resource for liquid-based biofuel production. Additionally, algae cultivation can also bring environmental benefits, as they can grow using nutrients in wastewater (especially nitrogen and phosphorus) and waste CO_2 from fossil energy powered plants and other high carbon emitting facilities [23]. Therefore, microalgae oil is considered a low-net carbon feedstock for liquid fuel production.

The lipids content of algae depends primarily on the algae species. For certain strains, the growth rate and lipid content, even lipid composition, mainly depend on growing conditions, including factors such as temperature, pH, light intensity, nutrients (organic C and N) source, dissolved oxygen level, salinity (for marine strains), and agitation speed [7]. Therefore, algae oil can be produced in a controlled manner, such as autotrophic cultures (light without additional nutrients), heterotrophic cultures (no light with additional nutrients) or mixotrophic cultures (with additional nutrients) in open raceways, photo-bioreactors or fermenters [7].

The oil extraction method and efficiency depend primarily on the algae species and their chemical compositions [23]. The extraction can be achieved through physical, chemical with organic solvents, and supercritical methods, with solvent extraction being the most practicably used approach for producing biofuels [24]. Algal oils are quite different from vegetable ones. Vegetable oils are primarily composed (□90–98 wt %) by neutral lipids (mostly triglycerides) with polar lipids being in a small amount that could be removed by simple aqueous degumming. While, crude algae oils contain neutral lipids (>30 wt %), a significant amount of polar lipids (depending on the algae strain, usually in the form of phospholipids of 10–40 wt %, and glycolipids of 10–40 wt %) and unidentified natural substances. Aqueous degumming has been proven to be ineffective in removing polar lipids in algae oil due to their high content. It is worth noting that, polar lipids contain heteroatoms (P, N, and S) that can cause the deactivation of catalysts, reducing their longevity. Currently, a wet extraction technology (patented by Valicor Renewables, LLC Inc) has been demonstrated to be effective to avoid the catalyst poisoning, hence separating these polar lipids from the oil to make the algae oil better suited for catalytic conversion to fuels

[25]. As summarized in Table **4**, fatty acids with even-number of carbon are present in microalgae oils, with C16 and C18 fatty acids being dominant. It also shows that the same algae strains, which were not grown under the same conditions but went through the same extraction process, would produce algae oils with different fatty acid profiles.

Table 4. Fatty acid profile of oils from different Scenedesmus obliquus strains.

C15:0	C16:0	C16:1	C16:2	C16:3	C16:4	C18:0	C18:1	C18:2	C18:3	
1.05	20.31					5.02	28.91	30.51	14.2	[26]
1.48	21.8	5.95	3.96	0.68	0.43	0.45	17.9	21.7	3.76	[27]

Summary of Lipid-Based Biomass Feedstocks

Based on the discussion above, the second generation lipid-based feedstocks are considered as wastes, the disposal of which is associated with economic and environmental costs, which makes their utilization as biofuel feedstock both cost-effective and eco-friendly. In addition, compared with the first generation feedstocks, the utilization of second generation feedstocks will not compete with food production. However, it needs to be noted that their water (or moisture), FFA and other contaminants contents are usually high, which may affect the biofuel quality and cause problems in the downstream upgrading process, in the form of catalyst deactivation and equipment corrosion. The development of upgrading processes with less sensitivity to high contents of FFA and water, and systematic control of the fuel quality are highly required. In conclusion, each lipid-based feedstock has its own advantages and limitations, and the most suitable biofuel feedstock for each community varies geographically and should be determined based on its productivity, local availability and manageability [6].

Non-Lipid-Based Biomass Feedstocks

In addition to lipid-based biomass feedstocks, the other two main classes of biomass feedstocks being currently investigated for transportation fuel production are lignocellulosics and amorphous sugars [5].

Lignocellulose refers to the structural materials constituting the essential part of the cell walls, leaves, stems, stalks and woody portions of biomass [28]. Cellulose, hemicellulose and lignin are the main components of lignocellulose, in which lignin and hemicellulose are cross linked with the rigid cellulose fibers embedded in the matrix [28]. Cellulose is a very large polymer molecule comprising glucose molecules, while hemicellulose is composed of short but highly branched chains of sugars. Through hydrolysis, these two components can be broken down to produce fermentable simple sugars [28]. However, lignin, a polymer made up of non-carbohydrate alcohol units, is not fermentable, and it needs to be removed from lignocellulose before subjecting lignocellulose to biorefining process [28]. Lignocellulosic biomass feedstocks, such as, corn stover, crop residuals, switchgrass, wood chips, and spent grains, vary in their compositions, with the typical biomass containing 40% - 60% cellulose, 20% - 40% hemicellulose, and 10% - 25% lignin [29]. Lignocellulosic biomass is usually available at low cost, as it is often a waste material of the food processing and forest products industries [28]. However, its energy density is low, thus, costs for upgrading to biofuel is relatively high, which makes it desirable to have dispersed upgrading facilities close to the biomass sources. However, economies of scale, which is positively affected by the size of production unit, should also be considered.

Amorphous sugars, such as starches, simple sugars, are the primary feedstock for producing bioethanol in the United States. The sugars are obtained from food crops, examples are sucrose, extracted from sugar cane or beet; and starch produced from corn, wheat, potatoes *etc*. Their upgrading process is simple and low-cost. However, the main challenge of utilizing sugars as feedstocks for biofuel production, exemplified by the food versus fuel debate, cannot be resolved currently. The competition with food production will decrease the farmable land for population and livestock supplies, and cause the shortage of crops for food production, thus increase the food price.

BIOFUEL PRODUCTION

Biofuels can be produced through mainly 4 types of process: (1) Gasification, to produce syngas (mainly CO and H_2), which can be further upgraded to liquid hydrocarbons through Fischer-Tropsch synthesis (2) Pyrolysis or thermal liquefaction, to produce crude bio-oils usually with high oxygen content that require further upgrading process (3) Chemical process, referring to homogeneous or heterogeneous catalytic processes, such as hydrotreating and (4) Biochemical process *via* enzymatic catalysis, such as fermentation [30].

Based on socio-economic, environmental characteristics and origin of the feedstock, the final produced biofuels can be designated into 4 groups, namely, first (1G), second (2G), third (3G) and fourth (4G) generation biofuels [31].

First generation biofuels refer to fuels produced from edible food crops and vegetable oils. Bioethanol is the most representative and successful 1G biofuel commercially available, and it's produced from sugarcane in Brazil, and corn in US [32]. Biodiesel production with edible vegetable oils as feedstocks has been commercialized in some countries, in Germany with oilseed rape, and in Malaysia with palm oil [32]. The main drawback of its production is the competition with food production, which may lead to the increase in food prices and pressure on land use, with associated social, environmental and economic challenges [31]. Therefore, the search for alternative feedstocks is the primary obstacle needed to be overcome in the development of 1G biofuels.

Second generation biofuels are mainly derived from non-edible feedstocks, including wood and wood waste, animal fats, non-food crops and waste cooking oil, with lignocellulosic biomass being predominant [31]. Lignocellulosic biofuels are produced through thermochemical process, mainly gasification and pyrolysis, as well as the biochemical process. Biodiesel is the primary product from non-edible oleaginous feedstocks, mainly animal fat and waste cooking oil. The production process exploits renewable low-cost feedstocks, and produces sustainable, affordable fuels with net zero carbon emissions.

Third generation biofuels, or the so called oilgae, are produced from microalgae biomass [32]. Algae are reported to have higher energy productivity per unit acre of land than conventional crops [33]. The typical biofuels are biodiesel and green diesel, which are produced *via* transesterification and hydrotreating, respectively. As a significant source of lipids, algae's lipid content and composition are the key factors that can directly affect the properties of the resulting biofuels.

Fourth generation biofuels refer to fuels that are derived from carbon captured from the environment. It can be produced either by geo-sequestering the captured CO_2 in old oil and gas fields or saline aquifers, or by capturing industrially emitted CO_2 into water *via* electrochemical process to produce liquid methanol, in which renewable energy sources, including solar energy, wind, hydro, and geothermal, are utilized to provide heat and electricity [34]. The production process is carbon negative, and the produced biofuel is a sulfur-free, aromatic rich high cetane number fuel that is compatible with petroleum diesel [31].

Biodiesel Production

The American Society for Testing and Materials (ASTM) defines biodiesel as a "fuel which comprises mono-alkyl esters of long chain fatty acids derived from renewable lipid feedstock, such as vegetable oils or animal fats, designated B100" [35]. The main component of vegetable oil and animal fat is acylglycerides, such as TAG, DAG and MAG, and depending on the source, some amount of FFA may be also present.

A conceptual flow diagram for the biodiesel production process is shown in Fig. (**2**). As indicated in Fig. (**3**), biodiesel can be produced from either the esterification of FFA or the transesterification of acylglycerides with alcohol. Methanol is the most frequently used alcohol due to its low cost and its advantageous physical and chemical properties (polar and shortest chain alcohol) [37]. The reaction can be catalyzed by alkali, acid and enzyme [37]. It has been reported that higher concentration of methanol will inhibit the lipase activity [38]. Therefore, lipase is only applicable to systems with diluted methanol, which will lower the reaction rate. Homogeneous strong acid or base catalysts, such as H_2SO_4, H_3PO_4, NaOH, and KOH are primarily used in biodiesel production. At room temperature, base catalyst has much higher activity than acid catalyst. When reaction temperature is increased to a value in the range of 100-150 °C, competitive activity could be obtained over acid catalyst. The direct use of base catalyst for feedstocks with high water and FFA content will cause saponification, which will consume the base catalyst, complicate the product purification step, and decrease the biodiesel yield [2]. A two-step process, in which, FFA is first pre-esterified with methanol using liquid acid catalyst, and then the unconverted acylglycerides transesterified using base catalyst, has been proposed to process feedstocks with high FFA content [39 - 41]. Acid catalyst can catalyze esterification and transesterification simultaneously [38]. So, compared with homogeneous base catalyst, it is more adaptable to raw oil feedstocks, which most often contain FFA. However, this homogeneous catalyzed biodiesel production has some major obstacles, such as the separation and reuse of catalysts, and environmental and corrosion-related problems [2]. Heterogeneous catalysts can provide an economical solution to the catalyst removal step. Various heterogeneous base and acid catalysts have been investigated by researchers. For base catalyst, a calcium oxide material was found to be of both scientific and industrial interest in transesterification reaction because of its high basicity, low solubility in methanol, and ease of availability [42]. Its stability and durability could be further enhanced when appropriate support, such as ZnO, which enabled uniform dispersion of CaO, was used [42]. For acid catalyst, more accomplishments have been reported. For example, a sulfated titanium oxide (SO_4^{2-}/TiO_2) was shown to have superior catalytic activity and stability when catalyzing esterification and transesterification simultaneously [43]. Ordered mesoporous silica, such as SBA-15 and MCM-41 [44], and carbon-based materials have both been considered as excellent support for acid catalysts [42]. However, none of the reported heterogeneous catalysts has been applied in the commercial production of biodiesel mainly limited by the high catalyst cost.

With the advantages of being biodegradable, renewable, non-toxic and of low polluting, especially SO_X emission [45, 46], biodiesel can be used either in pure form or blended with petroleum-based diesel at different concentrations. The blending technique not only helps to lower the viscosity of biodiesel and improve its cold-flow properties, but it also improves air quality for engines without selective catalytic reduction (SCR) systems. In terms of air quality, it produces less GHG than conventional diesel fuel that is roughly commensurate with the blend level [47]. In addition, biodiesel has better lubricity than petroleum diesel, which helps to reduce friction of moving parts [48]. Furthermore, the higher the biodiesel content, the higher the fuel price, and the lower the energy content. Therefore, the blending level should be a balance between fuel economy, fuel properties and emission benefits.

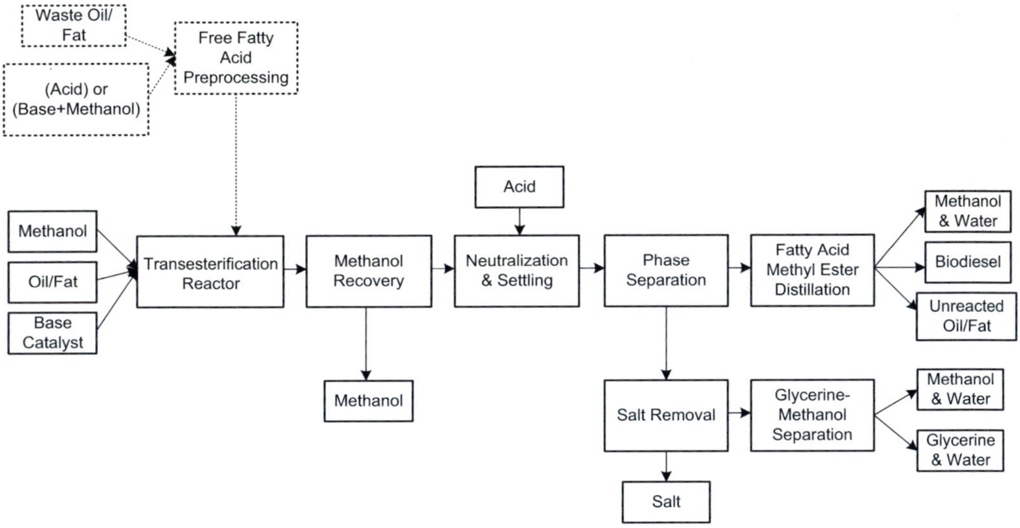

Fig. (2). Biodiesel production process (Adapted from [36]).

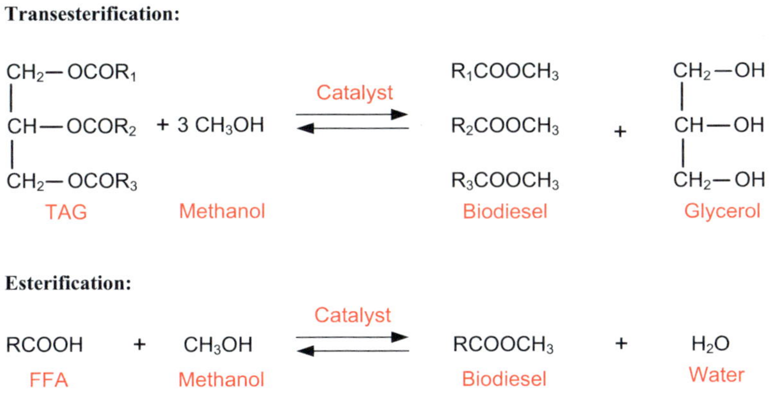

Fig. (3). Biodiesel production mechanism.

The most commonly used are low-level blends (up to 20% biodiesel), like B2, B5, and B20 where the B stands for biodiesel and the digit stands for the weight percentage of biodiesel in the diesel mixture. According to the ASTM D6751 specifications, low-level blends can also be called diesel fuel, which does not require separate labeling, and can directly be used in existing diesel engines without any modifications required [47]. B20 is the most popular since it represents a good balance of fuel properties and application concerns, involving cost, emissions, cold weather performance, materials compatibility and ability to act as a solvent [47]. Some studies have shown that the B20 has 1% to 2% less energy content than petroleum diesel, but no noticeable difference in performance or fuel economy has been found according to report from most B20 consumers [47]. However, it should be noted that not all engines are compatible with B20.

High-level blends and B100 are rarely used as a transportation fuel due to its poor cold-flow properties and solvent effect, which require special handling and an engine built with biodiesel-compatible material. Moreover, the utilization of B100 could increase the nitrogen oxides emissions, although other toxic emissions are greatly reduced [47].

Green Diesel Production

Green diesel is essentially a mixture of hydrocarbons, with the same chemical properties as petroleum diesel. It is a fully deoxygenated renewable alternative diesel fuel, which is compatible with all existing engines, pipelines, and infrastructure for application and distribution [25, 49]. Compared with biodiesel, green diesel has a higher cetane value, lower API gravity, better cold flow properties and excellent storage stability, with fuel properties independent of feed origin; and its production process is compatible with existing operations within petroleum refineries [25, 49]. Moreover, life-cycle analysis results also reveal the effectiveness of the GHG reduction and the advantages of less fossil energy consumption by green diesel compared with biodiesel [50]. Therefore, green diesel is an ideal substitute for petroleum diesel, being considered a good alternative for large-scale industrial production [25, 50].

Green Diesel Production Pathways

Green diesel can be produced by removing oxygen from lipid-based biomass feedstocks through the following routes:

Catalytic Cracking

Catalytic cracking is an efficient method for light liquid fuel production from lipids, with very low operational cost [51]. As shown in Fig. (**4**), many reactions, including cracking, hydrolysis, isomerization, dehydrogenation, aromatization and coking, are involved in this process [52, 53]. Its product is not feedstock source dependent, with diverse product components, including linear and cyclic paraffin, olefins, and oxygenated compounds, such as aldehydes, ketones and carboxylic acids [52]. The upgraded oils can be further fractionated *via* distillation processing according to the requirements of the boiling range of different fuels, such as, gasoline and diesel.

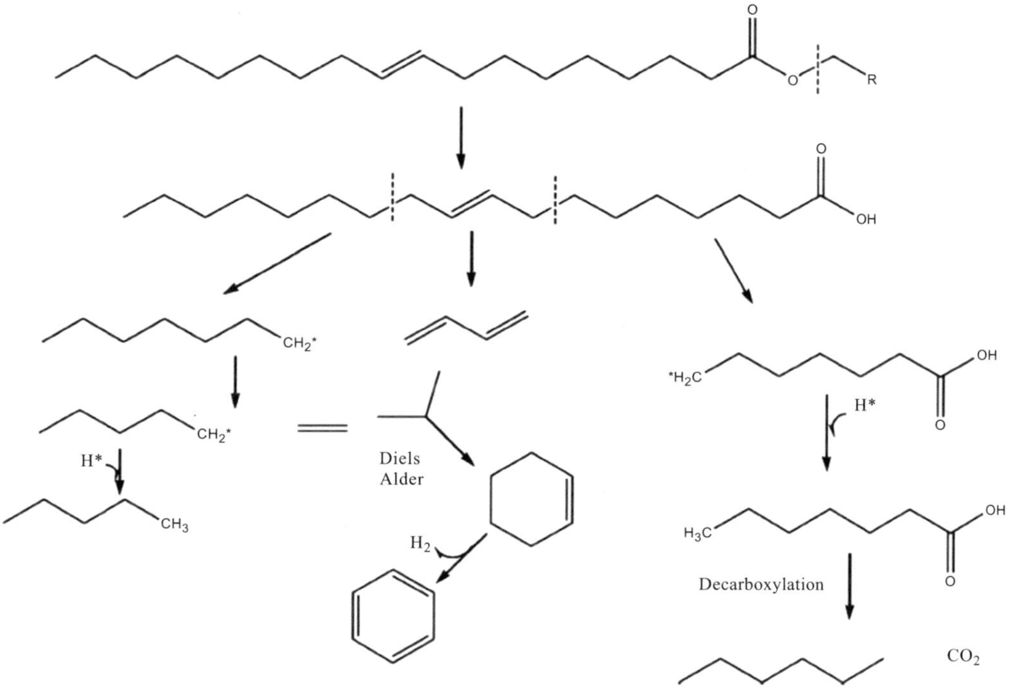

Fig. (4). Mechanism for fatty acids cracking [53].

The application of catalysts for thermal cracking of lipids benefits the process in lowering required reaction temperature, speeding up reaction rate, shortening residence time and improving the quality of the derived product [54]. The commonly investigated catalysts are metal oxides, molecular sieves, activated alumina and sodium carbonate [51]. The zeolite catalyst, HZSM-5, is most frequently studied in the cracking of vegetable oils, the product of which was reported to be similar to petrol with the octane number between 90 and 96 and high degree of conversion to aromatics [55 - 59] while the MCM-type mesoporous catalysts, with higher Al_2O_3 content and higher acidity were reported to have increased selectivity with respect to formation of linear hydrocarbons within diesel fuel range [60].

The major drawback of this cracking route is the wide product profiles, with substantial amount of oxygenated compounds, the separation of which is effected through a costly distillation step thereby impacting negatively the economic viability of the fuel production route. Therefore, the development of heterogeneous catalysts with high selectivity to desired fuel compounds is the main challenge for the lipids cracking process.

Deoxygenation

Deoxygenation is usually conducted under high pressure with low H_2 pressure or even in the complete absence of H_2. Decarboxylation, in which oxygen is removed in the form of CO_2, and decarboxylation, in which oxygen is removed in the form of CO are the two pathways of deoxygenation. Most of the studies that have been reported were conducted in batch mode using carbon supported catalysts. Many studies reported that the precious metal-based catalysts showed high activity and selectivity for fatty acids deoxygenation, with palladium supported on activated carbon being the most effective.

Deoxygenation has the advantage of low or no consumption of H_2. However, it should be noted that the primary deoxygenation products of fatty acids are unsaturated hydrocarbons [61]. Moreover, the insufficient hydrogen supply will lead to a high proportion of aromatic compounds in the products, which have been shown to play a key role in catalyst deactivation [62, 63].

Hydrotreating

The hydrotreating route mainly refers to oxygen removal, as lipids usually contains low levels of S, N, and P. Based on the nature of oxygen bonding in triglycerides, oxygen can be removed *via* two pathways (as listed below): (1) hydrodehydration (DHYD), in which oxygen is removed in the form of H_2O and (2) hydrodecarboxylation (DCO2) or hydrodecarbonylation (DCO), in which oxygen is removed in the form of CO_2 and CO, respectively. Due to the production of CO and CO_2, water gas shift (WGS) and methane formation are the main reactions in gas phase.

Liquid Phase:

$$DHYD\text{: } R - COOH + 3H_2 \rightarrow R - CH_3 + 2H_2O$$

$$DCO_2\text{: } R - COOH \rightarrow R - H + CO_2$$

$$DCO\text{: } R - COOH + H_2 \rightarrow R - H + H_2O + CO$$

Gas Phase:

$$Methanation\text{: } CO + 3H_2 \rightarrow CH_4 + H_2O$$

$$WGS: CO + H_2O \rightarrow CO_2 + H_2$$

One triglyceride molecule produces three hydrocarbon chains. Based on the reaction pathways, the hydrogen demand for each step of the reaction is listed in Table **5** [64]. Oxygen removal through the hydrodehydration route involves reactions 1–5, while that through the hydrodecarbonylation route involves reactions 1–3 and 6. The hydrogen demand for hydrotreating of fatty acids increases in the following order: DCO2 < DCO < DHYD.

Table 5. Hydrogen Demand in Hydrodeoxygenation of Acylglycerides and Free Fatty Acids [64].

Reaction	Hydrogen Consumption Per Mole FA Chain (mol H_2/mol FA)
1. Saturation of C=C double bond	1 mol H_2 demanded for per mol double bond
2. Hydrogenolysis to produce FFA	1
3. FFA → R-CHO (Fatty aldehyde)	1
4. R-CHO → R-CH_2OH (Fatty alcohol)	1
5. R-CH_2OH → HC(2n)	1
6. R-CHO → HC(2n-1)	0

Naturally available lipids feedstocks contain mostly triglycerides with even-carbon number fatty acids chains therefore even carbon-number hydrocarbons are produced through DHYD route, while odd carbon-number hydrocarbons are produced from DCO/DCO2 route. It should be noted that no catalyst will have a 100% selectivity to a certain pathway above, and the product is always a mixture from both pathways. The ratio of even-numbered carbon hydrocarbon to odd-numbered carbon hydrocarbon is a reflection of the catalyst selectivity to the two pathways. According to reviewed literature, precious metal-based catalysts were generally found to favor the DCO/DCO2 route. Ni favors DCO/DCO2 route, and Mo favors DHYD route, while conventional NiMo catalyst favors DHYD route.

Green Diesel Production from Microalgae Oil

Comparing Table **4** with Tables **1-3**, the fatty acids profile of microalgae is much more diverse than that of the other lipid-based biomass feedstocks. Moreover, the high content of PUFAs in microalgae oil makes it not particularly suitable for biodiesel production. If processed for biodiesel production, the PUFAs in algae oil will only go through transesterification reaction with the C-C double bond unchanged. The formation of polyunsaturated fatty acid methyl esters (PUFAMEs) will degrade the quality of produced biodiesel, namely viscosity, oxidation stability and cold flow properties [7]. In addition, the changing conditions during algae harvesting may lead to the rapid hydrolysis, through which lipids are converted to FFAs, which is not suitable to be converted to biodiesel *via* conventional process [7]. Therefore, microalgae oil is suitable for insertion as a middle distillate feed stream into the hydrotreating unit of a petroleum refinery, in which C-C double bond will be first saturated.

1. Microalgae growth in ponds
2. Biomass harvesting
3. Microalgae oil extraction from the biomass, and
4. Green diesel production from microalgae oil *via* hydrotreating

The first three steps are generally conducted at an aquatic farm exploiting both biological and agricultural engineering principles, while the last step is usually achieved in a hydrotreater in a refinery [48].

TECHNO-ECONOMIC ANALYSIS OF MICROALGAL ROUTE TO GREEN DIESEL

The following discussion is a condensation of that given in ref [65].

INTRODUCTION

To meet the US mandate for renewable fuels, algal based liquid fuels have been attracting interest [66 - 79]. Hydrotreating of algal derived oil can produce refinery ready paraffinic oil suitable for insertion in the middle distillate stream of a refinery's atmospheric distillation column [70]. However, there are significant challenges to commercialization which reside in processing demands and sustainability [80].

Economic analysis for open ponds and photo bio-reactors include refs [67, 73, 77, 81, 82]. The work discussed here examines (1) the dependence of area productivity on the pond operating parameters of algal doubling time, residence time, pond depth, and initial concentration, (2) this allows a unique sensitivity analysis, (3) the co-products of omega-3 and animal feed are treated in some detail, (4) commercial scale production is assumed.

Process description

Since open pond raceways are generally acknowledged to be more economical than photo bio-reactors, *e.g.*, Richardson *et al.* [83] and Davis, Aden, Pienkos [67], it is this configuration that forms the basis for this discussion. A block flow diagram, of the section to grow, harvest, and extract algal oil is shown in Fig. (**5**). Fig. (**5**) includes the production of animal feed. A very small stream of algal oil is diverted to the production of nutraceuticals (omega-3 fatty acids). Parameters associated with Fig. (**5**) are shown in Table **6**. The hydrotreater section, not discussed here is discussed in ref [65].

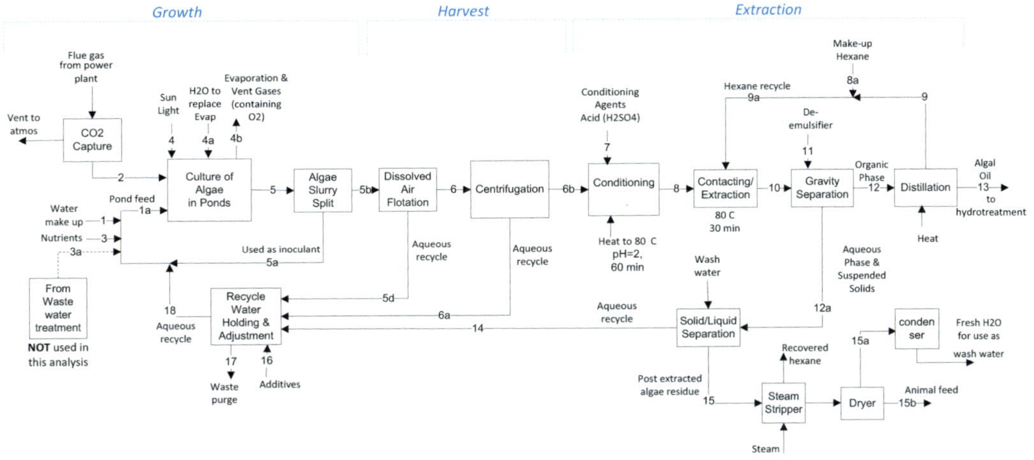

Fig. (5). Algae Growth-Harvesting-Extraction and Production of Algal Oil and Animal Feed - Section 1 [65].

Design Basis and Base Case

The following is assumed:

1. Production rate of 10,000 bbl/d (153 Mgal/yr) of HTAO. For the parameters of this study, this production rate requires ~140,000 acres of ponds.
2. Co-product production of animal feed and nutraceuticals, *e.g.,* omega-3 oils.
3. Production of nutraceuticals is 0.05% of the algal oil (AO) production. This represents ~25% of the global algal omega-3 market. All of the post extracted algal residue (PEAR) goes to animal feed; this being only 1.4% of its US market.
4. Water is obtained from a non-potable saline source such as a saline aquifer.
5. Disposal of wastewater is on land owned by the enterprise. This is permitted, for example, in Texas. Waste disposal cost was neglected. This situation is subject to change and could prove to be a significant problem.
6. Capital cost for ponds, harvesting, and extraction is based on Lundquist *et al.* [84]. For 10,000 bbl/day production we estimate CAPEX for ponds, harvesting, extraction, hydrotreating and co-product production to be ~ $B 5.2.
7. Operating and cost parameters are based on current reasonable values rather than future projected improved values. This more readily permits determination of areas of greatest potential improvement.
8. For the parameters chosen, the average annualized area productivity is 15.2 g (algae dry wt)$\cdot m^{-2} \cdot d^{-1}$. This value was computed from eq. (4) for 5 day

doubling time, 5 day retention time, pond depth of 10 inches, and an initial algal concentration of 0.3 g/L. This is considered to be an annual average value incorporating seasonal variation and is not unreasonable based on reported data and estimates [75, 77, 85]. Seasonality will impact economics by virtue, for example, of unemployed CAPEX during winter months [78].

9. AO composed only of acylglycerides and free fatty acids.
10. Oil content of algae of 25%.
11. Hydrotreater efficiency of 90%.
12. A carbon reduction credit of \$5/T is taken. However, this will be shown to have only a small effect being about \$ 0.01/gal. "T" represents 2,000 lb.
13. On-stream time of 90%. This is more reflective of n-th plant [81].
14. Return on investment of 10%.
15. Income tax rate of 38%.

Table 6. Parameters for Fig. (5) [65].

Algae Growth, Harvesting, & AO Extraction			
Total flow rate thru pond or reactor, gpm	4,729,466	Conditioning H2SO4 dosage, lb 100% acid/lb slurry	0.005
Feed broth algae concentration, g (dry wt)/L	0.300	Wt ratio of hexane to algae (dry wt)	5
Retention time in pond or reactor, hr	120	Efficiency of oil recovery, %	95
Algae doubling time, hr	120	Moisture content of biomass, strm 15, wt%	85
Pond depth, ft	0.83	Moisture content of dryed biomass, strm 15b, wt%	10
Algae gross composition oil, % of dry wt	25	H2O usage, lb H2O/lb algae	0.333
water solubles, % of dry wt	29	Algal Exit conc, g/L	0.60
Insolubles, % dry wt	46	Algal recycle ratio (stm5a/stm5b)	1.00
Efficiency of CO2 utilization, %	95	Fraction of total feed flow that is recycled	0.50
CO2 usage, lb CO2/lb algae	2.0	Total Pond volume, gal	3.405E+10
Oxygen production, lb O2/lb algae(dry)	1.33	Algal area productivity, g(dry wt algae)/m^2/day	15.2
Net Evaportion rate, inch/yr	43	Algal volumetric productivity, g(dry wt algae)/L/day	0.06
Theor N nutrient required, wt% of aglae dry wt	9.2	Total Area of pond, ac	125,915
Theor P nutrient required, wt% of aglae dry wt	1.3	Annual production, ton algae/ac/y	25
% excess nutrient fed to pond	15	Algal oil production rate, gal/yr/ac	1,734
Purge rate (of strm#17) % of strms 5d,6a & 14	10	Algal oil production rate, Mgal/yr	218.3
Conc of algae in floc thickener exit (strm 6), wt %	2	Algal oil production rate, bbl/day/ac	0.11
Conc of algae in centrifuge exit (strm 6b), wt %	12	HTAO oil to refinery rate, bbl/day	10,000

Flow rates are "calendar rates". Also, pond size etc. stated for 100% on-stream time.

Growth, Harvest and Extraction

Park *et al.* [86] give a typical algal empirical formula as $C_{106}H_{181}O_{45}N_{16}P$. Using this formula to calculate weight percent composition of dry algae, the nutrient requirements per unit weight of dry algae can be estimated, see Table **7**.

Table 7. Estimated Algae Nutrient Requirements [65].

Nutrient	**Weight of Nutrient per unit Weight of Dry Algae**
CO_2	1.91
H_2O	0.33
N as NH_3	0.11
P as P	0.013

Many pond configurations are possible. Consider that shown in Fig. (**6**). Inoculation ponds which could consist of photobioreactors followed by covered ponds then open plastic-lined ponds [84] are not shown.

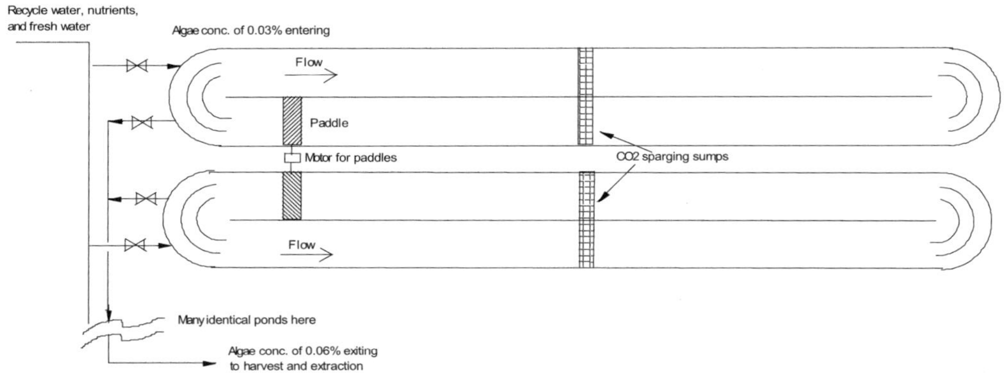

Fig. (6). High Rate Algal Ponds [65].

The ponds are raceways in which water flow is induced by paddle wheels and CO_2 is sparged into the flowing streams at various points. This design is known as a high rate algal pond. For the given design, the order of 140,000 acres of ponds at 90% on-stream time would be needed.

Calculation of Pond Area Productivity and Production Rate

Under idealized assumptions, algal area productivity can be related to four parameters: (1) initial algal concentration, (2) depth of pond, (3) residence time in pond, and (4) doubling time of algae. As shown in [65], the pond area productivity, P_A, in the units of grams of dry algae per square meter per day can be calculated from [65]:

$$P_A = \frac{454\rho^* c_i q}{4046A}\left[2^{\frac{\theta_r}{\theta_d}} - 1\right] = \frac{0.1122 c_i q}{A}\left[2^{\frac{\theta_r}{\theta_d}} - 1\right] \quad \textbf{(1)}$$

Symbol definitions and the required units are given in the Nomenclature. The flow into and out of the pond, q, should not be confused with the flow rate within the raceway which is a completely independent parameter.

The residence time in the pond is given by:

$$\theta_r = Ah/q \quad \textbf{(2)}$$

and the density of the algal slurry by:

$$\rho^* = \rho(62.4)(43,560) = 2.72*10^6 \rho \text{ lb/acre-ft} \quad \textbf{(3)}$$

Taking ρ = 1.0 and eliminating 'q' by using Equation 2, Equation 1 can be rewritten as:

$$P_A = \frac{3.05*10^5 c_i h}{\theta_r}\left[2^{\frac{\theta_r}{\theta_d}} - 1\right] \quad \textbf{(4)}$$

From this equation, it can be seen that productivity is directly proportional to the inlet algal concentration and to the depth of the pond. A deeper pond and high inlet concentration are favorable to the extent that light absorption does not overweight this effect. For example, if the initial algal concentration expressed as weight fraction is 0.0003 (*i.e.,* 0.3 $g \cdot L^{-1}$), pond depth is $h = 0.83$ ft, residence time in pond is $\theta_r = 5$ days, and doubling time is $\theta_d = 5$ days, the area productivity of the pond is calculated from Equation 4 being 15.2 gram dry algae per day per square meter (or 25 $T \cdot acre^{-1} \cdot yr^{-1}$). Under field conditions the doubling time of 5 days is significantly longer than that experienced for the ideal conditions of the laboratory. In the real world, the doubling time of algae is linked to the depth. For example, as the depth of the pond increases, the doubling time may also increase due to light shading.

Calculation of the production rate of algae, Q ($lb \cdot d^{-1}$) can be carried out from:

$$Q = \rho^* q c_i \left(2^{Ah/(q\theta_d)} - 1\right) = \rho^* q c_i \left(2^{\frac{\theta_r}{\theta_d}} - 1\right) \quad \textbf{(5)}$$

From Equation 5, for a fixed pond area and depth, we have the slightly paradoxical situation suggesting that if the flow rate into the pond is increased, reducing residence time the production rate goes down! Conversely, reducing the flow rate increases daily production, assuming that the shading effects do not come into play. For example, if residence time is reduced from 5 days to 2 by increasing the in and out flow rate, production rate is reduced by ~25%. Therefore, as great as possible a residence time is indicated (compatible with the adverse shading effects) for maximal use of land area. This suggests that it may be beneficial to expend more power to increase pond velocity to reduce the shading effect, thereby allowing greater algal density.

Pond volumetric productivity (in units of mg·L^{-1}·d^{-1}) can be calculated from area productivity by the following equation:

$$P_V = 3.28 P_A / h \quad (6)$$

where the units are given in the Nomenclature. Using Equation 4, we have:

$$P_V = \frac{1.00*10^6 c_i}{\theta_r}\left[2^{\frac{\theta_r}{\theta_d}} - 1\right] \quad (7)$$

For the base case area productivity of 15.2 g·m^{-2}·d^{-1} and a pond depth of 0.83 ft, the volumetric productivity is calculated to be 60 mg·L^{-1}·d^{-1}.

Harvesting and Extraction

Harvesting consists of dissolved air flotation followed by centrifugation. A variety of ways of extracting algal oil from the thickened slurry have been explored [70]. The method of oil extraction, illustrated in Fig. (**5**), is based on Valicor Separation Technologies LLC as discussed in [87]. The resulting bottom product is the isolated algal oil which requires a subsequent de-gumming step [79]. The algal oil is then sent to the hydrotreatment section. This step is discussed in [65].

ECONOMICS

Detailed economics have been treated in a number of recent studies, notably, refs [67, 73 - 75, 77, 79, 83 - 85, 88]. A recent paper by Davis *et al.* [78] addressed large (national) scale production of diesel from microalgae *via* hydrothermal liquefaction with a co-product of naptha. This reference shows the impact of seasonality on economics. The reader is referred to these papers for comparative capital and operating inputs and outputs.

Capital Cost

Growth-Harvesting-Extraction step

A number of capital cost estimates have been published. Some of these are listed in Table **8**.

Capital cost for the growth, harvest and extraction section was estimated from Lundquist *et al.* [84] adjusted to 2014 dollars with a scaling exponent of 0.75. The choice of a 0.75 exponent rather than the more conventional 0.60 accounts for ponds number up rather than scaling up. Since we are considering both ponds and harvesting and extraction, the latter two benefiting from economy of scale; a value of 0.75 was taken for the scale up exponent.

Table 8. Some Literature Capital Cost Estimates [65].

Estimate	Year of Estimate	Pond lined?	Harvest equipment?	Capital Cost $M/1000 acres
Lundquist *et al.* [84]	2010	no	yes	100
Davis *et al.* [67]	2011	no	yes	40
Sun *et al.* [73]	2011	no	yes	85 (estimated)
Junker & Faaij [85]	2013	yes	no	90
Davis *et al.* [78]	2014	yes	yes	88 (estimated)

Hydrotreater Step

The capital cost for hydrotreatment was taken from ref [89]. which suggests $M70 for a 30,000 bbl/day capacity. This value is scaled for capacity by the 0.6 exponent. For 10,000 bbl/day the hydrotreater plant is estimated to be $M 36.2.

Animal Feed Plant

Post extracted algae residue (PEAR) usually contains high levels of water which has significant cost implications. Further residual hexane in the algal-residue requires steam stripping or the use of a dryer which is vapor tight. Dewatering equipment and solar drying should be considered, see [90]. Capital cost for a drying unit (including off gas scrubbing) of animal feed is estimated to be $M 3 for each 12,000 $lb \cdot h^{-1}$ of wet cake containing 85% moisture [91]. This represents 278 drying units. For 10,000 $bbl \cdot d^{-1}$ production, CAPEX for animal feed drying is $M 839. This represents about 16% of the total CAPEX.

Nutraceutical Plant

An intimation of what could be involved in processing nutraceuticals is afforded by [92]. As a first pass estimate we take its capital to be 30% of the hydrotreater capital.

Summary

A summary of the capital estimates along with those for labor are given in Table **9**. Estimated total capital cost for a biorefinery to produce 10,000 bbl/day of HTAO is $B 5.2.

Table 9. Some Economic Input Parameters [65].

On-stream time, %	90
Base size for scale factor, acres	1,000
Capital cost for base size, $M	106
Scale exponent on capacity	0.75
Estimated ponds & harvest & extraction capital cost, $M	4,312
Pond operator per 1000 ac, shift-man	0.5
Harvesting, extraction, & distillation operators, shift-man	10
Hydrotreater base capacity, bbl/day	30,000
Hydrotreater base capital cost, $M	70
Estimated hydrotreater capital cost for HTAO to refinery rate, $M	38.6
Hydrotreater operators, shift-man	5
Animal feed & Nutraceutical operators, shift-man	7
Installed CAPEX for animal feed dryer for 12,000 lb/h, $M	3
Factor for capital cost of nutraceutical plant as frac of Hydrotreater cap,fra	0.3
Capital cost of animal feed plant, $M	839
Capital cost of nutraceutal plant, $M	11.6
Total capital cost, M$	5,201

The total capital cost of $B 5.2 gives a unit CAPEX cost of $24/(gal/yr) which compares with an estimate from ref [78]. of $30 to 40/(gal/y) for a much smaller scale production of ~3 Mgal/y using hydrothermal liquefaction. The effect of CAPEX is discussed in the economics section.

Variable Costs

Variable costs are calculated from usage and price of the raw material and utilities, see Table **10**. The usages for CO_2, NH_3 and DAP were derived from Table 7. Catalyst cost for hydrotreating was estimated at $0.09/gal. Utility costs for the hydrotreatment plant, were regarded as relatively small and were not included.

CO_2 is the major raw material cost contributing $1.62/gal to the sales price. It is remarked, however, that if a saline aquifer is being used it may contain dissolved sodium bicarbonate which would provide some of the carbon, thereby reducing CO_2 cost. Also, not accounted for in this analysis, flue gas from the animal feed dryer could be used to reduce CO_2 expense. The sum of nitrogen and phosphorous raw materials, *i.e*., the nutrients, contribute another $1.32/gal to the sales price. Thus, raw material cost is already at $3.14/gal.

Table 10. Usages and Variable Costs for Sections 1 and 2 [65].

Growth-harvest-extraction:		Price	Usage	Cost	Cost
Raw Materials	Units	$/Unit	Unit/bbl	$/bbl	$/gal
CO2	lb	0.02	3,402	68.0	1.62
nutrient: nitrogen as NH3	lb	0.20	171	34.3	0.82
nutrient: phosphate as DAP	lb	0.22	95	21.0	0.50
make up water (strms 1 & 4a)	k gal	0.05	1.25	0.06	0.00
flocculant	lb			0.0	0.00
H2SO4	lb	0.12	70.9	8.5	0.20
make up hexane	lb			0.0	0.00
deemulsifier	lb			0.0	0.00
TOTAL Raw Mat. Cost				131.9	3.14
Utilities & other costs					
Electric	KWH	0.08	610.5	48.8	1.16
Steam	k lb	8.00	8.8	70.3	1.67
Cooling water	k gal				
TOTAL Utilities cost				119.1	2.84
Hydrotretment:		Price	Usage	Cost	Cost
Raw Materials	Units	$/Unit	Unit/bbl	$/bbl	$/gal
H2	lb	0.68	11.3	7.7	0.18
Catalyst	lb			3.8	0.09
amine solvent	lb			0.0	
TOTAL Raw Mat. Cost				11.5	0.27

Although the saline water as a raw material has minimal economic impact, water use is estimated to be 4.5 billion gallons per year at a 10% purge rate for a 10,000 bbl/day plant. Thus possible depletion of the aquifer must be considered.

Utility Usages

These are shown in Table **11** for the growth, harvesting and extraction section and have been included in Table **10**.

Credits

Animal Feed

Animal feed is shown as stream 15b of Fig. (**5**) and can require washing especially if a saline water source is used to grow the algae. Water condensed from the dryer can be used for washing. Animal feed could be a very interesting co-product in that its market size is large. For example, in the US corn and soybean meal are consumed as livestock feeds at the rate of 158 million tons per in the 2011/2012 market year [88]. The amount of animal feed produced by a 10,000 bbl/day plant is about 2.2 million T/y which is only 1.4% of the US market. We take a sale price of $225/T for animal feed.

Table 11. Utility Usages and Costs for Section 1 [65].

Growth-harvest-extraction:					
Electric requirement	unit	value of unit	KWH/bbl	$/bbl	$/gal
Flow thru ponds	KW/ac	0.81	244.8	19.58	0.47
CO2 distribution & sparging	KW/ac	0.415	125.4	10.03	0.24
Dissolved air flotation	KW/gpm	0.012	69.2	5.53	0.13
Centrifuge of algae	KW/gpm	0.84	143.5	11.48	0.27
Contacting/Extraction	KW/gpm	0.15	4.3	0.34	0.01
Centrifuge of biomass (PEAR)	KW/gpm	0.84	23.4	1.87	0.04
		Total	610.5	48.8	1.16
Steam or Heat Requirment			MBTU/bbl	$/bbl	$/gal
Heat required for conditioning (strm8)			1.41	11.28	0.27
Heat req'ed for hexane distill'n (strm9)			1.34	10.68	0.25
Heat req'ed for drying wet biomass for animal feed(strm15			6.04	48.34	1.15
		Total	8.79	70.3	1.67

Our spreadsheet incorporates the drying requirement to remove water from 85% to 10% from the wet biomass in stream 15 of Fig. (**5**). As mentioned, this is a utility expense of $1.15/gal as well as a reduction in the CO_2 reduction credit. Dewatering equipment should be considered to reduce the moisture content of the wet cake. For example, if wet cake moisture content could be reduced from 85% to 45% before drying to 10%, a savings of $1.12/gal could be realized when both utility cost and dryer CAPEX are considered. Possibly sun drying may be an option. We will find that a sales price for animal feed of $225/T will reduce the sales price of HTAO by $3.24 /gal. CAPEX for drying animal feed is an expense of $0.94/gal; while operating cost is another $1.15/gal for a total of $2.09/gal expense. This, roughly (excluding labor), yields a net sale price reduction of $3.24 - $2.09 = $1.15/gal for animal feed production. Therefore, production of animal feed, per se, should be beneficial. It is noted that increasing the oil content of the algae will generally reduce the amount of animal feed produced.

Digestion of PEAR is an alternative to producing animal feed. This would generate methane gas which could be used to dry wet cake or generate electricity.

The resulting solids, which are said to be easier to de-water than non-digested solids, could possibly be sold as organic fertilizer.

Nutraceuticals

Anchovy and tuna are the major sources of omega-3 oils. Global annual consumption (human and pet foods) of omega-3 oils is around 89,000 metric tons (not including aquaculture and cattle feed). However, consumption of algal omega-3 oils is small, being about 1,500-1,600 metric tons/yr [93]. Average price of algal omega-3 is about $43/lb [93]. We have taken the price of algal omega-3 as $30/lb. The price of refined fish oil is about $3.00/lb. Global growth rate of algal omega-3 oils is about ~6%/yr [93]. Use of omega-3s in infant formulas, especially in China, is the largest and fastest growing segment of the omega-3 market [94]. Algal derived omega-3 compared to fish derived oil has a number of important advantages [95].

It is essential to prevent toxic materials from entering the production system if human and animal consumption is involved. This requires the water supply and nutrients, including CO_2, must be free of toxic materials. This requires use of purified CO_2 from an absorption-desorption process. Possibly flue gas from a natural gas fired power station might be acceptable, but the operating cost of distribution and compression might counterbalance the cost of purified CO_2. The present analysis assumes purified CO_2 at $40/T.

It is assumed that only 0.05% of the AO production is diverted to nutraceuticals. This volume would be about 25% of the global algal omega-3 market, which is a relatively large fraction. As the HTAO plant capacity increases the fraction of AO going to nutraceuticals would have to be further reduced to suit market demand so as not to disrupt sale price. The processing cost for production of EPA from algal biomass indicated that about 40% of the cost was associated with the biomass and 60% with the recovery/purification steps [96]. Our calculations assumes the cost of processing from AO into nutraceutical product to be 20% of the selling price. The nutraceuticals credit for these assumptions lowers the price of HTAO by only $0.14/gal.

Carbon Dioxide Reduction Credit

The social cost of carbon (SCC) is a measure of the incurred expense throughout the world for the discharge of a ton of CO_2 into the atmosphere. Therefore, a credit could be given to a process which results in CO_2 reduction compared to a conventional process. Current carbon credits appear to be about $5/T of CO_2 reduction. However, there are estimates of the SCC in the of about $50-100/T. Note that the cost of carbon capture from power plants is on the order of $50 –

150 per ton of CO_2 [97, 98]. Hence a carbon credit of 10 to 30 times higher than what is given in this analysis could be justified, *i.e.*, \$0.10 to \$0.30/gal rather than the \$0.01/gal, see Table **12**.

Table 12. Summary of Usages and Co-Product Credits [65].

Credits	Units	Price $/Unit	Usage Unit/bbl	Cost $/bbl	Cost $/gal
Waste biomass (animal feed)	ton	225	-0.60	-136	-3.24
Hydrotreatment Purge gas fuel credit (strm#11, Sec 2)	MBTU	3.5	-0.02	-0.1	-0.002
Nutraceuticals	lb	30	-0.2	-6.1	-0.14
Cost of processing nutraceutical	lb	6	0.2	1.2	0.03
CO2 reduction credit	T	5	-0.12	-0.6	-0.01
TOTAL credits				-141	-3.37

In capturing of 1 lb of purified CO_2 from flue gas, ~0.25 lb of CO_2 is emitted to the atmosphere by the capture process due to energy required for heating and cooling of the amine absorbent. Therefore, only ~75% of the purified CO_2 used by the algae process can be credited as being removed from the atmosphere. Reference [65] provides a life cycle assessment (LCA) comparison of CO_2e or GHG emissions between the algae based process of this paper and conventional crude oil to diesel.

A summary of usages (a negative usage means production) and credits is given in Table **12**. It is seen that co-products reduce sale price by \$3.37/gal. The main contributor being animal feed, nutraceuticals contributing a reduction of only ~\$0.14/gal.

Fixed Costs

These are shown in Table **13**. We have assumed 0.5 shift-man per 1000 acres, see Table **9**. A "shift-man" represents 4 people (3 to cover each of 3 daily shifts and an additional person for weekends and vacations). Each operator will manage a square ~1.8 mile on a side. For harvesting, extraction and hexane distillation it is estimated that 10 shift operators will be needed. Fixed costs contribute about \$ 2.08/gal to the sales price. The estimate (see Table **13**) of a work force compliment of 605 people is probably on the high side due to the estimate of 0.5 shift-man per 1000 acres.

Case Studies

Bear in mind that HTAO is nearly the quality of diesel. The sales price of HTAO for the base case (see Table **16**) is calculated to be \$9.89/gal for a 10% return on investment. The revenue brought in by each of the three products is shown in

Table **14**. Animal feed is seen to contribute a significant portion to the economics, while nutraceuticals are a minor factor (due to its necessarily small volume).

Table 13. Fixed Costs [65].

FIXED COSTS		Cost, $K/	Cost	Cost
A. Labor (incl PAC)	Shift-man	shft-man/y	$/bbl	$/gal
Operating	85.0	300	6.98	0.17
Oper Superv	6.2	420	0.71	0.02
Lab'tory	20.0	280	1.53	0.04
Yard	30.0	250	2.05	0.05
Packaging	10.0	320	0.88	0.02
Maintenance (1.5% of capital)	Tot op labor = 605		21.38	0.51
B. Supplies				
Operating (10% of total operating labor)			3.35	0.08
Maintenance (1.5% of capital)			21.38	0.51
C. Indirect				
Admin (Tech,acc,safety: 40%of tot labor)			13.41	0.32
Taxes & Ins (1.1% of capital)			15.68	0.37
TOTAL Fixed Cost			87.36	2.08

Table 14. Revenue Streams at 10,000 bbl/day Production [65].

Revenue Streams			$M/yr
HTAO	**@ $9.89/gal**		1,516.1
Animal feed			496.2
Nutraceuticals			22.1
		Total=	2,034.5

The effects of various conditions and parameters on the economics of the enterprise are now considered. For some values of the parameters, for example algae doubling time, the pond area requirement for a fixed production will change thereby causing the capital cost to change. This effect is incorporated into the economic calculations.

Base Case

Table **15** shows the contribution of each category of the enterprise for the base case parameters (see Table **16**). It is seen that costs associated with CAPEX is the major contributor to sale price, followed by raw materials and co-product credit.

Table 15. Summary of Base Case Economics [65].

Cost Component	$/gal
Raw materials	3.41
Utilities	2.84
Fixed costs	2.08
Capital cost recovery	4.93
Co-product credits	-3.37
Sales Price=	9.89

Table 16. Sensitivity Analysis of Base Case to 10% Increase in Parameter [65].

Production rate, bbl/day	10,000		
On-stream time, %	90		
Base sales price of HTAO, $/gal (ex tax)	9.89		
Base case capital cost, $M	5,201		

PARAMETERS	Base Parm Value	10% increase in Parm Value	Delta SP $/gal(ex tx)
Algae doubling time, hr	120	132	0.92
Algae gross composition: oil, % of dry wt	25	27.5	-0.69
Capital cost, $M	5,201	5,721	0.64
Moisture in post extracted algae biomass*, %	85	76.5*	-0.63
Feed broth algae concentration, g (dry wt)/L	0.3	0.33	-0.47
Return on investment, %	10.0	11	0.46
Pond depth, ft	0.83	0.91	-0.44
Retention time in pond or reactor, hr	120	132	-0.41
Sale price of animal feed, $/T	225	247.5	-0.33
Conc of algae in centrifuge exit (strm 6b), wt %	12	13.20	-0.09
Conc of algae in DAF floc thickener exit (strm 6), wt %	2.0	2.20	-0.03
Percent of AO going to nutraceuticals,%	0.050	0.055	-0.01
Sale price of nutraceuticals, $/lb	30	33.00	-0.01
CO2 creduction credit, $/T	5.0	5.5	0.00

Delta SP = change in sales price for a 10% change in parameter all other parameters remaining fixed, $/gal
* moisture in post extracted algae residue was reduced by 10%
note: area of pond and capital cost may change as a function of the changed parameter.

Sensitivity Analysis

Table **16** tabulates base case values of key parameters. These base case parameters generate a sale price of $9.89/gal. For a 10% increase in a given parameter, all other parameters remaining fixed, Table **16** indicates the change in sales price. Parameters are listed in the order of greatest effect, either beneficial or detrimental, to least effect. The doubling time of the algae causes the greatest effect. A 10% increase in the base case algae doubling time, going from 5 days to 5.5 days, increases the sales price by $0.92/gal. Conversely, a 10% reduction in doubling time reduces sale price by ~ $0.92/gal. This suggests research into reducing doubling time in the field operation versus the lab would be worthwhile. The next tier of importance includes: % oil content in algae, capital cost and moisture content of stream 15 of Fig. (**5**). Next in importance are feed broth algae concentration, % return on investment, pond depth, retention time in the pond and sale price of animal feed.

A sales price of $6.29/gal could be achieved if all of the 10% changes in parameters were made in the right direction. We mention that individual changes are not necessarily additive as there can be an interactive effect if all the changes were made simultaneously. None the less, this indicates that a significant beneficial effect on economics could be achieved by relatively modest improvements in a large number of areas.

Effect of Capital Cost, Oil Content and Productivity

The strict effect of capital cost for two combinations of algal oil content and productivity, *viz.,* (a) 25% oil & 15.2 g/m^2/day, and (b) 30% oil & 25.3 g/m^2/day is shown in Fig. (**7**). It is seen that at a given capital cost, the difference in sales price between the two cases, (a) and (b), is only $0.69/gal. This is due to the dominate effect of capital. The slope of the lines gives an effect of $1.20/gal per billion dollars of capital cost.

If we now let the capital cost be a function of oil content and of productivity, as it in fact will be, then the effect is amplified, see Table **17**. Variable cost is also given. It has been seen that oil content and productivity can have a major effect on sales price by virtue of their impacting capital cost primarily and also variable cost.

Table 17. Effect of Oil Content and Productivity on Capital and Sales Price [65].

% Oil	Productivity, g/m^2/d	CAPEX, $M	Variable Cost, $/gal	HTAO Sale Price, $/gal
25	15.2	5,201	2.88	9.89 (base)
25	25.3	3,832	2.54	7.75
30	15.2	4,440	2.60	8.62
30	25.3	3,246	2.31	6.77

If open ponds could be operated at a productivity of 25.3 g/m^2/day rather than 15.2 g/m^2/day with an algal oil content of 30% rather than 25%, then sale price is reduced to $6.77/gal.

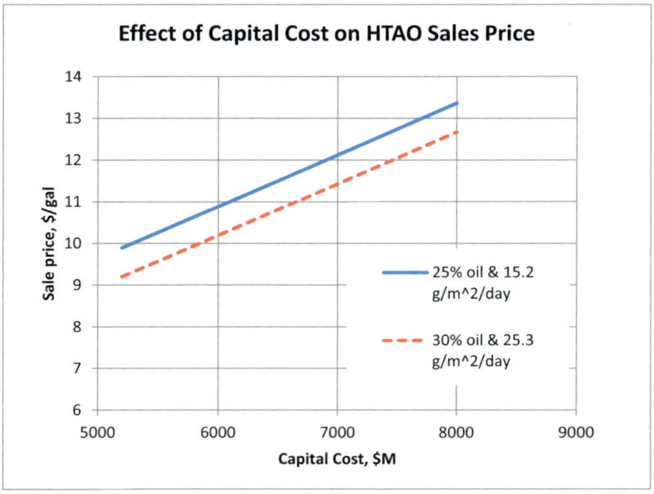

Fig. (7). Effects of capital on HTAO sales price [65].

Trade-off between Oil Content and Weight Productivity

A nitrogen starvation diet at the end of the growth cycle which adversely affects weight productivity, gr(dw)/m^2/day but elevates oil content in the algal biomass. This is done in an effort to increase oil area production rate (gal/ac/yr). Superficially, one might think it doesn't matter that much since oil area production rate (gal/ac/yr) is obtained by the multiplying oil content by weight productivity. Economically, however, it does matter. It is in fact preferable to enhance oil content at the expense of weight productivity provided the gal/ac/yr is unchanged. This results from the fact that increased weight productivity gr/m^2/day will require increased nutrients, greater animal feed production and drying, and the sale of more low priced animal feed at the expense of higher priced fuel oil. In Table **18**, the calculated sale price at a *constant* oil area production rate of 1734 gal/ac/yr (or, equivalently, 3.8 g (of oil)/m^2/day) is given for different oil contents and weight productivities. It is seen that lower sale price is obtained from the higher oil content providing the oil area productivity remains constant. There is a sharp reduction is sale price from 12.5 to 25% oil content. For oil contents greater than 25% the improvement is more gradual.

Table 18. Sale Price at Constant Oil Area Productivity vs Wt. Productivity and Oil Content [65].

Oil Content	Weight Area Productivity	Oil Area Productivity		HTAO Sale Price
% oil	g/m^2/day	g/m^2/day	gal/ac/yr	$/gal
12.5	30.8	3.8	1734	12.32
25	15.2	3.8	1734	9.89

(Table 18) contd.....

Oil Content	Weight Area Productivity	Oil Area Productivity		HTAO Sale Price
50	7.6	3.8	1734	8.67
100	3.8	3.8	1734	7.90

In the important limiting case of 100% oil content, that is, with no production of animal feed, the HTAO sale price is \$7.90/gal. Note that CAPEX is reduced by \$B0.8 due to elimination of the animal feed plant. This case of 100% algal oil content, *i.e.*, no co-product, seems to imply an inherently high cost of producing an algal derived oil due the high capital cost.

Solar Energy Conversion Efficiency

For the base case, the solar efficiency for the conversion to HTAO based on total solar energy input at a Texas location (6.0 KWH/m^2/day) is estimated to be ~ 0.5%. For the improved case of 30% oil content and 25 g/m^2/day area productivity, the efficiency rises to ~ 1.0%.

CONCLUSION

Capital cost, raw materials and credit for the co-product are the most significant costs contributors to HTAO sale price are. If relatively small improvements on the order of 10% in numerous parameters are made, significant reduction in sale price could be achievable. Algal doubling time is the most impactful parameter on economics. Increased agitation by increasing pond flow velocity or installation of turbulence inducing baffles in order to reduce doubling time should be examined. Also finding an economic means of dewatering algal residue for animal feed is important. The major energy sinks were indicated inviting improvement. Nutraceuticals reduced HTAO sales price by only \$0.14/gal. It was shown that, at constant oil area productivity (gal/acre/yr or g/m^2/day) the trade-off between oil content and weight productivity favors oil content. In the limit of 100% oil content a "lower bound" of the sale price is estimated to be \$7.90/gal. This is still a relatively high sale price and is somewhat reflective to the large capital cost contribution. Possible depletion of the water source must be considered as water use (although saline) is estimated to be 4.5 billion gallons per year at a 10% purge rate for a 10,000 bbl/day plant. Economic implications only were discussed here. Of great importance however are the sustainability and life cycle implications of fuel from microalgae. There appears to be in general a critical need for future work on sustainability and life cycle analysis. The material balance and economic spreadsheets developed here can be a useful tool in further sharpening economics as more accurate data become available.

NOMENCLATURE

A Area of pond(s), acre

c_i Initial concentration of algae, wt fraction

h Depth of pond, ft

n Number of algal doublings in pond, i.e., θr/ θd, or Ah/(qθd), dimensionless

P_A Pond area productivity, gram dry algae per square meter per day

P_V Pond volumetric productivity, mg/L/day

q Volumetric flow rate (time averaged over the duration of the batch) into and out of pond (assumed equal due to replacement of evaporation losses), acre-ft/day

Q Production rate of algae, lb(dry wt)/day

θ_d Doubling time of algae, day

θ_r Residence time in pond, *i.e.*, Ah/q, day

ρ Specific gravity (~1.0), dimensionless

ρ* Density of algal slurry, lb/ac-ft

CONFLICT OF INTEREST

The authors declare no conflict of interest, financial or otherwise.

ACKNOWLEDGEMENTS

We benefited from discussions with Sam Couture and David Hart of Qualitas-Health; Emilie Slaby of The Scoular Co., David Vaccari and Aspasia Kalomoiri, of Stevens Institute of Technology; Robert B. Levine and Kiran Kadam of Valicor Renewables LLC; Philip Pienkos and Ryan Davis of NREL, Warren Seider of Penn, Adam Ismail of GEOD, and Tom Capehart of the USDA; Jess Ornum and Kerry Kovacs of Komline-Sanderson; Aldo Bernasconi of GOED. We gratefully acknowledge that this work was funded in part by DOE contract DE-EE0006063.

REFERENCES

[1] Energy Information Administration, International Energy Outlook 2016, U.S. Department of Energy, May 2016.

[2] B-X. Peng, "Biodiesel production from waste oil feedstocks by solid acid catalysis", *Process Saf. Environ. Prot.,* vol. 86, no. 6, pp. 441-447, 2008. [http://dx.doi.org/10.1016/j.psep.2008.05.003]

[3] R.A. Lee, and J-M. Lavoie, "From first-to third-generation biofuels: Challenges of producing a commodity from a biomass of increasing complexity", *Anim. Front.,* vol. 3, no. 2, pp. 6-11, 2013. [http://dx.doi.org/10.2527/af.2013-0010]

[4] B. Dale, "Biofuels: thinking clearly about the issues", *J. Agric. Food Chem.,* vol. 56, no. 11, pp. 3885-3891, 2008. [http://dx.doi.org/10.1021/jf800250u] [PMID: 18422330]

[5] A.T. Bell, *Basic research needs: catalysis for energy. No. PNNL-17214. Pacific Northwest National Laboratory.* PNNL: Richland, WA, US, 2008.
[http://dx.doi.org/10.2172/927492]

[6] Gerhard Knothe, Jon Van Gerpen, and Jürgen Krahl, *The Biodiesel Handbook,* 2005.
[http://dx.doi.org/10.1201/9781439822357]

[7] R. Sawangkeaw, and S. Ngamprasertsith, "A review of lipid-based biomasses as feedstocks for biofuels production", *Renew. Sustain. Energy Rev.,* vol. 25, pp. 97-108, 2013.
[http://dx.doi.org/10.1016/j.rser.2013.04.007]

[8] R.K. Devappa, and P.S. Harinder, "Makkar, and Klaus Becker. "Jatropha toxicity—a review." Journal of toxicology and environmental health", *Part B,* vol. 13, no. 6, pp. 476-507, 2010.

[9] L. Li, "Catalytic hydrothermal conversion of triglycerides to non-ester biofuels", *Energy Fuels,* vol. 24, no. 2, pp. 1305-1315, 2010.
[http://dx.doi.org/10.1021/ef901163a]

[10] Achten, Wouter MJ, et al. "Jatropha biodiesel fueling sustainability?" Biofuels, bioproducts and biorefining 1.4 (2007): 283-291.

[11] A.A. Refaat, "Different techniques for the production of biodiesel from waste vegetable oil", *Int. J. Environ. Sci. Technol.,* vol. 7, no. 1, pp. 183-213, 2010.
[http://dx.doi.org/10.1007/BF03326130]

[12] M.K. Lam, K.T. Lee, and A.R. Mohamed, "Homogeneous, heterogeneous and enzymatic catalysis for transesterification of high free fatty acid oil (waste cooking oil) to biodiesel: a review", *Biotechnol. Adv.,* vol. 28, no. 4, pp. 500-518, 2010.
[http://dx.doi.org/10.1016/j.biotechadv.2010.03.002] [PMID: 20362044]

[13] B.Q. Araújo, "Synthesis and characterization of beef tallow biodiesel", *Energy Fuels,* vol. 24, no. 8, pp. 4476-4480, 2010.
[http://dx.doi.org/10.1021/ef1004013]

[14] M. Canakci, and J. Van Gerpen, "Biodiesel production from oils and fats with high free fatty acids", *Trans. ASAE,* vol. 44, no. 6, pp. 1429-1436, 2001.
[http://dx.doi.org/10.13031/2013.7010]

[15] K. White, "Production of biodiesel fuel from tall oil fatty acids via high temperature methanol reaction", *Fuel,* vol. 90, no. 11, pp. 3193-3199, 2011.
[http://dx.doi.org/10.1016/j.fuel.2011.06.017]

[16] M.J. Haas, "The general applicability of in situ transesterification for the production of fatty acid esters from a variety of feedstocks", *J. Am. Oil Chem. Soc.,* vol. 84, no. 10, pp. 963-970, 2007.
[http://dx.doi.org/10.1007/s11746-007-1119-4]

[17] A. Singh, S.I. Olsen, and P.S. Nigam, "A viable technology to generate third□generation biofuel", *J. Chem. Technol. Biotechnol.,* vol. 86, no. 11, pp. 1349-1353, 2011.
[http://dx.doi.org/10.1002/jctb.2666]

[18] R.A. Andersen, "Diversity of eukaryotic algae", *Biodivers. Conserv.,* vol. 1, no. 4, pp. 267-292, 1992.
[http://dx.doi.org/10.1007/BF00693765]

[19] X. Meng, "Biodiesel production from oleaginous microorganisms", *Renew. Energy,* vol. 34, no. 1, pp. 1-5, 2009.
[http://dx.doi.org/10.1016/j.renene.2008.04.014]

[20] B. Peng, Y. Yao, C. Zhao, and J.A. Lercher, "Towards quantitative conversion of microalgae oil to diesel-range alkanes with bifunctional catalysts", *Angew. Chem. Int. Ed. Engl.,* vol. 51, no. 9, pp. 2072-2075, 2012.
[http://dx.doi.org/10.1002/anie.201106243] [PMID: 22086827]

[21] T.M. Mata, A.A. Martins, and N.S. Caetano, "Microalgae for biodiesel production and other

applications: a review", *Renew. Sustain. Energy Rev.,* vol. 14, no. 1, pp. 217-232, 2010. [http://dx.doi.org/10.1016/j.rser.2009.07.020]

[22] P. Spolaore, C. Joannis-Cassan, E. Duran, and A. Isambert, "Commercial applications of microalgae", *J. Biosci. Bioeng.,* vol. 101, no. 2, pp. 87-96, 2006. [http://dx.doi.org/10.1263/jbb.101.87] [PMID: 16569602]

[23] L. Brennan, and P. Owende, "Biofuels from microalgae—a review of technologies for production, processing, and extractions of biofuels and co-products", *Renew. Sustain. Energy Rev.,* vol. 14, no. 2, pp. 557-577, 2010. [http://dx.doi.org/10.1016/j.rser.2009.10.009]

[24] M.A. Borowitzka, N.R. Moheimani, Ed., *Algae for biofuels and energy.* Springer, 2013. [http://dx.doi.org/10.1007/978-94-007-5479-9]

[25] L. Zhou, and A. Lawal, "Evaluation of Presulfided NiMo/γ-Al2O3 for Hydrodeoxygenation of Microalgae Oil To Produce Green Diesel", *Energy Fuels,* vol. 29, no. 1, pp. 262-272, 2014. [http://dx.doi.org/10.1021/ef502258q]

[26] H-W. Yen, P-W. Chen, and L-J. Chen, "The synergistic effects for the co-cultivation of oleaginous yeast-Rhodotorula glutinis and microalgae-Scenedesmus obliquus on the biomass and total lipids accumulation", *Bioresour. Technol.,* vol. 184, pp. 148-152, 2015. [http://dx.doi.org/10.1016/j.biortech.2014.09.113] [PMID: 25311189]

[27] S. Mandal, and N. Mallick, "Microalga Scenedesmus obliquus as a potential source for biodiesel production", *Appl. Microbiol. Biotechnol.,* vol. 84, no. 2, pp. 281-291, 2009. [http://dx.doi.org/10.1007/s00253-009-1935-6] [PMID: 19330327]

[28] Wisconsin Biorefining Development initiative, *Fermentation of lignocellulosic biomass. http://www.biomassinnovation.ca/pdf/fermlig.pdf.*

[29] U.S. Department of Energy, *Understanding Biomass as a Source of Sugars and Energy. http://www.eere.energy.gov/biomass/understanding_biomass.html#biomass.*

[30] E. Sari, Green diesel production via catalytic hydrogenation/decarboxylation of triglycerides and fatty acids of vegetable oil and brown grease. Diss. Wayne State University, Detroit, 2013.

[31] A.K. Azad, "Prospect of biofuels as an alternative transport fuel in Australia", *Renew. Sustain. Energy Rev.,* vol. 43, pp. 331-351, 2015. [http://dx.doi.org/10.1016/j.rser.2014.11.047]

[32] M.F. Demirbas, "Biorefineries for biofuel upgrading: a critical review", *Appl. Energy,* vol. 86, pp. S151-S161, 2009. [http://dx.doi.org/10.1016/j.apenergy.2009.04.043]

[33] Energy from Waste and Wood, *Generations of biofuels,* 2013.

[34] G.A. Olah, A. Goeppert, and G.K. Prakash, "Chemical recycling of carbon dioxide to methanol and dimethyl ether: from greenhouse gas to renewable, environmentally carbon neutral fuels and synthetic hydrocarbons", *J. Org. Chem.,* vol. 74, no. 2, pp. 487-498, 2009. [http://dx.doi.org/10.1021/jo801260f] [PMID: 19063591]

[35] ASTM. ASTM D6751-biodiesel blend stock specification (B100),

[36] Biodiesel Processes in CHEMCAD, http://www.chemstations.com/content/documents/Technical_ Articles/BiodieselWhitePaper.pdf, Chemstations, PDF file.

[37] F. Ma, and M.A. Hanna, "Biodiesel production: a review", *Bioresour. Technol.,* vol. 70, no. 1, pp. 1-15, 1999. [http://dx.doi.org/10.1016/S0960-8524(99)00025-5]

[38] Y. Watanabe, "Conversion of degummed soybean oil to biodiesel fuel with immobilized Candida antarctica lipase", *J. Mol. Catal., B Enzym.,* vol. 17, no. 3, pp. 151-155, 2002. [http://dx.doi.org/10.1016/S1381-1177(02)00022-X]

[39] S. Zullaikah, C.C. Lai, S.R. Vali, and Y.H. Ju, "A two-step acid-catalyzed process for the production of biodiesel from rice bran oil", *Bioresour. Technol.*, vol. 96, no. 17, pp. 1889-1896, 2005. [http://dx.doi.org/10.1016/j.biortech.2005.01.028] [PMID: 16084368]

[40] L. Talens, G. Villalba, and X. Gabarrell, "Exergy analysis applied to biodiesel production", *Resour. Conserv. Recycling,* vol. 51, no. 2, pp. 397-407, 2007. [http://dx.doi.org/10.1016/j.resconrec.2006.10.008]

[41] M. P. Dorado, *An alkali–catalyzed transesterification process for high free fatty acid waste oils.,* 2002.

[42] M.R. Avhad, and J.M. Marchetti, "Innovation in solid heterogeneous catalysis for the generation of economically viable and ecofriendly biodiesel: A review", *Catal. Rev.*, vol. 58, no. 2, pp. 157-208, 2016. [http://dx.doi.org/10.1080/01614940.2015.1103594]

[43] S. Furuta, H. Matsuhashi, and K. Arata, "Biodiesel fuel production with solid superacid catalysis in fixed bed reactor under atmospheric pressure", *Catal. Commun.,* vol. 5, no. 12, pp. 721-723, 2004. [http://dx.doi.org/10.1016/j.catcom.2004.09.001]

[44] A. Galarneau, "True microporosity and surface area of mesoporous SBA-15 silicas as a function of synthesis temperature", *Langmuir,* vol. 17, no. 26, pp. 8328-8335, 2001. [http://dx.doi.org/10.1021/la0105477]

[45] A. Srivastava, and R. Prasad, "Triglycerides-based diesel fuels", *Renew. Sustain. Energy Rev.,* vol. 4, no. 2, pp. 111-133, 2000. [http://dx.doi.org/10.1016/S1364-0321(99)00013-1]

[46] J. Van Gerpen, "Biodiesel processing and production", *Fuel Process. Technol.,* vol. 86, no. 10, pp. 1097-1107, 2005. [http://dx.doi.org/10.1016/j.fuproc.2004.11.005]

[47] Biodiesel blends, http://www.afdc.energy.gov/fuels/biodiesel_blends.html, Alternative fuels Data center, U.S. Department of Energy, Last accessed June 20, 2016.

[48] B.J. Gallagher, "The economics of producing biodiesel from algae", *Renew. Energy,* vol. 36, no. 1, pp. 158-162, 2011. [http://dx.doi.org/10.1016/j.renene.2010.06.016]

[49] Tom Kalnes, Terry Marker, and David R. Shonnard, *Green diesel: a second generation biofuel.,* . [http://dx.doi.org/10.2202/1542-6580.1554]

[50] T.N. Kalnes, "A technoeconomic and environmental life cycle comparison of green diesel to biodiesel and syndiesel", *Environ. Prog. Sustain. Energy,* vol. 28, no. 1, pp. 111-120, 2009. [http://dx.doi.org/10.1002/ep.10319]

[51] Z.D. Yigezu, and K. Muthukumar, "Catalytic cracking of vegetable oil with metal oxides for biofuel production", *Energy Convers. Manage.,* vol. 84, pp. 326-333, 2014. [http://dx.doi.org/10.1016/j.enconman.2014.03.084]

[52] B. Peng, Transformation of triglycerides and fatty acids into biofuels with sulfur-free catalysts., 2012.

[53] A.W. Schwab, "Diesel fuel from thermal decomposition of soybean oil", *J. Am. Oil Chem. Soc.,* vol. 65, no. 11, pp. 1781-1786, 1988. [http://dx.doi.org/10.1007/BF02542382]

[54] Y.K. Ong, and S. Bhatia, "The current status and perspectives of biofuel production via catalytic cracking of edible and non-edible oils", *Energy,* vol. 35, no. 1, pp. 111-119, 2010. [http://dx.doi.org/10.1016/j.energy.2009.09.001]

[55] "Katikaneni, Sai PR, John D. Adjaye, and Narendra N. Bakhshi. “Catalytic conversion of canola oil to fuels and chemicals over various cracking catalysts", *Can. J. Chem. Eng.,* vol. 73, no. 4, pp. 484-497, 1995. [http://dx.doi.org/10.1002/cjce.5450730408]

[56] P.B. Weisz, W.O. Haag, and P.G. Rodewald, "Catalytic production of high-grade fuel (gasoline) from biomass compounds by shape-selective catalysis", *Science,* vol. 206, no. 4414, pp. 57-58, 1979. [http://dx.doi.org/10.1126/science.206.4414.57] [PMID: 17812451]

[57] P. Tamunaidu, and S. Bhatia, "Catalytic cracking of palm oil for the production of biofuels: optimization studies", *Bioresour. Technol.,* vol. 98, no. 18, pp. 3593-3601, 2007. [http://dx.doi.org/10.1016/j.biortech.2006.11.028] [PMID: 17208441]

[58] J.T. Kloprogge, L.V. Duong, and R.L. Frost, "A review of the synthesis and characterisation of pillared clays and related porous materials for cracking of vegetable oils to produce biofuels", *Environmental Geology,* vol. 47, no. 7, pp. 967-981, 2005. [http://dx.doi.org/10.1007/s00254-005-1226-1]

[59] T.A. Milne, R.J. Evans, and N. Nagle, "Catalytic conversion of microalgae and vegetable oils to premium gasoline, with shape-selective zeolites", *Biomass,* vol. 21, no. 3, pp. 219-232, 1990. [http://dx.doi.org/10.1016/0144-4565(90)90066-S]

[60] F.A. Twaiq, A.R. Mohamed, and S. Bhatia, "Liquid hydrocarbon fuels from palm oil by catalytic cracking over aluminosilicate mesoporous catalysts with various Si/Al ratios", *Microporous Mesoporous Mater.,* vol. 64, no. 1, pp. 95-107, 2003. [http://dx.doi.org/10.1016/j.micromeso.2003.06.001]

[61] M. Chiappero, P.T. Do, S. Crossley, L.L. Lobban, and D.E. Resasco, "Direct conversion of triglycerides to olefins and paraffins over noble metal supported catalysts", *Fuel,* vol. 90, no. 3, pp. 1155-1165, 2011. [http://dx.doi.org/10.1016/j.fuel.2010.10.025]

[62] P. Mäki-Arvela, "Catalytic deoxygenation of tall oil fatty acid over palladium supported on mesoporous carbon", *Energy Fuels,* vol. 25, no. 7, pp. 2815-2825, 2011. [http://dx.doi.org/10.1021/ef200380w]

[63] E. Santillan-Jimenez, and M. Crocker, "Catalytic deoxygenation of fatty acids and their derivatives to hydrocarbon fuels via decarboxylation/decarbonylation", *J. Chem. Technol. Biotechnol.,* vol. 87, no. 8, pp. 1041-1050, 2012. [http://dx.doi.org/10.1002/jctb.3775]

[64] L. Zhou, and A. Lawal, "Hydrodeoxygenation of microalgae oil to green diesel over Pt, Rh and presulfided NiMo catalysts", *Catal. Sci. Technol.,* vol. 6, no. 5, pp. 1442-1454, 2016. [http://dx.doi.org/10.1039/C5CY01307K]

[65] J.L. Manganaro, A. Lawal, and B. Goodall, "Techno-economics of microalgae production and conversion to refinery-ready oil with co-product credits", *Biofuels Bioprod. Biorefin.,* vol. 9, no. 6, pp. 760-777, 2015. [http://dx.doi.org/10.1002/bbb.1610]

[66] J.L. Manganaro, and A. Lawal, "CO2 Life-Cycle Assessment of the Production of Algae-Based Liquid Fuel Compared to Crude Oil to Diesel", *Energy Fuels,* vol. 30, no. 4, pp. 3130-3139, 2016. [http://dx.doi.org/10.1021/acs.energyfuels.6b00207]

[67] R. Davis, A. Aden, and P.T. Pienkos, "Techno-economic analysis of autotrophic microalgae for fuel production", *Appl. Energy,* vol. 88, no. 10, pp. 3524-3531, 2011. [http://dx.doi.org/10.1016/j.apenergy.2011.04.018]

[68] A. Demirbas, and M. Fatih Demirbas, "Importance of algae oil as a source of biodiesel", *Energy Convers. Manage.,* vol. 52, no. 1, pp. 163-170, 2011. [http://dx.doi.org/10.1016/j.enconman.2010.06.055]

[69] E.H. Dunlop, "An energy□limited model of algal biofuel production: Toward the next generation of advanced biofuels", *AIChE J.,* vol. 59, no. 12, pp. 4641-4654, 2013. [http://dx.doi.org/10.1002/aic.14251]

[70] John Ferrell, and Valerie Sarisky-Reed, "National algal biofuels technology roadmap", *US Dept.*

Energy, Office of Energy Efficiency and Renewable Energy, Biomass Program, 2010. [http://dx.doi.org/10.2172/1218560]

[71] E.D. Frank, *Life-cycle analysis of algal lipid fuels with the greet model.* Center for Transportation Research, Energy Systems Division, Argonne National Laboratory: Oak Ridge, 2011, pp. 11-15.

[72] A.L. Stephenson, "Life-cycle assessment of potential algal biodiesel production in the United Kingdom: a comparison of raceways and air-lift tubular bioreactors", *Energy Fuels,* vol. 24, no. 7, pp. 4062-4077, 2010. [http://dx.doi.org/10.1021/ef1003123]

[73] A. Sun, "Comparative cost analysis of algal oil production for biofuels", *Energy,* vol. 36, no. 8, pp. 5169-5179, 2011. [http://dx.doi.org/10.1016/j.energy.2011.06.020]

[74] J.R. Benemann, and W.J. Oswald, *Systems and economic analysis of microalgae ponds for conversion of CO {sub 2} to biomass. Final report. No. DOE/PC/93204--T5. California Univ.* Dept. of Civil Engineering: Berkeley, CA, United States, 1996. [http://dx.doi.org/10.2172/493389]

[75] Y. Chisti, "Biodiesel from microalgae", *Biotechnol. Adv.,* vol. 25, no. 3, pp. 294-306, 2007. [http://dx.doi.org/10.1016/j.biotechadv.2007.02.001] [PMID: 17350212]

[76] L. Brennan, and P. Owende, "Biofuels from microalgae—a review of technologies for production, processing, and extractions of biofuels and co-products", *Renew. Sustain. Energy Rev.,* vol. 14, no. 2, pp. 557-577, 2010. [http://dx.doi.org/10.1016/j.rser.2009.10.009]

[77] R. Davis, *Renewable diesel from algal lipids: an integrated baseline for cost, emissions, and resource potential from a harmonized model.* Argonne National Laboratory, 2012. [http://dx.doi.org/10.2172/1044475]

[78] R.E. Davis, D.B. Fishman, E.D. Frank, M.C. Johnson, S.B. Jones, C.M. Kinchin, R.L. Skaggs, E.R. Venteris, and M.S. Wigmosta, "Integrated evaluation of cost, emissions, and resource potential for algal biofuels at the national scale", *Environ. Sci. Technol.,* vol. 48, no. 10, pp. 6035-6042, 2014. [http://dx.doi.org/10.1021/es4055719] [PMID: 24749989]

[79] D. Klein□Marcuschamer, "Technoeconomic analysis of renewable aviation fuel from microalgae, Pongamia pinnata, and sugarcane", *Biofuels Bioprod. Biorefin.,* vol. 7, no. 4, pp. 416-428, 2013. [http://dx.doi.org/10.1002/bbb.1404]

[80] Y. Chisti, "Constraints to commercialization of algal fuels", *J. Biotechnol.,* vol. 167, no. 3, pp. 201-214, 2013. [http://dx.doi.org/10.1016/j.jbiotec.2013.07.020] [PMID: 23886651]

[81] R. Davis, C Kinchin, J Markham, ECD Tan, LML Laurens, and D Dsexton, Knorr, P Schoen, J Lukas, Process design and economics for the conversion of algal biomass to biofuels: algal biomass fractionation to lipid- and carbohydrate-derived fuel products Tech. Rpt. NREL/TP-5100-62368, Sept. 2014.

[82] L.Y. Batan, G.D. Graff, and T.H. Bradley, "Techno-economic and Monte Carlo probabilistic analysis of microalgae biofuel production system", *Bioresour. Technol.,* vol. 219, pp. 45-52, 2016. [http://dx.doi.org/10.1016/j.biortech.2016.07.085] [PMID: 27475330]

[83] J. Richardson, J.L. Outlaw, and M. Allison, "The economics of microalgae oil", *AgBioForum,* vol. 13, no. 2, pp. 119-130, 2010.

[84] T.J. Lundquist, *A realistic technology and engineering assessment of algae biofuel production.* Energy Biosciences Institute, 2010, pp. 1-178.

[85] J.G. Jonker, and A.P. Faaij, "Techno-economic assessment of micro-algae as feedstock for renewable bio-energy production", *Appl. Energy,* vol. 102, pp. 461-475, 2013. [http://dx.doi.org/10.1016/j.apenergy.2012.07.053]

[86] J.B. Park, R.J. Craggs, and A.N. Shilton, "Wastewater treatment high rate algal ponds for biofuel production", *Bioresour. Technol.,* vol. 102, no. 1, pp. 35-42, 2011. [http://dx.doi.org/10.1016/j.biortech.2010.06.158] [PMID: 20674341]

[87] T.J. Czartoski, "Algae biomass fractionation", *U.S. Patent No. 9,296,985,* 2016.

[88] H.L. Bryant, "The value of post-extracted algae residue", *Algal Res.,* vol. 1, no. 2, pp. 185-193, 2012. [http://dx.doi.org/10.1016/j.algal.2012.06.001]

[89] L. Harwell, "Study outlines optimum ULSD hydrotreater design", *Oil Gas J.,* vol. 101, no. 29, pp. 50-57, 2003.

[90] K.L. Kadam, "Environmental implications of power generation via coal-microalgae cofiring", *Energy,* vol. 27, no. 10, pp. 905-922, 2002. [http://dx.doi.org/10.1016/S0360-5442(02)00025-7]

[91] Komline-Sanderson, *Personal communication,* 2014.

[92] L.R. Lynd, "The role of biomass in America's energy future: framing the analysis", *Biofuels Bioprod. Biorefin.,* vol. 3, no. 2, pp. 113-123, 2009. [http://dx.doi.org/10.1002/bbb.134]

[93] *Global Market for EPA/DHA Omega-3 Products, R.* Editor: Packaged Facts, MD, 2012.

[94] Report, Global Market for EPA/DHA Omega-3 Products. September 2012, Packaged Facts, 11200 Rockville Pike, Rockville, Maryland 20852.

[95] http://www.algaeindustrymagazine.com/aim-interview-dr-isaac-berzin/.

[96] E. Molina Grima, E.H. Belarbi, F.G. Acién Fernández, A. Robles Medina, and Y. Chisti, "Recovery of microalgal biomass and metabolites: process options and economics", *Biotechnol. Adv.,* vol. 20, no. 7-8, pp. 491-515, 2003. [http://dx.doi.org/10.1016/S0734-9750(02)00050-2] [PMID: 14550018]

[97] G.T. Rochelle, "Amine scrubbing for CO2 capture", *Science,* vol. 325, no. 5948, pp. 1652-1654, 2009. [http://dx.doi.org/10.1126/science.1176731] [PMID: 19779188]

[98] D. Singh, "Techno-economic study of CO 2 capture from an existing coal-fired power plant: MEA scrubbing vs. O2/CO2 recycle combustion", *Energy Convers. Manage.,* vol. 44, no. 19, pp. 3073-3091, 2003. [http://dx.doi.org/10.1016/S0196-8904(03)00040-2]

CHAPTER 16

Greenhouse Gas Balances of Microalgal Biofuels

Lucas Reijnders*

IBED, University of Amsterdam, Science Park 904, PO box 94248, 10 GE Amsterdam, the Netherlands

Abstract: Life cycle emissions of greenhouse gases associated with microalgal biofuels reported in peer- reviewed papers vary widely. For microalgal biodiesel, they range from -75 g CO_{2eq} to + 534 g CO_{2eq} MJ^{-1} biodiesel. Available studies on life cycle dealing with greenhouse gas emissions are consequential and subject to relatively large uncertainties. Thus, the results of life cycle assessments of specific combinations of product(s) and process(es) should be expressed as ranges rather than as one-point values. Choices about the application of credits, system boundaries, expected microalgal yields, allocation and the decarbonization of energy supply matter substantially to the results of life cycle assessments. The way emissions of methane are dealt with may also be important for estimated greenhouse gas emissions linked to microalgal biofuel lifecycles. Substantially reducing uncertainties in the outcomes of life cycle assessments awaits the availability of attributional life cycle assessments which deal with well-monitored existing commercial microalgal fuel production-consumption chains.

Keywords: Microalgae, Biofuel, Greenhouse gas, Life cycle assessment, Credits, System boundaries, Decarbonization, Allocation, Uncertainty, CO_2, CH_4, N_2O.

ESTIMATING GREENHOUSE GAS BALANCES

Greenhouse gases are gases which absorb infrared radiation emitted by the Earth and thus may have an upward effect on tropospheric temperature [1, 2]. They include CO_2, N_2O and CH_4 [1, 2]. Greenhouse gas emissions are often expressed as CO_{2eq}, assuming a time frame of 100 years [1, 2]. Assuming this time frame, the emission of one molecule of CH_4 is equivalent to the emission of 25 molecules CO_2 and the emission of one molecule N_2O is equivalent to the emission of 298 molecules of CO_2 [1, 2].

Environmental life cycle assessment (LCA) can be used to estimate greenhouse gas balances, and life cycle greenhouse gas emissions, associated with the produc-

* **Corresponding author Lucas Reijnders:** IBED, University of Amsterdam, Science Park 904, PO box 94248, 10 GE Amsterdam, the Netherlands; Tel +31205256269; Fax +31205257431; E-mail: l.reijnders@uva.nl

José Carlos Magalhães Pires (Ed.)

tion and consumption of microalgal biofuels and compare them with the greenhouse gas emissions linked to the production and consumption of *e.g.* mineral oil-based fuels [1 - 4]. Life cycle assessments may regard existing production-consumption chains. Such assessments are called attributional [3, 4]. They may also deal with production-consumption chains which do not (yet) exist, and may involve novel scales of production, novel products and/or novel production processes. Such life cycle assessments are called consequential [3, 4].

Estimates of greenhouse gas emissions are generally uncertain [2, 4, 5]. This in part follows from uncertainty in the input data (*e.g.* the amount of electricity needed). The uncertainty of input data tends to be larger in consequential life cycle assessments than in attributional assessments [2].

The choices made in life cycle assessment may impact the results. These choices in part also influence the inputs. Flue gas derived from fossil fuel burning used for the production of microalgae may be valued at zero (which is common for wastes [2]), but may also be valued in terms of a positive or negative emission. Other choices that may impact results regard system boundaries. System boundaries demarcate what is included and what is excluded in life cycle assessment. For instance, change of the land used, the production of capital goods and waste water treatment can be included or excluded in life cycle assessments of microalgal biofuels. Furthermore, not all life cycle assessments necessarily cover all stages of the life cycle ('cradle to grave', or in the case of fuels for automotive transport: 'cradle-to-wheel'). There are also 'cradle-to-(factory)-gate', 'cradle-to-pump', and 'cradle-to-de-watered-biomass' assessments, which are used for comparative assessment of other fuels, covering a similar part of the life cycle. Choices regarding system boundaries are substantially important for the outcomes of life cycle assessments [2 - 4]. Furthermore, there is usually more than one output from biofuel production processes. The life cycle greenhouse gas emissions should be allocated to all outputs. This can be done in several ways. Allocation can take place on the basis of monetary units (prices), on the basis of physical units (energy, mass) or on the basis of substitution (greenhouse gas emissions linked to a similar output) [2 - 4]. Different ways of allocation may lead to different estimates of life cycle greenhouse gas emissions of microalgal biofuels [2 - 4].

EVALUATION OF AVAILABLE PEER-REVIEWED LIFE CYCLE ASSESSMENTS OF MICROALGAL BIOFUELS

There is a set of relatively well documented peer-reviewed papers dealing with the assessment of life cycle greenhouse gas emissions linked to microalgal (including cyanobacterial) biofuel production [1, 6 - 26]. This set is briefly summarized in Table **1**. All LCAs in Table **1** are consequential. The life cycle greenhouse gas emissions MJ^{-1} microalgal biodiesel estimated in these papers range between -75 g CO_{2eq} (cradle-to-pump) [8] and +534 g CO_{2eq} [11] (cradle-to-gate). Other studies regarding microalgal biodiesel presented in Table **1** show outcomes between these extremes. For comparison, the cradle-to-wheel emission for conventional (mineral oil-based) diesel is estimated to be (+)79-96 g CO_{2eq} MJ^{-1} [15].

Table 1. Peer-reviewed life cycle assessments of greenhouse gas emissions associated with microalgal biofuels.

Authors/reference	Microlgal biofuel product	Life cycle coverage
Lardon *et al.* [6]	biodiesel from algae in pond	cradle-to wheel
Clarens *et al.* [7]	biodiesel from algae in pond	cradle-to-wheel
Batan *et al.* [8]	biodiesel from algae in pond	cradle-to-pump
Stephenson *et al.* [9]	biodiesel from algae in pond	cradle-to-wheel
Luo *et al.* [10]	ethanol from algae in bioreactor	cradle-to-gate
Brentner *et al.* [11]	biodiesel from algae in pond or bio- reactor	cradle-to-gate
Clarens *et al.* [12]	biodiesel from algae in pond	cradle-to-wheel
Khoo *et al.* [13]	biodiesel from algae in bioreactor/ pond combination	cradle-to-gate
Collet *et al.* [14]	methane from algae in pond	cradle-to-grave
Shirvani *et al.* [15]	biodiesel from algae in pond	cradle-to-gate
Campbell *et al.* [16]	biodiesel from algae in pond	cradle-to-gate
Vasudevan *et al.* [17]	algal oil from algae in pond	cradle-to-wheel
Itoiz *et al.* [18]	de-watered algal biomass from pond	cradle-to-de-watered-biomass
Frank *et al.* [1]	biodiesel from algae in pond	cradle-to-grave
Ferreira *et al.* [19]	bio-H_2 from bioreactor-derived algal biomass	cradle-to-grave
Quinn *et al.* [20]	biodiesel from algae in pond	cradle-to-pump
Handler *et al.* [21]	a variety of fuels from algae in pond	cradle-to-grave
Lokesh *et al.* [22]	bio-jet fuel from algae in pond	cradle-to-grave
Beal *et al.* [23]	biocrude from algae in pond	cradle to gate
Mu *et al.* [24]	biofuel from algae in pond or bio- reactor	cradle-to-grave
Conelly *et al.* [25]	biofuels from algae in pond	cradle-to-grave
Orfield *et al.* [26]	biodiesel from algae in pond	cradle-to-gate

Studies regarding other microalgal biofuels than biodiesel reported the following results. Collet *et al.* [14] calculated similar life cycle greenhouse gas emissions for conventional diesel and CH_4 generated from microalgal biomass. Lokesh *et al.* [22] compared the life cycle greenhouse gas emissions of micoroalgal jet fuel and conventional jet fuel A (based on mineral oil) and estimated a value of (+)40-42 g CO_{2eq} MJ^{-1} for microalgal jet fuel and (+)106.1 g CO_{2eq} MJ^{-1} for conventional jet fuel A. Luo *et al* [10], who considered ethanol from cyanobacteria, reported a greenhouse gas emission of (+)12.3-28.9 g CO_{2eq} MJ^{-1} ethanol. Ferreira *et al.* [19], considering only CO_2 as greenhouse gas, reported for algal biomass-derived H_2 a life cycle CO_2 emission of (+)670 10^3 g CO_2 MJ^{-1}.

The variation in outcomes regarding the life cycle assessment of microalgal biodiesel is large. Important determinants of this variation and problematic aspects of the set of life cycle assessments summarized in Table **1** are discussed below.

The Use of Credits

An important determinant of life cycle assessment outcomes is the use of credits. A first type of credits is carbon credits (negative CO_{2eq} emissions). For instance, studying jet fuel, Lokesh *et al.* [22] apply a carbon credit to microalgal jet fuel of about 71 g CO_{2eq} MJ^{-1}. Without such a credit the cradle-to-grave life cycle emission would have been about 112 g CO_{2eq} MJ^{-1}, higher than the cradle-to-grave life cycle greenhouse gas emission of conventional jet fuel (about 106 g CO_{2eq} MJ^{-1}), but with this credit the life cycle biofuel greenhouse gas emission decreases to 40-42 g CO_{2eq} MJ^{-1}. Similarly, studying bioreactor-based biodiesel production, Batan *et al.* [8] give an input of 0.93 MJ energy MJ^{-1} cradle-to-pump biodiesel, but estimate a negative cradle-to-pump emission of -75g CO_{2eq} MJ^{-1}. Batan *et al.* [8] state that the negative value is due to the capture of CO_2 intrinsic to the production of biomass during photosynthesis and the displacement of petroleum. Frank *et al.* [1] estimate in a cradle-to-grave life cycle assessment the energetic input in biodiesel production to be 2.5 times the energetic output, but found the life cycle greenhouse gas emission to be substantially less than for low-sulphur diesel based on mineral oil 'because the carbon in algal biofuel was treated as biogenic', which is equated by Frank *et al.* [1] to carbon sequestration. There is also the view that carbon credits can be claimed when flue gas generated by fossil fuel-fired installations is used as an input in microalgal production [16, 25, 26]. Credits similar to those used by Frank *et al.* [1] are in the studies of Lardon *et al.* [6], Clarens *et al.* [7] and Stephenson *et al.* [9], Campbell *et al.* [16], Connelly *et al.* [25] and Orfield *et al.* [26].

The choice to use carbon credits (negative CO_2 emissions) in estimating cradle-to-grave life cycle greenhouse gas emissions is debatable. The CO_2 initially sequestered in biofuel or taken from fossil fuel fired installations is released when biofuel is used for its purpose: combustion. This might be argued to favor valuing the CO_2 emitted by algal biofuel combustion at zero CO_2 emission because it is biogenic and the CO_2 taken from fossil fuel-fired installations at zero CO_2 emission because it is a waste. It might also be argued that the CO_2 from fossil fuel is rather to be valued in terms of a positive CO_2 emission, as such CO_2 used in the production of algae is apparently a useful co-product.

Another type of credit is used by Handler *et al.* [21] in their cradle-to-grave life cycle assessments of microalgal biofuel. Handler *et al.* [21] use wastewater treatment credits. This is also found in the study of Mu *et al.* [24]. In a scenario in which the effluent from a wastewater treatment plant is used as an input, Handler *et al.* [21] calculate the life cycle input of energy for pond-based microalgal pyrolytic oil at 1.98-2.6 MJ MJ^{-1} biofuel. To estimate the associated greenhouse gas emissions they apply a waste water treatment credit of 2.2 MJ MJ^{-1} biofuel, because they state that, as waste water is used in biofuel production, the need for waste water treatment is eliminated. This in turn leads to and overall energy balance of -0.22 to +0.4 MJ MJ^{-1} bio-oil, with correspondingly negative or low life cycle greenhouse gas emissions. It is uncommon to apply such a credit when waste water is used as an input [2 - 4]. Commonly, when nutrients from waste water are used as an input these are valued as zero, whereas the input of synthetic nutrients is valued at their life cycle greenhouse gas emissions [2, 4]. Also, there may be a case to value waste water treatment-derived nutrients positively in terms of greenhouse gas emissions (by allocation of a part of the greenhouse gas emissions of the process generating wastewater) as the nutrients are apparently a useful output, which can be used in the production of biofuels [2]. Furthermore, the output of biofuel treatment plants has to be subjected also to waste water treatment and Handler *et al.* [22] do not include such waste water treatment within their system boundaries.

System Boundaries

As noted before, the choice of system boundaries may impact life cycle assessment outcomes. It has been found [3] that in life cycle assessments water supply management, waste water treatment and upgrading of algal biodiesel are often outside system boundaries, which may lead to an underestimate of life cycle greenhouse gas emissions. There are exceptions to this exclusion. For instance, Mu *et al.* [24] included biofuel upgrading, whereas Lardon *et al.* [6] and Clarens *et al.* [7, 11] included waste water treatment. The emissions associated with producing the production facilities for biofuel production are often outside system boundaries. Canter *et al.* [27] estimate for microalgal biodiesel production in open ponds a contribution of 8.9 g CO_{2eq} MJ^{-1} to the life cycle greenhouse gas emissions, assuming a baseline design productivity of 13.2 g microalgal biomass m^{-2} day^{-1}. Also land use change is often outside the system boundaries. This may impact the outcomes of life cycle assessments, especially when the land use change needed for microalgal biofuel production facilities such as ponds regards forested land and/or peatland [2].

Dealing with Methane

Collet *et al.* [14] have studied biogas (methane), generated by anaerobic digestion of microalgae, as the main product of microalgal biofuel production. Frank *et al.* [1], Brentner *et al.* [11], Clarens *et al.* [12],Vasudevan *et al.* [17] and Orfield *et al.* [26] have studied methane, generated by anaerobic digestion, as a co-product of microalgal biofuel production. Frank *et al.* [1] have pointed out the importance of fugitive methane emissions in the overall life cycle emission of greenhouse gases linked to anaerobic digestion, which was not noted in the previous studies of Brentner *et al.* [11], Clarens *et al.* [12], Collet *et al.* [14], and Vasudevan *et al.* [17]. Indeed, the fugitive methane emissions may be an important determinant of greenhouse gas emissions linked to the biofuel methane [1]. However, Frank *et al.* [1] have only given a one-point value for these fugitive emissions, whereas there would seem to be substantial uncertainty about their actual size.

The fugitive methane emissions are in part related to the reactor, pipes, valves and fittings of the anaerobic digestion facility. Estimates of these emissions by existing anaerobic digesters range between 1 and 10% of the methane produced in these facilities [28]. When facilities operate relatively well a loss of 1.4% has been reported [28]. When there is on-site combustion of biogas by lean-burn gas engines there may be a relatively large methane slip [29].

Furthermore, fugitive methane emissions are related to the management of biogas digestate, a liquid co-product of biogas production. An option is the agricultural application of digestate. Conventional agricultural management for biogas digestate is storage in open tanks and field application (as fertilizer) with a splash plate [28]. Depending on factors such as hydraulic retention time and organic loading, digestate may contain substantial to relatively large amounts of methane [30, 31]. There are also treatment options to reduce the amount of methane in digestate [32]. The presence of methane in digestate may give rise to fugitive emissions during storage and application, which so far have not been well quantified. Limited efforts to quantify methane losses during digestate storage [33 - 35] suggest a range from 0.5-15% of the methane produced at the anaerobic digestion facilities [34, 35]. No data have been found regarding methane losses during application. After application methane losses are apparently relatively low [34].

Assumptions about Microalgal Yields from Ponds

Rather often in life cycle assessments of pond-derived microalgal biofuels relatively low inputs of energy in microalgal biofuel production are associated with relatively high yields of microalgal biomass (~90 Mg ha^{-1} $year^{-1}$) [3]. In practice growth of microalgae in ponds is susceptible to organisms consuming microalgae, to infections by viruses and bacteria and to competing organisms [3]. Reducing such susceptibility is possible by growing microalgae under extreme conditions (high salt concentrations or high alkalinity) but these conditions are not favorable to high yields [3]. It remains uncertain whether dry-weight microalgal yields of ~90 Mg ha^{-1} $year^{-1}$ can be achieved in widespread commercial practice.

Decarbonization of Energy Supply

Shirvani *et al.* [15] have shown that life cycle greenhouse gas emissions of algal biofuel can be much reduced when energy supply for microalgal fuel production is decarbonized (*e.g.* as fossil fuels are replaced by energy sources such as solar and wind energy). In their study the life cycle input of energy was about 2.5 MJ MJ^{-1} microalgal biofuel. When co-products were considered as wastes, life cycle greenhouse gas emissions might be up to 450 g CO_{2eq} MJ^{-1} biofuel, but with decarbonized energy supply life cycle greenhouse gas emissions could be < 20 g CO_{2eq} MJ^{-1} biofuel. However, it should be noted that to be fit for a significant role in future energy supply the life cycle input of energy in microalgal biofuel production should be < 0.2 MJ MJ^{-1} biofuel [3]. Thus decarbonization at the input of 2.5MJ MJ^{-1} microalgal biofuel as considered by Shirvani *et al.* [15] is unlikely to save the internal combustion engine (see also [36]).

Allocation of Greenhouse Gas Emissions Based on Prices

In consequential life cycle assessments allocation based on current prices may be problematic. There are substances which have currently a high value and may be generated as co-product to algal biofuel production. Misra *et al.* [37] refer in this context for instance to substances such as beta-carotene, astaxanthin and phycoerythrin. Allocation of greenhouse gas emissions to such co-products on the basis of current prices may substantially improve greenhouse gas balances of microalgal biofuels, if compared to allocation on the basis of the physical unit mass. However, experience with terrestrial biofuel production has shown that, even with modest shares in the overall fuel market, the price of the co-product glycerol has collapsed [2]. Major reductions in the prices of current high value microalgal products would also seem likely when there would be large scale production of biofuels from microalgae which can supply substances such as beta-carotene, astaxanthin and phycoerythrin.

Dealing with Uncertainty

The life cycle assessments summarized in Table 1 tend to contain one-point estimates of life cycle greenhouse gas emissions for specific combinations of product(s) and process(es). Such one-point estimates neglect the uncertainties which are present in the emission inventories for existing processes and products used in LCAs [38, 39]. These have been illustrated in the section **Dealing with methane**. The uncertainties which pertain to new products and processes are even larger [2]. For this reason the results of life cycle assessments regarding greenhouse gas emissions linked to specific combinations of product(s) and process(es) generating microalgal biofuels should be expressed in terms of ranges [*e.g.* 40]. Uncertainties in these results may be substantially reduced by the emergence of life cycle assessments that deal with well-monitored existing commercial microalgal fuel production-consumption chains.

CONCLUSION

All in all, apart from the conclusion that the production of H_2 from microalgal biomass is associated with excessive greenhouse gas emissions [19], the set of life cycle assessments discussed in this chapter does not allow for definitive conclusions as to the greenhouse gas balances of microalgal biofuels. Choices regarding the use of credits, system boundaries and allocation, and the handling of uncertainty should be carefully considered in life cycle assessment. The availability of (attributional) life cycle assessments that deal with well-monitored existing commercial microalgal fuel production-consumption chains would allow for more certainty in estimates of greenhouse gas emissions linked to microalgal biofuel life cycles.

CONFLICT OF INTEREST

The authors declare no conflict of interest, financial or otherwise.

ACKNOWLEDGEMENTS

Declared none.

REFERENCES

[1] E.D. Frank, J. Han, I. Palu-Rivera, A. Elgowainy, and M.Q. Wang, "Methane and nitrous oxide emissions affect the life cycle analysis of algal biofuels", *Environ. Res. Lett.*, vol. 7, no. 014030, 2013.

[2] L. Reijnders, and M.A. Huijbregts, *Biofuels for road transport.* Springer: Heidelberg, 2009.

[3] L. Reijnders, "Lipid-based liquid biofuels from autotrophic microalgae: energetic and environmental performance", *WIRE's Energy Environ,* vol. 2, pp. 73-85, 2013. [http://dx.doi.org/10.1002/wene.29]

[4] L. Reijnders, "Life Cycle Assessment of Greenhouse Gas Emissions", In: *Handbook of Climate Change Mitigation and Adaptation,* W.-Y. Chen, Ed., 2-2. Springer Part 1, 2017, pp. 63-91. [http://dx.doi.org/10.1007/978-3-319-14409-2_2]

[5] D.L. Sills, V. Paramita, M.J. Franke, M.C. Johnson, T.M. Akabas, C.H. Greene, and J.W. Tester, "Quantitative uncertainty analysis of Life Cycle Assessment for algal biofuel production", *Environ.*

Sci. Technol., vol. 47, no. 2, pp. 687-694, 2013.
[http://dx.doi.org/10.1021/es3029236] [PMID: 23237457]

[6] L. Lardon, A. Hélias, B. Sialve, J.P. Steyer, and O. Bernard, "Life-cycle assessment of biodiesel production from microalgae", *Environ. Sci. Technol.,* vol. 43, no. 17, pp. 6475-6481, 2009. [http://dx.doi.org/10.1021/es900705j] [PMID: 19764204]

[7] A.F. Clarens, E.P. Resurreccion, M.A. White, and L.M. Colosi, "Environmental life cycle comparison of algae to other bioenergy feedstocks", *Environ. Sci. Technol.,* vol. 44, no. 5, pp. 1813-1819, 2010. [http://dx.doi.org/10.1021/es902838n] [PMID: 20085253]

[8] L. Batan, J. Quinn, B. Willson, and T. Bradley, "Net energy and greenhouse gas emission evaluation of biodiesel derived from microalgae", *Environ. Sci. Technol.,* vol. 44, no. 20, pp. 7975-7980, 2010. [http://dx.doi.org/10.1021/es102052y] [PMID: 20866061]

[9] A.L. Stephenson, E. Kazamis, J.S. Dennis, C.J. Howe, S.A. Satt, and A.G. Smith, "Life cycle assessment of potential algal biodiesel production in the United Kingdom: a comparison of raceways and air lift tubular bioreactors", *Energy Fuels,* vol. 24, pp. 4062-4077, 2010. [http://dx.doi.org/10.1021/ef1003123]

[10] D. Luo, Z. Hu, D.G. Choi, V.M. Thomas, M.J. Realff, and R.R. Chance, "Life cycle energy and greenhouse gas emissions for an ethanol production process based on blue-green algae", *Environ. Sci. Technol.,* vol. 44, no. 22, pp. 8670-8677, 2010. [http://dx.doi.org/10.1021/es1007577] [PMID: 20968295]

[11] L.B. Brentner, M.J. Eckelman, and J.B. Zimmerman, "Combinatorial life cycle assessment to inform process design of industrial production of algal biodiesel", *Environ. Sci. Technol.,* vol. 45, no. 16, pp. 7060-7067, 2011. [http://dx.doi.org/10.1021/es2006995] [PMID: 21662987]

[12] A.F. Clarens, H. Nassau, E.P. Resurreccion, M.A. White, and L.M. Colosi, "Environmental impacts of algae-derived biodiesel and bioelectricity for transportation", *Environ. Sci. Technol.,* vol. 45, no. 17, pp. 7554-7560, 2011. [http://dx.doi.org/10.1021/es200760n] [PMID: 21774477]

[13] H.H. Khoo, P.N. Sharatt, P. Das, R.K. Balasubramanian, P.K. Naraharisetti, and S. Shaik, "Life cycle energy and CO_2 analysis of micro-algae-to-biodiesel: preliminary results and comparisons", *Bioresour. Technol.,* vol. 102, pp. 5800-5807, 2011. [http://dx.doi.org/10.1016/j.biortech.2011.02.055] [PMID: 21435867]

[14] P. Collet, A. Hélias, L. Lardon, M. Ras, R.A. Goy, and J.P. Steyer, "Life-cycle assessment of microalgae culture coupled to biogas production", *Bioresour. Technol.,* vol. 102, no. 1, pp. 207-214, 2011. [http://dx.doi.org/10.1016/j.biortech.2010.06.154] [PMID: 20674343]

[15] T. Shirvani, X. Yan, O.R. Inderwildi, P.P. Edwards, and D.A. King, "Life cycle energy and greenhouse gas analysis for algae-derived biodiesel", *Energy Environ. Sci.,* vol. 4, pp. 3773-3778, 2011. [http://dx.doi.org/10.1039/c1ee01791h]

[16] P.K. Campbell, T. Beer, and D. Batten, "Life cycle assessment of biodiesel production from microalgae in ponds", *Bioresour. Technol.,* vol. 102, no. 1, pp. 50-56, 2011. [http://dx.doi.org/10.1016/j.biortech.2010.06.048] [PMID: 20594828]

[17] V. Vasudevan, R.W. Stratton, M.N. Pearlson, G.R. Jersey, A.C. Beyene, J.C. Weissman, M. Rubino, and J.I. Hileman, "Environmental performance of algal biofuel technology options", *Environ. Sci. Technol.,* vol. 46, pp. 2431-2439, 2012.

[18] E.S. Itoiz, C. Fuentes-Grünewald, C.M. Gasol, E. Garces, E. Alacid, S. Rossi, and J. Rieradevall, "Energy balance and environmental impact analysis of marine microalgal biomass production for biodiesel generation in a photobioreactor pilot plant", *Biomass Bioenergy,* vol. 39, pp. 324-335, 2012. [http://dx.doi.org/10.1016/j.biombioe.2012.01.026]

[19] A.F. Ferreira, J. Ortigueira, L. Alves, L. Gouveia, P. Moura, and C.M. Silvia, "Energy requirement and CO2 emissions of bioH2 production form microalgal biomass", *Biomass Bioenergy,* vol. 49, pp. 249-259, 2013.
[http://dx.doi.org/10.1016/j.biombioe.2012.12.033]

[20] J.C. Quinn, T.G. Smith, C.M. Downes, and C. Quinn, "Microalgae to biofuels life cycle assessment - multiple pathway evaluation", *Algal Res.,* vol. 4, pp. 116-122, 2014.
[http://dx.doi.org/10.1016/j.algal.2013.11.002]

[21] R.M. Handler, D.R. Shonnard, T.N. Kanles, and F.S. Lupton, "Life cycle assessment of algal biofuels: influence of feedstock cultivation systems and conversion platforms", *Algal Res.,* vol. 4, pp. 105-115, 2013.
[http://dx.doi.org/10.1016/j.algal.2013.12.001]

[22] K. Lokesh, V. Sethi, T.T. Nikolaides, and E. Goodger, "Life cycle greenhouse gas analysis of biojet fuels with a technical investigation into their impact on jet engine performance", *Biomass Bioenergy,* vol. 77, pp. 26-44, 2015.
[http://dx.doi.org/10.1016/j.biombioe.2015.03.005]

[23] C.M. Beal, L.N. Gerber, D.L. Sills, M.E. Huntley, S.C. Machesky, M.J. Walsh, J.W. Tester, I. Archibald, J. Granados, and C.H. Greene, "Algal biofuel production for fuels and feed in a 100-ha facility: a comprehensive techno-economic analysis and life cycle assessment", *Algal Res.,* vol. 10, pp. 266-279, 2015.
[http://dx.doi.org/10.1016/j.algal.2015.04.017]

[24] D. Mu, M. Min, B. Krohn, K.A. Mullins, R. Ruan, and J. Hill, "Life cycle environmental impacts of wastewater-based algal biofuels", *Environ. Sci. Technol.,* vol. 48, no. 19, pp. 11696-11704, 2014.
[http://dx.doi.org/10.1021/es5027689] [PMID: 25220843]

[25] E.B. Connelly, L.M. Colosi, A.F. Clarens, and J.H. Lambert, "Life cycle assessment of biofuels from algae hydrothermal liquefaction: the upstream and downstream factors affecting regulatory compliance", *Energy Fuels,* vol. 29, pp. 1653-1661, 2015.
[http://dx.doi.org/10.1021/ef502100f]

[26] N.D. Orfield, R.B. Levine, G.A. Keoleian, S.A. Miller, and P.E. Savage, "Growing algae for biodiesel on direct sunlight or sugars: a comparative life cycle assessment", *ACS Sustain. Chem.& Eng.,* vol. 3, pp. 386-395, 2015.
[http://dx.doi.org/10.1021/sc5004117]

[27] C.E. Canter, R. Davis, M. Urgun-Demirtas, and E.D. Frank, "Infrastructure associated emissions for renewable diesel production from microalga", *Algal Res.,* vol. 5, pp. 195-203, 2014.
[http://dx.doi.org/10.1016/j.algal.2014.01.001]

[28] N.D. Miranda, H.L. Tuomisto, and M.D. McCulloch, "Meta-analysis of greenhouse gas emissions from anaerobic digestion processes in dairy farms", *Environ. Sci. Technol.,* vol. 49, no. 8, pp. 5211-5219, 2015.
[http://dx.doi.org/10.1021/acs.est.5b00018] [PMID: 25790272]

[29] J. Møller, A. Boldrin, and T.H. Christensen, "Anaerobic digestion and digestate use: accounting of greenhouse gases and global warming contribution", *Waste Manag. Res.,* vol. 27, no. 8, pp. 813-824, 2009.
[http://dx.doi.org/10.1177/0734242X09344876] [PMID: 19748957]

[30] T. Rehl, and J. Müller, "Life cycle assessment of biogas digestate processing technologies", *Resour. Conserv. Recycling,* vol. 56, pp. 92-104, 2011.
[http://dx.doi.org/10.1016/j.resconrec.2011.08.007]

[31] C. Rico, J.L. Rico, I. Tejero, N. Muñoz, and B. Gómez, "Anaerobic digestion of the liquid fraction of dairy manure in pilot plant for biogas production: residual methane yield of digestate", *Waste Manag.,* vol. 31, no. 9-10, pp. 2167-2173, 2011.
[http://dx.doi.org/10.1016/j.wasman.2011.04.018] [PMID: 21612905]

[32] C. Sambusiti, F. Monau, E. Ficara, A. Musatti, M. Rollini, A. Barakat, and M. Malpei, "Comparison of various post-treatments for recovering methane from agricultural digestate", *Fuel Process. Technol.*, vol. 137, pp. 359-365, 2015.
[http://dx.doi.org/10.1016/j.fuproc.2015.04.028]

[33] M. Hrad, M. Piringer, L. Kamarad, K. Baumann-Stanzer, and M. Huber-Humer, "Multisource emission retrieval within a biogas plant based on inverse dispersion calculations--a real-life example", *Environ. Monit. Assess.*, vol. 186, no. 10, pp. 6251-6262, 2014.
[http://dx.doi.org/10.1007/s10661-014-3852-0] [PMID: 24869953]

[34] S. Siegl, M. Laaber, and P. Holubar, "Green electricity from biomass, Part I: Environmental impacts from direct life cycle emissions", *Waste Biomass Valoriz.*, vol. 2, pp. 267-284, 2011.
[http://dx.doi.org/10.1007/s12649-011-9077-3]

[35] H. Baldé, A.C. VanderZaag, S.D. Burtt, C. Wagner-Riddle, A. Crolla, R.L. Desjardins, and D.J. MacDonald, "Methane emissions from digestate at an agricultural biogas plant", *Bioresour. Technol.*, vol. 216, pp. 914-922, 2016.
[http://dx.doi.org/10.1016/j.biortech.2016.06.031] [PMID: 27323243]

[36] J.B. van Beilen, "Why microalgal biofuels won't save the internal combustion engine", *Biofuels Bioprod. Biorefin.*, vol. 4, pp. 41-52, 2010.
[http://dx.doi.org/10.1002/bbb.193]

[37] N. Misra, P.K. Panda, B.K. Parida, and B.K. Mishra, "Way forward to achieve sustainable and cost-effective biofuel production form microalgae: a review", *Int. J. Environ. Sci. Technol.*, vol. 13, pp. 2735-2756, 2016.
[http://dx.doi.org/10.1007/s13762-016-1020-5]

[38] J.S. Cooper, M. Noon, and E. Kahn, "Parametrization in life cycle assessment inventory data: review of current use and the representation of uncertainty", *Int. J. Life Cycle Assess.*, vol. 17, pp. 689-695, 2012.
[http://dx.doi.org/10.1007/s11367-012-0411-1]

[39] S. Muller, P. Lesage, A. Ciroth, C. Mutel, B.P. Weidema, and R. Samson, "The application of the pedigree approach to the distributions foreseen in Ecoinvent v3", *Int. J. Life Cycle Assess.*, vol. 21, pp. 1327-1337, 2016.
[http://dx.doi.org/10.1007/s11367-014-0759-5]

[40] L.N. Gerber, J.W. Tester, C.M. Beal, M.E. Huntley, and D.L. Sills, "Target cultivation and financing parameters for sustainable production of fuel and feed from microalgae", *Environ. Sci. Technol.*, vol. 50, no. 7, pp. 3333-3341, 2016.
[http://dx.doi.org/10.1021/acs.est.5b05381] [PMID: 26942694]

CHAPTER 17

Triple Bottom Line Assessment of Algae Bio-crude Production

Arunima Malik*

ISA, School of Physics A28, The University of Sydney, NSW 2006, Australia

Abstract: Triple bottom line (TBL) accounting is a framework for assessing the sustainability performance of a product or an organization. The aim of this chapter is to present the results of a TBL assessment of algae bio-crude production. Algae are considered an ideal feedstock for biofuel production owing to their ability to grow on marginal land without the need for freshwater resources. This chapter presents an input-output based life-cycle assessment of algae bio-crude production. The following stages are considered in the analysis - cultivation of algae, extraction of bio-crude from algae, and transportation of algae bio-crude to a refinery for processing. A region in Australia ideal for the production of algae was chosen for this study. The results show that the supply chain of algae bio-crude production is more sustainable than that of crude oil production.

Keywords: Algae biofuel, Sustainability assessment, Input-output analysis, Energy use, Employment, Microalgae, Bio-crude, Footprint, Carbon-negative, Hybrid life-cycle assessment, Triple bottom line.

INTRODUCTION

The demand for fossil fuels has continued to increase since the industrial revolution. These fuels have become the key energy resources for many economies around the globe. Whilst these fuels have provided humankind with electricity for lighting their homes and offices, obtaining fuels for transportation, and many everyday products such as medicines, cosmetics, lubricants and plastics, this has come at a price. An increase in the greenhouse gas emissions resulting from the combustion of these fuels has become a major environmental concern in today's world. Fossil fuels release carbon dioxide, a major greenhouse gas responsible for radiative forcing resulting in warming of the Earth's climate. World's countries acknowledge this as a global problem that needs to be tackled in unison.

* **Corresponding author Arunima Malik:** ISA, School of Physics A28, The University of Sydney, NSW 2006, Australia; Tel: +61 2 9351 5451; Fax: +61 2 9351 7726; E-mail: arunima.malik@sydney.edu.au

José Carlos Magalhães Pires (Ed.)

In December 2015, world leaders gathered in Paris, France to come up with an agreement for limiting the increase of temperature to below 2 °C, and to initate efforts to limit the warming to 1.5 °C [1, 2]. One of the solutions suggested at the meeting was to increase the renewable energy capacity of countries in order to reduce the consumption of fossil fuels [3, 4]. Out of the renewable energy sources available at our disposal, biofuels are gaining increasing attention. As opposed to wind and solar energy technologies that rely on the availability of wind and sun, respectively, biofuels can be produced all year round using a range of feedstocks. Research published on biofuels so far suggests that sugarcane, crops, vegetable oil, forestry biomass and algae are potential feedstocks for the production of ethanol and biodiesel [5 - 12]. Algae are considered an ideal feedstock for biofuel production [13 - 17].

Algae biofuels are classified as third-generation biofuels, and have many advantages over the first- and second-generation biofuels. Algae are photosynthetic organisms that can grow on lands unsuitable for the production of crops. These organisms do not require fresh water for growth and reproduction, instead can grow in saline water thus reducing the load on freshwater resources. Algae are able to sequester carbon dioxide for growth, a characteristic that is advantageous in situations where an algae production plant is located close to a coal-fired power plant [5, 16 - 19]. The processes for growing, harvesting and ultilising microalgae are explained in the initial chapters of this book. This chapter focusses on the life cycle assessment of algae biofuels. Put simply, is algae biofuel production economically, socially and environmentally sustainable? To this end, I will introduce and describe the methodology underlying hybrid life cycle assessment using input-output (IO) analysis. Following this, I will present results for IO-based life cycle assessment of algae biofuel production.

METHODOLOGY

In this section, I will describe the methodology underlying input-output based hybrid life cycle assessment.

A Brief Introduction to Input-Output (IO) Analysis

Input-output analysis is a macroeconomic technique that relies on economic input-output tables [20, 21]. These tables document the flow of money, in return of goods and services, between various industry sectors. In essence, IO tables provide a snapshot of the inter-industrial structure of an economy [22]. Often termed as input-output matrices, the size of these matrices varies depending on the number of sectors that the IO table features. Single-region IO tables represent monetary transactions between the industry sectors of a single region, for example Australia's national input-output table. Multi-regional input-output (MRIO) tables come in two forms – sub-national MRIO tables contain data on different regions of a single country, and global MRIO tables contain data on many countries. See SI3.1 of [23] for an introduction to multi-regional input-output analysis.

Hybird Life-Cycle Assessment

A combination of input-output analysis and process analysis is termed as hybrid life-cycle assessment (LCA) [24 - 27]. Literature on the process LCA of algae biofuels is abound in comparison to hybrid LCA of the same. LCA of algae biofuel production has been undertaken based on a number of regions and countries. Gao *et al.* [16] summarised these details in Table 1 of their article. Hybrid LCA offers many advantages over conventional process LCAs. In conventional LCAs, selection of a system boundary results in truncation errors that lead to an incomplete quantification of the impacts [28]. In Input-output based LCA, however, all infinite supply chains are taken into account thus reducing the need for boundary selection. Hybrid LCA is holistic as all supply chains are included in the analysis. Furthermore, it offers specificity with the inclusion of bottom-up process data [28, 29]. Fig. (**1**) shows the process of hybridizing bottom-up processing of data with top-down processing of data. This process is also called hybridization, and is explained in the next section.

Hybridisation of Process Data with Input-Output Data

The process of hybridizing bottom-up process data with input-output data is explained in Fig. (**1**). As a first step, two sets of data were collected for the inputs needed for cultivating, harvesting and processing microalgae – monetary data [30 - 32] and data on physical indicators (employment, economic stimulus, energy use and greenhouse gas emissions) [33]. The data collection steps are described in detail in Section 2.2 of [34]. The data were then arranged into a column vector of inputs (see supplementary information of [34]), which was then integrated into a comprehensive multi-regional input-output (MRIO) table for Australia [35]. The MRIO table distinguishes 19 regions and 344 sectors [35]. For the purposes of this case study, it was considered that the algae bio-crude production industry will be established in Western Australia. Hence, the column vector of inputs was inserted into the Western Australia region of the MRIO table (see use and value-added blocks in Fig. **1**). Furthermore, it was considered that the algae bio-crude will be transported to the closest refinery that refines crude oil. The row for the sales structure of crude oil was therefore replicated, and scaled according to the total input of algae-biocrude (see use and final demand blocks in Fig. **1**). The augmented table was used for undertaking TBL footprint assessments using standard Leontief input-output equations [22, 34].

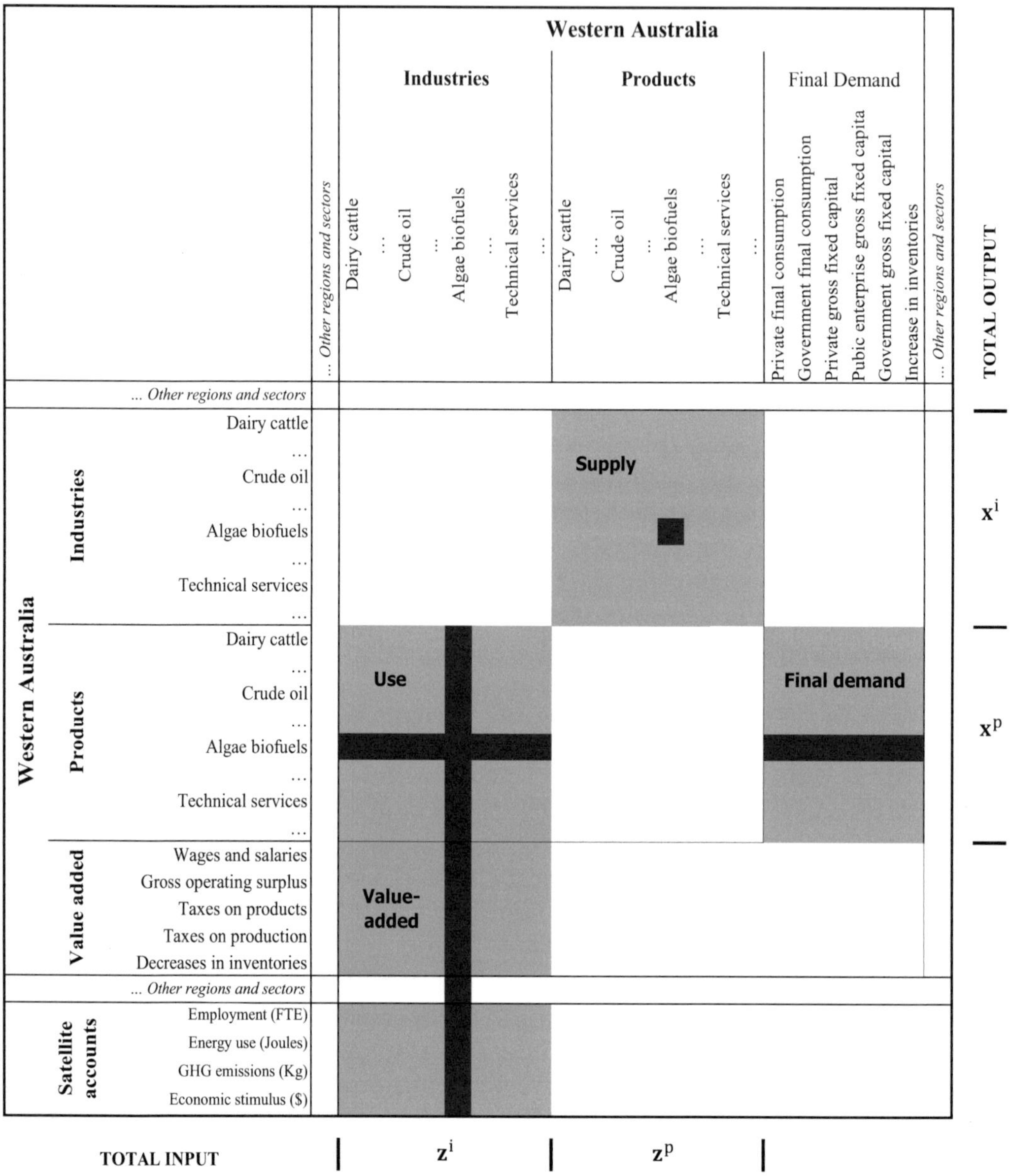

Fig. (1). Hybridisation of algae biofuels data with an Australian multi-regional input-output (MRIO) table. The table is arranged in a basic supply-use format. This figure has been adapted from [34].

RESULTS AND DISCUSSION

Comparison: Algae Bio-crude Supply Chain *vs.* Crude Oil Supply Chain

The sustainability performance of algae bio-crude and crude oil supply chains was assessed for four indicators – energy use and carbon dioxide (CO2-e) emissions (environmental indicators), employment (social indicator) and economic stimulus (economic indicator). The performance of the two supply chains was compared using an spider diagram (Fig. **2**). The results for the indicators are calculated on a common basis, for example total CO2-e emissions resulting from algae-bio crude supply chain are presented in kg per dollar of the total output of the bio-crude sector. Similarly, the impacts of the bio-crude and crude oil supply chains are calculated per dollar of output of the respective sectors. Thus, the spider diagram allows the indicators to be represented in an integrated way, allowing comparisons to be made. The spider is drawn such that if the performance of a sector is better than that of the benchmark sector, the line falls inside the inner benchmark polygon, and the line falls outside the inner benchmark polygon if the performance of a sector is worse than that of the benchmark sector. Spider diagrams are explained further here [36]. In short, the spikes are considered "bad", and dips are considered "good" in a spider diagram. In Fig. (**2**), the blue line is the performance line for algae bio-crude production, and the inner polygon is that of the benchmark sector (in this case study, crude oil production is taken as the benchmark sector – see also Fig. **1**). The spider diagram shows that the performance of the algae bio-crude industry is better than that of the crude oil industry. The supply chain of algae bio-crude production results in less carbon dioxide emissions, uses less energy and generates more employment and economic stimulus than the crude oil supply chain.

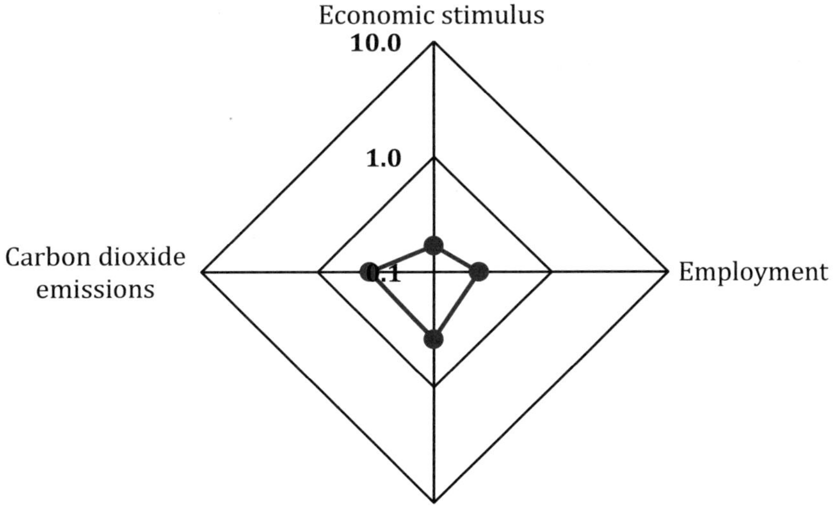

Fig. (2). Spider diagram showing the performance of the algae bio-crude production against the benchmark sector – crude oil production. The blue line represents the performance of the algae bio-crude sector. The spider diagram is drawn on a logarithmic scale. This figure is based on results from [34].

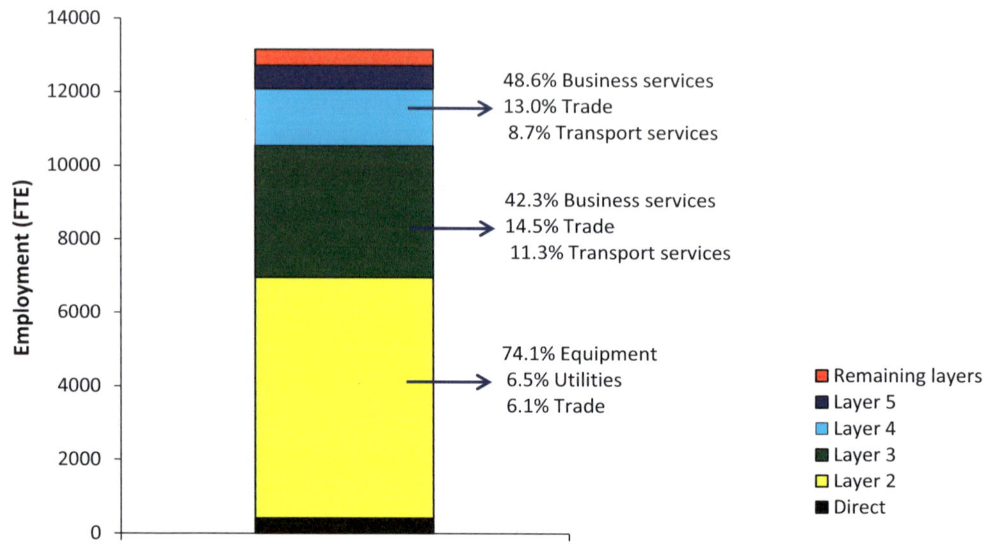
14000
12000
10000
8000
6000
4000
2000
0
Employment (FTE)
48.6% Business services
13.0% Trade
8.7% Transport services
42.3% Business services
14.5% Trade
11.3% Transport services
74.1% Equipment
6.5% Utilities
6.1% Trade
Remaining layers
Layer 5
Layer 4
Layer 3
Layer 2
Direct

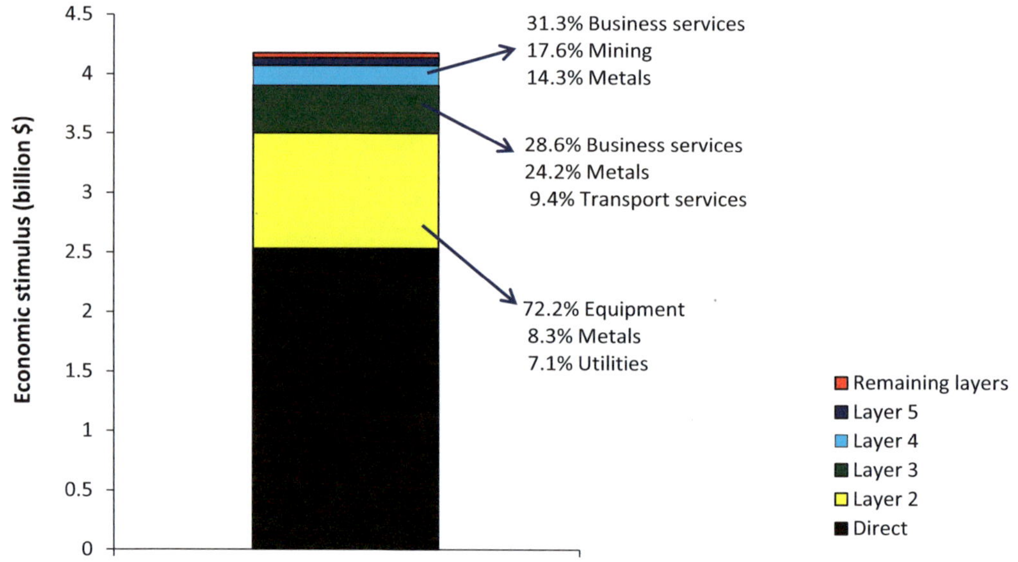
4.5
4
3.5
3
2.5
2
1.5
1
0.5
0
Economic stimulus (billion $)
31.3% Business services
17.6% Mining
14.3% Metals
28.6% Business services
24.2% Metals
9.4% Transport services
72.2% Equipment
8.3% Metals
7.1% Utilities
Remaining layers
Layer 5
Layer 4
Layer 3
Layer 2
Direct

Fig. 3 contd.....

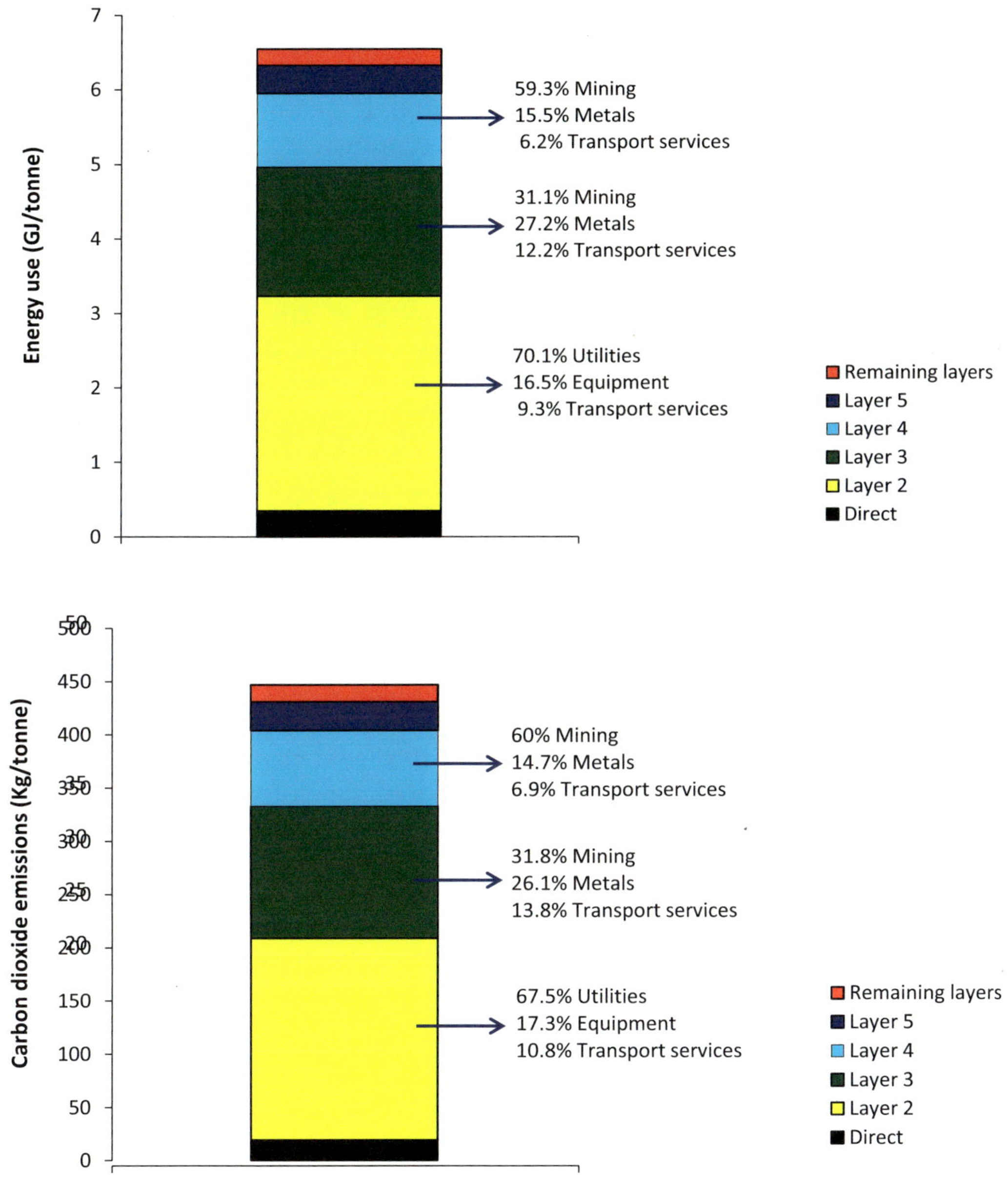

Fig. (3). Production Layer Decomposition of triple bottom line impacts of algae bio-crude production. The figures show the employment, economic stimulus, energy use and carbon dioxide emissions across various production layers of the algae bio-crude supply chain. Figures generated using data from [34].

Triple Bottom Line Footprints of Algae Bio-crude Production

Fig. (**3**) shows the results of production layer decomposition (PLD) of the indicators analysed in this study. A PLD analysis considers the upstream supply chain of an industry as various layers of production. If a company is considered as layer 1 of the supply chain, then the company's suppliers are layer 2, suppliers' of suppliers' of the company are layer 3, suppliers of suppliers of suppliers' of the company are layer 4 and so on. Using input-output analysis, a production layer decomposition can be carried out to quantify the impacts at each of the different production layers [34, 37, 38]. Fig. (**3**) shows such impacts for all four indicators. Layer 1 is also referred to as direct or on-site impacts, as these impacts refer to the operations carried out by the industry itself, for example if the algae bio-crude industry burns fuels on-site its premises, then the emissions from fuel burning are considered as direct emissions. The direct (on-site) or Layer 1 impacts are marked "black" in Fig. (**3**). The impacts resulting from bio-crude industry buying inputs from its suppliers (Layer 2) are marked yellow. As an example, the energy use impacts at layer 2 are primarily due to the utilities (70.1%), equipment (16.5%) and transport services (9.3%) sectors. The extraction of bio-crude from algae requires substantial input of energy, resulting in energy use impacts and also carbon dioxide emissions. Whilst these are considered negative impacts, the production of bio-crude also results in positive impacts such as the generation of jobs and economic stimulus in the utilities sector (see yellow marked impacts for the employment and stimulus indicators, Fig. **3**). The impacts from layers 1-5 are illustrated in Fig. **3**, along with main sectors that contribute to those impacts. Since the fundamental input-output equation used for constructing these PLD graphs is based on the famous Leontief inverse (*i.e.* inverting the input-output matrix), in most cases, layers 6 and beyond have a minor contribution to the overall impacts of a company or an industry. Therefore, the impacts of layer 6 and beyond are aggregated into one category termed "Remaining layers". The reader can refer to the comprehensive book on input-output analysis for an explanation of Leontief inverse [22].

CONCLUDING REMARKS

In this chapter, the performance of algae bio-crude supply chain was compared with that of the crude oil supply chain using hybrid life cycle assessment. Results suggest that the supply chain of algae bio-crude production is more sustainable in terms of generation of employment and economic stimulus, use of energy and production of carbon dioxide emissions. Further examination of the algae bio-crude supply chain reveals that energy use impacts and carbon dioxide emissions are largely due to the use of electricity for harvesting algae and extraction of bio-crude from algae. The production layer decomposition results presented in this chapter are valuable for understanding areas of concern, and where efforts should be directed to reduce negative environmental impacts such as energy use and greenhouse gas emissions. Previous research on algae bio-crude also points to the considerable use of energy for the production of bio-crude [39, 40]. Efforts are underway to develop technologies for increasing the efficiency of algae bio-crude production. Once efficient algae harvesting and bio-crude extraction technologies are developed, algae have a significant potential to contribute to the world's future energy mix.

CONFLICT OF INTEREST

The authors declares no conflict of interest, financial or otherwise.

ACKNOWLEDGEMENTS

This work was supported by the National eResearch Collaboration Tools and Resources project (NeCTAR) through its Industrial Ecology Virtual Laboratory. NeCTAR is an Australian Government project conducted as part of the Super Science initiative and financed by the Education Investment Fund.

REFERENCES

[1] J. Rogelj, M. den Elzen, N. Höhne, T. Fransen, H. Fekete, H. Winkler, R. Schaeffer, F. Sha, K. Riahi, and M. Meinshausen, "Paris Agreement climate proposals need a boost to keep warming well below 2 °C", *Nature,* vol. 534, no. 7609, pp. 631-639, 2016. [http://dx.doi.org/10.1038/nature18307] [PMID: 27357792]

[2] UNFCCC. (2016, 15 September). The Paris Agreement. Available at: , http://unfccc. int/paris_agreement/items/9485.php.

[3] J-Y. Moon, J. Song, and S. Lee, "Catalyzing Investment for Renewable Energy in Developing Countries", *KIEP Research Paper World Economy Update,* 2016. [http://dx.doi.org/10.2139/ssrn.2801434]

[4] H.J. Schellnhuber, S. Rahmstorf, and R. Winkelmann, "Why the right climate target was agreed in Paris", *Nat. Clim. Chang.,* vol. 6, pp. 649-653, 2016. [http://dx.doi.org/10.1038/nclimate3013]

[5] L. Brennan, and P. Owende, "Biofuels from microalgae—a review of technologies for production, processing, and extractions of biofuels and co-products", *Renew. Sustain. Energy Rev.,* vol. 14, pp. 557-577, 2010. [http://dx.doi.org/10.1016/j.rser.2009.10.009]

[6] A. Demirbas, "Political, economic and environmental impacts of biofuels: A review", *Appl. Energy,* vol. 86, pp. S108-S117, 2009. [http://dx.doi.org/10.1016/j.apenergy.2009.04.036]

[7] E. Gnansounou, A. Dauriat, J. Villegas, and L. Panichelli, "Life cycle assessment of biofuels: energy and greenhouse gas balances", *Bioresour. Technol.,* vol. 100, no. 21, pp. 4919-4930, 2009. [http://dx.doi.org/10.1016/j.biortech.2009.05.067] [PMID: 19553106]

[8] *Towards Sustainable Production and Use of Resources: Assessing Biofuels.* UNEP/Earthprint, 2009.

[9] J.F. Sanz Requena, A.C. Guimaraes, S. Quiros Alpera, E. Relea Gangas, S. Hernandez-Navarro, L.M. Navas Gracia, J. Martin-Gil, and H. Fresneda Cuesta, "Life Cycle Assessment (LCA) of the biofuel

production process from sunflower oil, rapeseed oil and soybean oil", *Fuel Process. Technol.*, vol. 92, pp. 190-199, 2011.
[http://dx.doi.org/10.1016/j.fuproc.2010.03.004]

[10] A. Malik, M. Lenzen, and A. Geschke, "Triple bottom line analysis of a lignocellulosic biofuel industry", *Glob. Change Biol. Bioenergy,* vol. 8, pp. 96-110, 2016.
[http://dx.doi.org/10.1111/gcbb.12240]

[11] A. Malik, M. Lenzen, R.N. Ely, and E. Dietzenbacher, "Simulating the impact of new industries on the economy – the case of biorefining in Australia", *Ecol. Econ.*, vol. 107, pp. 84-93, 2014.
[http://dx.doi.org/10.1016/j.ecolecon.2014.07.022]

[12] S. Naik, V.V. Goud, P.K. Rout, and A.K. Dalai, "Production of first and second generation biofuels: a comprehensive review", *Renew. Sustain. Energy Rev.*, vol. 14, pp. 578-597, 2010.
[http://dx.doi.org/10.1016/j.rser.2009.10.003]

[13] M.A. Borowitzka, and N.R. Moheimani, "Sustainable biofuels from algae", *Mitig. Adapt. Strategies Glob. Change,* vol. 18, pp. 13-25, 2013.
[http://dx.doi.org/10.1007/s11027-010-9271-9]

[14] P.K. Campbell, T. Beer, and D. Batten, "Life cycle assessment of biodiesel production from microalgae in ponds", *Bioresour. Technol.*, vol. 102, no. 1, pp. 50-56, 2011.
[http://dx.doi.org/10.1016/j.biortech.2010.06.048] [PMID: 20594828]

[15] P.K. Campbell, T. Beer, and D. Batten, *Greenhouse gas sequestration by algae: energy and greenhouse gas life cycle studies.* CSIRO Energy Transformed Flagship, 2009.

[16] X. Gao, Y. Yu, and H. Wu, "Life Cycle Energy and Carbon Footprints of Microalgal Biodiesel Production in Western Australia: A Comparison of Byproducts Utilization Strategies", *ACS Sustain. Chem.& Eng.*, vol. 1, pp. 1371-1380, 2013.
[http://dx.doi.org/10.1021/sc4002406]

[17] M.K. Lam, and K.T. Lee, "Microalgae biofuels: A critical review of issues, problems and the way forward", *Biotechnol. Adv.*, vol. 30, no. 3, pp. 673-690, 2012.
[http://dx.doi.org/10.1016/j.biotechadv.2011.11.008] [PMID: 22166620]

[18] G.B. Leite, A.E. Abdelaziz, and P.C. Hallenbeck, "Algal biofuels: challenges and opportunities", *Bioresour. Technol.*, vol. 145, pp. 134-141, 2013.
[http://dx.doi.org/10.1016/j.biortech.2013.02.007] [PMID: 23499181]

[19] Y. Li, M. Horsman, N. Wu, C.Q. Lan, and N. Dubois-Calero, "Biofuels from microalgae", *Biotechnol. Prog.*, vol. 24, no. 4, pp. 815-820, 2008.
[PMID: 18335954]

[20] J. Murray, and R. Wood, *The Sustainability Practitioner's Guide to Input-Output Analysis.* Common Ground Pub., 2010.
[http://dx.doi.org/10.18848/978-1-86335-748-7/CGP]

[21] R. Dixon, *Inter-Industry Transactions and Input-Output Analysis*, 1996.

[22] R.E. Miller, and P.D. Blair, *Input-Output Analysis:Foundations and Extensions.* 2nd ed. Cambridge University Press: Cambridge, 2009.
[http://dx.doi.org/10.1017/CBO9780511626982]

[23] A. Oita, A. Malik, K. Kanemoto, A. Geschke, S. Nishijima, and M. Lenzen, "Substantial nitrogen pollution embedded in international trade", *Nat. Geosci.*, vol. 9, pp. 111-115, 2016.
[http://dx.doi.org/10.1038/ngeo2635]

[24] R. Heijungs, and S. Suh, *The computational structure of life cycle assessment.* vol. Vol. 11. Kluwer Academic Publishers: Dordrecht, Netherlands, 2002.
[http://dx.doi.org/10.1007/978-94-015-9900-9]

[25] S. Suh, M. Lenzen, G.J. Treloar, H. Hondo, A. Horvath, G. Huppes, O. Jolliet, U. Klann, W. Krewitt, Y. Moriguchi, J. Munksgaard, and G. Norris, "System boundary selection in life-cycle inventories using hybrid approaches", *Environ. Sci. Technol.,* vol. 38, no. 3, pp. 657-664, 2004. [http://dx.doi.org/10.1021/es0263745] [PMID: 14968848]

[26] S. Suh, and G. Huppes, "Methods for life cycle inventory of a product", *J. Clean. Prod.,* vol. 13, pp. 687-697, 2005. [http://dx.doi.org/10.1016/j.jclepro.2003.04.001]

[27] S. Suh, and S. Nakamura, "Five years in the area of input-output and hybrid LCA", *Int. J. Life Cycle Assess.,* vol. 12, pp. 351-352, 2007. [http://dx.doi.org/10.1065/lca2007.08.358]

[28] M. Lenzen, "Errors in Conventional and Input□Output—based Life—Cycle Inventories", *J. Ind. Ecol.,* vol. 4, pp. 127-148, 2000. [http://dx.doi.org/10.1162/10881980052541981]

[29] M. Lenzen, and C. Dey, "Truncation error in embodied energy analyses of basic iron and steel products", *Energy,* vol. 25, pp. 577-585, 2000. [http://dx.doi.org/10.1016/S0360-5442(99)00088-2]

[30] J. Sheehan, T. Dunahay, J. Benemann, and P. Roessler, *A Look Back at the U.S. Department of Energy's Aquatic Species Program—Biodiesel from Algae.* National Renewable Energy Laboratory: Colorado, 1998. [http://dx.doi.org/10.2172/15003040]

[31] X. Liu, B. Saydah, P. Eranki, L.M. Colosi, B. Greg Mitchell, J. Rhodes, and A.F. Clarens, "Pilot-scale data provide enhanced estimates of the life cycle energy and emissions profile of algae biofuels produced via hydrothermal liquefaction", *Bioresour. Technol.,* vol. 148, pp. 163-171, 2013. [http://dx.doi.org/10.1016/j.biortech.2013.08.112] [PMID: 24045203]

[32] U. Jena, and K. Das, "Comparative evaluation of thermochemical liquefaction and pyrolysis for bio-oil production from microalgae", *Energy Fuels,* vol. 25, pp. 5472-5482, 2011. [http://dx.doi.org/10.1021/ef201373m]

[33] WorleyParsons, *Pilbara Algae Industry Study.* Perth, Western Australia, 2012.

[34] A. Malik, M. Lenzen, P.J. Ralph, and B. Tamburic, "Hybrid life-cycle assessment of algal biofuel production", *Bioresour. Technol.,* vol. 184, pp. 436-443, 2015. [http://dx.doi.org/10.1016/j.biortech.2014.10.132] [PMID: 25465782]

[35] M. Lenzen, A. Geschke, T. Wiedmann, J. Lane, N. Anderson, T. Baynes, J. Boland, P. Daniels, C. Dey, J. Fry, M. Hadjikakou, S. Kenway, A. Malik, D. Moran, J. Murray, S. Nettleton, L. Poruschi, C. Reynolds, H. Rowley, J. Ugon, D. Webb, and J. West, "Compiling and using input-output frameworks through collaborative virtual laboratories", *Sci. Total Environ.,* vol. 485-486, pp. 241-251, 2014. [http://dx.doi.org/10.1016/j.scitotenv.2014.03.062] [PMID: 24727042]

[36] B. Foran, M. Lenzen, and C. Dey, *Balancing Act - A Triple Bottom Line Account of the Australian Economy.* CSIRO Resource Futures and The University of Sydney: Canberra, ACT, Australia, 2005.

[37] T. Wiedmann, M. Lenzen, and J. Barrett, "Companies on the scale - Comparing and benchmarking the sustainability performance of businesses", *J. Ind. Ecol.,* vol. 13, pp. 361-383, 2009. [http://dx.doi.org/10.1111/j.1530-9290.2009.00125.x]

[38] A.M. Rodríguez-Alloza, A. Malik, M. Lenzen, and J. Gallego, "Hybrid input-output life cycle assessment of warm mix asphalt mixtures", *J. Clean. Prod.,* vol. 90, pp. 171-182, 2015. [http://dx.doi.org/10.1016/j.jclepro.2014.11.035]

[39] L. Lardon, A. Hélias, B. Sialve, J.P. Steyer, and O. Bernard, "Life-cycle assessment of biodiesel production from microalgae", *Environ. Sci. Technol.,* vol. 43, no. 17, pp. 6475-6481, 2009. [http://dx.doi.org/10.1021/es900705j] [PMID: 19764204]

[40] K. Sander, and G.S. Murthy, "Life cycle analysis of algae biodiesel", *Int. J. Life Cycle Assess.*, vol. 15, pp. 704-714, 2010.
[http://dx.doi.org/10.1007/s11367-010-0194-1]

SUBJECT INDEX

A

B

C

D

E

I

K

L

N

O

P

R

S

T

U

V

W

Z

Made in the USA
Las Vegas, NV
04 November 2026

57424421R10266